ISBN 978-0-483-18391-9
PIBN 10585552

NORTH OF ENGLAND INSTITUTE OF MINING AND MECHANICAL ENGINEERS.

TRANSACTIONS

VOL. XXXII.

1887-8

NEWCASTLE-UPON-TYNE:
ANDREW REID, PRINTING COURT BUILDINGS, AKENSIDE HILL.

LONDON OFFICE:
4 AND 5, QUEEN'S HEAD PASSAGE, PATERNOSTER ROW, E.C.

1888.

NORTH OF ENGLAND INSTITUTE OF MINING AND MECHANICAL ENGINEERS.

///

TRANSACTIONS

VOL. XXXVII.

1887-8.

NEWCASTLE-UPON-TYNE:
ANDREW REID, PRINTING COURT BUILDINGS, AKENSIDE HILL.

LONDON OFFICE:
4 AND 5, QUEEN'S HEAD PASSAGE, PATERNOSTER ROW, E.C.

1888.

NEWCASTLE-UPON-TYNE:
ANDREW REID, PRINTING COURT BUILDINGS, AKENSIDE HILL.

CONTENTS OF VOL. XXXVII.

GENERAL MEETINGS.

Report.

IN presenting their Report for the year 1887–1888, the Council of the North of England Institute of Mining and Mechanical Engineers may once more fairly congratulate the members upon the continued prosperity and usefulness of the Institute.

Notwithstanding unusual claims upon its resources the accounts still show a balance in favour of the Institute, its Transactions have not fallen below their accustomed standard, and the Committees appointed to carry out special investigations have been of an important character.

The number of new members elected during the year is 15, or, counting honorary members, 24. During the same period there was a loss by death of 10 members, and by resignation of 23 members, whilst 10 were removed from the list for non-payment of their annual contributions, making a total loss of 43 members.

The total number of members of all classes was 721 at the end of the year 1886–1887 and is now 702. These are distributed as follows :—

	1886-87.	1887-88.
Honorary members	21	29
Original members	412	390
Ordinary members	42	41
Associate members	168	173
Students	63	53
Collieries	15	16

The increase in the number of honorary members is due to the action of the Council, which, it is hoped, that the present meeting, by adopting this Report, will ratify, in appointing all H.M. Chief Inspectors of Mines honorary members of the Institute during their tenure of office.

The necessarily constant decrease in the class of original members is considerably less than in past years.

The number of ordinary members remains small and stationary, whilst there is an increase in that of associate members. The Council have under consideration a scheme by which the evident comparative popularity of the latter class may be still further heightened.

The number of contributing collieries shows a slight increase.

The Exhibition, in the theatre of which the last Annual Meeting was held, and the inception of which was entirely due to this Institute, may be claimed as a great success; and it is with much satisfaction that the Council are able to announce that the guarantee of £1,000 will, in all probability, remain untouched in the hands of the Institute.

The proposals for a Federation of Mining Institutes made last year, in a remarkable and carefully considered paper read by Mr. T. W. Bunning—the last public act of a devoted servant of the Institute—has borne fruit. A meeting of members of this Council and representatives of several of the other Institutes in the country was held in London in June last, the proceedings of which will reach members very shortly. Another meeting, for the further discussion of the matter from a more definite and practical standpoint, is to be held before long at Sheffield. In this matter, therefore, satisfactory progress may be reported.

In the secretarial department of the Institute material changes have taken place, owing to the regretted resignation through ill-health of the Secretary, Mr. T. W. Bunning, whose sudden and lamented death was to occur so soon after. These changes, the Council trust, will tend towards economy without loss of efficiency. Professor G. A. Lebour, M.A., of the Durham College of Science, was appointed Secretary; and now that the College has migrated to its handsome new buildings near the Leazes, may serve as a remaining link between the two Institutions. The rooms freed by the removal of the College will, it is thought, be capable of being used so as to add to the comfort and convenience of members.

The Institute has again been placed on the list of Corresponding Societies of the British Association, and the Council have appointed the Secretary to act as delegate to the forthcoming meeting at Bath.

In connection with the meeting of the Association at Newcastle in 1889, the following six gentlemen have been nominated to represent the Institute on the Local General Committee formed to carry out the necessary arrangements in this city :—Messrs. John Marley, W. H. Hedley, J. G. Weeks, M. Walton Brown, Thomas Heppell, and the Secretary.

The Committee appointed some years ago to investigate the causes of the Ryhope receiver explosion has terminated its labours and presented a valuable report which will very shortly be in the hands of the members.

Two new and important committees have been nominated, and are now in full working order. One, to report on fan ventilation, is a Joint

Committee, in the work of which gentlemen from South Wales and the Midlands will take part. The other, equally important, is to investigate the nature of flameless explosives.

The papers read during the year are the following :—

"A further attempt for the Correlation of the Coal Seams of the Carboniferous Formation of the North of England; with some Notes on the Probable Duration of the Coal-field." By Mr. M. Walton Brown.

"On the Pyrites Deposits of Huelva." By Mr. John Allan.

"Report of the Committee on Earth Tremors."

"The Endless Chain in Spain." By Mr. George Lee.

"On a Safety-Lamp designed to meet the Requirements of the Mines Act, 1887." By Mr. Emerson Bainbridge.

"On the Coal-field of Tkiboulli (Caucasus)." By Mr. C. J. Murton.

"On an Improved form of Seismoscope." By Professor A. S. Herschel, D.C.L., F.R.S.

"On Bornêt's Hand Boring Machine." By Mr. E. L. Dumas.

"On Ackroyd and Best's Patent Safety-Lamp Cleaning Machine." By Mr. William Ackroyd.

"On the Use of Iron Supports in the Main Roads of Mines, instead of Masonry or Timbering." By Mr. W. J. Bird, A.Sc.

"On Coal Nodules from the Bore-Hole Seam at Newcastle, New South Wales." By Mr. T. E. Forster, M.A.

"Notes on the Horizon of the Low Main Seam in a portion of the Durham Coal-field." By Mr. Hugh Bramwell.

"Report of the Ryhope Receiver Explosion Committee."

"Timber v. Steel in Mining." By Mr. A. L. Steavenson.

"A Contribution to our Knowledge of Coal-Dust." By Professor P. P. Bedson, D.Sc.

The close of the session was spent in an excursion to Scotland, which may be regarded as having been very successful in spite of somewhat inclement weather.

Finance Report.

THE Income for the year 1887-88 amounted to £1,586 7s. 5d., showing a decrease, when compared with that of the preceding year, of £167 14s. 11d., caused principally by the non-payment by the Institute and Coal Trade Chambers Company of the second half-yearly dividend, the estimated amount of which has been treated as an asset.

The Expenditure was £1,767 1s. 9d., being £200 9s. more than that of last year, and £180 14s. 4d. above the Income, due to some exceptional payments of large amount.

The total receipts for subscriptions and arrears were £1,307 7s., and the arrears of subscriptions amount to £642 12s.

G. B. FORSTER.
J. B. SIMPSON.

July 21st, 1888.

ADVERTISEMENT.

TREASURER IN ACCOUNT WITH THE NORTH OF ENGLAND

Dr.   *July 20th, 1887,*

	£	s.	d.	£	s.	d.
July 20, 1887.						
To Balance at Bankers	395	1	10			
„ „ in Cashier's-hands	14	10	2			
				409	12	0
July 17, 1888.						
To Dividend of 8½ % per annum on 134 Shares of £20 each in the Institute and Coal Trade Chambers Company, Ltd., for the Half-Year ending December, 1887 ...	113	18	0			
To Dividend for Second Half-Year, unpaid...	„	„	„			
To Interest on Investments with the River Tyne Commissioners	77	13	2			
				191	11	2
To Rent of College Class Rooms				48	8	8
To Subscriptions for 1887–88, as follows :—						
307 Original Members	644	14	0			
29 Ordinary do.	89	5	0			
118 Associate do.	247	16	0			
1 Do. paid as Life Member	20	0	0			
37 Students	38	17	0			
1 Do. paid as an Associate	1	1	0			
1 New Ordinary Member	3	3	0			
10 New Associate Members	21	0	0			
1 Do. paid as Life Member ...	20	0	0			
1 New Student	1	1	0			
	1,086	17	0			
To Subscribing Collieries, &c., namely :—						
Ashington £2 2 0						
Birtley Iron Company 6 6 0						
Haswell 4 4 0						
Hetton 10 10 0						
Lambton 10 10 0						
Londonderry 10 10 0						
Marquis of Bute 10 10 0						
North Hetton 6 6 0						
Ryhope 4 4 0						
Seghill 2 2 0						
South Hetton and Murton 4 4 0						
Stella 2 2 0						
Throckley 2 2 0						
Victoria Garesfield 2 2 0						
Wearmouth 4 4 0						
The Bridgewater Trustees 6 6 0						
	88	4	0			
	1,175	1	0			
To Members' Arrears	123	18	0			
To Students' do.	4	4	0			
To Arrears considered as Irrecoverable, but since Paid	4	4	0			
	132	6	0			
				1,307	7	0
To Sale of Publications, per Andrew Reid, less 10 % Commission	28	17	7			
To Sale of Publications per Secretary	10	3	0			
				39	0	7
				£1,995	19	5

INSTITUTE OF MINING AND MECHANICAL ENGINEERS.
to July 17th, 1888.   Cʀ.

1888.						£	s.	d.	£	s.	d.
July 17.											
By Andrew Reid:—											
Publishing Accounts	...	...	...	...	...	449	12	0			
Covers for Parts, Folding and Stitching		...	...			33	9	3			
Binding and Sewing Volumes		...	...	...		26	2	2			
Borings and Sections	...	...	...	...	...	10	3	0			
Library	...	...	...	...	...	15	14	8			
Stationery and Circulars		...	...	...	...	67	1	2			
Postage	...	...	...	...	...	40	6	1			
									642	8	4
By Books for Library, in addition to amount paid A. Reid					...	58	8	7			
By Printing and Stationery do. do.					...	3	15	1			
By Abstracts of Foreign Papers	...	...	...	...		54	15	10			
By Secretary's Incidental Expenses and Postages	...		...		...	66	9	4			
By Sundry Accounts and Payments	...	...	...		...	28	9	0			
By Travelling Expenses	...	...	...	...	...	14	5	0			
By Secretary's Salary	..	...	...	...	...	283	6	8			
By Cashier's do.	...	...	...	...	...	75	0	0			
By Clerks' Wages	...	...	...	...	...	164	18	9			
By Reporter's Salary	...	...	...	...	...	12	12	0			
By Rent	...	...	...	...	...	77	15	3			
By Rates and Taxes	...	...	...	...	...	17	8	0			
By Fire Insurance	...	...	...	...	...	8	14	11			
By Furnishing, Repairs, &c.	...	...	...	...		2	16	2			
By Coals, Gas, and Water	...	...	...	...		7	0	11			
									875	15	6
By British Association Meeting—Delegate's Expenses					...	15	0	0			
By Testimonial to Mr. T. W. Bunning		...	...		...	110	0	0			
By Alteration of Wood Memorial Hall Drains			...		...	29	8	4			
By Expenses in connection with the Visit of Engineers,											
August, 1887	...	...	...	...	...	94	9	7			
									248	17	11
By Balance at Bankers	...	...	...	...		199	17	7			
By Do. in Cashier's hands		...	...	...		29	0	1			
									228	17	8

Audited and found correct,

JOHN G. BENSON,

Cʜᴀʀᴛᴇʀᴇᴅ Aᴄᴄᴏᴜɴᴛᴀɴᴛ.

Newcastle-upon-Tyne,

1st August, 1888.

£1,995 10 5

DR.   THE TREASURER IN ACCOUNT

			£	s.	d.

To 412 Original Members. as per List, 1887-88,
10 of whom are Life Members.

63 @ £2 2s. ... ...

402 @ £2 2s. 844 4 0

To 42 Ordinary Members, as per List, 1887-88,
3 of whom are Life Members.

39 (36 @ £3 3s. ; 3 @ £2 2s) 119 14 0

To 168 Associate Members, as per List, 1887-88,
7 of whom are Life Members.

161

1 paid as Life Member 20 0 0

160 @ £2 2s. 336 0 0

To 63 Students, as per List, 1887-88.

63 @ £1 1s. 66 3 0

To 1 Student paid extra as an Associate Member. 1 1 0

To 2 New Ordinary Members, @ £3 3s.... ... ...   6 6 0

To 12 New Associate Members,
1 paid as Life Member 20 0 0

11 @ £2 2s. 23 2 0

To 1 New Student @ £1 1s. 1 1 0

To Subscribing Collieries, &c. 88 4 0

                                                             1,525 15 0

To Arrears. as per Balance Sheet, 1886-87... ... ... £519 15 0
  Deduct—Irrecoverable 99 15 0

                                                 420 0 0
  Add—Considered as irrecoverable, but since paid ...   4 4 0
                                                 ————————   424 4 0

                                                             £1,949 19 0

WITH SUBSCRIPTIONS, 1887-88.   CR.

			PAID. £ s. d.	UNPAID. £ s. d.
By 307 Original Members paid	@ £2 2s. ...	...	614 14 0	......
By 82 Do.	unpaid	,, ...	......	172 4 0
By 6 Do.	dead	,, ...	......	12 12 0
By 2 Do.	resigned	,, ...	......	4 4 0
By 2 Do.	no address	,, ...	......	4 4 0
By 3 Do.	struck off	,, ...	......	6 6 0
402				
By 27 Ordinary Members paid	@ £3 3s. ...	...	85 1 0	......
By 2 Do. ,,	@ £2 2s. ...	...	4 4 0	......
By 7 Do. unpaid	@ £3 3s. ...	...	......	22 1 0
By 1 Do. ,,	@ £2 2s. ...	...	......	2 2 0
By 1 Do. resigned	@ £3 3s. ...	...	......	3 3 0
By 1 Do. struck off	@ £3 3s. ...	...	......	3 3 0
39				
By 118 Associate Members paid	@ £2 2s. ...	...	247 16 0	......
By 36 Do.	unpaid	,, ...	......	75 12 0
By 4 Do.	resigned	,, ...	......	8 8 0
By 2 Do.	no address	,, ...	......	4 4 0
160				
1 Do.	paid as Life Member ...	...	20 0 0	......
By 37 Students paid	@ £1 1s.		38 17 0	......
By 17 Do. unpaid	,, ...		......	17 17 0
By 4 Do. resigned	,, ...		......	4 4 0
By 4 Do. no address	,, ...		......	4 4 0
By 1 Do. struck off	,, ...		......	1 1 0
63				
By 1 Student paid extra as an Associate Member		...	1 1 0	......
By 1 New Ordinary Member paid	@ £3 3s. ...	...	3 3 0	......
By 1 Do. do. unpaid	,, ...	...	......	3 3 0
2				
By 10 New Associate Members paid	@ £2 2s. ...	...	21 0 0	......
By 1 Do. do. unpaid	,, ...	...	......	2 2 0
11				
By 1 Do. do. paid as Life Member		...	20 0 0	......
By 1 New Student paid @ £1 1s.	... ...		1 1 0	......
By Subscribing Collieries paid	... ...	...	88 4 0	......
			1,175 1 0	350 14 0
By Members' Arrears	... ...	...	123 18 0	250 19 0
By Students' do.	... ...	...	4 4 0	40 19 0
By considered as irrecoverable, but since paid	...	...	4 4 0	......
			1,307 7 0	642 12 0
				1,307 7 0
				£1,949 19 0

GENERAL STATEMENT, JULY 17TH, 1888.

DR.

Liabilities.

	£	s.	d.
None			
Capital	11,709	6	8
	£11,709	**6**	**8**

Audited and found correct.
(Share Certificates and Bonds produced.)

JOHN G. BENSON,
CHARTERED ACCOUNTANT.

Newcastle-on-Tyne,
1st August, 1888.

CR.

Assets.

	£	s.	d.	£	s.	d.
Balance of Account at Bankers	199	17	7			
Do. in Cashier's hands	29	0	1	228	17	8
134 Shares of £20 each in the Institute and Coal Trade Chambers Company, Limited...				2,680	0	0
Invested with the River Tyne Commissioners				2,000	0	0
Arrears of Subscriptions				642	12	0
Value of 492 Bound Volumes of Transactions, @ 11s. 6d.	282	18	0			
Value of 4,455 Sewn Copies of Transactions, @ 9s.	2,004	15	0			
Value of Sundry Unbound Parts of Transactions	80	0	0			
Value of 35 Copies of Mr. T. F. Brown's Map, @ 5s.	8	15	0			
Value of 382 Copies of General Index, @ 3s.	57	6	0			
Value of 760 Copies of Fossil Illustrations, @ 12s. 6d.	475	0	0			
Value of 852 Copies of Catalogue of Fossils, @ 5s.	213	0	0			
Value of 313 Copies of Borings and Sinkings, Vol. I., @ 5s.	78	5	0			
Value of 328 Copies Do. Do. Vol. II., @ 5s.	82	0	0			
Value of 350 Copies Do. Do. Vol. III., @ 5s.	87	10	0			
Value of 471 Copies Do. Do. Vol. IV., @ 5s.	117	15	0			
Value of 1,500 Copies of Borings and Sinkings, Vol. I., in Sheets	300	0	0			
Value of Sheets of Borings and Sinkings, Vol. V., unpublished at date	25	0	0			
Value of 261 Copies of Library Catalogue, @ 5s.	65	5	0	3,877	9	0
Value of Office Furniture and Fittings				450	0	0
Value of Books and Maps in Library				1,750	0	0
Second Half-Year's Dividend on Shares in the Institute and Coal Trade Chambers Company, Limited, estimated at 3 %				80	8	0
				£11,709	**6**	**8**

Patrons.

His Grace the DUKE OF NORTHUMBERLAND.
His Grace the DUKE OF CLEVELAND.
The Most Noble the MARQUESS OF LONDONDERRY
The Right Honourable the EARL OF LONSDALE.
The Right Honourable the EARL OF DURHAM.
The Right Honourable the EARL GREY.
The Right Honourable the EARL OF RAVENSWORTH.
The Right Honourable the EARL OF WHARNCLIFFE.
The Right Reverend the LORD BISHOP OF DURHAM.
The Very Reverend the DEAN AND CHAPTER OF DURHAM.
WENTWORTH B. BEAUMONT, Esq., M.P.

Honorary Members.

Honorary Members during term of office only.

	ELECTED	
---	MEM	HON.
The Right Honourable the EARL OF RAVENSWORTH, Ravensworth Castle, Gateshead-on-Tyne...		1877
* PROF. P. PHILLIPS BEDSON, D.Sc. (Lond.), F.G.S., Durham College of Science, Newcastle-on-Tyne ...		1883
* THOMAS BELL, Esq., Inspector of Mines, Durham		
M. DE BOUREUILLE, Commandeur de la Légion d'Honneur, Conseiller d'état, Inspecteur Général des Mines, Paris		1853
* PROF. G. S. BRADY, M.D., F.R.S., F.L.S., Durham College of Science, Newcastle-on-Tyne		1875
DR. BRASSERT, Berghauptmann, Bonn-am-Rhein, Prussia		1883
DR. H. VON DECHEN, Berghauptmann, Bonn-am-Rhein, Prussia...		1853
JOSEPH DICKINSON, Esq., F.G.S., Inspector of Mines, Manchester		1853
* C. LE NEVE FOSTER, Esq., Inspector of Mines, Llandudno		1888
* PROF. WILLIAM GARNETT, M.A., D.C.L., Principal of the Durham College of Science, Newcastle-on-Tyne		1884
* HENRY HALL, Esq., Inspector of Mines, Rainhill, Prescott		1876
* PROF. A. S. HERSCHEL, M.A., D.C.L., F.R.S., F.R.A.S., Durham College of Science, Newcastle-on-Tyne		1872
THE VERY REV. DR. LAKE, Dean of Durham		1872
* PROF. G. A. LEBOUR, M.A., F.G.S., Durham College of Science, Newcastle-on-Tyne	1873	1879
J. A. LONGRIDGE, Esq., Grève d'Ayetté, Jersey		1886
* J. S. MARTIN, Esq., Inspector of Mines, Clifton		1888
* RALPH MOORE, Esq., Inspector of Mines, Glasgow		1866
* A. E. PINCHING, Esq., Inspector of Mines, Stoke, Devonport		1888
* JOSEPH T. ROBSON, Esq., Inspector of Mines, Swansea ...		1888

	ELECTED.	
	MEM.	HON.
* J. M. RONALDSON, Esq , Inspector of Mines, 44, Athole Gardens, Glasgow		1888
* W. B. SCOTT, Esq., Inspector of Mines, Parkdale, Wolverhampton ..		1888
SIR WARINGTON W. SMYTH, M.A., F.R.S., F.G.S., F.R.G.S., 28, Jermyn Street, London		1869
* A. H. STOKES, Esq., Inspector of Mines, Greenhill, Derby		1888
M. E. VUILLEMIN, Mines d'Aniche, Nord, France		1878
* FRANK N. WARDELL, Esq., F.G.S., Inspector of Mines, Wath-on-Dearne, near Rotherham	1864	1868
* JAMES WILLIS, Esq., Inspector of Mines, 14, Portland Terrace, Newcastle-on-Tyne	1857	1871
THOMAS WYNNE, Esq., F.G.S., Inspector of Mines, Manor House, Gnosall, Stafford		1853

Life Members.

	ELECTED.	
	MEM.	LIFE.
C. W. BARTHOLOMEW, Esq., Blakesley Hall, near Towcester ...	1875	1875
THOS. HUGH BELL, Esq., Middlesbro'-on-Tees	1882	1882
DAVID BURNS, Esq., C.E., F.G.S., Canal Bank, Carlisle	1877	1877
T. E. CANDLER, Esq., F.G.S., Hong Kong Club, Hong Kong, China	1875	1885
E. B. COXE, Esq., Drifton, Jeddo, P.O., Luzerne Co., Penns., U.S. ...	1873	1874
JAMES S. DIXON, Esq., 170, Hope Street, Glasgow	1878	1880
ERNEST HAGUE, Esq., Castle Dyke, Sheffield	1872	1876
G. C. HEWITT, Esq., Coal Pit Heath Colliery, near Bristol ...	1871	1879
JAMES HILTON, Esq., Wigan Coal and Iron Co., Limited, Wigan ..	1867	1883
THOS. E. JOBLING, Esq., Croft Villa, Blyth, Northumberland ...	1876	1882
ROBERT KNOWLES, Esq., Arncliffe, Cheetham Hill, Manchester...	1886	1886
HENRY LAPORTE, Esq., M.E., Aciéries de France, Aubin, Aveyron, France	1877	1877
W. MERIVALE, Esq., District Engineer's Office, Indian Midland Railway, Sangur, India	1881	1884
NATHAN MILLER, Esq., 31, Hyde Lane, Hyde, near Manchester	1878	1878
H. J. MORTON, Esq., 2, Westbourne Villas, South Cliff, Scarborough	1856	1861
RUDOLPH NASSE, Esq., Oberbergrath, Saarbrücken, Prussia ...	1869	1880
ARTHUR PEASE, Esq., Darlington	1882	1882
EDWARD G. PRIOR, Esq., Victoria, British Columbia	1880	1883
WALTER SAISE, D.Sc. (London), M. Inst. C.E., Manager E.I.R. Collieries, Giridi, Bengal, India	1877	1887
R. CLIFFORD SMITH, Esq., F.G.S., Ashford Hall, Bakewell ...	1874	1874
JOHN EMANUEL TYERS, Mohpani Coal Mines, Gadawarra, India, C.P.		1887
T. H. WARD, Esq., F.G.S., Assistant Manager, East Indian Railway Collieries, Giridi, Bengal, India	1882	1882

OFFICERS, 1888-89.

President.

JOHN MARLEY, Esq., Thornfield, Darlington.

Vice-Presidents.

CUTHBERT BERKLEY, Esq., Marley Hill, Whickham, R.S.O., Co. Durham.
T. J. BEWICK, Esq., M.I.C.E., F.G.S., Suffolk House. Laurence Pountney Hill. near London, E.C.
WM. COCHRANE, Esq., Grainger Street West, Newcastle-on-Tyne.
THOMAS DOUGLAS, Esq., Peases' West Collieries, Darlington.
A. L. STEAVENSON, Esq., Durham.
JAMES WILLIS, Esq., 14, Portland Terrace, Newcastle-on-Tyne.

Council.

WM. ARMSTRONG, Jun., Esq., Wingate, County Durham.
J. B. ATKINSON, Esq., Stocksfield-on-Tyne.
T. W. BENSON, Esq., 11, Newgate Street, Newcastle-on-Tyne.
R. F. BOYD, Esq., Houghton-le-Spring, Fence Houses, County Durham.
M. WALTON BROWN, Esq , 3, Summerhill Terrace, Newcastle-on-Tyne.
S. C. CRONE. Esq., Killingworth Hall, Newcastle-on-Tyne.
W. H. HEDLEY, Esq., Consett Collieries, Medomsley, Newcastle-on-Tyne.
HENRY LAWRENCE, Esq., Grange Iron Works, Durham.
T. LISHMAN, Esq., Hetton Colliery, Fence Houses.
W. LISHMAN, Esq., Bunker Hill, Fence Houses.
G. MAY, Esq., Harton Colliery Offices, South Shields.
PROF. J. H. MERIVALE, M.A., 2, Victoria Villas, Newcastle-on-Tyne.
R. ROBINSON, Esq., Howlish Hall, near Bishop Auckland.
J. B. SIMPSON, Esq.. F.G.S., Hedgefield House, Blaydon-on-Tyne.
T. H. M. STRATTON. Esq., Cramlington House, Northumberland.
J. G. WEEKS, Esq., Bedlington, R.S.O., Northumberland.
W. H. WOOD, Esq., Coxhoe Hall, Coxhoe, County Durham.
W. O. WOOD, Esq., South Hetton, Sunderland.

Ex-officio {
LORD ARMSTRONG, C.B., LL.D., F.R.S., D.C.L., Jesmond, Newcastle.
SIR LOWTHIAN BELL, Bart., F.R.S., F.C.S., Rounton Grange, Northallerton.
E. F. BOYD, Esq., F.G.S., Moor House, Leamside, Fence Houses.
JOHN DAGLISH, Esq., F.G.S., Marsden, South Shields.
SIR GEORGE ELLIOT, Bart., M.P., D.C.L., Houghton Hall, Fence Houses.
G. B. FORSTER, Esq., M.A., F.G.S., Lesbury, R.S.O., Northumberland.
G. C. GREENWELL, Esq., F.G.S., Elm Tree Lodge, Duffield, Derby.
LINDSAY WOOD, Esq., The Hermitage, Chester-le-Street.
} Past-Presidents.

WM. ARMSTRONG, Sen., Esq., F.G.S., Pelaw House, Chester-le-Street. } Retiring Vice-President.

Secretary.

PROF. G. A. LEBOUR, M.A., F.G.S.. Neville Hall, Newcastle-on-Tyne.

List of Members.

AUGUST, 1888.

Original Members.

Marked * are Life Members.

ELECTED.

1 AITKIN, HENRY, Falkirk, N.B. ... ... Mar. 2, 1865
2 ANDERSON, C. W., Belvedere, Harrogate ... ... Aug. 21, 1852
3 ANDREWS, HUGH, Swarland Hall, Felton, Northumberland ... ... Oct. 5, 1872
4 ARCHER, T., Dunston Engine Works, Gateshead ... ... July 2, 1872
5 ARMSTRONG, LORD, C.B., LL.D., F.R.S., D.C.L., Jesmond, Newcastle-on-Tyne (PAST-PRESIDENT, Member of Council) ... ... May 3, 1866
6 ARMSTRONG, WM., F.G.S., Pelaw House, Chester-le-Street (RETIRING VICE-PRESIDENT, Member of Council) ... ... Aug. 21, 1852
7 ARMSTRONG, W., Jun., Wingate, Co. Durham (Member of Council) April 7, 1867
8 ARMSTRONG, W. L., Newton Lane Colliery, Victoria Coal and Coke Co., Limited, near Wakefield ... ... Mar. 3, 1864
9 ARTHUR, D., M.E., Sherfin House, Baxenden, nr. Accrington, Manchester Aug. 4, 1877
10 ASHWORTH, JAMES, Stanley Hall, near Derby ... ... Feb. 5, 1876
11 ASQUITH, T. W., Harperley, Lintz Green, Newcastle-on-Tyne ... Feb. 2, 1867
12 ATKINSON, J. B., Stocksfield-on-Tyne (Member of Council) ... ... Mar. 5, 1870
13 ATKINSON, W. N., Shincliffe Hall, Durham ... ... June 6, 1868
14 AUBREY, R. C., Wigan Coal & Iron Co., Ld., Standish, near Wigan ... Feb. 5, 1870
15 AUSTINE, JOHN, Cadzow Coal Co., Glasgow ... ... Nov. 4, 1876
16 AYNSLEY, WM., Chilton Colliery, Ferry Hill ... ... Mar. 3, 1873

17 BAILES, GEORGE, Murton Colliery, Sunderland ... ... Feb. 3, 1877
18 BAILES, T., Jesmond Gardens, Newcastle ... ... Oct. 7, 1858
19 BAILEY, SAMUEL, Perry Barr, Birmingham ... ... June 2, 1859
20 BAIN, R. DONALD, 85, Pembroke Road, Clifton, Bristol ... ... Mar. 3, 1873
21 BAINBRIDGE, E., Nunnery Colliery Offices, Sheffield ... ... Dec. 3, 1863
22 BANKS, THOMAS, 60, King Street, Manchester ... ... Aug. 4, 1877
23 BARTHOLOMEW, C., Castle Hill House, Ealing, London, W. ... ... Aug. 5, 1853
24 *BARTHOLOMEW, C. W., Blakesley Hall, near Towcester ... ... Dec. 4, 1875
25 BATES, MATTHEW, ... Mar. 3, 1874
26 BATES, W. J., Winlaton, Blaydon-on-Tyne ... ... Mar. 3, 1874
27 BATEY, JOHN, Newbury Collieries, Coleford, Bath ... ... Dec. 5, 1868
28 BEANLANDS, ARTHUR, M.A., Palace Green, Durham ... ... Mar. 7, 1867
29 BELL, SIR LOWTHIAN, Bart., D.C.L., F.R.S., F.C.S., Rounton Grange, Northallerton, (PAST-PRESIDENT, Member of Council) ... ... July 6, 1854
30 BENSON, J. G., Accountant, 12, Grey Street, Newcastle-on-Tyne ... Nov. 7, 1874
31 BENSON, T. W., 11, Newgate Street, Newcastle (Member of Council) Aug. 2, 1866

ELECTED.

32 BERKLEY, C., Marley Hill, Whickham, R.S.O., Co. Durham (VICE-
PRESIDENT) ... ... Aug. 21, 1852
33 BEWICK, T. J., M.I.C.E., F.G.S., Suffolk House, Laurence Pountney
Hill, near London (VICE-PRESIDENT) ... ... April 5, 1860
34 BIGLAND, J., Bedford Lodge, Bishop Auckland ... ... June 4, 1857
35 BIRAM, B., Peaseley Cross Collieries, St. Helen's, Lancashire ... 1856
36 BLACK, W., Hedworth Villa, South Shields ... ... April 2, 1870
37 BOLTON, H. H., Newchurch Collieries, near Manchester ... ... Dec. 5, 1868
38 BOOTH, R. L., Ashington Colliery, near Morpeth 1864
39 BOYD, E. F., F.G.S., Moor House, Leamside, Fence Houses (PAST-
PRESIDENT, *Member of Council)* ... ... Aug. 21, 1852
40 BOYD, R. F., Houghton-le-Spring, Fence Houses, County Durham
(Member of Council) ... ... Nov. 6, 1869
41 BOYD, WM., North House, Longbenton, Newcastle-on-Tyne ... ... Feb. 2, 1867
42 BRECKON, J. R., 41, Fawcett Street, Sunderland ... ... Sept. 3, 1864
43 BRETTELL, T., Mine Agent, Dudley, Worcestershire ... ... Nov. 3, 1866
44 BROMILOW, WM., Preesgweene, near Chirk, North Wales ... ... Sept. 2, 1876
45 BROWN, JOHN, Priory Place, 155, Bristol Road, Birmingham ... Oct. 5, 1854
46 BROWN, THOS. FORSTER, F.G.S., Guildhall Chambers, Cardiff ... 1861
47 BROWNE, SIR BENJAMIN C., M.I.C.E., Westacres, Benwell, Newcastle-
on-Tyne ... ... Oct. 1, 1870
48 BRYHAM, WILLIAM, Rosebridge Colliery, Wigan ... ... Aug. 1, 1861
49 BRYHAM, W., Jun., Douglas Bank Collieries, Wigan ... ... Aug. 3, 1865
50*BURNS, DAVID, C.E., F.G.S., Canal Bank, Carlisle ... ... May 5, 1877
51 BURROWS, J. S., Yew Tree House, Atherton, near Manchester ... Oct. 11, 1873

52 CAMPBELL, W. B., Consulting Engineer, Grey Street, Newcastle ... Oct. 7, 1876
53 CARR, WM. COCHRAN, South Benwell, Newcastle-on-Tyne ... ... Dec. 3, 1857
54 CHAMBERS, A. M., Thorncliffe Iron Works, near Sheffield ... ... Mar. 6, 1869
55 CHEESMAN, I. T., Throckley Colliery, Newcastle-on-Tyne ... ... Feb. 1, 1873
56 CHEESMAN, W. T., Wire Rope Manufacturer, Hartlepool ... ... Feb. 5, 1876
57 CLARK, C. F., Garswood Coal and Iron Co., Limited, near Wigan ... Aug. 2, 1866
58 CLARK, R. B., Springwell Colliery, Gateshead ... ... May 3, 1873
59 CLARK, W., M.E., The Grange, Teversall, near Mansfield ... ... April 7, 1866
60 CLARKE, WILLIAM, Victoria Engine Works, Gateshead ... ... Dec. 7, 1867
61 COCHRANE, B., Aldin Grange, Durham ... ... Dec. 6, 1866
62 COCHRANE, C., Green Royde, Pedmore, near Stourbridge ... ... June 3, 1857
63 COCHRANE, W., St. John's Chambers, Grainger Street West, Newcastle
(VICE-PRESIDENT) 1859
64 COLE, RICHARD, Walker Colliery, near Newcastle-on-Tyne . ... April 5, 1873
65 COLE, ROBERT HEATH, Endon, Stoke-upon-Trent ... ... Feb. 5, 1876
66 COLLIS, W. B., Swinford House, Stourbridge, Worcestershire ... June 6, 1861
67 COOK, J., Jun., Washington Iron Works, Gateshead .. .. May 8, 1869
68 CORBETT, V. W., Chilton Moor, Fence Houses ... ... Sept. 3, 1870
69 CORBITT, M., Wire Rope Manufacturer, Teams, Gateshead ... ... Dec. 4, 1875
70 COULSON, F., 10. Victoria Terrace, Durham ... . Aug. 1, 1868
71 COULSON, W., High Coniscliffe, Darlington . . .. Oct. 1, 1852

ELECTED.

72 COWEY, JOHN, Wearmouth Colliery, Sunderland	Nov.	2, 1872
73 COX, JOHN H., 10, St. George's Square, Sunderland	Feb.	6, 1875
74*COXE, E. B., Drifton, Jeddo, P.O. Luzerne Co., Penns., U.S. ...	Feb.	1, 1873
75 CRAWFORD, T., 3, Grasmere Street, Gateshead-on-Tyne	Sept.	3, 1864
76 CRAWSHAY, E., Gateshead-on-Tyne	Dec.	4, 1869
77 CRAWSHAY, G., Gateshead-on-Tyne	Dec.	4, 1869
78 CRONE, E. W., Killingworth Hall, near Newcastle-on-Tyne	Mar.	5, 1870
79 CRONE, J. R., Tudhoe House, via Spennymoor	Feb.	1, 1868
80 CRONE, S. C., Killingworth Hall, Newcastle (Member of Council) ...		1853
81 CROSS, JOHN, 77, King Street, Manchester	June	5, 1869
82 CROUDACE, C. J., Bettisfield Colliery Co., Limited, Bagillt, N. Wales	Nov.	2, 1872
83 CROUDACE, JOHN, West House, Haltwhistle	June	7, 1873
34 CROUDACE, THOMAS, 16, Lower Park Field, Putney, London... ...		1862
85 DAGLISH, JOHN, F.G.S., Marsden, South Shields (PAST-PRESIDENT.		
Member of Council)	Aug.	21, 1852
86 DAGLISH, W. S., Solicitor, Newcastle-on-Tyne	July	2, 1872
87 DALE, DAVID, West Lodge, Darlington	Feb.	5, 1870
88 D'ANDRIMONT, T., Liége, Belgium	Sept.	3, 1870
89 DANIEL, W., Steam Plough Works, Leeds	June	4, 1870
90 DARLING, FENWICK, South Durham Colliery, Darlington ...	Nov.	6, 1875
91 DARLINGTON, JAMES, Black Park Colliery, Ruabon, North Wales ...	Nov.	7, 1874
92 DAVEY, HENRY, C.E., 3 Princes Street, Westminster. London, S.W.	Oct.	11, 1873
93 DEES, R. R., Solicitor, Newcastle-on-Tyne	Oct.	7, 1871
94 DIXON, D. W., Lumpsey Mines, Brotton, Saltburn-by-the-Sea ...	Nov.	2, 1872
95 DIXON, NICH., Dudley Colliery, Dudley, Northumberland ...	Sept.	1, 1877
96 DIXON, R., Wire Rope Manufacturer, Teams, Gateshead ...	June	5, 1875
97 DODD, B., Bearpark Colliery, near Durham	May	3, 1866
98 DODDS, JOSEPH, M.P., Stockton-on-Tees	Mar.	7, 1874
99 DOUGLAS, C. P., Parliament Street, Consett, Co. Durham ...	Mar.	6, 1869
100 DOUGLAS, T., Peases' West Collieries, Darlington (VICE-PRESIDENT) ..	Aug.	21, 1852
101 DOVE, G., Viewfield, Stanwix, Carlisle	July	2, 1872
102 DOWDESWELL, H., Butterknowle Colliery, via Darlington ...	April	5, 1873
103 DYSON, GEORGE, Middlesbro'	June	2, 1866
104 ELLIOT, SIR GEORGE, Bart., M.P., D.C.L., Houghton Hall, Fence		
Houses, (PAST-PRESIDENT, Member of Council)	Aug.	21, 1852
105 ELSDON, ROBERT, 76, Manor Road, Upper New Cross, London	Nov.	4, 1876
106 EMBLETON, T. W., The Cedars, Methley, Leeds	Sept.	6, 1855
107 EMBLETON, T. W., Jun., The Cedars, Methley, Leeds	Sept.	2, 1865
108 EMINSON, J. B., Londonderry Offices, Seaham Harbour ...	Mar.	2, 1872
109 EVERARD, J. B., M.E., 6, Millstone Lane, Leicester	Mar.	6, 1869
110 FARMER, A., Seaton Carew, near West Hartlepool	Mar.	2, 1872
111 FAVELL, THOMAS M., F.G.S., Etruria Iron Works, near Stoke-on-Trent	April	5, 1873
112 FENWICK, BARNABAS, 84, Osborne Road, Newcastle-on-Tyne ...	Aug.	2, 1866
113 FERENS, ROBINSON, Oswald Hall, near Durham	April	7, 1877

ELECTED.

114 FLETCHER, H., Ladyshore Coll., Little Lever, Bolton, Lancashire ... Aug. 3, 1865
115 FLETCHER, JAS., Manager, Co-operative Collieries, Wallsend, near
    Newcastle, New South Wales ... ... Sept. 11, 1875
116 FLETCHER, JOHN, Rock House, Ulverstone ... . July 2, 1872
117 FOGGIN, WM., North Biddick Coll., Washington Station. Co. Durham Mar. 6, 1875
118 FORSTER, G. B., M.A., F.G.S., Lesbury, R.S.O., Northumberland
    (PAST-PRESIDENT, *Member of Council*) ... ... Nov. 5, 1852
119 FORSTER, J. R., Water Company's Office, Newcastle-on-Tyne ... July 2, 1872
120 FORSTER, J. T., Burnhope Colliery. near Lanchester, Co. Durham . Aug. 1, 1868
121 FORSTER, R., 25. Old Elvet, Durham ` .. .. Sept. 5, 1868
122 FOSTER, GEORGE, Osmondthorpe Colliery, near Leeds... ... ... Mar. 7, 1874
123 FRANCE, FRANCIS, St. Helen's Colliery Co., Ld., St. Helen's, Lancashire Sept. 1, 1877

124 GALLOWAY, T. LINDSAY, M.A., Argyll Colliery, Campbeltown, N.B. Sept. 2, 1876
125 GERRARD, JOHN, Westgate, Wakefield .. .. Mar. 5, 187⁰
126 GILLETT, F. C., 20, Midland Road, Derby ... ... July 4, 1861
127 GILROY, G., Woodlands, Parbold, near Wigan ... ... Aug. 7, 1856
128 GILROY, S. B., Mining Engineer, Hednesford, Stafford ... ... Sep'. ·5· 1868
129 GJERS, JOHN, Southfield Villas, Middlesbro' . ... June 7, 1873
130 GODDARD, F. R., Accountant, Newcastle-on-Tyne . ... Nov. 7, 1874
131 GORDON, JAMES N., c/o W. Nicolson, 5, Jeffrey's Square, St. Mary
    Axe, London, E.C. ... . Nov. 6, 1875
132 GRACE, E. N., Dhadka, Asansol, Bengal, India ... ... Feb. 1, 1868
133 GREAVES, J. O., St. John's, Wakefield .. .. Aug. 7, 1862
134 GREEN, J. T., Mining Engineer, Ty Celyn, Abercarne, Newport, Mon. Dec. 3, 1870
135 GREENER, JOHN, General Manager, Vale Coll., New Glasgow, Pictou,
    Nova Scotia ... ... Feb. 6, 1875
136 GREENWELL, G. C., F.G.S., Elm Tree Lodge, Duffield, Derby (PAST-
    PRESIDENT, *Member of Council*) ... ... Aug. 21, 1852
137 GREENWELL, G. C., Jun., Poynton, near Stockport ... ... Mar. 6, 1869
138 GREY, C. G.. Ballycourcy, Enniscorthy, County Wexford ... ... May 4, 1872
139 GRIEVES, D., Brancepeth Colliery, Willington, County Durham ... Nov. 7, 1874
140 GRIFFITH, N. R., Wrexham 1866
141 GRIMSHAW, E. J., 23, Hardshaw Street, St. Helen's, Lancashire ... Sept. 5, 1868

142 HAGGIE, D. H., Wearmouth Patent Rope Works, Sunderland ... Mar. 4, 1876
143*HAGUE, ERNEST, Castle Dyke, Sheffield ... ... Mar. 2, 1872
144 HAINES, J. RICHARD, Adderley Green Colliery, near Longton ... Nov. 7, 1874
145 HALES, C., Nercuis Cottage, Nercuis, near Mold, Flintshire 1865
146 HALL, M., Lofthouse Station Collieries, near Wakefield . ... Sept. 5, 1868
147 HALL, M. S., 8, Victoria Street, Bishop Auckland ... ... Feb. 14, 1874
148 HALL, WM., Murton Colliery, *via* Sunderland .. .. Dec. 4, 1875
149 HALL, WILLIAM F., Haswell Colliery, Haswell, *via* Sunderland ... May 13, 1858
150 HANN, EDMUND, Aberaman, Aberdare ... ... Sept. 5, 1868
151 HARBOTTLE, W. H., Orrell Coal and Cannel Co., near Wigan ... Dec. 4, 1875
152 HARGREAVES, WILLIAM, Rothwell Haigh, Leeds ... ... Sept. 5, 1868
153 HARLE, RICHARD, Browney Colliery, Durham .. .. April 7, 1877

ELECTED.

154 HARLE, WILLIAM, Pagebank Colliery, near Durham	Oct.	7, 1876
155 HARRISON, R., Eastwood, near Nottingham		1861
156 HARRISON, W. B., Brownhills Collieries, near Walsall	April	6, 1867
157 HAY, J., Jun., Widdrington Colliery, Acklington	Sept.	4, 1869
158 HEDLEY, J. J., Consett Collieries, Leadgate, County Durham ...	April	6, 1872
159 HEDLEY, J. L., Flooker's Brook, Chester	Feb.	5, 1870
160 HEDLEY, W. H., Consett Collieries, Medomsley, Newcastle-on-Tyne		
(Member of Council)		1864
161 HENDERSON, H., Pelton Colliery, Chester-le-Street	Feb.	14, 1874
162 HEPPELL, T., Leafield House, Birtley, Chester-le-Street	Aug.	6, 1863
163 HEPPELL, W., Western Hill, Durham	Mar.	2, 1872
164 HERDMAN, J., Park Crescent, Bridgend, Glamorganshire	Oct.	4, 1860
165 HESLOP, C., Upleatham & Lingdale Mines, Upleatham, R.S.O., Yorks ...	Feb.	1, 1868
166 HESLOP, GRAINGER, Whitwell Coal Company, Sunderland	Oct.	5, 1872
167 HETHERINGTON, D., Coxlodge Colliery, Newcastle-on-Tyne		1859
168*HEWITT, G. C., Coal Pit Heath Colliery, near Bristol	June	3, 1871
169 HEWLETT, A., Haseley Manor, Warwick	Mar.	7, 1861
170 HIGSON, JACOB		1861
171*HILTON, J., Wigan Coal and Iron Co., Limited, Wigan	Dec.	7, 1867
172 HILTON, T. W., F.G.S., Wigan Coal and Iron Co., Limited, Wigan ...	Aug.	3, 1865
173 HOLLIDAY, MARTIN F., Langley Grove, Durham	May	1, 1875
174 HOLMES, C., Grange Hill, near Bishop Auckland	April	11, 1874
175 HOMER, CHARLES J., Mining Engineer, Stoke-on-Trent	Aug.	3, 1865
176 HOOD, A., 6, Bute Crescent, Cardiff	April	18, 1861
177 HOPE, GEORGE, Success House, Fence Houses	Feb.	3, 1877
178 HORNSBY, H., Rodridge House, Wingate, R.S.O., Co. Durham ...	Aug.	1, 1874
179 HORSLEY, W., Whitehill Point, Percy Main, Newcastle-on-Tyne ...	Mar.	5, 1857
180 HOSKOLD, H. D., C. and M.E., F.R.G.S., F.G.S., M. Soc. A., &c.,		
Inspector General of Mines of the Argentine Republic, and		
Director of the National Department of Mines and Geology,		
Casilla, Correos, 900, Buenos Ayres	April	1, 1871
181 HOWARD, W. F., 13, Cavendish Street, Chesterfield	Aug.	1, 1861
182 HUMBLE, JOHN, West Pelton, Chester-le-Street	Mar.	4, 1871
183 HUMBLE, JOS., Staveley Works, near Chesterfield	June	2, 1866
184 HUNTER, J., Waratah Coal Co., Charlestown, N.S. Wales, Australia ..	Mar.	6, 1869
185 HURST, T. G., F.G.S., Osborne Road, Newcastle-on-Tyne	Aug.	21, 1852
186 JACKSON, C. G., Chamber Colliery Co., Limited, Hollinwood	June	4, 1870
187 JACKSON, W. G., Loscoe Grange, Normanton, Yorkshire	June	7, 1873
188 JARRATT, J., Houghton Main Colliery, near Barnsley	Nov.	2, 1867
189 JEFFCOCK, T. W., 18, Bank Street, Sheffield	Sept.	4, 1869
190 JENKINS, W., M.E., Ocean Collieries, Treorky, Glamorgan	Dec.	6, 1862
191 JENKINS, WM., Consett Iron Works, Consett, Durham	May	2, 1874
192 JOHNSON, J., Carlton Main Colliery, Barnsley	Mar.	7, 1874
193 JOHNSON, R. S., Sherburn Hall, Durham	Aug.	21, 1852
194 JOICEY, J. G., Forth Banks West Factory, Newcastle-on-Tyne ...	April	10, 1869
195 JOICEY, W. J., Urpeth Lodge, Chester-le-Street	Mar.	6, 1869

ELECTED.

196 KENDALL, JOHN D., Roper Street, Whitehaven ... ... Oct. 3, 1874
197 KIMPTON, J. G., 40, St. Mary's Gate, Derby ... ... Oct. 5, 1872
198 KIRKBY, J. W., Kirkland, Leven, Fife ... ... Feb. 1, 1873
199 KNOWLES, A., Swinton Old Hall, Manchester ... ... Dec. 5, 1856
200 KNOWLES, JOHN, Westwood, Pendlebury, Manchester ... ... Dec. 5, 1856

201 LAMB, R., Bowthorn Colliery, Cleator Moor, near Whitehaven ... Sept. 2, 1865
202 LAMB, R. O., The Lawn, Ryton-on-Tyne ... ... Aug. 2, 1866
203 LAMB, RICHARD W., 29, Great Cumberland Place, London, W. ... Nov. 2, 1872
204 LANCASTER, JOHN, Anfield House, Leamington ... ... Mar. 2, 1865
205 LANDALE, A., Comely Park Place, Dunfermline ... ... Dec. 2, 1858
206*LAPORTE. HENRY, M.E., Aciéries de France, Aubin, Aveyron, France May 5, 1877
207 LAVERICK, ROBT., West Rainton, Fence Houses ... ... Sept. 2, 1876
208 LAWRENCE, HENRY, Grange Iron Works, Durham (Mem. of Council) Aug. 1, 1868
209 LAWS, H., Grainger Street W., Newcastle-on-Tyne ... ... Feb. 6, 1869
210 LEBOUR, G. A., M.A., F.G.S., Durham College of Science, Newcastle
    (SECRETARY) ... ... Feb. 1, 1873
211 LEVER, ELLIS, Bowdon, Cheshire 1861
212 LEWIS, SIR WILLIAM THOMAS, Mardy, Aberdare 1864
213 LIDDELL, G. H., Somerset House, Whitehaven ... ... Sept. 4, 1869
214 LINSLEY, R., Cramlington Colliery, Northumberland ... ... July 2, 1872
215 LINSLEY, S. W., Whitburn Colliery, South Shields ... ... Sept. 4, 1869
216 LISHMAN, T., Jun., Hetton Colliery, Fence Houses (Mem. of Council) . Nov. 5, 1870
217 LISHMAN, WM., Holly House, Witton-le-Wear... 1857
218 LISHMAN, WM., Bunker Hill, Fence Houses (Member of Council) ... Mar. 7, 1861
219 LIVESEY, C., Bradford Colliery, near Manchester ... ... Aug. 3, 1865
220 LIVESEY, T., Bradford Colliery, near Manchester ... ... Nov. 7, 1874
221 LLEWELYN, L., Abersychan House, Abersychan ... ... May 4, 1872
222 LOGAN, WILLIAM, Langley Park Colliery, Durham .. .. Sept. 7, 1867
223 LONGBOTHAM, J., Barrow Collieries, Barnsley, Yorkshire ... ... May 2, 1868
224 LUPTON, A., F.G.S., 6, De Grey Road, Leeds ... ... Nov. 6, 1869

225 MALING, C. T., Ellison Place, Newcastle-on-Tyne ... ... Oct. 5, 1872
226 MAMMATT, J. E., C.E., St. Andrew's Chambers, Leeds 1864
227 MARLEY, JOHN, Thornfield, Darlington (PRESIDENT) ... ... Aug. 21, 1852
228 MARLEY, J. W., Marley, Pinching, & Marley, 41, Threadneedle Street,
    London ... ... Aug. 1, 1868
229 MARSHALL, F. C., Messrs. R. & W. Hawthorn, St. Peter's, Newcastle Aug. 2, 1866
230 MARSTON, W. B., Leeswood Vale Oil Works, Mold ... ... Oct. 3, 1868
231 MARTEN, E. B., C.E., Pedmore, near Stourbridge ... ... July 2, 1872
232 MATTHEWS, R. F., Marske Hall, Richmond, Yorkshire ... ... Mar. 5, 1857
233 MAUGHAN, J. A., Manager of the Government Central Provinces'
    Collieries, Umaria, via Katni, India, C.P. .. .. Nov. 7, 1863
234 MAY, GEO., Harton Colliery Offices, near South Shields (Member of
    Council) ... ... Mar. 6, 1862
235 McCREATH, J., 95, Bath Street, Glasgow ... ... Mar. 5, 1870
236 McCULLOCH, DAVID, Beech Grove, Kilmarnock, N.B. ... ... Dec. 4, 1875

ELECTED.

237 McMURTRIE, J., Radstock Colliery, Bath	... ...	Nov. 7, 1863
238 MERIVALE, PROF. J. H., M.A., 2, Victoria Villas, Newcastle-on-Tyne		
(Member of Council)	... ...	May 5, 1877
239 MILLER, ROBERT, Silkstone and Worsbro' Park Collieries, Locke Park,		
near Barnsley	... ...	Mar. 2, 1865
240 MILLS, M. H., Kirklye Hall, Alfreton ...	... ...	Feb. 4, 1871
241 MITCHELL, CHAS., Jesmond, Newcastle-on-Tyne	... ...	April 11, 1874
242 MITCHELL, JOSEPH, Mining Offices, Eldon Street, Barnsley ...	...	Feb. 14, 1874
243 MITCHINSON, R., Jun., Pontop Coll., Lintz Green Station, Co. Durham		Feb. 4, 1865
244 MONKHOUSE, JOS., Gilcrux, Cockermouth	... ...	June 4, 1863
245 MOOR, T., Cambois Colliery, Blyth	... ...	Oct. 3, 1868
246 MOOR, WM., Jun., Hetton Colliery, Fence Houses	... ...	July 2, 1872
247 MOORE. R. W., Somerset House, Whitehaven ...	... ...	Nov. 5, 1870
248 MORRIS, W., Waldridge Colliery, Chester-le-Street	... ...	1858
249*MORTON, H. J., 2, Westbourne Villas, South Cliff, Scarborough	...	Dec. 5, 1856
250 MORTON, H. T., Lambton, Fence Houses	... ...	Aug. 21, 1852
251 MUNDLE, ARTHUR, St. Nicholas' Chambers, Newcastle-on-Tyne	...	June 5, 1875
252 MUNDLE, W., Redesdale Mines, Bellingham	... ...	Aug. 2, 1873
253*NASSE, RUDOLPH, Oberbergrath, Saarbrücken, Prussia	... ...	1869
254 NEVIN, JOHN, Dunbottle House, Mirfield, Normanton	...	May 2, 1868
255 NEWALL, R. S., D.C.L., F.G.S., Ferndene, Gateshead-on-Tyne	...	May 2, 1863
256 NICHOLSON, E., Jun., Beamish Colliery, Chester-le-Street	...	Aug. 7, 1869
257 NICHOLSON, MARSHALL, Middleton Hall, Leeds	... ...	Nov. 7, 1863
258 NOBLE, CAPTAIN, C.B, F.R.S., F.R.A.S., F.C.S., Jesmond, New-		
castle-on-Tyne	... ...	Feb. 3, 1866
259 NORTH, F. W., F.G.S., Rowley Hall Colliery, Dudley, Staffordshire ...		Oct. 6, 1864
260 OGDEN, JOHN M., Solicitor, Sunniside, Sunderland	... ...	Mar. 5, 1857
261 OGILVIE, A. GRAEME, 8, Grove End Road, St. John's Wood, London		Mar. 3, 1877
262 OLIVER, ROBERT, Charlaw Colliery, near Durham	.. ...	Nov. 6, 1875
263 PALMER, A. S., Usworth Hall, Washington Station, Co. Durham	...	July 2, 1872
264 PALMER, SIR CHARLES MARK, Bart., M.P., Quay, Newcastle-on-Tyne		Nov. 5, 1852
265 PAMELY, C., Springfield, Berw Road. Pontypridd, South Wales	...	Sept. 5, 1868
266 PANTON, F. S., Silksworth Colliery, Sunderland	... ...	Oct. 5, 1867
267 PARRINGTON, M. W., Wearmouth Colliery, Sunderland	...	Dec. 1, 1864
268 PARTON, T., F.G.S., Hill Top, West Bromwich ..	... ...	Oct. 2, 1869
269 PEACE, M. W., Wigan, Lancashire	... ...	July 2, 1872
270 PEARCE, F. H., Bowling Iron Works, Bradford	... ...	Oct. 1, 1857
271 PEASE, Sir J. W., Bart., M.P., Hutton Hall, Guisbro', Yorkshire	...	Mar. 5, 1857
272 PEEL, JOHN, Wharncliffe Silkstone Collieries, near Barnsley	...	Nov. 1, 1860
273 PEEL, JOHN, Leasingthorne Colliery, Bishop Auckland	...	Mar. 3, 1877
274 PEILE, WILLIAM, Cartgate, Hensingham, Whitehaven	...	Oct. 1, 1863
275 PICKUP, P. W., 71, Preston New Road, Blackburn	... ...	Feb. 6, 1875
276 POTTER, ADDISON, C.B., Heaton Hall, Newcastle-on-Tyne	...	Mar. 6, 1869
277 POTTER, A. M., Shire Moor Colliery, Earsdon, Newcastle	...	Feb. 3, 1872

ELECTED.

321 SIMPSON, R., Moor House, Ryton-on-Tyne ... ... Aug. 21, 1852
322 SIMPSON, ROBT., Drummond Coll., Westville, Pictou, Nova Scotia ... Dec. 4, 1875
323 SLINN, T., 2, Choppington Street, Westmorland Road, Newcastle ... July 2, 1872
324 SMITH, G. F., Grovehurst, Tunbridge Wells ... ... Aug. 5, 1853
325 *SMITH, R. CLIFFORD, F.G S., Ashford Hall, Bakewell ... ... Dec. 5, 1874
326 SMITH, T. E., Phœnix Foundry, Newgate Street, Newcastle-on-Tyne Dec. 5, 1874
327 SOPWITH, A., Cannock Chase Collieries, near Walsall ... ... Aug. 1, 1868
328 SOPWITH, THOS., 6, Great George St., Westminster, London, S.W. ... Mar. 3, 1877
329 SOUTHERN, R., Burleigh House, The Parade, Tredegarville, Cardiff ... Aug. 3, 1865
330 SOUTHWORTH, THOS., Hindley Green Collieries, near Wigan ... ... May 2, 1874
331 SPENCER, JOHN, Westgate Road, Newcastle-on-Tyne ... ... Sept. 4, 1869
332 SPENCER, M., Newburn, near Newcastle-on-Tyne ... ... Sept. 4, 1869
333 SPENCER, T., Ryton, Newcastle-on-Tyne .. .. Dec. 6, 1866
334 SPENCER, W., Southfields, Leicester ... ... Aug. 21, 1852
335 STEAVENSON, A. L., Durham (VICE-PRESIDENT) ... ... Dec. 6, 1855
336 STEPHENSON, G. R., 9 Victoria Chambers, Westminster, London, S.W. Oct. 4, 1860
337 STOBART, W., Pepper Arden, Northallerton ... ... July 2, 1874
338 STOREY, THOS. E., Clough Hall Iron Works, Kidsgrove, Staffordshire Feb. 5, 1872
339 STRAKER, J. H., Stagshaw House, Corbridge-on-Tyne ... ... Oct. 3, 1876
340 STRATTON, T. H. M., Cramlington House, Northumberland *(Mem. of*
    Council) ... ... Dec. 3, 1870
341 SWALLOW, J., Bushblades House, Lintz Green, Newcastle-on-Tyne ... May 2, 1874
342 SWALLOW, R. T., Springwell, Gateshead-on-Tyne 1862
343 SWAN, H. F., Shipbuilder, Newcastle-on-Tyne ... ... Sept. 2, 1871
344 SWAN, J. G., Upsall Hall, near Middlesbro' ... ... Sept. 2, 1871
345 SWANN, C. G., Sec., General Mining Asso. Ld., Blomfield House, Lon-
    don Wall, and New Broad Street., London, E.C. ... ... Aug. 7, 1875

346 TATE, SIMON, Trimdon Grange Colliery, Co. Durham ... ... Sept. 11, 1875
347 TAYLOR, HUGH, King Street, Quay, Newcastle-on-Tyne ... ... Sept. 5, 1856
348 TAYLOR, T., Quay, Newcastle-on-Tyne ... ... July 2, 1872
349 TAYLOR-SMITH, THOMAS, Greencroft Park, Durham ... ... Aug. 2, 1866
350 THOMPSON, R., Jun., 19, The Crescent, Gateshead ... ... Sept. 7, 1867
351 THOMSON, JOHN, Eston Mines, by Middlesbro' ... ... April 7, 1877
352 THOMSON, JOS. F., Manvers Main Colliery, Rotherham ... ... Feb. 6, 1875
353 TINN, J., C.E., Ashton Iron Rolling Mills, Bedminster, Bristol ... Sept. 7, 1867
354 TYZACK, D., c/o Mr. Donnison, 71, Westgate Road, Newcastle-on-Tyne Feb. 14, 1874
355 TYZACK, WILFRED, So. Medomsley Colliery, Lintz Green, Newcastle... Oct. 7, 1876

356 VIVIAN, JOHN, Diamond Boring Company, Whitehaven ... ... Mar. 3, 1877

357 WADHAM, E., C. and M.E., Millwood, Dalton-in-Furness ... ... Dec. 7, 1867
358 WALKER, J. S., Pagefield Iron Works, Wigan, Lancashire ... ... Dec. 4, 1869
359 WALKER, W., Hawthorns, Saltburn-by-the-Sea ... ... Mar. 5, 1870
360 WALLACE, HENRY, Trench Hall, Gateshead ... ... Nov. 2, 1872
361 WARD, H., Rodbaston Hall, near Penkridge, Stafford... ... ... Mar. 6, 1862
362 WARDELL, S. C., Doe Hill House, Alfreton ... ... April 1, 1865

Ordinary Members.

Marked * are Life Members.

ELECTED.

11 DACRES, THOMAS, Dearham Colliery, *via* Carlisle ... ... May 4, 1878
12 DAVIES, JOHN, Hartley House, Coundon, Bishop Auckland ... ... April 10, 1886
13 DEES, J. GIBSON, Floraville, Whitehaven ... ... Oct. 13, 1883
14*DIXON, JAMES S., 97, Bath Street, Glasgow ... ... Aug. 3, 1878

15 ELLIS, W. R., F.G.S., Wigan .. .. June 1, 1878

16 FORREST, B. J., c/o G. Scoville, Calle Piedad 481, Buenos Ayres,
    Argentine Republic ... ... April 12, 1884
17 FORREST, J. C., Witley Coal Co., Limited. Halesowen, Birmingham .. April 12, 1884

18 GALLOWAY, WM., Mining Engineer, Cardiff ... ... April 23, 1887
19 GEDDES, GEORGE H., 142 Princes Street, Edinburgh ... ... Oct. 1, 1881
20 GOUDIE. J. H., Ironwood, by Watersmeet, Michigan, U.S.A. ... Sept. 7, 1878

21 JOHNSON, H., Jun., Mining Offices, Trindle Road, Dudley. So. Staff. Feb. 10, 1883
22 JOHNSON, WILLIAM, Radcliffe Colliery, Acklington, Northumberland Dec. 9, 1882

23 KELLETT, WILLIAM, Wigan ... ... June 1, 1878
24 KNOWLES, I., Wigan .. .. Oct. 13, 1883
25*KNOWLES. ROBERT, Arncliffe, Cheetham Hill, Manchester ... ... April 10, 1886

26 LANCASTER JOHN, Auchinheath, Southfield and Fence Collieries,
    Lesmahagow ... ... Sept. 7, 1878
27 LAWS, W. G. Town Hall, Newcastle-on-Tyne ... ... Oct. 2, 1880
28 LEACH, C. C., Seghill Colliery, Northumberland ... ... Mar. 7, 1874
29 LLEWELLIN, DAVID MORGAN, F.G.S., Glanwern Offices, Pontypool ... May 14, 1881

30 MARTIN, TOM PATTINSON, Allhallows Colliery, Mealsgate, Carlisle ... Feb. 15, 1879

31 POTTS, JOS., Jun., North Cliff, Roker, Sunderland ... ... Dec. 6, 1879
32*PRIOR, EDWARD G., Victoria, British Columbia... ... ... Feb. 7, 1880

33 RHODES, C. E., Carr House, Rotherham ... ... Aug. 4, 1883
34 RUSSELL, ROBERT, Coltness Iron Works, Newmains, N.B. ... ... Aug. 3, 1878

35 SELBY, ATHERTON, Leigh, near Manchester ... ... Oct. 13, 1883
36 SPENCER, JOHN W., Newburn, near Newcastle-on-Tyne ... ... May 4, 1878
37 STEVENS. JAMES. M.E., Kaiping Mines, c/o H.B.M.'s Consulate,
    Tientsin, North China ... ... Feb. 14, 1885

38 TOPPING, WALTER, Messrs. Cross, Tetley, & Co., Bamfurlong, nr. Wigan Mar. 2, 1878

39 VARTY, THOMAS, Skelton Park Mines, Skelton, R.S.O., Cleveland ... Feb. 12, 1887

40 WALKER, SYDNEY FERRIS, 196. Severn Road. Canton. Cardiff ... Dec. 9, 1882
41 WALKER, WILLIAM EDWARD, Lowther Street, Whitehaven ... ... Nov. 19, 1881
42 WINSTANLEY ROBT., M.E., 28, Deansgate, Manchester ... ... Sept. 7, 1878

Associate Members.

Marked * are Life Members.

ELECTED

1 AGNIEL. S., Mines de Vicoigne (Nord), Nœux (P. de C.), France ... April 23, 1887

2 ALLAN, JOHN F., La Carolina, Provincia de St. Luis, Argentine
    Republic ... ... Feb. 10, 1883

3 ALLISON, J. J. C., Hedley Hill Colliery, Waterhouses, Durham ... Feb. 13, 1886

4 ANDERSON, R. S., Elswick Colliery, Newcastle-on-Tyne ... ... June 9, 1883

5 ARMSTRONG, HENRY, Pelaw House, Chester-le-Street ... ... April 14, 1883

6 ARMSTRONG, J. H., St. Nicholas' Chambers, Newcastle-on-Tyne ... Aug. 1, 1885

7 ARMSTRONG, T. J., Hawthorn Terrace, Newcastle-on-Tyne ... ... Feb. 10, 1883

8 ARNOLD, THOS., Mineral Surveyor, Castle Hill, Greenfields, Llanelly Oct. 2, 1880

9 ATKINSON, A. A., New Brancepeth Colliery, near Durham ... ... Aug. 3, 1878

10 ATKINSON, FRED., Maryport ... ... Feb. 14, 1874

11 AUDUS, T., Mineral Traffic Manager, N.E. Railway, Newcastle-on-Tyne Aug. 7, 1880

12 AYTON, HENRY, Cowpen Colliery, Blyth, Northumberland ... ... Mar. 6, 1875

13 BAILES, E. T., Wingate, Ferryhill ... ... June 7, 1879

14 BALL, ALFRED F., 14, Landsdowne Terrace, Gosforth ... ... Dec. 11, 1886

15 BELL, GEO. FRED., 2, Belmout Crescent, Hillhead, Glasgow . . ... Sept. 6, 1879

16*BELL, THOMAS HUGH, Middlesbro'-on-Tees ... ... Dec. 11, 1882

17 BENNETT, ALFRED H., Dean Lane Collieries, Bedminster, Bristol ... April 10, 1886

18 BERKLEY, FREDERICK, Murton Colliery, near Sunderland ... ... Dec. 11, 1882

19 BERKLEY, R. W., Marley Hill, Whickham, R.S.O., Co. Durham ... Feb. 14, 1874

20 BEWICK, T. B., Hebburn, Newcastle-on-Tyne ... ... Mar. 7, 1874

21 BIRD, W. J., Wingate, Co. Durham ... ... Nov. 6, 1875

22 BLACKETT, W. C., Jun., Kimblesworth Colliery, Chester-le-Street ... Nov. 4, 1876

23 BOUCHER, A. S., La Salada puerto Bertio, E de Antioguia, United
    States of Colombia, S.A. ... ... Aug. 4, 1883

24 BRAMWELL, HUGH, Mining Offices, Marsden, South Shields ... ... Oct. 4, 1879

25 BROUGH, BENNETT H., F.G.S., 5, Robert Street, Adelphi, London, W.C. Dec. 10, 1887

26 BROUGH, THOMAS, Seaham Colliery, Sunderland ... ... Feb. 1, 1873

27 BROWN, C. GILPIN, Hetton Colliery, Fence Houses ... ... Nov. 4, 1876

28 BROWN, M. WALTON, 3, Summerhill Terrace, Newcastle-on-Tyne
    (Member of Council) ... ... Oct. 7, 1871

29 BROWN, ROBERT M., Norwood Colliery, via Darlington ... ... April 10, 1886

30 BRUCE, JOHN, Port Mulgrave, Hinderwell, R.S.O., Yorkshire ... Feb. 14, 1874

31 BULMAN, H. F., Broomside Colliery, near Durham ... ... May 2, 1874

32 BURDON, A. E., Hartford House. Cramlington, Northumberland ... Feb. 10, 1883

33 CABRERA, FIDEL, c/o H. Kendall & Son, 12, Great Winchester Street,
    London... ... ... Oct. 6, 1877

34*CANDLER, T. E., F.G.S., Hong Kong Club, Hong Kong. China ... May 1, 1875

35 CHARLTON, W. A. (of Tangyes Ltd.), 8, Richmond Terrace, Gateshead Nov. 6. 1880

36 CHILDE, HENRY S., Mining Engineer, Wakefield ... ... Feb. 12, 1887

ELECTED.

37 CLOUGH, JAMES, Willow Bridge, Choppington, Morpeth ... ... April 5, 1873
38 COCHRANE, RALPH D., Hetton Colliery Offices, Fence Houses ... June 1, 1878
39 COCKBURN, W. C., 1, St. Nicholas' Buildings, Newcastle-on-Tyne .. Oct. 8, 1887
40 COCKSON, CHARLES, Ince Coal and Cannel Co., Ince, Wigan ... ... April 22, 1882
41 COOPER, R. W., Solicitor, Newcastle-on-Tyne ... ... Sept. 4, 1880
42 CRAWFORD, T. W., 32, Poultry, London, E.C. ... ... Dec. 4, 1875
43 CRONE, F. E., Killingworth, Newcastle-on-Tyne ... ... Sept, 2, 1876
44 CURRY, W. THOS., Chelvey, Backwell, Somerset ... ... Sept. 4, 1880

45 DAKERS, W. R., Croxdale Colliery, Durham ... ... Oct. 14, 1882
46 DENNISTON, ROBERT B., Stuart Street, Dunedin, New Zealand ... Dec. 11, 1886
47 DOBINSON, LANCELOT, Durham Place House, Murton Colliery,
   Sunderland ... ... Feb. 11, 1888
48 DODD, M., Lemington, Scotswood-on-Tyne ... ... Dec. 4, 1875
49 DONKIN, WM., Warora Colliery, Wardha Coal State Railway, C.P., India Sept. 2, 1876
50 DOUGLAS, A. S., Stanley Villa, near Crook, via Darlington ... ... June 1, 1878
51 DOUGLAS, JOHN, Seghill Colliery, Dudley, Northumberland ... ... April 22, 1882
52 DOUGLAS, JOHN, Jun., Seghill Colliery, Dudley, Northumberland ... April 22, 1882
53 DOUGLAS, M. H., Marsden Colliery, South Shields ... ... Aug. 2, 1879
54 DOYLE, PATRICK, C.E., F.M.S., F.L.S., M.R.A.S., F.G.S., M.S.I.,
   Bengal E.I.R.. Chord Line, Sitarampur, India ... ... Mar. 1, 1879
55 DU PRE, F. B., 13, Old Elvet, Durham .. .. Oct. 9, 1886
56 DUNN, A. F., Poynton, Stockport, Cheshire ... ... June 2, 1877
57 DURNFORD, H. ST. JOHN, Swaithe Colliery, near Barnsley ... ... June 2, 1877

58 EDGE, JOHN H., Coalport Wire Rope and Chain Works, Shifnal, Salop Sept. 7, 1878
59 EDWARDS, F. H., Forth House, Bewick Street, Newcastle-on-Tyne ... June 11, 1887

60 FAIRLEY, JAMES, Craghead and Holmside Collieries, Chester-le-Street Aug. 7, 1880
61 FARROW, JOSEPH, Brotton Mines, Brotton, R.S.O. ... . Feb. 11, 1882
62 FERENS, FREDERICK J., Silksworth Colliery, Sunderland ... ... Dec. 4, 1880
63 FERGUSON, D., Cadzow Colliery, Hamilton, N.B. ... ... Dec. 8, 1883
64 FISHER, EDWARD R., Naut Glas, Cross Hands, near Llanelly, So. Wales Aug. 2, 1884
65 FLETCHER, LANCELOT, Marsden Colliery, South Shields ... ... April 14, 1888
66 FLETCHER, W., Brigham Hill, via Carlisle ... ... Oct. 13, 1883
67 FORSTER, THOMAS E., Lesbury, R.S.O., Northumberland ... ... Oct. 7, 1876
68 FRYAR, MARK, Denby Colliery, Derby ... ... Oct. 7, 1876

69 GERRARD, JAMES, 19, King Street, Wigan ... ... Mar. 3, 1873
70 GILCHRIST, J. R., Durham Main Colliery, Durham ... ... Feb. 3, 1877
71 GREENER, HENRY, South Pontop Colliery, Annfield Plain ... ... Dec. 11, 1882
72 GREENER, T. Y., Hucknall Torkard Collieries, near Nottingham ... July 2, 1872
73 GRESLEY, W. S., F.G.S., Assoc. Inst. C.E., Overseile, Ashby-de-la-Zouch Oct. 5, 1878

74 HADDOCK, W. T., Jun., Ryhope Colliery, Sunderland ... ... Oct. 7, 1876
75 HAGGIE, PETER SINCLAIR, Gateshead-on-Tyne ... ... April 14, 1883
76 HALLAS, G. H., Wigan and Whiston Coal Co., Limited, Prescot . Oct. 7, 1876

ELECTFD.

77 HALSE, EDWARD, 15, Clarendon Road, Notting Hill, London, W. ... June 13, 1885
78 HAMILTON, E., Rig Wood, Saltburn-by-the-Sea ... ... Nov. 1, 1873
79 HARRIS, W. S., Kibblesworth, Gateshead-on-Tyne ... ... Feb. 14, 1874
80 HEDLEY, E., Rainham Lodge, The Avenue, Beckenham, Kent ... Dec. 2, 1871
81 HEDLEY, SEPT. H., Bank Chambers, Wakefield ... ... Feb. 15, 1879
82 HEDLEY, T. F. Jun., Valuer, Sunderland ... ... April 23, 1887
83 HENDERSON, C. W. C., The Riding, Hexham .. .. Dec. 11, 1882
84 HENDY, J. C. B., Stanton Iron Co.'s Collieries, Pleasley, near Mansfield, Notts ... ... Sept. 2, 1876
85 HESLOP, SEPTIMUS, Belrooi Colliery, Sitarampur, E.I.R., Bengal, India Dec. 4, 1880
86 HESLOP, THOMAS, Storey Lodge Colliery, Cockfield, via Darlington ... Oct. 2, 1880
87 HILL, WILLIAM, Carterthorne Colliery Offices, Witton-le-Wear .. June 9, 1883
88 HOLME, JAMES, Engineer's Department, Canadian Pacific Railway,
Winnipeg, Canada ... ... June 12, 1886
89 HOOPER, FRED. G., South Derwent Coll., Annfield Plain, Lintz Green Feb. 14, 1885
90 HUMBLE, JOICEY .. .. Mar. 3, 1877
91 HUMBLE, ROBERT .. .. Sept. 2, 1876
92 HUMBLE, STEPHEN, 5, Westminster Chambers, Victoria St., London, S.W. Oct. 6, 1877

93 IRVINE, JOSEPH R., Hendon Ropery, Sunderland ... ... Dec. 10, 1887

94 JEPSON, H., 10, Crossgate, Durham ... ... July 2, 1872
95*JOBLING, THOS. E., Croft Villa, Blyth, Northumberland ... ... Oct. 7, 1876
96 JOHNSON, F. D., Aykleyheads, Durham ... ... Feb. 10, 1883
97 JOHNSON, W., Abram Colliery, Wigan .. .. Feb. 14, 1874

98 KIRKUP, PHILIP, Cornsay Colliery Office, Esh, near Durham ... Mar. 2, 1878
99 KIRTON, HUGH, Waldridge Colliery, Chester-le-Street ... ... April 7, 1877

100 LAVERICK, JOHN WALES, Tow Law Colliery Office, Tow Law, R.S.O.,
Co. Durham ... ... Dec. 11, 1882
101 LEE, JOHN F., Castle Eden Colliery, County Durham... ... ... June 13, 1885
102 LEE, WILLIAM, Felling Colliery, Newcastle-on-Tyne ... ... Dec. 10, 1887
103 LIDDELL, J. M., 3, Victoria Villas, Newcastle-on-Tyne ... ... Mar. 6, 1875
104 LISLE, J., Washington Colliery, County Durham ... ... July 2, 1872
105 LIVEING, E. H., 52, Queen Anne Street, Cavendish Square, London, W. Sept. 1, 1877

106 MACCABE, H. O., Russell Vale, Wollongong, New South Wales ... Sept. 7, 1878
107 MACKINLAY, T. B., West Pelton Colliery, Chester-le-Street ... Nov. 1, 1879
108 MADDISON, THOS. R., The Knowle, Mirfield ... ... Mar. 3, 1877
109 MAKEPEACE, H. R., Cwmaman Colliery, Aberdare ... ... Mar. 3, 1877
110 MARKHAM, G. E., Howlish Offices, Bishop Auckland ... ... Dec. 4, 1875
111 MATTHEWS, J., Messrs. R. & W. Hawthorn, Newcastle-on-Tyne April 11, 1885
112 McCARTHY, EDWARD T., A.R.S.M., c/o Col. Pigott, Higbury,
Eastbourne ... ... Oct. 8, 1887
113 McLAREN, R., Heddon Coal and Fire Brick Co., Wylam-on-Tyne ... Dec. 10, 1883
114*MERIVALE, W., District Engineer's Office, Indian Midland Railway,
Sangur, India ... ... Mar. 5, 1881

ELECTED.

115 MILLER, D. S., Cheadle, Staffordshire . ... Nov. 7, 1874

116*MILLER, N., 31, Hyde Lane, Hyde, near Manchester ... . Oct. 5, 1878

117 MOORE, WILLIAM, Loftus Mines, Loftus in Cleveland, R.S.O. ... Nov. 19, 1881

118 MOREING, C. A., Suffolk House, Laurence Pountney Hill, London, E.C. Nov. 7, 1874

119 MORISON, JOHN, Newbattle Collieries, Dalkeith, N.B. ... ... Dec. 4, 1880

120 MULHOLLAND, M. L., 74, Weardale Street, Mount Pleasant, Spennymoor, County Durham ... ... Dec. 11, 1886

121 MURTON, CHARLES J., Delaval Benwell Colliery, Newcastle-on-Tyne Mar. 6, 1880

122 MUSGRAVE, HENRY, Havercroft Main Colliery, Wakefield ... ... June 12, 1886

123 NICHOL, WM., Boldon Colliery, Newcastle-on-Tyne ... ... Oct. 9, 1886

124 ORNSBY, R. E., Seaton Delaval Colliery, Northumberland ... ... Mar. 6, 1875

125 PALMER, HENRY, East Howle Colliery, near Ferryhill ... ... Nov. 2, 1878

126 PARSONS, HON. CHARLES ALGERNON, Elvaston Hall, Ryton-on-Tyne June 12, 1886

127 PEAKE, C. E., Eskell Chambers, Nottingham ... ... Nov. 3, 1877

128 PEAKE, R. C., Stoke Lodge, Bletchley, Bucks. ... ... Feb. 7, 1880

129*PEASE, ARTHUR, Darlington ... ... Dec. 11, 1882

130 PREST, J. J., Kimblesworth Colliery, Chester-le-Street ... ... May 1, 1875

131 PREST, T., Bedlington Colliery, R.S.O , Northumberland ... .. June 14, 1884

132 PRICE, S. R., Cottam Colliery, Barlbro, near Chesterfield ... ... Nov. 3, 1877

133 PROCTOR, C. P., Shibden Hall Collieries, near Halifax, Yorkshire ... Oct. 7, 1876

134 PROUD, JOSEPH, South Hetton Colliery Offices, Sunderland ... ... Oct. 14, 1882

135 RATHBONE, EDGAR P., 2, Great George Street, Westminster, London Mar. 7, 1878

136 RICH, WILLIAM, Minas de Rio Tinto, Provincia de Huelva, Spain ... June 9, 1888

137 RIDLEY, Sir MATTHEW WHITE, Bart., M.P., Blagdon, Northumberland Feb. 10, 1883

138 ROBINSON, FRANK, The Nunnery, Orrell Mount, Wigan ... ... Sept. 2, 1876

139 ROBSON, T. O., Bensham Crescent, Gateshead-on-Tyne ... ... Sept. 11, 1875

140 ROUTLEDGE, W. H., The Rhyd, Tredegar, Mon., Wales ... ... Oct. 7, 1876

141*SAISE, WALTER, D.Sc. (Lond.), F.G.S., M.Inst.C.E., Manager E.I.R.
Collieries, Giridi, Bengal, India ... ... Nov. 3, 1877

142 SAWYER, A. R., Ass. R.S.M., Newcastle, Staffordshire ... ... Dec. 6, 1873

143 SCURFIELD, GEO. J., Hurworth-upon-Tees, Darlington ... ... Dec. 11, 1882

144 SHIPLEY, T., Woodland Colliery Office, Woodland, Butterknowle,
R.S.O., Co. Durham ... ... Aug. 2, 1884

145 SIMPSON, F. L. G., Mohpani Coal Mines, Gadawarra, C.P., India ... Dec. 13, 1884

146 SMITH, EUSTACE, Wire Rope Manufacturer and Shipbuilder, Newcastle June 11, 1887

147 SNOWBALL, JOSEPH, Seaton Burn House, Dudley, Northumberland ... Feb. 10, 1883

148 SOUTHERN, E. O., Ashington Colliery, near Morpeth .. .. Dec. 5, 1874

149 SPENCE, R. F., Cramlington, R.S.O., Northumberland ... ... Nov. 2, 1878

150 STOBART, F., Pensher House, Fence Houses ... ... Aug. 2, 1873

151 STOBART, H. T., Wearmouth Colliery, Sunderland ... ... Oct. 2, 1880

152 STOBBS, FRANK, 1, Queen Street, Newcastle-on-Tyne .. .. Oct. 1, 1881

153 STOKER, ARTHUR P., Birtley, near Chester-le-Street ..' ... . Oct. 6, 1877

ELECTED.

154 TELFORD, W. H., Hartford Coll., Cramlington, R.S.O., Northumberland Oct. 3, 1874
155 THOMPSON, CHARLES LACY, Milton Hall, Carlisle ... ... Feb. 10, 1883
156 TODD, JOHN T., Hamsteels, near Durham ... ... Nov. 4, 1876
157*TYERS, JOHN E., Mohpani Coal Mines, Gadawarra, C.P., India ... Dec. 10, 1887

158 VITANOFF, GEO. N., Sophia, Bulgaria ... ... April 22, 1882

159 WAIN, WM. HOLT, Podmore Hall Collieries, Newcastle-under-Lyne .. Feb. 12, 1887
160 WALLAU, JACOB, Messrs. Black, Hawthorn & Co., Gateshead ... Dec. 10, 1887
161 WALTERS, HARGRAVE, Birley Collieries, near Sheffield ... ... June 4, 1881
162 WALTON, J. COULTHARD, Writhlington Collieries, Radstock, via Bath Nov. 7, 1874
163*WARD, T. H., F.G.S., Assistant Manager, E.I.R. Collieries, Giridi,
Bengal, India ... ... Aug. 7, 1882
164 WARDLE, EDWARD, Craghead Colliery, Chester-le-Street ... ... Feb. 5, 1881
165 WATKYN-THOMAS, W., M.E., Mineral Office, Cockermouth Castle ... Feb. 10, 1883
166 WEARS, W. G., M.E., 28 and 29, St. Swithin's Lane, London, E.C. ... June 9, 1888
167 WEBSTER, H. INGHAM, Morton House, Fence Houses ... ... April 14, 1883
168 WEEKS, R. L., Willington, Co. Durham ... ... June 10, 1882
169 WHITE, C. E., Hebburn Colliery, near Newcastle-on-Tyne ... ... Nov. 4, 1876
170 WIGHT, EDWD. S., c/o R. M. Wight, Askam-in-Furness, Lancashire Dec. 12, 1885
171 WILSON, J. D., Ouston House, Chester-le-Street ... ... Sept. 11, 1873
172 WILSON, JOHN R., Swaithe, near Barnsley ... ... June 9, 1883
173 WORMALD, C. F., Mayfield Villa, Saltwell, Gateshead-on-Tyne ... Dec. 8, 1885

174 YOUNG, JOHN A., 7, Tyne Vale Terrace, Gateshead ... ... Dec. 10, 1887

Students.

1 BARRASS, M., Tudhoe Colliery, Spennymoor ... ... Dec. 10, 1883
2 BAUMGARTNER, W. O., Houghton-le-Spring, Fence Houses, Co. Durham Sept. 6, 1879
3 BLAKELEY, A. B., Soothill Wood Colliery Co., Limited, near Batley... Feb. 15, 1879
4 BROWN, WESTGARTH F., Alston House, Cardiff ... ... Oct. 9, 1886

5 CHANDLEY, CHARLES, Atherton Collieries, near Manchester ... ... Nov. 6, 1880
6 COLE, COLLIN, Broomfield, Newcastle-on-Tyne ... ... Oct. 18, 1882
7 CRAWFORD, JAMES MILL, Murton Colliery, near Sunderland ... Dec. 11, 1882

8 FORSTER, C. W., Lesbury, R.S.O., Northumberland ... ... June 10, 1882
9 FORSTER, GEORGE W., Heworth Colliery, near Newcastle-on-Tyne ... Oct. 8, 1887
10 FUTERS, THOMAS 97, Stanhope Street, Newcastle-on-Tyne ... ... Feb. 12, 1887

11 GALLWEY, A. P., Ruby and Dunderburg Mining Co., Eureka,
Nevada, U.S. ... ... Oct. 2, 1880
12 GREIG, J., Eston Mines, Middlesbro'-on-Tees ... ... Feb. 5, 1881

ELECTED.

13 HAGGIE, DOUGLAS, Thorncliffe Iron Works, Sheffield... ... ... April 14, 1883
14 HARE, SAMUEL, Brymbo Co., near Wrexham, North Wales ... ... Aug. 2, 1879
15 HARRISON, R. W., Leicestershire Club, Leicester ... ... Mar. 3, 1877
16 HAY, W., Jun., Nostell Colliery, Wakefield ... ... Dec. 10, 1883
17 HILL, LEONARD, Newport Wire Mills, Middlesbro' ... . Oct. 6, 1877
18 HOOPER, EDWARD, c/o J. H. Hooper, College Precincts, Worcester .. June 4, 1881
19 HOWARD, WALTER, c/o F. W. Schwager, Coronel, Chili ... ... April 13, 1878
20 HURST, GEO., Seaton Delaval Colliery, Northumberland ... ... April 14, 1883
21 HUTT, E. H., Medomsley, near Newcastle-on-Tyne ... ... Aug. 4, 1883

22 KAYLL, A. C., Gosforth, Newcastle-on-Tyne ... ... Oct. 7, 1876
23 KIRKHOUSE, E. G., 1, Edith Street, Consett, Co. Durham ... ... Aug. 3, 1878

24 LISHMAN, R. R., Celynen Colliery, Abercarne, via Newport, Mon. ... June 9, 1883

25 McMURTRIE, G. E. J., 7, Clifton Bank, Rotherham ... ... Aug. 2, 1884
26 MITTON, A. D., Hetton Colliery, Fence Houses ... ... June 9, 1883

27 NICHOLSON, A. D., Eldon Colliery, Co. Durham ... ... June 13, 1885
28 NICHOLSON, J. H., North Seaton Colliery Office, Newbiggin-by-the-Sea Oct. 1, 1881

29 OATES, ROBERT J. W., E.I.R. Collieries, Giridi, Bengal, India ... Feb. 10, 1883

30 PEART, A. W., 70, Caeharris, Dowlais, South Wales ... ... Nov. 4, 1876
31 PEASE, J. F., Pierremont, Darlington ... ... June 9, 1883
32 POTTER, E. A., Cramlington, Northumberland ... ... Feb. 6, 1875
33 PRINGLE, H. A., Barrow Collieries, Barnsley, Yorkshire ... ... Oct. 2, 1880
34 PRINGLE, HY. GEO., The Southern Coal Co., Ltd., Wollongong, New
        South Wales ... ... Dec. 4, 1880

35 REDMAYNE, R. A. S., Hetton Collieries, near Fence Houses... ... Dec. 13, 1884
36 RICHARDSON, RALPH, Field House, West Rainton, Fence Houses ... June 9, 1883
37 RIDLEY, WM., So. Tanfield Coll., Stanley, R.S.O., Newcastle-on-Tyne Dec. 11, 1882

38 SCOTT, JOSEPH SAMUEL, East Hetton Colliery, Coxhoe, Co. Durham Nov. 19, 1881
39 SCOTT, WALTER, Cornsay Colliery, Lanchester... ... ... Sept. 6, 1879
40 SCOTT, WM., Brancepeth Colliery Offices, Willington, Co. Durham ... Mar. 4, 1876
41 SHUTE, WM. ASHLEY, Westoe, South Shields ... ... April 11, 1885
42 SIMPSON, F. R., Hedgefield House, Blaydon-on-Tyne ... ... Aug. 4, 1883
43 SMITH, THOS., Leadgate, Co. Durham ... ... Feb. 15, 1879
44 SMITH, T. F., Jun., c/o Mr. Parry, Grocer and Draper, Littledean,
        Newnham ... ... May 5, 1877
45 STEAVENSON, C. H., Durham ... ... April 14, 1883
46 SYKES, FRANK K., Peases' West Collieries, Crook, by Darlington ... Feb. 13, 1886

47 WAUGH, C. L., Ffalda Steam Coal Colliery, Garw Valley, near Bridgend Nov. 19, 1881

48 YEOMAN, THOMAS, Willington Hall, Willington, Co. Durham ... Feb. 14, 1885

Subscribers under Bye-law 9.

1 Ashington Colliery, Newcastle-on-Tyne.

2 Birtley Iron Company, Birtley.

3 Bridgewater Trustees.

4 Haswell Colliery, Sunderland.

5 Hetton Collieries, Fence Houses.

6 Lambton Collieries, Fence Houses.

7 Londonderry Collieries, Seaham Harbour.

8 Marquess of Bute.

9 North Hetton Colliery, Fence Houses.

10 Ryhope Colliery, near Sunderland.

11 Seghill Colliery, Northumberland.

12 South Hetton and Murton Collieries.

13 Stella Colliery, Hedgefield, Blaydon-on-Tyne.

14 Throckley Colliery, Newcastle-on-Tyne.

15 Victoria Garesfield Colliery, Lintz Green.

16 Wearmouth Colliery, Sunderland.

CHARTER

OF

THE NORTH OF ENGLAND

𝕴𝖓𝖘𝖙𝖎𝖙𝖚𝖙𝖊 𝖔𝖋 𝕸𝖎𝖓𝖎𝖓𝖌 𝖆𝖓𝖉 𝕸𝖊𝖈𝖍𝖆𝖓𝖎𝖈𝖆𝖑 𝕰𝖓𝖌𝖎𝖓𝖊𝖊𝖗𝖘.

FOUNDED 1852.

INCORPORATED NOVEMBER 28TH, 1876.

𝖁𝖎𝖈𝖙𝖔𝖗𝖎𝖆, by the Grace of God, of the United Kingdom of Great Britain and Ireland, Queen, Defender of the Faith, TO ALL TO WHOM THESE PRESENTS SHALL COME, GREETING:

WHEREAS it has been represented to us that NICHOLAS WOOD, of Hetton, in the County of Durham, Esquire (since deceased); THOMAS EMERSON ·FORSTER, of Newcastle-upon-Tyne, Esquire (since deceased); SIR GEORGE ELLIOT, Baronet (then George Elliot, Esquire), of Houghton Hall, in the said County of Durham, and EDWARD FENWICK BOYD, of Moor House, in the said County of Durham, Esquire, and others of our loving subjects, did, in the year one thousand eight hundred and fifty-two, form themselves into a Society, which is known by the name of THE NORTH OF ENGLAND INSTITUTE OF MINING AND MECHANICAL ENGINEERS, having for its objects the Prevention of Accidents in Mines and the Advancement of the Sciences of Mining and Engineering generally, of which Society LINDSAY WOOD, of Southill, Chester-le-Street, in the County of Durham, Esquire, is the present President. AND WHEREAS it has been further represented to us that the Society was not constituted for gain, and that neither its projectors nor Members derive nor have derived pecuniary profit from its prosperity; that it has during its existence of a period of nearly a quarter of a century steadily devoted itself to the preservation of human life and the safer development of mineral property; that it has contributed substantially and beneficially to the prosperity of the country and the welfare and happiness of the working members of the community; that the Society has since its establishment diligently pursued its aforesaid objects, and in so doing has made costly experiments

and researches with a view to the saving of life by improvements in the
ventilation of mines, by ascertaining the conditions under which the safety
lamp may be relied on for security; that the experiments conducted by
the Society have related to accidents in mines of every description, and
have not been limited to those proceeding from explosions; that the various modes of getting coal, whether by mechanical appliances or otherwise,
have received careful and continuous attention, while the improvements
in the mode of working and hauling belowground, the machinery employed for preventing the disastrous falls of roof underground, and the
prevention of spontaneous combustion in seams of coal as well as in cargoes, and the providing additional security for the miners in ascending
and descending the pits, the improvements in the cages used for this purpose, and in the safeguards against what is technically known as "over-
winding," have been most successful in lessening the dangers of mining,
and in preserving human life; that the Society has held meetings at stated
periods, at which the results of the said experiments and researches have
been considered and discussed, and has published a series of Transactions
filling many volumes, and forming in itself a highly valuable Library of
scientific reference, by which the same have been made known to the
public, and has formed a Library of Scientific Works and Collections of
Models and Apparatus, and that distinguished persons in foreign countries
have availed themselves of the facilities afforded by the Society for communicating important scientific and practical discoveries, and thus a useful
interchange of valuable information has been effected; that in particular,
with regard to ventilation, the experiments and researches of the Society,
which have involved much pecuniary outlay and .personal labour,
and the details of which are recorded in the successive volumes of
the Society's Transactions, have led to large and important advances
in the practical knowledge of that subject, and that the Society's researches have tended largely to increase the security of life; that the
Members of the Society exceed 800 in number, and include a large proportion of the leading Mining Engineers in the United Kingdom. AND
WHEREAS in order to secure the property of the Society, and to extend its
useful operations, and to give it a more permanent establishment among
the Scientific Institutions of our Kingdom, we have been besought to grant
to the said LINDSAY WOOD, and other the present Members of the Society,
and to those who shall hereafter become Members thereof, our Royal
Charter of Incorporation. Now KNOW YE that we, being desirous of
encouraging a design so laudable and salutary of our special grace, certain knowledge, and mere motion, have willed granted, and declared, and

do, by these presents, for us, our heirs, and successors, will, grant, and
declare, that the said LINDSAY WOOD, and such others of our loving subjects as are now Members of the said Society, and such others as shall
from time to time hereafter become Members thereof, according to such
Bye-laws as shall be made as hereinafter mentioned, and their successors,
shall for ever hereafter be, by virtue of these presents, one body, politic and
corporate, by the name of "THE NORTH OF ENGLAND INSTITUTE OF
MINING AND MECHANICAL ENGINEERS," and by the name aforesaid shall
have perpetual succession and a Common Seal, with full power and
authority to alter, vary, break, and renew the same at their discretion, and
by the same name to sue and be sued, implead and be impleaded, answer
and be answered unto, in every Court of us, our heirs and successors, and
be for ever able and capable in the law to purchase, acquire, receive, possess, hold, and enjoy to them and their successors any goods and chattels
whatsoever, and also be able and capable in the law (notwithstanding the
statutes and mortmain) to purchase, acquire, possess, hold and enjoy to
them and their successors a hall or house, and any such other lands, tenements, or hereditaments whatsoever, as they may deem requisite for the
purposes of the Society, the yearly value of which, including the site
of the said hall or house, shall not exceed in the whole the sum of three
thousand pounds, computing the same respectfully at the rack rent which
might have been had or gotten for the same respectfully at the time of
the purchase or acquisition thereof. AND WE DO HEREBY GRANT our
especial licence and authority unto all and every person and persons and
bodies politic and corporate, otherwise competent, to grant, sell, alien,
convey or devise in mortmain unto and to the use of the said Society and
their successors, any lands, tenements, or hereditaments not exceeding
with the lands, tenements or hereditaments so purchased or previously acquired such annual value as aforesaid, and also any moneys,
stocks, securities, and other personal estate to be laid out and disposed of
in the purchase of any lands, tenements, or hereditaments not exceeding
the like annual value. AND WE FURTHER will, grant, and declare, that the
said Society shall have full power and authority, from time to time, to
sell, grant, demise, exchange and dispose of absolutely, or by way of
mortgage, or otherwise, any of the lands, tenements, hereditaments and
possessions, wherein they have any estate or interest, or which they shall
acquire as aforesaid, but that no sale, mortgage, or other disposition of any
lands, tenements, or hereditaments of the Society shall be made, except
with the approbation and concurrence of a General Meeting. And our will
and pleasure is, and we further grant and declare that for the better rule

and government of the Society, and the direction and management of the concerns thereof, there shall be a Council of the Society, to be appointed from among the Members thereof, and to include the President and the Vice-Presidents, and such other office-bearers or past office-bearers as may be directed by such Bye-laws as hereinafter mentioned, but so that the Council, including all *ex-officio* Members thereof, shall consist of not more than forty or less than twelve Members, and that the Vice-Presidents shall be not more than six or less than two in number. AND WE DO HEREBY FURTHER will and declare that the said LINDSAY WOOD shall be the first President of the Society, and the persons now being the Vice-Presidents, and the Treasurer and Secretary, shall be the first Vice-Presidents, and the first Treasurer and Secretary, and the persons now being the Members of the Council shall be the first Members of the Council of the Society, and that they respectfully shall continue such until the first election shall be made at a General Meeting in pursuance of these presents. AND WE DO HEREBY FURTHER will and declare that, subject to the powers by these presents vested in the General Meetings of the Society, the Council shall have the management of the Society, and of the income and property thereof, including the appointment of officers and servants, the definition of their duties, and the removal of any of such officers and servants, and generally may do all such acts and deeds as they shall deem necessary or fitting to be done, in order to carry into full operation and effect the objects and purposes of the Society, but so always that the same be not inconsistent with, or repugnant to, any of the provisions of this our Charter, or the Laws of our Realm, or any Bye-law of the Society in force for the time being. AND WE DO FURTHER will and declare that at any General Meeting of the Society, it shall be lawful for the Society, subject as hereinafter mentioned, to make such Bye-laws as to them shall seem necessary or proper for the regulation and good government of the Society, and of the Members and affairs thereof, and generally for carrying the objects of the Society into full and complete effect, and particularly (and without its being intended hereby to prejudice the foregoing generality), to make Bye-laws for all or any of the purposes hereinafter mentioned, that is to say: for fixing the number of Vice-Presidents, and the number of Members of which the Council shall consist, and the manner of electing the President and Vice-Presidents, and other Members of the Council, and the period of their continuance in office, and the manner and time of supplying any vacancy therein; and for regulating the times at which General Meetings of the Society and Meetings of the Council shall be held, and for convening the same and regulating the proceedings thereat, and for regulating the manner of admitting persons to be Members of the Society, and of removing or expelling Members from the Society, and for imposing reasonable fines or penalties for non-performance of any such Bye-laws, or for disobedience thereto, and from time to time to annul, alter, or change any such Bye-laws so always that all Bye-laws to be made as aforesaid be not repugnant to these presents, or to any of the laws of our Realm. AND WE DO FURTHER will and declare that the present Rules and Regulations of the Society, so far as they are not inconsistent with these presents, shall continue in force, and be deemed the Bye-laws of the Society until the same shall be altered by a General Meeting, provided always that the present Rules and Regulations of the Society and any future Bye-laws of the Society so to be made as aforesaid shall have no force or effect whatsoever until the same shall have been approved in writing by our Secretary of State for the Home Department. IN WITNESS WHEREOF WE HAVE CAUSED THESE OUR LETTERS TO BE MADE PATENT.

Witness Ourself at our Palace, at Westminster, this 28th day of November, in the fortieth year of our reign.

By Her Majesty's Command.

CARDEW.

THE NORTH OF ENGLAND INSTITUTE

OF

MINING AND MECHANICAL ENGINEERS.

BYE-LAWS

PASSED AT A GENERAL MEETING ON THE 16TH JUNE. 1877.

1.—The members of the North of England Institute of Mining and Mechanical Engineers shall consist of four classes, viz.:—Original Members, Ordinary Members, Associate Members, and Honorary Members, with a class of Students attached.

2.—ORIGINAL MEMBERS shall be those who were Ordinary Members on the 1st of August, 1877.

3.—ORDINARY MEMBERS.—Every candidate for admission into the class of Ordinary Members, or for transfer into that class, shall come within the following conditions :—He shall be more than twenty-eight years of age, have been regularly educated as a Mining or Mechanical Engineer, or in some other recognised branch of Engineering, according to the usual routine of pupilage, and have had subsequent employment for at least five years in some responsible situation as an Engineer, or if he has not undergone the usual routine of pupilage, he must have practised on his own account in the profession of an Engineer for at least five years, and have acquired a considerable degree of eminence in the same.

4.—ASSOCIATE MEMBERS shall be persons practising as Mining or Mechanical Engineers, or in some other recognised branch of Engineering, and other persons connected with or interested in Mining or Engineering.

5.—HONORARY MEMBERS shall be persons who have distinguished themselves by their literary or scientific attainments, or who have made important communications to the Society.

6.—Students shall be persons who are qualifying themselves for the profession of Mining or Mechanical Engineering, or some other of the recognised branches of Engineering, and such persons may continue Students until they attain the age of twenty-three years.

7.—The annual subscription of each Original Member, and of each Ordinary Member who was a Student on the 1st of August, 1877, shall be £2 2s., of each Ordinary Member (except as last mentioned) £3 3s., of each Associate Member £2 2s., and of each Student £1 1s., payable in advance, and shall be considered due on election, and afterwards on the first Saturday in August of each year.

8.—Any Member may, at any time, compound for all future subscriptions by a payment of £25, where the annual subscription is £3 3s., and by a payment of £20 where the annual subscription is £2 2s. All persons so compounding shall be Original, Ordinary, or Associate Members for life, as the case may be ; but any Associate Member for life who may afterwards desire to become an Ordinary Member for life, may do so, after being elected in the manner described in Bye-law 13, and on payment of the further sum of £5.

9.—Owners of Collieries, Engineers, Manufacturers, and Employers of labour generally, may subscribe annually to the funds of the Institute, and each such subscriber of £2 2s. annually shall be entitled to a ticket to admit two persons to the rooms, library, meetings, lectures, and public proceedings of the Society ; and for every additional £2 2s., subscribed annually, two other persons shall be admissible up to the number of ten persons ; and each such Subscriber shall also be entitled for each £2 2s. subscription to have a copy of the Proceedings of the Institute sent to him.

10.—In case any Member, who has been long distinguished in his professional career, becomes unable, from ill-health, advanced age, or other sufficient cause, to carry on a lucrative practice, the Council may, on the report of a Sub-Committee appointed for that purpose, if they find good reason for the remission of the annual subscription, so remit it. They may also remit any arrears which are due from a member, or they may accept from him a collection of books, or drawings, or models, or other contributions, in lieu of the composition mentioned in Bye-law 8, and may thereupon constitute him a Life Member, or permit him to resume his former rank in the Institute.

11.—Persons desirous of becoming Ordinary Members shall be proposed and recommended, according to the Form A in the Appendix, in which form the name, usual residence, and qualifications of the candidate shall be distinctly specified. This form must be signed by the proposer and at least five other Members certifying a personal knowledge of the candidate. The proposal so made being delivered to the Secretary, shall be submitted to the Council, who on approving the qualifications shall determine if the candidate is to be presented for ballot, and if it is so determined, the Chairman of the Council shall sign such approbation The same shall be read at the next Ordinary General Meeting, and afterwards be placed in some conspicuous situation until the following Ordinary General Meeting, when the candidate shall be balloted for.

12.—Persons desirous of being admitted into the Institute as Associate Members, or Students, shall be proposed by three Members; Honorary Members shall be proposed by at least five Members, and shall in addition be recommended by the Council, who shall also have the power of defining the time during which, and the circumstances under which, they shall be Honorary Members. The nomination shall be in writing, and signed by the proposers (according to the Form B in the Appendix), and shall be submitted to the first Ordinary General Meeting after the date thereof. The name of the person proposed shall be exhibited in the Society's room until the next Ordinary General Meeting, when the candidate shall be balloted for.

13.—Associate Members or Students, desirous of becoming Ordinary Members, shall be proposed and recommended according to the Form C in the Appendix, in which form the name, usual residence, and qualifications of the candidate shall be distinctly specified. This form must certify a personal knowledge of the candidate, and be signed by the proposer and at least two other Members, and the proposal shall then be treated in the manner described in Bye-law 11. Students may become Associate Members at any time after attaining the age of twenty-three on payment of an Associate Member's subscription.

14.—The balloting shall be conducted in the following manner :— Each Member attending the Meeting at which· a ballot is to take place shall be supplied (on demand) with a list of the names of the persons to be balloted for, according to the Form D in the Appendix, and shall strike out the names of such candidates as he desires shall not be elected, and return the list to the scrutineers appointed by the presiding Chairman for the purpose, and such scrutineers shall examine the lists so returned, and inform the meeting what elections have been made. No candidate shall be elected unless he secures the votes of two-thirds of the Members voting.

15.—Notice of election shall be sent to every person within one week after his election, according to the Form E in the Appendix, enclosing at the same time a copy of Form F, which shall be returned by the person elected, signed, and accompanied with the amount of his annual subscription. or life composition, within two months from the date of such election, which otherwise should become void.

16.—Every Ordinary Member elected having signed a declaration in the Form F, and having likewise made the proper payment, shall receive a certificate of his election.

17.—Any person whose subscription is two years in arrear shall be reported to the Council, who shall direct application to be made for it, according to the Form G in the Appendix, and in the event of its continuing one month in arrear after such application, the Council shall have the power, after remonstrance by letter, according to the Form H in the Appendix, of declaring that the defaulter has ceased to be a member.

18.—In case the expulsion of any person shall be judged expedient by ten or more Members, and they think fit to draw up and sign a proposal requiring such expulsion, the same being delivered to the Secretary, shall be by him laid before the Council for consideration. If the Council, after due inquiry, do not find reason to concur in the proposal, no entry thereof shall be made in any minutes, nor shall any public discussion thereon be permitted, unless by requisition signed by one-half the Members of the Institute ; but if the Council do find good reason for the proposed expulsion, they shall direct the Secretary to address a letter, according to the Form I in the Appendix, to the person proposed to be expelled, advising him to withdraw from the Institute. If that advice be followed, no entry on the minutes nor any public discussion on the subject shall be permitted ; but if that advice be not followed, nor an explanation given which is satisfactory to the Council, they shall call a General Meeting for the purpose of deciding on the question of expulsion ; and if a majority of the persons present at such Meeting (provided the number so present be not less than forty) vote that such person be expelled, the Chairman of that Meeting shall declare the same accordingly, and the Secretary shall communicate the same to the person, according to the Form J in the Appendix.

19.—The Officers of the Institute, other than the Treasurer and the Secretary, shall be elected from the Original, Ordinary and Associate Members, and shall consist of a President, six Vice-Presidents, and eighteen Councillors, who, with the Treasurer and the Secretary (if Members of the Institute) shall constitute the Council. The President, Vice-Presidents, and Councillors shall be elected at the Annual Meeting in August (except in cases of vacancies) and shall be eligible for re-election, with the exception of any President or Vice-President who may have held office for the three immediately preceding years, and such six Councillors as may have attended the fewest Council Meetings during the past year; but such Members shall be eligible for re-election after being one year out of office.

20.—The Treasurer and the Secretary shall be appointed by the Council, and shall be removable by the Council, subject to appeal to a General Meeting. One and the same person may hold both these offices.

21.—Each Original, Ordinary, and Associate Member shall be at liberty to nominate in writing, and send to the Secretary not less than eight days prior to the Ordinary General Meeting in June, a list, duly signed, of Members suitable to fill the offices of President, Vice-Presidents, and Members of Council, for the ensuing year. The Council shall prepare a list of the persons so nominated, together with the names of the Officers for the current year eligible for re-election, and of such other Members as they deem suitable for the various offices. Such list shall comprise the names of not less than thirty. The list so prepared by the Council shall be submitted to the General Meeting in June, and shall be the balloting list for the annual election in August. (See Form K in the Appendix.) A copy of this list shall be posted at least seven days previous to the Annual Meeting, to every Original, Ordinary, and Associate Member; who may erase any name or names from the list, and substitute the name or names of any other person or persons eligible for each respective office; but the number of persons on the list, after such erasure or substitution, must not exceed the number to be elected to the respective offices. Papers which do not accord with these directions shall be rejected by the scruti-neers. The Votes for any Members who may not be elected President or Vice-Presidents shall count for them as Members of the Council. The Chairman shall appoint four scrutineers, who shall receive the balloting papers, and, after making the necessary scrutiny, destroy the same, and sign and hand to the Chairman a list of the elected Officers. The balloting papers may be returned through the post, addressed to the Secretary, or be handed to him, or to the Chairman of the Meeting, so as to be received before the appointment of the scrutineers for the election of Officers.

22.—In case of the decease or resignation of any Officer or Officers, the Council, if they deem it requisite that the vacancy shall be filled up, shall present to the next Ordinary General Meeting a list of persons whom they nominate as suitable for the vacant offices, and a new Officer or Officers shall be elected at the succeeding Ordinary General Meeting.

23.—The President shall take the chair at all meetings of the Institute, the Council, and Committees, at which he is present (he being *ex-officio* a member of all), and shall regulate and keep order in the proceedings.

24.—In the absence of the President, it shall be the duty of the senior Vice-President present to preside at the meetings of the Institute, to keep order, and to regulate the proceedings. In case of the absence of the President and of all the Vice-Presidents, the meeting may elect any Member of Council, or in case of their absence, any Member present, to take the chair at the meeting.

25.—The Council may appoint Committees for the purpose of transacting any particular business, or of investigating specific subjects connected with the objects of the Institute. Such Committees shall report to the Council, who shall act thereon as they see occasion.

26.—The Treasurer and the Secretary shall act under the direction and control of the Council, by which body their duties shall from time to time be defined.

27.—The Funds of the Society shall be deposited in the hands of the Treasurer, and shall be disbursed or invested by him according to the direction of the Council.

28.—The Copyright of all papers communicated to, and accepted for printing by the Council, and printed within twelve months, shall become vested in the Institute, and such communications shall not be published for sale or otherwise, without the written permission of the Council.

29.—An Ordinary General Meeting shall be held on the first Saturday of every month (except January and July) at two o'clock, unless otherwise determined by the Council; and the Ordinary General Meeting in the month of August shall be the Annual Meeting, at which a report of the proceedings, and an abstract of the accounts of the previous year, shall be presented by the Council. A Special General Meeting shall be called whenever the Council may think fit, and also on a requisition to the Council, signed by ten or more Members. The business of a Special Meeting shall be confined to that specified in the notice convening it.

30.—At meetings of the Council, five shall be a quorum. The minutes of the Council's proceedings shall be at all times open to the inspection of the Members.

31.—All Past-Presidents shall be *ex-officio* Members of the Council so long as they continue Members of the Institute, and Vice-Presidents who have not been re-elected or have become ineligible from having held office for three consecutive years, shall be *ex-officio* Members of the Council for the following year.

32.—Every question, not otherwise provided for, which shall come before any Meeting, shall be decided by the votes of the majority of the Original, Ordinary, and Associate Members then present.

33.—All papers shall be sent for the approval of the Council at least twelve days before a General Meeting, and after approval, shall be read before the Institute. The Council shall also direct whether any paper read before the Institute shall be printed in the Transactions, and notice shall be given to the writer within one month after it has been read, whether it is to be printed or not.

34.—All proofs of reports of discussions, forwarded to Members for correction, must be returned to the Secretary within seven days from the date of their receipt, otherwise they will be considered correct and be printed off.

35.—The Institute is not, as a body, responsible for the statements and opinions advanced in the papers which may be read, nor in the discussions which may take place at the meetings of the Institute.

36.—Twelve copies of each paper printed by the Institute shall be presented to the author for private use.

37.—Members elected at any meeting between the Annual Meetings shall be entitled to all papers issued in that year, so soon as they have signed and returned Form F, and paid their subscriptions.

38.—The Transactions of the Institute shall not be forwarded to Members whose subscriptions are more than one year in arrear.

39.—No duplicate copies of any portion of the Transactions shall be issued to any of the Members unless by written order from the Council.

40.—Invitations shall be forwarded to any person whose presence at the discussions the Council may think advisable, and strangers so invited shall be permitted to take part in the proceedings but not to vote. Any Member of the Institute shall also have power to introduce two strangers (see Form L) to any General Meeting, but they shall not take part in the proceedings except by permission of the Meeting.

41.—No alteration shall be made in the Bye-laws of the Institute, except at the Annual Meeting, or at a Special Meeting for that purpose, and the particulars of every such alteration shall be announced at a previous Ordinary Meeting, and inserted in its minutes, and shall be exhibited in the room of the Institute fourteen days previous to such Annual or Special Meeting, and such Meeting shall have power to adopt any modification of such proposed alteration of the Bye-laws.

Approved,
R. ASSHETON CROSS.

Whitehall,
  2nd July, 1877.

APPENDIX TO THE BYE-LAWS.

[FORM A.]

A. B. [Christian Name, Surname, Occupation, and Address in full], being upwards of twenty-eight years of age, and desirous of being elected an -Ordinary Member of the North of England Institute of Mining and Mechanical Engineers, I recommend him from *personal knowledge* as a person in every respect worthy of that distinction, because—

[Here specify distinctly the qualifications of the Candidate, according to the spirit of Bye-law 3.]

On the above grounds, I beg leave to propose him to the Council as a proper person to be admitted an Ordinary Member.

Signed_____ Member.

Dated this   day of   18

We, the undersigned, concur in the above recommendation, being convinced that A. B. is in every respect a proper person to be admitted an ordinary Member.

FROM PERSONAL KNOWLEDGE.

Five Members.

[To be filled up by the Council.]

The Council, having considered the above recommendation, present A. B. to be balloted for as a   of the North of England Institute of Mining and Mechanical Engineers.

Signed_____ Chairman.

Dated this   day of   18

[FORM B.]

A. B. [Christian Name, Surname, Occupation, and Address in full], being desirous of admission into the North of England Institute of Mining and Mechanical Engineers, we, the undersigned, propose and recommend that he shall become [an Honorary Member, or an Associate Member, or a Student] thereof.

$$\left.\begin{array}{c}\text{———————————}\\\text{———————————}\\\text{———————————}\\\text{———————————}\end{array}\right\}\begin{array}{l}\text{Three}^*\\\text{Members.}\end{array}$$

* If an Honorary Member, five signatures are necessary, and the following Form must be filled in by the Council.

Dated this   day of   18

[*To be filled up by the Council.*]

The Council, having considered the above recommendation, present A. B. to be balloted for as an Honorary Member of the North of England Institute of Mining and Mechanical Engineers.

      Signed ————————————Chairman.

Dated   day of   18

[FORM C.]

A. B. [Christian Name, Surname, Occupation, and Address in full], being at present a   of the North of England Institute of Mining and Mechanical Engineers, and upwards of twenty-eight years of age, and being desirous of becoming an Ordinary Member of the said Institute, I recommend him, from *personal knowledge*, as a person in every respect worthy of that distinction, because—

[*Here specify distinctly the Qualifications of the Candidate according to the spirit of Bye-law 3.*]

On the above grounds, I beg leave to propose him to the Council as a proper person to be admitted an Ordinary Member.

      Signed — ————————————Member.

Dated this   day of   18

We, the undersigned, concur in the above recommendation, being convinced that A. B. is in every respect a proper person to be admitted an Ordinary Member.

FROM PERSONAL KNOWLEDGE.

———————————————— ⎫ Two
———————————————— ⎰ Members.

[*To be filled up by the Council.*]

The Council, having considered the above recommendation, present A. B. to be balloted for as an Ordinary Member of the North of England Institute of Mining and Mechanical Engineers.

Signed ————————————Chairman.

Dated   day of   18

[FORM D.]

List of the names of persons to be balloted for at the Meeting on , the   day of   18

ORDINARY MEMBERS:—

————————————————————————
————————————————————————
————————————————————————

ASSOCIATE MEMBERS:—

————————————————————————
————————————————————————

HONORARY MEMBERS:—

————————————————————————
————————————————————————

STUDENTS:—

————————————————————————
————————————————————————

Strike out the names of such persons as you desire should *not* be elected, and hand the list to the Chairman.

[FORM E.]

SIR,—I beg leave to inform you that on the   day of you were elected a   of the North of England Institute of Mining and Mechanical Engineers, but in conformity with its Rules your election cannot be confirmed until the enclosed form be returned to me with your signature, and until your first annual subscription be paid, the amount of which is £ , or, at your option, the life-composition of £

If the subscription is not received within two months from the present date, the election will become void under Bye-law 15.

I am, Sir,

Yours faithfully,

Secretary.

Dated 18

[FORM F.]

I, the undersigned, being elected a of the North of England Institute of Mining and Mechanical Engineers, do hereby agree that I will be governed by the Charter and Bye-laws of the said Institute for the time being ; and that I will advance the objects of the Institute as far as shall be in my power, and will not aid in any unauthorised publication of the proceedings, and will attend the meetings thereof as often as I conveniently can ; provided that whenever I shall signify in writing to the Secretary that I am desirous of withdrawing my name therefrom, I shall (after the payment of any arrears which may be due by me at that period) cease to be a Member.

Witness my hand this day of 18

[FORM G.]

Sir,—I am directed by the Council of the North of England Institute of Mining and Mechanical Engineers to draw your attention to Bye-law 17, and to remind you that the sum of £ of your annual subscriptions to the funds of the Institute remains unpaid. and that you are in consequence in arrear of subscription. I am also directed to request that you will cause the same to be paid without further delay, otherwise the Council will be under the necessity of exercising their discretion as to using the power vested in them by the Article above referred to.

I am, Sir,

Yours faithfully,

Secretary

Dated 18

[FORM H.]

SIR,—I am directed by the Council of the North of England Institute of Mining and Mechanical Engineers to inform you, that in consequence of non-payment of your arrears of subscription, and in pursuance of Bye-law 17, the Council have determined that unless payment of the amount £   is made previous to the   day of
next, they will proceed to declare that you have ceased to be a Member of the Institute.

But, notwithstanding this declaration, you will remain liable for payment of the arrears due from you.

I am, Sir,

Yours faithfully,

Secretary.

Dated   18

[FORM I.]

SIR,—I am directed by the Council of the North of England Institute of Mining and Mechanical Engineers to inform you that, upon mature consideration of a proposal which has been laid before them relative to you, they feel it their duty to advise you to withdraw from the Institute, or otherwise they will be obliged to act in accordance with Bye-law 18.

I am, Sir,

Yours faithfully,

Secretary.

Dated   18

[FORM J.]

SIR,—It is my duty to inform you that, under a resolution passed at a Special General Meeting of the North of England Institute of Mining and Mechanical Engineers, held on the   day of
18   , according to the provisions of Bye-law 18 you have ceased to be a Member of the Institute.

I am, Sir,

Yours faithfully,

Secretary.

Dated   18

[FORM K.]

BALLOTING LIST.

Ballot to take place at the Meeting of   18   at Two o'Clock.

PRESIDENT—ONE NAME only to be returned, or the vote will be lost.

——— President for the current year eligible for re-election.

———} New Nominations.

VICE-PRESIDENTS—SIX NAMES only to be returned, or the vote will be lost.

     The Votes for any Members who may not be elected as President or Vice-Presidents will count for them as other Members of the Council.

———} Vice-Presidents for the current year eligible for re-election.

———} New Nominations.

COUNCIL—EIGHTEEN NAMES only to be returned, or the vote will be lost.

———} Members of the Council for the current year eligible for re-election.

———} New Nominations

Any list returned with a greater number of Names than ONE PRESIDENT, SIX VICE-PRESIDENTS, EIGHTEEN COUNCILLORS, Will be rejected by the Scrutineers as informal, and the Votes will consequently be lost.

Extract from Bye-law 21.

Each Original, Ordinary, and Associate Member shall be at liberty to nominate in writing, and send to the Secretary not less than eight days prior to the Ordinary General Meeting in June, a list, duly signed, of Members suitable to fill the Offices of President, Vice-Presidents, and Members of Council, for the ensuing year. The Council shall prepare a list of the persons so nominated, together with the names of the Officers for the current year eligible for re-election, and of such other Members as they deem suitable for the various offices. Such list shall comprise the names of not less than thirty. The list so prepared by the Council shall be submitted to the General Meeting in June, and shall be the balloting list for the annual election in August. (See Form K in the Appendix.) A copy of this list shall be posted at least seven days previous to the Annual Meeting, to every Original, Ordinary, and Associate Member; who may erase any name or names from the list, and substitute the name or names of any other person or persons eligible for each respective office; but the number of persons on the list, after such erasure or substitution, must not exceed the number to be elected to the respective offices. Papers which do not accord with these directions shall be rejected by the Scrutineers. The Votes for any Members who may not be elected President or Vice-Presidents shall count for them as Members of the Council. The Chairman shall appoint four Scrutineers, who shall receive the balloting papers, and after making the necessary scrutiny destroy the same, and sign and hand to the Chairman a list of the elected Officers. The balloting papers may be returned through the post, addressed to the Secretary, or be handed to him, or to the Chairman of the Meeting, so as to be received before the appointment of the Scrutineers for the election of Officers.

Names substituted for any of the above are to be written in the blank spaces opposite those they are intended to supersede.

The following Members are ineligible from causes specified in Bye-law 19:—

As PRESIDENT _____

As VICE-PRESIDENT _____

As COUNCILLORS _____

[FORM L.]

Admit

of

to the Meeting on Saturday, the

(Signature of Member or Student)

The Chair to be taken at Two o'Clock.

I undertake to abide by the Regulations of the North of England Institute of Mining and Mechanical Engineers, and not to aid in any unauthorised publication of the Proceedings.

(Signature of Visitor)

Not transferable.

NORTH OF ENGLAND INSTITUTE

OF

MINING AND MECHANICAL ENGINEERS.

GENERAL MEETING, SATURDAY, OCTOBER 8TH, 1887, IN THE
WOOD MEMORIAL HALL.

Mʀ. T. J. BEWICK ɪɴ ᴛʜᴇ Cʜᴀɪʀ.

The Sᴇᴄʀᴇᴛᴀʀʏ read the minutes of the last meeting, and reported
the proceedings of the Council.

The following gentlemen were elected, having been previously nomin‑
ated :—

Oʀᴅɪɴᴀʀʏ Mᴇᴍʙᴇʀs—

Mr. Cʜᴀʀʟᴇs Z. Bᴜɴɴɪɴɢ, Warora, Central Provinces, India.
Mr. Jᴏʜɴ Cʀɪɢʜᴛᴏɴ, 2, Clarence Buildings, Booth Street, Manchester

Assᴏᴄɪᴀᴛᴇ Mᴇᴍʙᴇʀs—

Mr. W. C. Cᴏᴄᴋʙᴜʀɴ, 1, St. Nicholas' Buildings, Newcastle‑on‑Tyne.
Mr. E. McCᴀʀᴛʜʏ, A.R.S.M., 60, Sunniside Road, Ealing, W., London.

Sᴛᴜᴅᴇɴᴛ—

Mr. Gᴇᴏ. W. Fᴏʀsᴛᴇʀ, Heworth Colliery, near Newcastle‑on‑Tyne.

The following gentlemen were nominated for election :—

Hᴏɴᴏʀᴀʀʏ Mᴇᴍʙᴇʀ—

Mr. Wɪʟʟɪᴀᴍ Bᴇᴀᴛᴛɪᴇ Sᴄᴏᴛᴛ, Mines Inspector, Wolverhampton.

Oʀᴅɪɴᴀʀʏ Mᴇᴍʙᴇʀ—

Mr. Wɪʟʟɪᴀᴍ Sᴛᴇᴘʜᴇɴsᴏɴ Bʟᴀᴄᴋʙᴜʀɴ, Mining Engineer, Astley House.
Woodlesford, near Leeds.

––––––––

Mr. M. WALTON BROWN read the following paper on "A further attempt for the correlation of the Coal-seams of the Carboniferous Formation of the North of England, with some notes on the probable duration of the Coal-field":—

A FURTHER ATTEMPT FOR THE CORRELATION OF THE COAL SEAMS OF THE CARBONIFEROUS FORMATION OF THE NORTH OF ENGLAND, WITH SOME NOTES UPON THE PROBABLE DURATION OF THE COAL-FIELD.

By M. WALTON BROWN.

INTRODUCTION.

THE Carboniferous formation of the North of England is usually divided into two divisions :—

> A.—Upper Carboniferous, comprising the true Coal-Measures, Gannister beds, and Millstone Grit.
>
> B.—Lower Carboniferous, comprising the Carboniferous Limestone or Bernician series, Tuedian and Basement beds.

The great Northern coal-field of Northumberland and Durham consists of the Upper and Lower Carboniferous Measures lying against the coast line of the North Sea.

The Upper Carboniferous rocks extend from Staindrop, near the River Tees, on the south, to the mouth of the River Coquet on the north, a distance of about 52 miles, with a maximum width of about 20 miles.

This coal-field is of trough-like form, whose longer axis lies along the coast line, with the beds rising more or less regularly to the north and west.

The Lower Carboniferous rocks are found in the western and northern portions of the two counties, as they rise from under the denuded Upper Carboniferous rocks, and form the rolling moorlands and mountainous areas of the Pennine chain.

In other districts the Coal-Measures are divided into upper, middle, and lower, but in this district it is preferable to class all the strata above the Brockwell Seam as true Coal-Measures.

UPPER CARBONIFEROUS MEASURES.

A valuable "Synopsis of the several Seams of Coal in the Newcastle District" was read by Mr. John Buddle in 1830, before the Natural History Society of Northumberland, Durham, and Newcastle-upon-Tyne.*

* See Transactions, 1831, Vol. I., pp. 117–131.

This synopsis was revised in 1863, by Messrs. Nicholas Wood, J. Taylor, and J. Marley, and will be found in a paper on "Coal Mining, &c.," appearing in Vol. XII. of the Transactions of this Institute, page 153. A most valuable sheet diagram was arranged by Mr. J. B. Simpson in 1877, which shows the depth, thickness, and local names of the seams in several of the principal collieries of the various districts.

Upon the aforementioned basis the synopsis given in Table I., pages 6 and 7, shows the probable correlation of the seams of the Coal-Measures in the different districts.

Throughout the whole coal-field the coal is of bituminous qualities, with the exception of small areas in certain seams, where local deposits of anthracite and cannel coal have been discovered.

Various estimates have been made at different times as to the profitable duration of the Northumberland and Durham coal-field, which are summarised in the Table given on the opposite page.

The estimate of the Royal Commission in 1871 of the coal then remaining unworked was :—

	Area in Square Miles.		Tons of Coals in Millions.
Land	685	...	6,734
3½ miles under the sea	111	...	1,137
Totals	796	...	7,871
Deduct quantity worked out since 1871 ...	...	...	500
			7,371

This quantity would supply the present annual demand of about 36 million tons for about 200 years.

In the Gannister beds and Millstone Grit series (lying below the true Coal-Measures) some thin seams of coal are found, which have been worked from time to time. The chief of these are the Victoria and Marshall Green Seams. A workable seam of about 2 feet in thickness is occasionally found at about 12 fathoms below the Brockwell Seam, a second workable seam of similar thickness being found at a further depth of 15 fathoms. It is also known by borings that other and probably worthless seams exist within a further depth of 150 feet, or 300 feet in all, below the Brockwell Seam.

LOWER CARBONIFEROUS MEASURES.

The Lower Carboniferous coals are found in the Carboniferous Limestone or Bernician series, which are separated from the Brockwell Seam by a thickness of about 1,100 feet of strata.

......

......

......

......

......

Coal.

......

Coal.

Coal.

Radcliffe.

......

Prince Albert.

Queens.

Little Wonder.

Princess or Top.

......

......

......

Duke or Main.

......

Yard or Bottom.

Givins (?)

Coal.

Coal.

Coal.

TABLE I.—SYNOPSIS OF THE COAL SEAMS ; THE U|

TOW LAW.	CROOK.	WEARMOUTH TO BISHOP AUCKLAND.	CONSETT.	PELTON.	WALLSEN	GATESHEAD AND PRUD|
				......	Closing H|	
				......	Hebburn or Mon|	
		......	......	......	Coal ...	
	......	Splint ...	Crow Coal ...	Coal ...	Three-Qu|	
	......	Three-Quarter	Shield Row ...	Shield Row	High Mai|	
	......	Five-Quarter	Five-Quarter or Splint Coal ...	Hard Coal	Metal Co| Stone Co|	
	......	Main Coal ...	Brass Thill ...	Main Coal	Yard Coa|	
	......	Maudlin ...	Hutton ...	Maudlin ...	Bensham		Main Coal
				Low Main	Six-Quar	Main Coal
......	......	Low Main ...			Five-Qua|	
......	Brass Thill ...	Brass Thill ...	Little Coal ...	Brass Thill	......	
... .	Hutton ...	Hutton ...	Main Coal ...	Hutton ...	Low Mai	Five-Quart
		......	Coal	......	Two-Qua	
... .	......	.. ....	......	... .	......	
... Three-Quarter	Harvey or Constantine	Harvey ...	No. 1 Seam ...	Beaumont	Beaumon|	
					......	
		......	Stone Coal ...	......	......	
......	......	......	......		....	
Ballarat ...	Ballarat ...	Busty	Busty Bank ...	Upper / Lower	Busty ...	....
r Busty ... oal	Five-Quarter or Jet				Coal ...	
ter } Top Coal	"B" or Little Seam	......	Pasture Drift ...	......	...	
... } Main Coal	Main Coal ..	Brockwell ...	Brockwell ...	......	Brockw(	
... .	Victoria ...	Coal	Coal	......	..	
... .	Marshall Green	......	......	......	..	
			......	......		

THE UPPER CARBONIFEROUS MEASURES.

GARESFIELD AND PRUDHOE.	RYTON.	KENTON, CALLERTON, AND DENTON.	BLYTH AND HARTLEY.	ASHINGTON AND MORPETH.	WIDDRINGTON.
......	......				
		......	Five-Quarter	......	
	......	Three-Quarter or Seventy Fathoms	Black Close or Moorland	......	
	......	Kenton Main Coal or Seven-Quarter	Glebe ...	Longhirst Top Seam	......
	......	Hollywell Reins, Five-Quarter, or Newbiggen Stone Coal	Gray or Blake	Main ...	
	......				
	......	Yard Coal...	... Yard Coal ...	Yard	......
		Bensham ..	... Bensham ...	Coal	......
Main Coal ...	Grand Lease Main Coal	......	Six-Quarter	Quarry ...	Ulgham Top...
			Hartley Stone Coal	Band	Ulgham Main
......	Crow Coal...	Benwell Main Coal or Grove Seam	......	....	
Five-Quarter	Five-Quarter		Low Main ...	Yard Seam at Pegswood	Coal ... ..
Ruler ...	Ruler ...	Coal	Plessy ...	Plessy ...	Coal ... ...
......	......	Coal	Coal ...	......	......
Barlow Fell	Towneley Main	Beaumont or Engine	Beaumont ...	Beaumont ...	Widdrington Main Coal
......	Hodge or Splint	Hodge	......		
Tilley ...	Tilley ...	......			
Hand ...	Hand ...	Coal	......	...	....
Five-Quarter or Main Coal	Whickham Stone Coal	Stone Coal or Three-Quarter	Coal ...	Coal ...	Coal ... ...
Six-Quarter	Five-Quarter or Jet	Denton Low Main or Main Coal	......	Coal	......
Yard ...	Three-Quarter	Three-Quarter or Black	Coal ...	Coal	......
Brockwell ...	Horsley Wood	Denton Low Low Main or Splint	Coal ...	Coal ...	Coal ... ...
Coal	......	Coal	......		
				...	....
			......	Coal ...	Coal ... ...

GANNISTER

ESTIMATES OF THE DURATION OF THE NORTHUMBERLAND AND DURHAM COAL-FIELD.

Date.	Authority.	Where quoted from.	Area, Square Miles.	Coal Remaining Unwrought, Millions of Tons.	Yearly Demand, Tons.	Duration Years.
1610	Sir George Selby	Speech in the House of Commons	?	?	?	21
1789	John Williams	Natural History of the Mineral Kingdom, Vol. II., pp. 158, 161, 171 to 173	Points out that coal mines are rapidly being exhausted.			
1793	Dr. Macnab	Letters addressed to the Right Hon. William Pitt, pages 131–136	300	1,153	3,100,000 / 3,850,000	375 / 300
1796	J. Bailey and G. Culley	General View of the Agriculture of the county of Northumberland. Third Edition, pp. 14–16. The estimates vary in later editions	200	2,188	2,650,000	825
1811	Eneas Mackenzie	Historical and Descriptive View of the County of Northumberland, p. 128. The estimates vary in later editions	200	697	3,100,000	223
1816	J. H. H. Holmes	A Treatise on the Coal Mines of Northumberland and Durham, p. 20	300	1,161	3,100,000	387
1827	Dr. Thomson	Annals of Philosophy	?	?	?	1,000
1828	Robert Bakewell	Introduction to Geology, p. 182	?	?	3,000,000	360
1829	H. Taylor	An Estimate of the Extent and Produce of the Durham and Northumberland Coal Field. See Report of S lect Committee (Lords) on the state of the Coal Trade, pp. 124–5	837	6,046	3,500,000	1,727
1829	Rev. Adam Sedgwick	Report of Select Committee (Commons) on the State of the Coal Trade, pp. 232–239	540	?	?	90
1829	Rev. W. Buckland, D.D.	Report of Select Committee (Commons) on the State of the Coal Trade, p. 240	418	1,511	3,500,000	432
1838	Thomas John Taylor	Richardson's Descriptive Companion through Newcastle-upon-Tyne and Gateshead, pp. 268–9	924	7,589	5,200,000	1,450
1817	G. C. Greenwell	?	?	3,315	10,000,000	331
1854	T. Y. Hall	Transactions of North of England Institute of Mining and Mechanical Engineers. Vol. II., p. 226	750	5,122	14,000,000 / 20,000,000	363 / 256
1871	T. E. Forster / Sir George Elliot	Report of the Commissioners—Coal in the United Kingdom. Vol. I., pp. 22, 23, and 26	796	7,871	30,476,000	?
1882	M. A. Soubeiran	Sur la Géologie du Bassin Houiller de Newcastle. Annales des Mines, series 18, Mémoires Vol. I., pp. 409–448	765	7,000	35,000,000	200

Many papers upon the geology of the Lower Carboniferous Measures of Northumberland and Durham have been published from time to time.* Some of the most valuable of these appear in the Transactions of this Institute, written by the late Mr. Nicholas Wood, Mr. E. F. Boyd, Mr. E. Gibsone, Professor G. A. Lebour, Mr. T. J. Bewick, and others.

The synopsis of the coal-seams and limestones of the Bernician series given in Table II., pages 12 and 13, is brought forward with some hesitation; it is not to be considered as more than a tentative effort, and subject to the criticism of those persons who are well acquainted with the geological features of these measures in the two northern counties.

The qualities of the coals, the produce of these seams, render them suitable for household, gas-making, and manufacturing purposes, but at present their application depends more especially upon the facilities of working and sale.

These seams are not extensively worked as compared with the Upper Carboniferous Measures, but their commercial value is becoming enhanced, and, with greater facilities of transit, their exploration is well worthy the attention of capitalists. They appear to be of contemporaneous formation with the Carboniferous Limestone coals of Scotland (known there as the Lower Coal-Measures).

The resources of the Lower Carboniferous or Mountain Limestone districts of Northumberland were estimated by Mr. T. E. Forster, in 1871, for the Coal Commission, at 665 million tons. An examination of the synopsis of these seams tends to show that the available areas and thickness of the seams have been under-estimated. Thus in the Scremerston district the seams appear to be from 54 to 73 feet in aggregate thickness. This development is not found in other parts of the Mountain Limestone, as in the more southern districts not more than four seams are found, which are only of workable thickness in small areas.

In an area of 1,200 square miles in the northern parts of this coalfield (omitting the portions covered by the Upper Coal-Measures) it may be assumed that there is at least 10 feet in thickness of workable coal. The contents of one square mile of this thickness will be 9,600,000 tons, and of 1,200 square miles will, consequently, be 11,520 million tons.

After an ample allowance for dykes and other interruptions, there will probably remain an available supply of 8,000 million tons.

---

* N. J. Winch, "Observations on the Geology of Northumberland and Durham," Transactions of the Geological Society (London), 1816, Vol. IV., pages 1 to 100. "Remarks on the Geology of the Banks of the Tweed," Transactions of Natural History Society of Northumberland, Durham, and Newcastle-upon-Tyne, 1831, Vol. I., pp. 117–131. "On the Geology of a part of Northumberland and Cumberland," by Nicholas Wood, Killingworth, 1831. *Ibid.* Vol. I., 302-334. "A Geological Map of Northumberland and Durham," George Tate, 1867. *Ibid.* Vol. II.

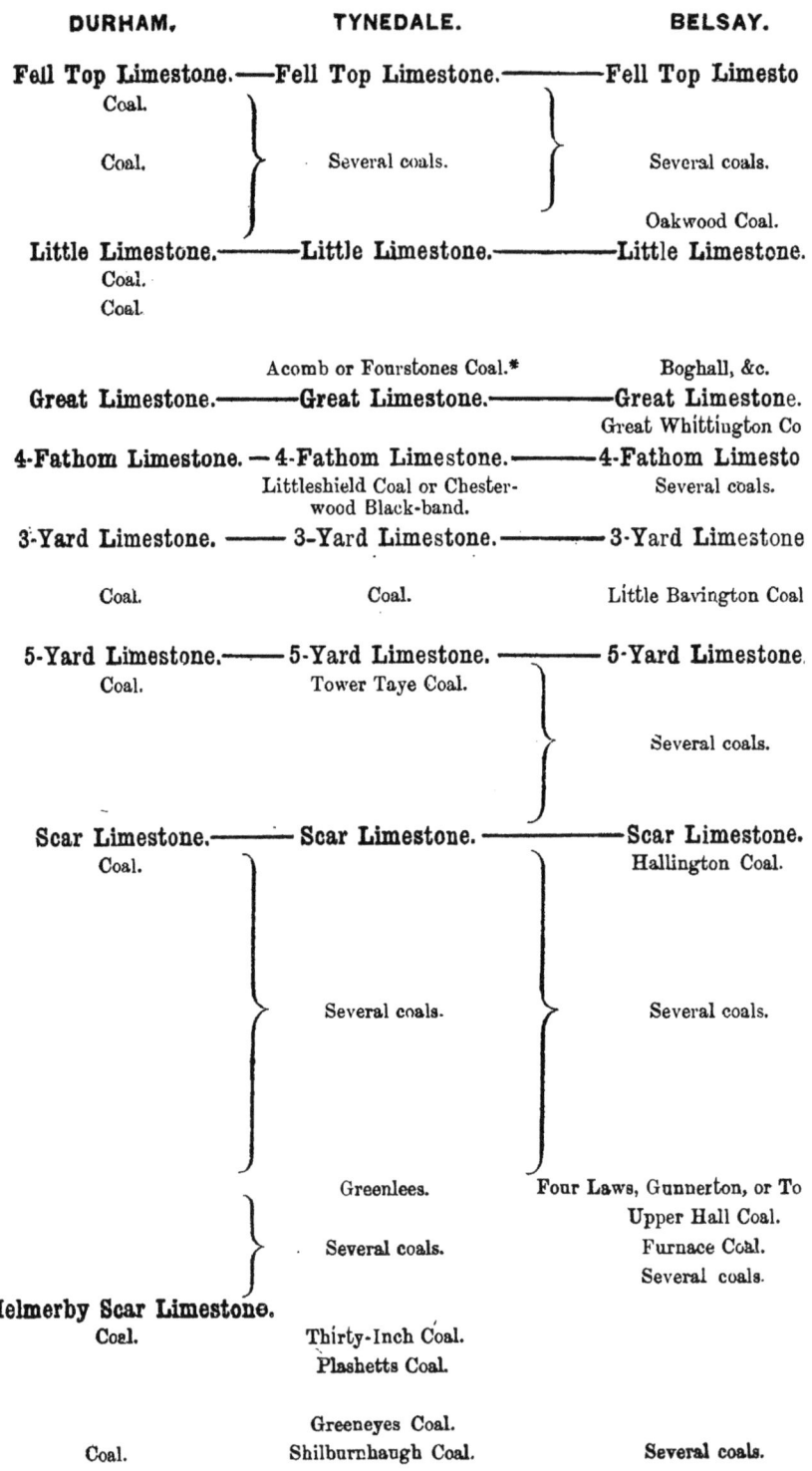

DURHAM.	TYNEDALE.	BELSAY.

Fell Top Limestone.——Fell Top Limestone.————Fell Top Limesto
Coal.

Coal.   Several coals.   Several coals.

                                              Oakwood Coal.
Little Limestone.———Little Limestone.————Little Limestone.
Coal.
Coal.

              Acomb or Fourstones Coal.*   Boghall, &c.
Great Limestone.———Great Limestone.————Great Limestone.
                                              Great Whittington Co
4-Fathom Limestone. — 4-Fathom Limestone.———4-Fathom Limesto
              Littleshield Coal or Chester-   Several coals.
              wood Black-band.
3-Yard Limestone. ——— 3-Yard Limestone. ————3-Yard Limestone

Coal.   Coal.   Little Bavington Coal

5-Yard Limestone.——— 5-Yard Limestone. ——— 5-Yard Limestone
Coal.   Tower Taye Coal.

                                              Several coals.

Scar Limestone.——— Scar Limestone. ———Scar Limestone.
Coal.   Hallington Coal.

                    Several coals.   Several coals.

                    Greenlees.   Four Laws, Gunnerton, or To
                                              Upper Hall Coal.
                    Several coals.   Furnace Coal.
                                              Several coals.
Melmerby Scar Limestone.
Coal.   Thirty-Inch Coal.
                    Plashetts Coal.

                    Greeneyes Coal.
Coal.   Shilburnhaugh Coal.   Several coals.

BEBNICIAN SERIES.

BRINKBURN.	ALNWICK.	SCREMERSTON.
‹ll Top Limestone. ———	Fell Top Limestone.	
‹wton Underwood Coal.		
Stanton Coal.		
‹ongwitton Top Coal.	Several coals.	Denuded off.
‹ngwitton Bottom Coal.		
Several coals.		
Chirm Limestone.———	Chirm Limestone. ———	Licker Limestone.
	Coal.	Coal.
Several coals.	Several coals.	Parrot Coal.
		Rough Coal.
Rothley or Chirm.	Shilbottle Hill Head Coal.	Licker Main Coal.
‹n-Yard Limestone.———	Ten-Yard Limestone.———	Dryburn Limestone.
Brinkburn Top Coal.	Shilbottle Townhead Coal.	Dryburn Coal.
3-Yard Limestone. ———	8-Yard Limestone. ———	Low Dean Limestone.
Several coals.	Coal.	Coal.
‹-Yard Limestone. ———	6-Yard Limestone.———	Acre Limestone.
	Coal.	Coal.
‹rinkburn Main Seam.	Shilbottle Main Coal.	Acre or Lowick Lime Coal.
		Coal.
-Yard Limestone.———	Beadnell Main Limestone. ———	Eelwell Limestone.
Coal.	Beadnell Main Coal, or Denwick Big Coal.	Eelwell Upper Coal.
	Beadnell Windmill Coal.	Eelwell Lower Coal.
		Coal.
Several coals.	Several coals.	Crosshill Limestone.
		Coal.
	Beadnell Stone Close Coal.	Coal.
	5-Yard Limestone. ———	Oxford or Greenses Limest‹
	Swinhoe Coal.	Greenses or Allerdean Coal.
	Fleetham Coal.	Muckle Howgate or Woodside Coa
Several coals.		Little Howgate Coal.
		Woodend or Biteabout.Limest
	Several coals.	Several coals.
		Biteabout Coal.
	Dun Limestone. ———	Dun Limestone.
Coal.	Dun Coal.	Dun Coal.
		Robies Coal.
Coal.	Hobberlaw Coal.	Fassett or Caldside Coal.
		Several coals.
Several coals.	Several coals.	Hetton Coal.
		Several coals.
	Craw Coal.	Scremerston Main Coal or Blackhill
		Hardy, Kiln, or Stony Coal.
	Several coals.	Diamond Coal.
		Coal.
Several coals.	Main Coal.	Bulman, Cancer, or Main Coal.
		Several coals.
		Three-Quarter Coal.

OUTLIERS OF THE UPPER CARBONIFEROUS MEASURES.

There is also a string of small detached trough-like formations of Coal-Measures extending from the Main Coal Field into Cumberland. They are found lying along the north and down-throw side of a large East and West Fault known as the Great Stublick Dyke. Their presence is due to two concurrent causes, the general flatness of the beds in the western districts and the disturbance caused by the East and West Fault. Had these five seams lying in the lower part of the Coal-Measures and the Gannister beds maintained their regular inclination, they would have outcropped about twenty miles from the sea, but the joint effect of the causes already mentioned allows the existence of the small and isolated coal-fields of Midgeholme, Coanwood, Stublick, etc., for a further distance of thirty miles.

	No.	Midgeholme.	Coanwood.	Plainmeller.	Stublick.
True Coal Measures.	4	Five-Quarter.	Five-Quarter.	Five-Quarter.	Cannel.
	3	Three-Quarter	Three-Quarter	Yard.	Yard.
	2	Wellsike.	Coomroof.	Coomroof.	Three-Quarter.
	1*	Seven-Quarter	Slag.	Main Coal.	Main Coal.
Gannister Beds and Millstone Grit Series.	3	Cannel.	Cannel.	Cannel.	Foot Coal.
	2	......	......	......	Little Coal.
	1	......	......	......	Stone Coal.

* This is probably upon the same horizon as the Brockwell Seam.

CONCLUSION: A NEW THEORY OF THE FORMATION OF THE COAL SEAMS OF THE UPPER CARBONIFEROUS MEASURES.

From a full consideration of the relative positions of the coal-seams in the true measures, it appears possible that they should be considered as portions of one and the same seam which was in continuous formation during a long period of time. If this theory be based upon fact, the Coal-Measures must be then considered as one seam of coal, with intercalated bands or beds of sandstone, shale, and other rocks. The various districts where coal-seams are brought into contact, with thin or no bands between them, are shown by the ⁓⁓⁓ in the synopsis.

It is safe, therefore, to assume that in the case of seams 13, 14, and 15, the formation of the coal was continuous, and that the varying thicknesses of strata intercalated in various areas are merely bands of greater than ordinary thickness. This is shown in Plate I.

Many other instances could be instanced in support of the theory but the one already quoted shows that during the formation of the Consett Hutton Seam it was divided by the intercalated strata A and B into three seams at Felling, and into two seams at Hebburn and Teams.

These ideas upon the mode of formation of coal-seams are strengthened by the following extract from a paper upon the "Magnesian Limestone of Durham," by Messrs. John Daglish and G. B. Forster, appearing in the Transactions of this Institute, Vol. XIII., page 212 :—"The roof of a seam of coal consists at one place of a hard sandstone, which, thinning out more or less abruptly, is replaced by soft shale, and at times the shale comes in as a wedge, without displacing the sandstone, and gradually increases to a thick bed. Even beds of coal themselves, commencing with a few inches, thicken to many feet, are separated by layers of shale into distinct seams, and again become one by the disappearance of the band of shale."

———

The following section of the Mountain Limestone has been added by the writer at the request of various members :—

Section of the Mountain Limestone Formation between the Tweed and Coquet, prepared by Messrs. William and John Wilson, of Shilbottle Colliery.

Reference.		Fs.	Ft.	In.	Fs.	Ft.	In.	Remarks
	Coarse sandstone rock ..	10	0	0				At Warkworth Hermitage and Hounden Dene.
	Blue metal	1	1	0				
1.	**COAL**	0	2	0				At Hounden Dene Burn and the Coquet.
					11	3	0	
	Black and grey metal ...	3	0	0				At Shortridge House.
	Coarse pebbly sandstone	14	0	0				
	Blue metal	1	4	0				
	Iron band	0	0	2				
	White metal	0	2	0				
	Blue metal	0	0	3				
	White metal	1	1	2				
	Blue metal	0	1	0				
	Hard sandstone	1	5	9				
	Blue metal	1	4	9				
	COAL	0	0	6				
					24	1	7	
	Black metal	0	1	6				
	White metal	0	3	8				
	Blue metal	0	5	6				
	Beddy freestone	1	0	9				
	White freestone	1	3	0				
	Blue metal	0	5	4				
	Sandstone	0	0	6				
	COAL	0	0	6				
					5	2	9	
	Grey metal	1	4	0				
	Bastard limestone	0	1	0				
	Grey metal	0	3	0				
	White metal	0	3	9				
	Blue metal	0	5	9				
	Beddy freestone	0	3	9				
	Carried forward	4	3	8	41	1	4	

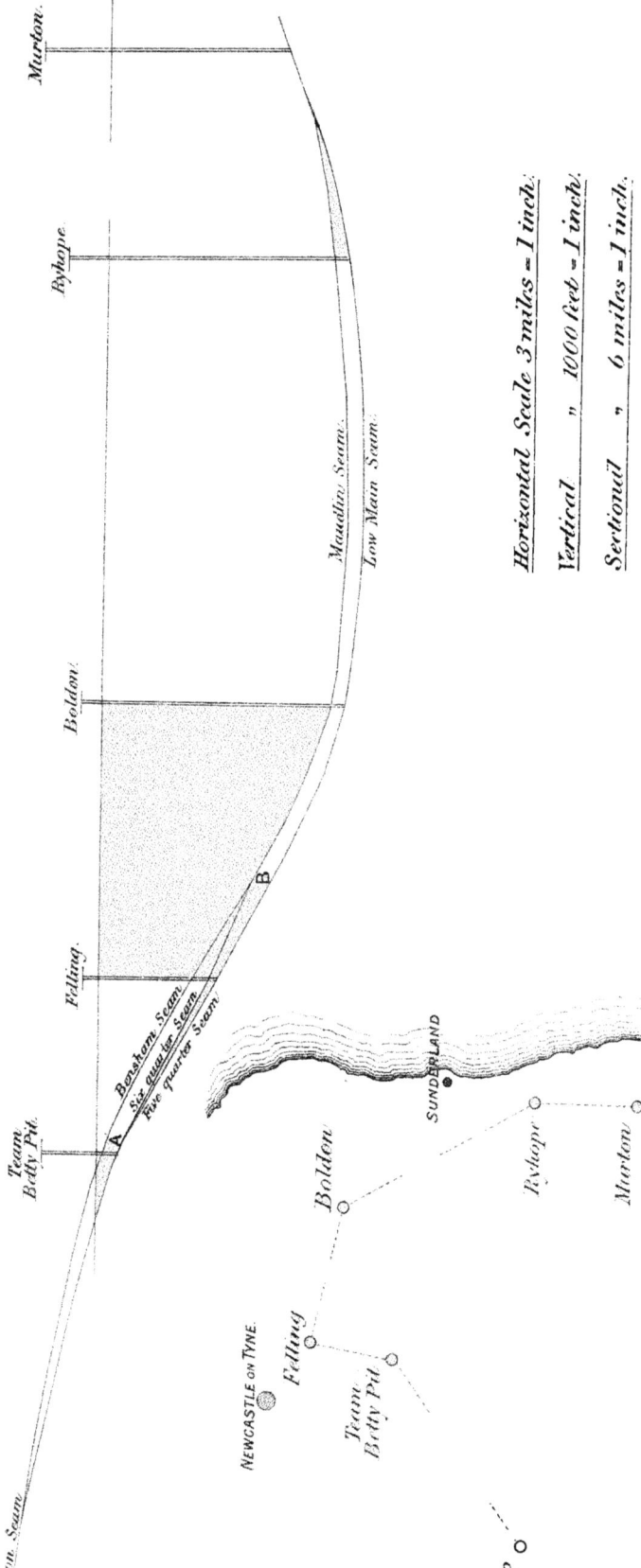

Horizontal Scale 3 miles = 1 inch.

Vertical " 1000 feet = 1 inch.

Sectional " 6 miles = 1 inch.

Murton.

Ryhope.

Boldon.

Felling.

Team
Betty Pit.

Maudlin Seam.

Low Main Seam.

Bensham Seam.

Six quarter Seam.

Five quarter Seam.

A

B

...on Seam.

NEWCASTLE on TYNE

Felling

Team
Betty Pit

Boldon

SUNDERLAND

Ryhope

Murton

P

SECTION OF THE MOUNTAIN LIMESTONE.—*Continued.*

Reference.		Fs.	Ft.	In.	Fs.	Ft.	In.	Remarks.
	Brought forward	4	3		41	1	4	
	Blue metal	0	3					
	Bastard limestone ...	0	0					
	Blue metal	0	1					
	Brown sandstone ...	0	1					
	Beddy sandstone ...	1	5					
	Pale blue metal ...	3	0					
	Soft limestone	3	5	8				
	Sandstone	0	3	0				
	Bastard limestone ...	0	3	3				
	Metal	0	2	9				
	Sandstone	0	0	6				
	Metal	0	1	6				
	White sandstone ...	1	5	0				
	Bastard limestone ...	0	1	4				
	White limestone ...	0	4	9				
	Dun sandstone	0	2	5				
	Blue metal	0	1	1				
	Grey sandstone... ...	0	3	11				
	Limestone	0	4	3				
	Grey sandstone... ...	1	0	10				
	Blue metal	4	2	0				
	White flinty stone, very hard	2	0	0				
	Metal	0	2	0				
	Sandstone	0	4	0				
	White sandstone ...	0	2	7				
	Dun metal	0	3	0				
	Grey sandstone ...	1	4	0				
	Metal	0	1	6				
	Hard dark stone ...	0	1	2				
	Blue metal	2	2	6				
A.	Limestone	0	4	0				*Little Lime* (?).
	Blue metal	0	1	3				
	Grey sandstone ...	0	1	6				
	Blue metal	0	1	3				
	Grey metal	0	2	7				
	Grey sandstone... ...	3	1	9				
	White sandstone ...	5	1	7				Full of water, running 20 gallons per minute at top of hole.
	Blue metal	0	2	2				
	Dun sandstone	0	4	1				
	Dun metal	0	3	2				
	Dun sandstone	1	4	5				
	Dun metal	1	0	6				
	Grey sandstone beds ...	2	0	9				
	Beddy metal ...	1	2	0				
	Grey sandstone ...	1	0	2				
	Blue metal	0	3	9				
	Sandstone and beds of metal	12	0	0				
	Coarse grey sandstone...	4	3	0				
	Sandstone and beds of metal	3	2					
B.	Limestone	0	3	0				
	Blue metal	1	2	0				

		Ft.	In.				
2. {	COAL, fine ...	0	9				
	COAL, coarse ..	1	0				Lying in valley to North-east from Woodhouse.
	COAL, fine ...	0	4				
		—		0	2	1	
					76	5	1

Carried forward   118   0   5

SECTION OF THE MOUNTAIN LIMESTONE.—*Continued.*

Reference.		Fs.	Ft.	In.	Fs.	Ft.	In.	Remarks.
	Brought forward				118	0	5	
	Blue metal	1	4	0				
	Dun sandstone	0	3	0				
	Coarse yellow sandstone	11	0	0				At Hill Head Dene Well.
	Grey metal	8	2	0				
	Beddy sandstone ...	4	3	0				
	Metal	0	5	0				
3.	**COAL**	0	2	0				*Shilbottle, Hill Head,* or *Licker Coal,* found South of Shilbottle, and at Hazon Lea, Dryburn, Berrington, Licker, Chirm, and Birkheads.
					27	1	0	
	Grey thilly stone ...	2	3	0				
·	Sandstone	3	5	6				Shilbottle Beacon Freestone Quarry. Fossil Plants.
	Blue metal	1	3	0				
C.	Limestone	0	4	0				At Beacon, above Shilbottle.
	Black metal	0	4	6				
	Limestone	0	2	6				
	Black band stone ...	0	3	3				
	Limestone	0	5	6				
	Beddy sandstone ...	2	1	0				
	Blue metal, with hard bands	9	3	0				
D.	Limestone	4	4	0				*Ten-Yard* or *Great Lime,* at Shilbottle Townhead Quarry, Whittle, Newton, Brinkburn, Beadnell, Spital, Ford, Dryburn and Berrington. At Ebsnook, near Beadnell, this limestone contains a portion of magnesia.
	Blue metal, with bands	1	0	6				
	Sandstone	5	5	6				
4.	**COAL**	0	3	6				At Shilbottle Townhead, Whittle, Newton, Brinkburn, Beadnell, Dryburn, Longframlington, Lowick, Embleton, Littlehoughton, Fenwick, Hawkhill, Tuggall, and Christon Bank.
					35	0	9	
	Fire clay and black dent	1	1	0				
	Coarse sandstone ...	12	2	0				At Longhoughton Station and Buston Quarry.
	Blue metal	6	3	0				
E.	Limestone, and metal about 30 inches thick	4	4	0				*Eight-Yard Stone,* at Shilbottle, Beadnell, Newton, Brinkburn, Tuggall, Embleton, Christon Bank. Rock, Denwick, Lowick, Beal, and Kyloe. With marine fossils.
	COAL and black dent ..	0	0	6				
					24	4	6	
	White hard sandstone...	1	4	0				
	Brown sandstone ...	2	3	0				
	Blue metal and ironstone	6	5	0				
F.	Limestone	3	0	0				*Six-Yard Stone,* at Beadnell, Denwick, Little Mill, Hawkbill, Shilbottle, Lowick, Tuggall, Scremerston, Holy Island, Newton, Brinkburn, Rock, and Whittle.
	COAL	0	0	6				
					14	0	6	
	Grey slaty sandstone ...	2	3	4				
	Blue metal	3	3	4				
	Limestone	0	2	0				
	Thill stone and sandstone	0	5	0				
	Grey metal	1	5	0				
5.	**COAL**	0	2	4				*Shilbottle Main Coal* or *Lowick Lime Coal,* at Brinkburn, Littlehoughton, Denwick, Hawkhill, Shilbottle, Beadnell, Tuggall, Embleton, Christon Bank, Rock, Lowick, Beal, Holy Island, and Newton.
					9	3	0	
	Grey freestone thill ...	1	2	0				
	Blue metal	2	1	0				
	Beddy sandstone ...	6	1	0				
*	Pale blue metal ...	1	4	0				
G.	Black limestone ...	1	1	0				
	Grey slaty sandstone ...	4	3	2				
H.	Limestone	4	3	0				*Nine-Yard* or *Main Lime,* at Beadnell Main Quarry, Shilbottle Tile Sheds, Brinkburn, Newhouses Woods, Tuggall North Sunderland, Beadnell, Falloden, Rock, Denwick, Lowick, Hetton Coal Houses, Oxford, and Greenses
	Sandstone	5	0	0				
	Blue metal	3	0	0				
I.	Limestone	0	5	0				
	Blue metal	2	0	0				
6.	**COAL**	0	3	2				*Beadnell Main Coal,* at North Sunderland Beadnell, Falloden, Rock, Denwick (big coal), Shilbottle Tile Sheds, Greenses, Hetton Coal Houses, Ancroft, and Oxford.
					32	5	4	
	Carried forward				261	3	6	

* Approximate horizon of whin sill.

SECTION OF THE MOUNTAIN LIMESTONE.—*Continued.*

Reference.		Fs.	Ft.	In.	Fs.	Ft.	In.	Remarks.
	Brought forward				261	3	6	
	Grey sandstone thill ...	0	5	0				
	Beddy sandstone ...	2	0	0				
7.	COAL	0	1	5				*Beadnell Windmill Coal.*
					3	0	5	
	Grey beddy sandstone...	7	3	0				
	Pale blue metal ...	1	1	6				
	Grey beddy sandstone...	1	3	0				
	Blue metal	1	4	0				
	Red sandstone	6	2	0				
	Grey metal	3	4	6				
	Limestone	0	2	6				
	Grey beddy sandstone...	1	0	6				
	Grey metal	4	5	6				
8.	COAL	0	1	4				*Stoneclose Coal,* at Beadnell and Stoneclose.
					28	3	10	
	Grey beddy sandstone ...	1	4	0				
	Blue metal	0	5	1				
	Beddy sandstone ...	1	3	6				
	Blue metal	3	0	8				
	Limestone brat... ...	0	2	6				
K.	Limestone	2	0	0				*Five-Yard* or *Stoneclose Limestone,* at Beadnell.
	COAL	0	0	4				
					9	4	1	
	Grey thill stone ...	1	3	0				
	Grey beddy sandstone...	2	0	0				
L.	Limestone	0	5	0				At Fleetham and Newton Mill.
9.	COAL	0	1	4				*Swinhoe Coal,* at Beadnell, Swinhoe. Fleetham, and Broomford.
					4	3	4	
	Grey sandstone... ...	4	3	4				
	Grey metal	1	3	2				
	White sandstone ...	2	0	6				
	Blue metal	6	0	0				
M.	Limestone	1	0	0				
	COAL	0	0	9				
					15	1	9	
	Grey metal	4	0	0				
10.	COAL	0	1	6				*Hetton, Fleetham,* or *Muckle Howgate Coal,* at Beadnell, Swinhoe, Fleetham, Coalybars. Doxford, Hickley, Rugly, Hetton, Chatton, Woodside, Fenwick, and Doddington Crops out in freestone quarry at Alnwick Moor. Alnwick Moor Quarry.
					4	1	6	
	Grey metal and sandstone	11	0	0				
	Blue metal	3	0	0				
N.	Limestone	2	3	0				*Hobberlaw Stone,* at Beadnell, Swinhoe, Coldrife, Hetton Coal Houses, Lowick, Barmoor, Fassett. Woodside, Chatton, Hobberlaw, Rugly, Ford Moss, and Botany.
	COAL	0	0	4				
	Beddy sandstone, with metal partings ...	10	0	0				
	COAL	0	0	10				
	Grey metal and dun sandstone	8	0	8				
O.	White limestone ...	1	0	0				*Dunstone,* at North of Beadnell, Fleetham, Coldrife, and Detchant.
	Sandstone	6	0	0				
	COAL	0	0	9				
					41	5	7	
	Grey sandstone... ...	7	0	0				
11.	COAL	0	2	0				*Hobberlaw, Coldside,* or *Fassett Coal,* at Hetton Coal Houses, Doddington, Ford Moss. Fassett, Scremerston. Chatton, Hobberlaw, Allerdene, and Botany.
	Grey posty metal ...	3	0	0				
12.	COAL	0	2	6				At Hetton Coal Houses.
					10	4	6	
	Carried forward				379	4	6	

SECTION OF THE MOUNTAIN LIMESTONE.—*Continued.*

Reference.		Fs.	Ft.	In.	Fs.	Ft.	In.	Remarks.
	Brought forward				379	4	6	
	Grey sandstone and metal bands	20	0	0				
P.	Yellow sandstone ...	19	0	0				At Reigham Quarry.
	Limestone	0	2	0				
	Grey sandstone and me- tals	9	0	0				
	COAL ... Ft. In. 0 10							
	Band stone ... 0 10							
13.	**COAL** ... 4 0							
		0	5	6				*Crow Coal. Scremerston Main Coal*, or *Blackhill Seam*, at Scremerston, Un-
					49	1	6	thank, Etal, Ford Moss, Detchant, Eglingham, Barmoor, Leamington,
	White metal	0	0	6				Chatton, Alnwick Moor, and Banna- moor.
	Grey ribby sandstone ...	3	0	0				
14.	**COAL**	0	3	0				*Stony* or *Hardy Coal*, at Etal.
					3	3	6	
	Sandstones	20	0	0				At Stonypeth.
	Limestone	0	2	6				
15.	**COAL**	0	4	0				*Main Coal, Cancer*, or *Bulman Seam*, at
					21	0	6	Spital, Unthank, Shoreswood, Green- lawalls, Etal, Ford Moss Barmoor,
	Grey sandstone	10	0	0				Doddington, Detchant, Eglingham, Leamington, Debden, Chatton, Bot-
	COAL	0	1	0				any, Clattery, and Bannamoor.
					10	1	0	
	Grey sandstone and metal	3	0	0				
	Limestone	0	2	0				
	Metal	1	0	0				
	Limestone	0	2	0				
	Grey sandstone and me- tals	10	0	0				
	Limestone	0	1	6				
16.	**COAL**	0	2	6				*Three-Quarter Coal*, at Etal and Felk- ington.
					15	2	0	
	Bastard sandstone or ironstone	0	3	0				
	Blue metal	0	2	0				
	Limestone	0	1	6				
17.	**COAL**	0	2	9				*Cooper Eye Coal*, at Berwick Hill, Mor- ton, Shoreswood, Greenlawalls, Etal,
					1	3	3	Ford Moss, Detchant, and Chatton.
	Grey sandstone thill ...	2	0	0				
	Limestone	0	4	0				
	Metals and thin lime- stones	7	0	0				
	Blue metals	2	0	0				
	COAL ... Ft. In. 1 3							
	Band stone ... 1 6							
18.	**COAL** ... 2 0							*Wester Coal Seam*, at Ford Moss and Doddington. This is the lowest coal
		0	4	9				worked.
					12	2	9	
	Grey thill stone ...	0	2	6				
	White sandstone ...	2	3	6				
	Pale blue metal ...	2	0	0				
	Sandstone	7	3	0				
	Beddy freestone and me- tal	10	4	0				
	Sandstone	18	4	0				
	Beddy sandstone ...	9	2	0				
	Blue metal and iron bands	13	0	0				
	Carried forward	64	1	0	493	1	0	

SECTION OF THE MOUNTAIN LIMESTONE.—*Continued.*

Reference.		Fs.	Ft.	In.	Fs.	Ft.	In	Remarks.
	Brought forward	64	1	0	493	1	0	
Metal and bands of sand-stone		8	0	0				
Sandstone		7	0	0				
Grey metal, with bands 4 inches to 12 inches thick		22	3	0				
Pale blue metal, with bands 4 inches to 24 inches thick		9	4	0				
Sandstone bands, with metal partings ...		10	2	0				
Soft green metal ...		1	3	0				
Beddy green sandstone..		7	3	0				
Pale blue metal ...		20	3	0				
Hard white metal ...		0	3	6				
Very hard band stone ...		0	2	6				
Hard bastard band stone		0	5	0				
Hard iron bands ...		0	4	6				
White metal		0	2	6				
White beddy sandstone		0	4	0				
Pale blue metal ...		0	4	0				
Hard flinty stone ...		0	5	0				
Beds of hard sandstone		1	0	0				
Sandstone		0	2	0				
Hard band beds ...		0	2	0				
Pale blue metal ...		1	3	0				
Hard sandstone bands, with partings ...		1	2	9				
Hard flinty stone ...		0	2	0				
Hard bands, with metal partings		1	3	0				
Sandstone		0	2	6				
White metal		0	4	0				
Hard bastard stone ...		0	1	10				
Green metal		0	4	0				
Hard bands, with soft partings		0	3	0				
Pale metal		0	3	0				
Sandstone bands ...		0	2	0				
Pale blue metal ...		0	4	0				
Beddy sandstone ...		3	1	0				
Hard bastard bands 3 inches and 4 inches thick		1	0	0				
Pale metal		1	2	0				
Hard white stone ...		1	0	0				
Hard band stone ...		1	5	8				
Pale metal		0	3	6				
Hard stone		0	2	6				
Grey metal		0	3	0				
Hard sandstone ...		0	2	0				
Hard iron or bastard limestone bands ...		0	2	0				
Grey metal		1	5	0				
Hard bands 3 inches and 4 inches thick ...		0	2	0				
Grey metal stone ...		0	0	10				
					179	4	7	
	Total				672	5	7	

Professor LEBOUR said, he would like to express his thanks, and he thought the thanks of almost everyone connected with the coal-field, to Mr. Brown for the great trouble he had taken in getting up his paper. Only those who had tried to correlate beds in limestone measures over a large extent of country knew the very great trouble Mr. Brown must have taken in that portion of his work. He (Professor Lebour) himself knew, from being so much in that building, the amount of trouble Mr. Brown had taken in this matter, and the pains he had taken to test the truth of the hypotheses he had brought before them. Never had a correlation of this sort been done more carefully, and he thought it would rank in time with the two other correlations which were the fathers of this kind of thing. One of these was Mr. Buddle's, who was the first to throw any light on the arrangement of the coal-seams in this district, and the second was Mr. Simpson's. It said a great deal for Mr. Buddle's work that it was a work much of which stood good to this day. Mr. Simpson's work was of a very much later date, and therefore much more perfect. In the Mountain Limestone in this district the correlation was more difficult; the seams were not so constant as in the Coal-Measures, and they split up a great deal more. Mr. Brown was right in saying that his correlation of the Bernician series must be taken as a tentative one; but he (Professor Lebour) thought, as a tentative one, it was bound to hold a very high rank indeed. Mr. Brown, with his usual brevity, made a most important statement in two or three lines in his paper; and as it was very likely the statement might be overlooked, he would specially point it out. It was this, that "from a full consideration of the relative positions of the coal-seams in the true measures, it appears possible that they should be considered as portions of one and the same seam which was in continuous formation during a long period of time." If this was true, then it was an absolutely new fact. He had known for some time that Mr. Brown was inclined to think this was likely, and such data as Mr. Brown had had at his disposal seemed to show it was probable as to some of the seams. If it were shown that each seam was connected with those above and below, then it was a new fact which would throw a great light upon the physical geography of the Carboniferous formation. This was a most important matter, and it was a pity that Mr. Brown put it into three lines, and hid it under a bushel in this manner. Other people might aid in throwing a light upon it. He moved a vote of thanks to Mr. Brown, and hoped Mr. Simpson would say something upon this subject. No one was more capable than Mr. Simpson to criticise the details. He (Professor Lebour) would himself like to say something of the details, but would defer his remarks until the next discussion.

Mr. J. B. SIMPSON seconded the vote of thanks to Mr. Brown, and said the subject which he had brought before them was a most interesting one.   He agreed with Professor Lebour that, to come to a correct conclusion on so difficult a subject as the correlation of seams, involved great labour.   The more papers they could get on the subject the more able should they be to come to a conclusion on the extended continuity of the seams.   He would have liked if Mr. Brown could in the latter synopsis have stated the thickness of the strata, in some form or other, in the different districts.   This would add much interest, even if it could be done only approximately.   He did not feel competent at present to go into the details in Mr. Brown's paper, but hoped, when the paper came up for discussion, that he might be able to criticise and approve some of Mr. Brown's propositions.

Professor MERIVALE said he would suggest that before the paper came up for discussion, Mr. Brown might be able to add to Coanwood, Plainmeller, and Stublick, particulars of that most interesting little coalfield, Midgeholme.   He, perhaps, could give Mr. Brown some information that would enable him to correlate the seams at that place.   In the paper, Mr. Brown mentioned various eminent authorities as to the duration of the coal-field.   There was one authority, however, perhaps the earliest one, whom he had not mentioned, and that was Sir George Selby, who, in 1610, announced in Parliament that the Newcastle coal-field would not last for 21 years.

The CHAIRMAN said that, if he mistook not, there was a later authority on the duration of coal than Mr. Brown had mentioned, and that was Sir William Armstrong, when president of the British Association meeting at Newcastle in 1863.

Mr. SIMPSON : Professor Jevons was later still.

Professor LEBOUR : Mr. Greenwell is a later authority than either Sir Wm. Armstrong or Professor Jevons.

The CHAIRMAN : Sir William Armstrong's statement as to the duration of the coal, made when the British Association met in Newcastle, created a great sensation all over the country.   He suggested that Mr. Brown should add to his paper the estimates of Sir Wm. Armstrong and the late Professor Jevons ; and he would further suggest that he should give the year in which each estimate was made.   He was quite recently astonished in London by a friend saying to him that in fifty years' time the whole of the coal-fields of Northumberland and Durham would be exhausted; and that any person contemplating buying property on the Tyne should very seriously consider how it would be affected by this exhaustion of the coal. The gentlemen present at this meeting knew a great deal more about the duration of coal than he did, and he would leave this part of the subject in their hands; but he must say that he did not think such a thing as his friend in London predicted was likely to happen, or that capitalists need be afraid of investing their money in such property, for they ought to get a good return long before the coal was exhausted. He was not sufficiently acquainted with the Scremerston coal-field to speak authoritatively, but it struck him that the 73 feet mentioned in the paper was a great thickness of coal, and there might be a mistake. With respect to what Professor Lebour had specially called attention to, as to the coal seams in the true measures being perhaps portions of one and the same seams, this was a matter which should be left over for discussion at another meeting. He could endorse what had been said by Professor Lebour and Mr. Simpson as to the great labours which Mr. Brown had undertaken, and of the paper that had been the issue. He knew from experience there was very great difficulty indeed in tracing out the different beds in various districts where they occurred.

The vote of thanks was unanimously agreed to.

Mr. WALTON BROWN returned thanks. He trusted that, on analysis, his paper would be found to be fairly correct. The estimates of Sir William Armstrong, Professor Jevons, and Mr. G. C. Greenwell, related to the coal-fields of the United Kingdom, whereas his estimates referred only to the duration of the coal of the carboniferous formation of the counties of Northumberland and Durham. With respect to the 73 feet at Scremerston he thought himself that it was incorrect when he stated it, and on subsequent investigation he had found it to be 53 feet.

Mr. STEAVENSON said, he was not at all so sanguine as Mr. Brown and others appeared to be with respect to the coal included in the return being worked. Long before the thin seams were available foreign competition would cut the North-country coal out of the market. Mr. Brown, in the synopsis, spoke of the Brockwell extending all the way from Monkwearmouth to Bishop Auckland. He (Mr. Steavenson) thought this was a mistake; his impression was that the Brockwell seam had not been seen, or been found or proved, in any pit east of Durham. He tried for it many times in the neighbourhood of Coxhoe. He would like some one to give them particulars of what had been recently learned on the subject of the southern boundary. He fancied that in late years there had been a great deal ascertained that was formerly unknown, for there had been shafts sunk which had been found to be useless.

The CHAIRMAN said that in the synopsis there was a heading " Durham." Durham was rather an indefinite term; it was both a county and a city.

Mr. WALTON BROWN said he meant the county.

The CHAIRMAN suggested that Mr. Brown should mention the district and not the county.

————

A paper by Mr. John Allan, on " The Pyrites Deposits of the Province of Huelva," was read as follows :—

THE PYRITES DEPOSITS OF THE PROVINCE OF HUELVA.

By JOHN ALLAN.

THE ever-increasing production of copper in this locality, in spite of low prices, has brought it prominently before the eyes of the financial world. The writer trusts that a few remarks about the mines, though given inadequately, may prove of some interest to the members of the Institute.

SITUATION AND GENERAL REMARKS.

The province of Huelva is situated in the south-west of Spain, is separated from Portugal by the river Guadiana, and is bounded south by the Atlantic. Huelva, the capital, lies at the confluence of the rivers Odiel and Tinto, about three miles from the mouth of the river, and forms the centre for railroads and shipping. The bar is passable at high water to ships of heavy tonnage, and the river is navigable as far as the piers built by the Rio Tinto and Tharsis companies for the shipment of their minerals.

Public railways connect the town with Seville and Zafra, whilst mining companies have constructed narrow gauge lines to Tharsis, Buitron, and Rio Tinto, the Tharsis and Buitron being open to the public.

The climate is mild in winter; in summer the thermometer ranges from 95 to 110 degs. F. Dry for the greater part of the year, heavy rains fall in early winter and spring, swelling rivers and watercourses, mostly dry in summer, into foaming torrents.

The country is mountainous. After ten or fifteen miles of flat and fertile ground near the sea coast, it rises in a series of hills to the foot of the Sierra Morena. These hills, almost denuded of soil, are thickly covered with brushwood, locally termed "sara," with here and there a sparse plantation of Spanish oak, the acorns of which during the season maintain numerous herds of swine.

Fuel, Labour, Material.—The people of the country are frugal, easily controlled, and as a rule very good workmen. The wages paid are as follows :—

				s.	d.
Miner, 8 hours shift on contract	...	...	...	3	4
„ 12 „ day's work	...	...	...	2	11
Labourer on surface, per day	...	...	...	2	1
Boys and girls „ „	...	...	...	1	3

An average workman will easily pick up any special work, such as plate-laying, and timbering, but requires supervision.

The nearest coal-fields being in the vicinity of Cordova, all coal has to be imported, and costs roughly 25s. per ton at the mines which have their own piers and railroads. Mines not in direct communication with the sea have to pay for transport on mules' backs, at the rate of 7d. to 1s. per mile per ton.

Timber is scarce; the native pine is, however, excellent for timbering purposes, the cost of round logs is roughly 34s. per cubic yard. Baltic pine costs roughly 46s. per cubic yard in square logs.

Mining Concessions, Customs, etc.—The Spanish Government offers every facility for mining enterprise. The subsoil belongs to the nation ; any person or company, whether foreign or Spanish, can register a claim or "Denuncio" at the Government Mining Office, situated in the capital of each province. After the lapse of time prescribed by law, the engineer deputed by Government lays off the claim on the ground, and after payment of dues the title deeds are delivered. The denouncer of "claim" thus becomes possessed of all mineral wealth, and has a right to start workings, expropriating the surface owner should he refuse to sell his ground at a fair price. Starting workings is not necessary, and the owner can allow his concession to remain unworked any length of time. The smallest claim laid off is a square of 110 yards square, or about 12,000 square yards. Should several of these not contiguous enclose a free space on which a square of the above-named dimensions cannot be marked off, the space, or "Demacia," becomes the property, free of charge, of any owner of circumscribing claims who may petition for it to the Government office.

The charges are very slight, two dollars per annum for a "pertenencia" of about 12,000 square yards, and a small original cost for marking off.

Mining companies, or private owners, in other respects, pay the same taxes as ordinary proprietors ; on the ore extracted, however, there is a tax of 1 per cent.

The customs duties are heavy—coal, 3s. 1½d. per ton ; iron, 15s. per ton ; and it is of advantage to use the products of native industry.

GEOLOGICAL FORMATION.

The pyrites deposits occur in a zone of clay slate, which traverses the province from east to west, beginning with the deposit at Asnalcollar, in the province of Seville, and ending with Aljustrel and Gràndola in Portugal.

These slates running W. 30 degs. N., with a dip of 70 degs. to 80 degs. N., vary in colour, hardness, and composition. They are often traversed in every direction by small veins of quartz, evidently of a later formation. A difference of opinion arises as to the relative age of the formation; it is indicated in the geological maps of "Maestra" and "De Verneuil" as Silurian.* Roëmer describes it as belonging to a low horizon of the "Culm" measures, Phillips,† as apparently, of Silurian, Devonian and Carboniferous age.

Fossils are of rare occurrence, and it would perhaps be rash to form an opinion until the ground has been more carefully studied.

The deposits occur parallel to the stratification of the slate, and are generally bounded north by porphyry.

This happens so frequently that a connection has been suggested between the two. The lodes, if such they can be called, are lenticular in shape, 110 to 330 yards in length. They vary in width from a few feet to 150 yards. Some dip north, at about the same angle as the slate, with parallel walls; these have never been tried in depth, continuing below the deepest workings; others vary considerably in width on lower or upper levels, the north wall advancing or receding; others again take the form of a boat, and are entirely cut off below. The ore is composed of iron pyrites, intimately mixed with copper pyrites. It is, as a rule, finely crystalline, and in this state is divided in a series of joints, at right angles to the stratification of the slate, and to a lesser degree parallel to it. The structure, colour, density, and lustre also vary. Its colour is a silvery white, with metallic lustre, when hard and poor; and dark green, granular, and earthy, when soft and rich. Its density varies from 3 to 4·85. The percentage in copper varies considerably, and although about 3 per cent. is the average, some parts contain 1 per cent. and under, whilst others vary from 3, 5 to 8 and 10 per cent.

Some deposits are traversed, though to a very subordinate extent, by strings of copper pyrites, grey copper, fählerz, galena, and quartz. These ores occur in the joints of cleavage, as a rule, but attain neither width nor depth. Quartz, however, forms part of the structure, and is found

* Zeitschrift der Geolog. Gesselschaft, 1876, p. 354.
† Phillips on Ore Deposits, p. 15.

up to 10 and 12 per cent.   Many other metals occur intimately ad-
mixed, such as gold, silver (15 to 30 dwt. per ton), lead, zinc, bismuth,
nickel, antimony, and copper, but merely as traces.   Seen from a distance,
these deposits present a striking appearance, from their outcrops of
ironstone, produced by the decomposition of the upper portions of
the mass.   This gossan covers the deposit with varying thickness, the
harder portions forming a series of ridges, whilst the softer have been
removed through denudation.   The part immediately above the mineral
is soft and crumbling, and the contact marked only by 1 or 2 inches of
soft decomposed mineral.   The composition of this ironstone varies
considerably, but the following analysis may be taken as an example :—

Water ...	6·06
Copper ...	traces
Iron ...	53·06
Sulphur...	1·40
Oxygen	22·74
Silicon ...	16·74
	100·00

Deposits of iron ore, posterior to the actual formation, can be seen
in the vicinity of every mine.   The iron salts resulting from the original
decomposition of the pyrites, have formed beds of iron ore, which
through denudation are left covering the slate.   An excellent example of
this is the Mesa de las Pinas, in the vicinity of the south lode, at the
Rio Tinto mines, where a variety of imprints of leaves, belonging to
trees still growing in the neighbourhood, have been found.   This class of
ore is easily identified, being stratified, and containing fragments of
quartz, slate, etc.   It has at various times been mistaken for the outcrop
of a lode, and workings actually started.

Deposits.—A complete description of the numerous deposits would
entail giving an idea of the mineral resources of the various companies;
as this is neither advisable, nor within the limits of this paper, the
author will confine himself to mentioning the names of deposits, roughly
classified according to their size, with a few brief remarks on those which
offer any striking features from a geological point of view.

   1.--Rio Tinto, Tharsis, Calañas, *Santo Domingo.

   2.—Cueva de la Mora, Lagunazo, Castillo de las Guardas, Sotiel
       Coronada, *Aljustrel.

   3.—Poderosa, Concepion, Carpio, San Telmo, San Miguel, Asnalcollar,*
       Grandola, El Tinto, Buitron, Peña de Hierro, Consessionarios
       (iron pyrites).

                  * In Portugal.

4.—Vuelta Falsa, Vulcano, Monte Rubio, Aguas Teñidas Romanera, Lomero, Chapparita, Barranco de los Bueyes.

Besides these mines there are a quantity of others some of which may be of importance.   Soya, Cabezas del Pasto, Herrerias, Sierra de los Veneros, Sierra Vicaria, Ternancia, Chanza Trimpacho, Carmen, Manolito, Mosquitos, Angostura, Tomas, San Nicolas, Santa Flora Campanario.

Rio Tinto, the most important, is composed, according to some plans, of the South Lode, and its continuation the San Dionisio Lode, the Middle Lode, and the North Lode.

The South Lode dips slightly south, and its South Wall is composed of slate.   These slates are dark and talcose in the part of the lode on which the opencast is situated, further west yellowish white, and decomposed by the acid salts, resulting from the decomposition of the pyrites.   The North Wall formed of porphyry, bulges in and out, the lode attaining, immense width in some places, and narrowing to a thin vein for a short distance, before reaching San Dionisio.

The porphyry immediately in contact with the lode is decomposed and soft, especially in the opencast veins.

Continning north after passing a dyke of porphyry, what may be called the Centre Lode is reached.

Its South Wall dips north, and is composed for 33 to 50 feet thick of a soft, friable, granular quartz, containing two to three per cent. of copper, and generally very rich when it touches the mineral.   This quartz gets gradually harder and smoother to the touch on nearing the porphyry; for this reason it has been suggested that it is the result of the decomposition of the porphyry, as veins of hard crystalline quartz, cutting through the decomposed ground, and continuing into the hard porphyry, have been met with.   The North Wall dips south, and is formed of hard quartz porphyry.   There is no distinct joint between the mineral and porphyry, the latter being for many yards partly metallised, with veins and impregnations of iron and copper pyrites.

Continuing north, through a band of quartz and porphyry, bearing strings and specks of pyrites, the North Lode appears.   This lode has so to speak, no distinct walls, and consequently no inclination, being bounded north and south by a band of impregnations, and gradually going into porphyry.

It is, however, impossible to lay a fixed rule, as the various deposits exhibit different shapes and structures in various places.

It may be interesting to note, that the joints in the mineral, well defined in the South Lode, on contact with the slate become less so in the Centre Lode, and are scarcely ever seen in the North Lode.

The direction of joints in the Centre Lode is confused ; in one part the ground is much cut up, and lumps of ore the size of a fist are found, polished through pressure, in four or five directions.

The Tharsis Mines next in importance to the Rio Tinto, are composed of the North Lode and its continuation, the Sierra Bullones, and Poca Pringue, the Centre Lode, the South Lode and its continuation, the Esperanza schist deposit.

The deposits can be taken as typical of the usual formation.

The North Lode, the most important, dips north with about the same inclination as the slate, both walls running parallel, and formed of black slate, porphyry being within a short distance of the North Wall. The slates and porphyries in the vicinity of the lode are decomposed in some places, but offer no peculiarity. The system of joints is highly developed, both in the direction and across the lode. The other deposits have the same dip and general features.

The Calañus Mine, next in importance, is also the property of the Tharsis Company. The lode can be traced for a long distance and attains great width in one place. The walls are parallel, and dip north, with the same inclination as the slate, the south wall being composed of slate, and the north in some places of impregnations, the wall being, however, clearly defined.

The San Miguel Mine, can be taken as an example of a lode dipping slightly south, with slate south and porphyry north, as is the Poyatos Mine an example of a boat-shaped deposit.

MINING.

History and Ancient Workings.—These mines were known to the ancients as far back as 1,000 B.C. Phœnicians and Carthagenians possessed trading establishments on the coast and worked the mines to a certain extent. The workings nearer the surface can probably be ascribed to them, as also a part of the slag found in large quantities in the vicinity of every mine. The Romans leave more unmistakable traces of their workings. Pliny, in the year 79, describes the mines as giving employment to 20,000 slaves ; and the course of their occupation can be traced by coins bearing the stamp of various emperors until the year 412. Every mine in the province has been more or less worked by them, as shown by Shafto, there being adit levels at great length, depressions formed by cavins in of workings, and deposits of slag, variously estimated at from 10 to 20 millions of tons.

The great centres of Roman industry, however, were at the Tharsis and Rio Tinto mines. Remains of houses and tombs are met with in great profusion, in which delicate glass articles, ornaments, surgical instruments, weights, and lamps are found daily. The mining works are on a gigantic scale, considering the means at their disposal, adits at various levels are still open, and large cavities are found at the lowest depths of modern workings. The largest workings are met with near the walls of the lodes, where the mineral is richest and easiest to work. Regular stopes, well timbered with oak, can still be seen, and at Rio Tinto in soft quartz, large areas have been worked longwall, extracting the strings of rich sulphides, the cavities being packed with the refuse. The administration and methods of mining and smelting must have, however, undergone many changes and improvements during the long period of the Roman occupation.

As regards mining, undoubtedly the first step was the driving of an adit; if this was successful in striking the mineral, galleries were driven in all directions skirting the lode in search of a soft working face. Numerous shafts were sunk, mostly in pairs, for the purpose of extraction, development, and ventilation; and as the extraction continued, more adits were driven, or the waters raised by means of a series of bucket water wheels, or other appliances, specimens of which have been found in most mines.

Not much is known about the metallurgical treatment. From the few remains of furnaces in existence, one side appears to be formed by excavation in the solid rock, with a semicircular wall about 7 feet in height built in front of it, leaving thus a circular section $2\frac{1}{2}$ feet diameter, with openings below for outflow of slag and admission of blast.

The slags, though varying in composition according to locality, contain only traces of copper. The following analysis may be taken as a fair average :—

Water	0·83
Copper	0·04
Iron	55·09
Aluminium	·93
Oxygen	15·83
Silicon	27·13
Magnesium and sulphur	traces
	99·85

At the Rio Tinto mines, a kind of speiss, locally termed "metal blanquillo," is found in large quantities; it contains roughly :—

D

Copper	...	...	...	...	...	...	2·3
Iron	...	...	...	...	...	...	0·50
Lead	...	...	...	...	...	... traces	0·16
Antimony	...	...	...	...	...	...	3·10
Arsenic	...	...	...	...	...	...	20·25
Sulphur	...	...	...	...	...	...	2·3
Silver	...	...	...	...	...	... traces	·03

Matte and refined copper are seldom found, though there are plenty of manufactured articles.

It is impossible to underrate the vast importance these works are to those engaged in modern mining, as a mine bearing no trace of Roman enterprise can, as a rule, be put down as worthless. Some years ago a large deposit was discovered, bearing every indication of being similar to those existing, the only objection that could be raised was that the Roman slag and workings bore the appearance of being investigatory and not extensive. The subsequent trials proved the deposit to be one of iron pyrites, containing no copper. Old workings may also prove a source of insecurity, if not danger, and in a great many cases may either upset arrangements or forward them.

The Romans seem to have left suddenly, probably at the alarm of the Gothic invasion. During a long lapse of time the mines were deserted, and bear no trace of other workings. In 1725 some works were started at the Rio Tinto mines, and also in 1840, when extraction of copper by the wet way was introduced. About this time the Concepcion, El Tinto, Chapparita, and San Miguel mines were worked. In 1853, Mr. Deligny, a French engineer, discovered and laid claims to the Tharsis, Calañas, Santo Domingo, Poyatos, and Curva de la Mora mines, and started investigatory works on some of them. It was not, however, till 1858 that a company was started, and 1866 that the present Tharsis Sulphur and Copper Company was formed. In 1873, the Rio Tinto Company bought their mines from the Government, in whose possession they had been since 1840. Almost every mine in the province was more or less worked about that time, though now, owing to the depression in the copper market, only large mines are able to work at a profit.

Modern Mining.—The systems of mining generally adopted have been invariably either by opencast, or by what has been termed pillar and stall. The former has been used, or, rather, misused, in every imaginable way. At the San Mionel Mines a hole without an exit was made, the overburden being taken up zig-zag paths on mules and donkeys, and the mineral actually laid bare at 130 feet from surface. At the El Tinto

Mines, large quantities of sterile were removed to lay bare a small vein a few yards thick.   An opencast was also made at the Buitron, the mineral uncovered, but through want of previous underground investigations a comparatively small amount of ore was extracted, owing to its hardness, a few yards from surface.   It would be useless to bring forward any more examples of the badly-planned haste which the early miners of these deposits displayed, wasting large sums of money for little visible result.

If the opencasts were badly planned, the underground workings fully corresponded.   Workings which, from a primary condition, ought to be regular in order to let pillars correspond on one floor and another, were allowed, through want of ordinary precautions, to lose levels and dircetions.   The result was, in many cases, a fine mass of mineral in a ruinous condition, riddled with galleries, impossible for future workings except by an opencast, and even then at a distinct disadvantage, as the cost of removing the overburden has to be borne by two-thirds of the mineral, one-third approximately being extracted by previous workings.

With the larger companies, however, all the most modern appliances came into play; the gigantic opencasts of Rio Tinto and Tharsis were planned and executed, and enormous quantities of pyrites were investigated and laid bare.

Pillar and Stall.—This method consists of working the lode in floors. Galleries of different sections, according to the distance between floors, are driven parallel to each other and afterwards cross-cut at stated distances, leaving thus a series of superposed pillars bound together at each floor by a roof of more or less thickness.   Should the extraction of ore be carried on by means of cages or skips, the mineral on each floor is simply taken to the engine-shaft by the quickest way, in hutches, on narrow gauge lines.   In the event of a tunnel at lower levels, shoots, for dropping the ore, consisting of perpendicular or inclined winzes, are cut at convenient distances according to requirement.   The laying out of workings, in order to take advantage of the joints in mineral, natural ventilation, etc., vary according to local circumstances.   In most mines the Roman adit was cleaned out and widened, or a new one cut, unwatering the lode at a given level.   After cross-cutting the mass and sinking one or two shafts from surface for ventilation, galleries termed "reales" were driven parallel with the lode and each other, and floors started as convenient.

The various galleries or driving ends are generally let out on contract to gangs of four or six men, who, amongst themselves, choose their foreman, all, however, having equal rights and dividing the profits equally. They are, as a rule, paid by the ton, have to provide their own explosives, and pay for the sharpening and repairs of their tools.   In ends which require timbering, or are of special importance, a single contractor employing twelve or fourteen men is sometimes employed, for the better supervision of the work.

Working ends at different sections have been tried.   No fixed rule can be laid down, as conditions vary ; but, in average pyrites, with the following result :—Ends about 13 feet square were found to be a convenient size in ordinary ground.   The section per lineal metre gives roughly 63 tons.   An average price paid is—

					s.	d.
Labour	...	...	...	...	1	10
Explosives	...	...	...	...		6½
Repairs—tools		...	...	...		0¼
		Total ...	...	...	2	5 per ton.

The face is generally attacked by a small end driven in the back of the level, leaving a bench about 8 feet high, which is afterwards stoped down, 16½ by 20 feet and 20 feet square galleries have also been tried, and were driven in the same way as the 13 feet square ones, with a small level in advance.   The section in one case yields 106 tons, and in the other 122 tons, per lineal yard of gallery ; the cost was found to be about 2s. per ton.   These workings have the disadvantage of offering too large a section, and, on jointy or soft ground presenting itself, become unsafe.   A better method is driving 13 feet square galleries, which, being of a convenient size, can be reduced in dimensions, at the slightest change of ground.   Should the sides and roof stand well, they can be widened out to 16½ by 20 feet, or 20 feet square, at an average price of 1s. 3d. per ton, thus reducing the cost of the whole gallery to about 2s. per ton.   6½ feet square workings have been driven, but generally for investigation and for the sake of rapidity.   The cost of winning ore, from such a confined section, generally rises to double the cost per ton in a 13 feet square end.   The original expense is, however, compensated when the gallery is afterwards widened out.

With the exception of very few deposits, every mine in the province has been more or less worked by the pillar and stall method, as it was found to be the quickest and, temporarily, the cheapest way of extracting mineral.

OPENCAST WORKINGS.

Working by opencast consists in removing the overburden from the surface of the mineral, and working the lode as an open quarry.

To illustrate Mr John Allan's paper on "The Pyrites deposits of the Province of Huelva."

FIG. 2.

FIG. 1.

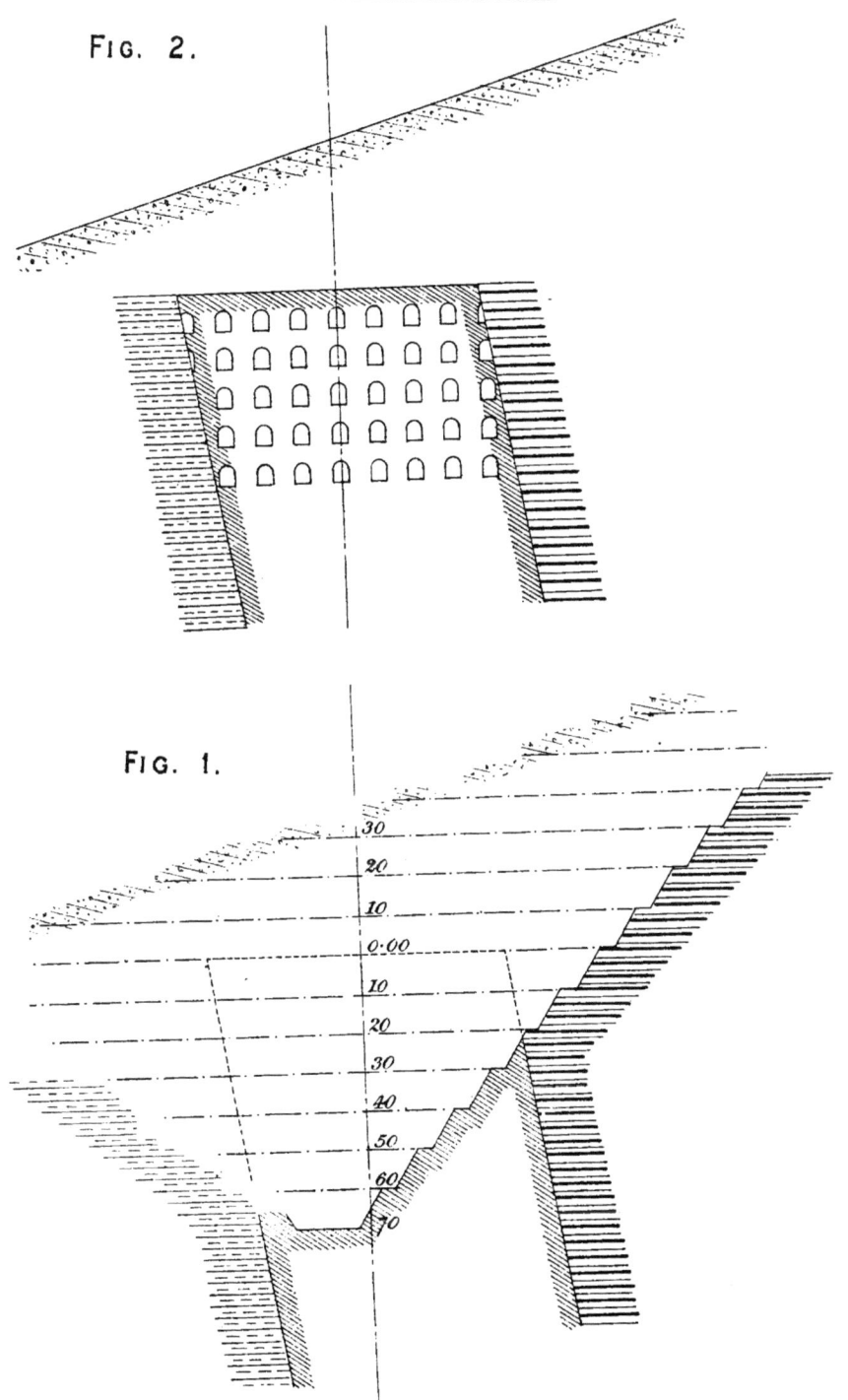

Removal of Overburden.—This is done in successive lifts (Plate II., Fig. 1) varying in height from 26 to 40 feet. Railways are laid according to the configuration of the ground, for the removal of the *débris* in wagons and their transport to a convenient tipping place. The mode of attack depends on the stratification and hardness of the ground to be removed. The removal of ground, to which no level lines can be laid—the cost of a cutting becoming too great—is carried on : 1st, by means of a vertical shaft or inclined plane to level outlet ; 2nd, through a tunnel ; 3rd, tipped down spouts to a tunnel driven into the lode at a lower level. As each bench approaches completion, it becomes necessary to settle the distance between the foot of one bench and top of the other. This varies considerably ; in porphyry, 10 or 15 degrees off the plane is considered sufficient, with a distance of 13 or 16½ feet ; in slate, its natural dip or more ; and in soft ground as much as 45 degrees. Generally the benches are only pushed far enough to discover mineral for one or two years' anticipated output, and afterwards keep pace with the winning of ore.

Winning of Ore.—The mineral once uncovered, benches are laid off similar to those for the removal of overburden. The work is commenced by a long cutting laid off parallel to the most pronounced system of joints, this is pushed on until sufficient space is left for another cutting below, and so on. The distance usually left between one floor and another is about 33 feet, but varies. The removal of the mineral to surface is carried on by the same means as the removal of the lower floors of the overburden, generally the same plan serves for both.

General Remarks.—The laying off of works of this kind depends entirely on local circumstances, and requires an intimate knowledge of the shape, depth, and richness of lode, so as to find out at first the amount of overburden that has to be paid for by each ton of ore made available by the uncoverings, and whether this is done at a profit. When once this is decided there remains to be seen in which way the removal of the part immediately above the lode, for which there is no level outlet, is to be carried on.

The construction of a tunnel of more or less length, entirely for the removal of overburden, may offer advantages in some cases, and it is a matter of calculation whether the original cost of tunnel and traction through it is cheaper than winding the same amount, either on an inclined plane or a vertical shaft, which can afterwards be used for hoisting mineral. Undoubtedly the most efficient arrangement is fixing on a method which can be utilised both for the removal of the overburden and minerals. Of the systems mentioned above, each one has its advantages.

Using a vertical shaft, with no other outlet, underground waters have to be raised, thus forming an item of expense, although the water usually met is easily coped with. The cost of raising one ton 100 yards may be roughly estimated at $3\frac{1}{4}$d.; add to this $\frac{1}{4}$d. for tipping at pit mouth—total $3\frac{1}{2}$d. Once raised, the ore may require classification, and means of traction to its destination.

A tunnel has the advantage that the ore, if necessary, can be classified on the spot before loading and taken to its destination at once. In many cases a Roman adit can be cleaned out and widened, thus cheapening the cost.

Lastly, if a tunnel is driven at a low level, say at the proposed limit of extraction by opencast, dropping the ore, or previously the overburden, down shoots is a great advantage, as filling into wagons at the shoot mouth scarcely costs one halfpenny per ton. The cost, however, of raising winzes from tunnel is heavy and not always practicable, though in some cases the ore won counterbalances this. After these primary conditions have been settled, the plan of attacking the upper floors, in order to have the working face parallel to the stratification of the rock, and get as large a working face as possible, becomes of great importance. This depends on local circumstances, as also does the possibility of laying off gradients in favour of the loaded train. Locomotive and mule traction have been used, and in the case of the tip being near, the full train was sent there by gravitation, the empties being taken back by a locomotive or a team of mules.

The organisation of the men employed has also varied. In order to prevent the men being idle whilst the train was unloaded, a double line was laid along the working face, and one train loaded whilst the other was being emptied. This, however, necessitates the stuff being carried, as it cannot be shovelled direct to the off line, and is consequently more expensive.

Cost of loading with spade	...	$2\frac{1}{2}$d. per cu. m.	1·9d. per cut yard.
„ on trays ...	...	4d. „	3d. „ „
Difference ...	...	$1\frac{1}{2}$d. „	1·1d. „ „

Loading with the spade increases the expense of plate-laying, as the line has to be continually shifted.

The usual arrangement is to provide two working faces not far distant, and load alternately at each face.

Steam navvies have been used; the cost compares most favourably with hand labour in soft irony ground, but in slate offers no advantage. The cost for loading, boring, and blasting in mineral varies considerably, and is carried on in the same way as in the sterile benches. The average cost might be put down roughly:—

> Boring and explosives ...   .   4d. to 8d. per ton.
> Loading ...   .   2½d.   „

The following are the names of the mines which are, or have been, partly worked by opencast:—

Rio Tinto.—Part of South Lode.

Tharsis.—North Lode, Centre Lode, Sierra Bulloms, Esperanza.

Calerñas.—Part of Lode.

Cueva de la Mora, Lasunuzo, Poderosa, Concepcion La Torya, El Tinto, San Miguel, Buitron, San Telmo, Payatus, Lomero, Chapparita, Vulcano Peña de Hierro.

These works exhibit the greatest variety of execution and offer striking examples of success and failure. The opencast recently made at Caluñas can, perhaps, be taken as an example of careful planning and economical execution.

CLASSIFICATION OF ORE.

The varied nature and copper contents of the ore render a classification necessary in mines which have their own railroads. Considering the low prices of sulphur and copper, it does not pay to export ore under about 3 per cent., the remainder of the ore, representing about 66 per cent. of the total, has to be treated on the spot. Accordingly, two separations are made, export ore and ore for local treatment, the latter being again subdivided according to the method of treatment adopted. The export ore, containing about :—

> Copper   ...   .   3 per cent.
> Sulphur   48   „
> Iron   45   „
> Silver, zinc, bismuth, lead, nickel, antimony, silicon   ..   4   „
> ———
> 100

is shipped direct to various parts of Europe. The ore for local treatment, containing, roughly :—

Copper	...	...	...	...	...	...	2 per cent.
Iron	...	...	...	...	...	...	38 .,
Sulphur	...	...	...	...	...	...	48 ,,
Arsenic, lead. zinc, and silicon			...		...	...	12 ..
							100

undergoes a rough process, known as the Rio Linto cementation process, introduced in 1840 by Don Felipe Pinto. This system is imperfect, as it only aims at the extraction of copper, 85 to 90 per cent. of which is extracted in two or three years. It consists in :—

1.—Calcining the ore in open heaps or tips, to convert the sulphide of copper into sulphate.

2.—In washing the ore thus calcined to dissolve the sulphate of copper.

3.—In decomposing with metallic iron the solution of sulphate of copper, and producing a precipitate of metallic copper.

The mode of selection at Rio Linto and Tharsis differs considerably. The Tharsis having works of their own in England, consume their own pyrites. The Rio Tinto Company mostly export to consumers, and have to undergo regulations as to the percentage of silica, dust, etc. They have also erected large blast-furnaces at their mines for the treatment of rich ores and quartz. This necessitates a careful classing and is mostly done by hand.

In mines not connected with railways all the ore undergoes local treatment, and is only classed according to size. In some cases the roughs are divided from the smalls by passing over screens about 1 inch clearance between bars, smalls and dust falling through to one wagon and the roughs passing on to another. Often this is dispensed with and done roughly by hand when the ore is being built in heaps for calcination.

GENERAL REMARKS.

The facility of winning ore in an opencast, compensates generally for the charge of the removal of the overburden, and compares favourably with the winning of ore from confined galleries by the pillar and stall method. By the former, there is a feasible and complete extraction of the ore, by the latter an incomplete and wasteful system, the only advantage of which is, that less capital is needed to start with, and that the ore can be extracted at once.

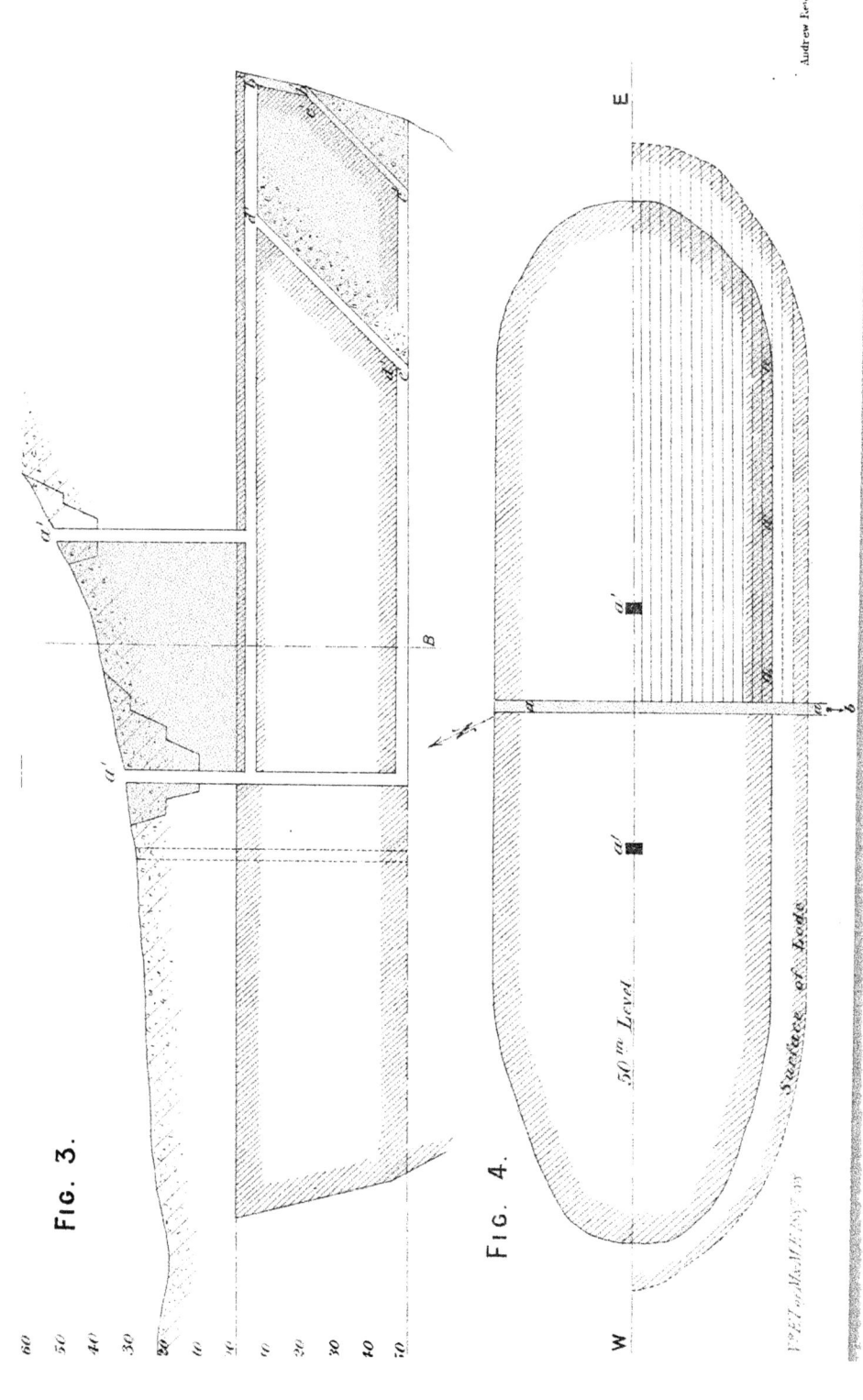

FIG. 3.

FIG. 4.

50ᵐ Level

Surface of Lode

W

E

Andrew Reid, Newcastle.

Once the lode worked by pillar and stall, and about one-third of the ore extracted, the rest being left in pillars and roof, the problem arises, how to move the remainder. In the ease of irregular workings this is almost impossible, in regular workings most dangerous and costly, as the ground covering the lode is in most cases composed of loose ironstone, clay, and boulders, which cannot stand without support, and which gather weight at the slightest filtration of surface waters. Filling up the cavity as pillars are removed has been tried but is costly, and alternate pillars and roofs have to be left to avoid a general crush.

By using care and working downwards, leaving safe ground behind the men, a good deal of ore can be won by stripping down workings to a size consistent with immediate safety, removing, perhaps between the first and ultimate workings, half of the whole mass.

Generally, an opencast is the solution of the following figures, taken from an ideal cross section :—Plate II., Fig. 1, through the line A B on longitudinal section, Plate III., Fig. 3, showing with what result.

Taking the section for a distance of 1·1 yards, leaving benches of 45 degrees in slate, and 60 degrees in porphyry and mineral, with slate on the South Wall and porphyry on the North, 8,528 cubic yards of slate and ironstone, and 5,689 cubic yards of porphyry have to be removed to uncover 5,428 cubic yards of mineral, or allowing 3·43 tons per cubic yard, 18,618 tons. Estimating the cost of removal of porphyry at 2s. 3d. per cubic yard, and slate and gossan at 1s. 1½d., the overburden account comes out at about 1s. 0¼d. per ton. Should the lode have been worked previously by pillar and stall, as shown on cross section, Fig. 2, Plate II., by 5½ feet by 6½ feet workings and 4½ feet roofs on a distance of 1·1 feet, namely, about ½ a yard on pillar section, and ¼ a yard on cross-cut section, it gives 2,043 cubic yards of ore extracted from a total of 5,240, or rather over a third. Estimating, however, the mineral extracted as a third of the whole, the overburden has to be paid by 12,450 tons, and the overburden account comes out at 1s. 6½d., or 6¼d. dearėr.

These figures distinctly prove the disadvantage of the system as regards pillar and stall workings, and a plan for removing all the ore when an opencast becomes too expensive is urgently needed.

As far back as 1860, the Government engineers, especially Don Felia Astiroz, spoke in favour of working the lodes partly by opencast where the least overburden had to be removed, and removing all the ore from the remaining part, packing the cavity with the sterile obtained from the opencast. This plan is advantageous at first sight, as the double advantage is obtained of removing a portion of the lode and thus obtaining sterile for filling in the underground workings, which otherwise would have to be quarried for that purpose.

This system was, however, never put into execution; difficulties occurred as to the method to be employed for working and filling in, and it is not until lately that this plan has been adopted for the working of a lode at the Tharsis mines.

The writer trusts that a few words on a system of mining, which he thinks would be feasible, cheap, and effective, would not be out of place.

PROPOSED METHOD.

Taking an ideal lode, Fig. 4 in plan, Fig. 3 in section (see Plate III.), 87 yards wide, with parallel walls dipping north, having shallow over-burden east of line A B and heavier west. The part west of A, B, would be advantageously worked by opencast, and east by underground workings, according to a new system.

This method consists in working with inclined stopes about 4 feet wide, either across the lode or longitudinally, as shown on sections, Fig. 3, Plate III., and Fig. 5, Plate IV., and filling in with sterile from the opencast workings, in measure as the stopes advance. Working is started on the 54 yards level by a $4\frac{1}{2}$ yards by $4\frac{1}{2}$ yards gallery $a, a, a,$ Fig. 4, in communication either with an adit or an engine shaft, as shown at b. Near the surface of the mineral, the corresponding level is driven in direct communication with shafts $a', a',$ leaving a slight roof of mineral. From this level an inclined winze, $b', b',$ Fig. 3, $3\frac{1}{4}$ yards wide is sunk, following the dip of the lode to the 54 yards level.

The actual working then commences by stripping down the back of winze about a yard at a time, as at $x, x,$ Plate IV., Fig. 6, and tipping sterile in measure as the stope advances, building a rough stone wall against the exposed face of mineral until the stope assumes the angle shown at $c', c',$ and, later on, $d', d',$ Plate III., Fig. 3, the slope to be determined by the angle formed by tipping. The sterile is loaded in end-tip wagons at shafts $a', a',$ which are partly used as shoots for tipping sterile from opencast, and partly for ventilation, being divided into two compartments for this purpose. The wagons are taken by the nearest way to the tip.

The ore after blasting rolls down the stope and is loaded into wagons at $e',$ Plate III., Fig. 3, and taken through gallery $a, a, a,$ Fig. 4, to engine shaft or adit. The work of filling and winning ore has to be kept distinct, and for this purpose the time can be divided proportionately, allowing a certain time for boring, blasting, and loading of ore, this work being stopped during the tipping of sterile.

VOL. XXXVII, PLATE IV.

FIG. 5.

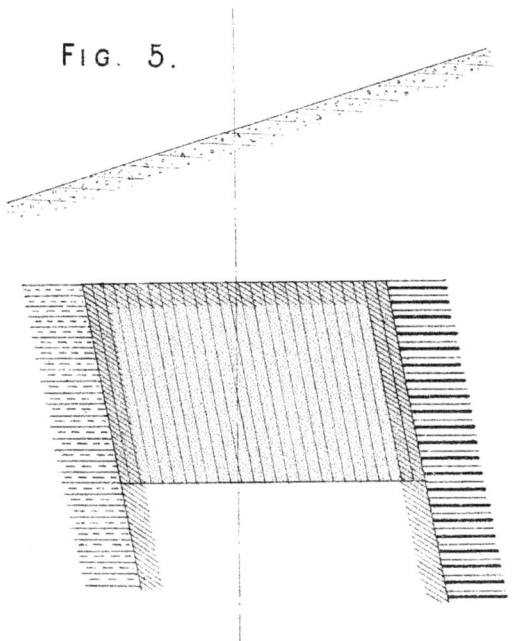

FIG. 6.

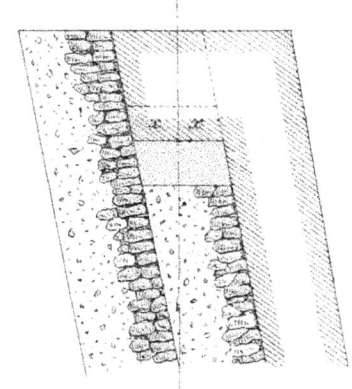

General Remarks.—It is thought that this mode of working is perfectly safe. Laying off the stopes longitudinally at the same angle as the lode the packing can exert no pressure, the inclination serving as a batter on the rough stone wall, which is only exposed $2\frac{3}{4}$ yards in height after the blasting has taken place. The mineral also stands perfectly for a width of $3\frac{1}{4}$ yards. Some dust and smalls would of course get mixed with the sterile after blasting, but with care this could to a great extent be avoided.

As regards expense, allowing 1s. 7d. per cubic metre for loading at shoot mouth, traction, tipping, and building of walls, the charge of packing per ton of ore would be 5d.; winning of ore and loading, say, 1s. 8d. per ton; total, 2s. 1d.; which compares favourably with both pillar and stall and opencast workings.

To conclude; in Plate III., Fig. 3, the depth of lode to be worked has been marked at 50m. or 54 yards; this distance can be modified according to requirement, and once the lode worked to this depth, the same operation can be repeated below. The system can also be put in operation by stoping across the lode, but the advantage of working in favour of the joints in mineral is lost.

Winning Ore in lodes worked by Pillar and Stall where a general fall has taken place.—In this case, the writer thinks that the system of mining used in the tin mines of Altenberg, Saxony, under similar circumstances, might be used. This method consists in driving a gallery at lower levels in firm ground, skirting the fall, and driving strongly timbered cross-cuts, $1\frac{1}{2}$ yards by $2\frac{1}{4}$ yards, into the loose ground. The workmen thus protected, draw the loose blocks of ore into the cross-cut at small expense. Care is taken to divide and localise the slips thus caused as much as possible, in order to avoid the ground taking a movement of any magnitude. After drawing a certain amount of ore through one of these cross-cuts, all work in it is stopped until the expected crush has taken place.

The cheapness of the method is proved, as it is carried on in ore containing only one-third per cent. of lassiterite, with considerable profit.

CATALOGUE OF FOSSILS EXHIBITED WHEN THE
PAPER WAS READ.

1.—Fossils in slate, from the same bed.
2.—Slates, from various localities.
3.—Porphyry ?
4.—Decomposed porphyry from walls of lode.
5.—Samples of partially metallised porphyry and quartzite (?) from walls of lode.
6.—Quartz.
7.—A. Ironstone from surface.
„   B. Azurite, fählerz, and quartz, in connection with pyrites.
„   C. Grey copper ore,   „   „
„   D. Peacock ore, pyrites, etc.,
„   E. Galena,
„   F. Grey copper ore,
„   G. Copper pyrites, galena,
„   K. Galena, etc.,
„   L. Galena, grey copper, ...
„   M. Peacock ore,
„   N. Native copper, quartz, ...
„   O. Fählerz,
„   P. Vein of peacock ore, pyrites,   „
„   Q. Pyrites, 6·50 per cent. copper,   „
8.—Joints in mineral, polished through pressure.
9.—Roman slag.
10.—Speiss (Roman), locally termed "metal blanquillo."
11.—A. Vein of slag (Roman), running between joints of two blocks of granite.
            forming part of Roman furnace.
„   B. Corner covered with slag, from same furnace.
12.—Roman implement, found four feet underground in ruins of house.
13.—Part of Roman miner's lamp.

MR. D. TYZACK said, in the absence of the writer of this most interesting paper, and as there are probably not many gentlemen present who like himself have been engaged in pyrites mining in province Huelva, the members perhaps may find interest in a few supplementary notes he had made.

As his experience was confined to the neighbourhood of the Rio Tinto Company's mines and works, some 40 miles from the town of Huelva, it would be understood that these notes applied only to that district and to the Rio Tinto mines.

Geology.—On the subject of the Geology of this district, he should say but little, leaving to more able men the solution of the phenomena of the occurrence of these immense deposits of pyrites, but the striking appearance of the neighbourhood for many miles around and about the mines demands some attention. Mr. Allan states that these deposits occur in a zone of clay slate; this clay slate is almost universally vertical in its cleavage, and in the mountainous part of the railway from Huelva to the mines and elsewhere, the fantastic pinnacles of the clay slate assume the appearance of church spires, huge pointed towers, and saw-like indentations, and though all on edge it never appears to lose its laminated structure which points directly to its original deposit, being due to the action of water settling it in a horizontal position, though here and there may be seen zig-zag contortions of strata evidently the result of lateral pressure, whilst in a more plastic condition than at present. He had been fortunate enough to obtain from the clay slate formation at Rio Tinto, the fossil impression of a pair of bivalves, which he believed Professor Lebour, from specimens in his possession, would be able to show are commonly found in the Lower Carboniferous system.

Pyrites Lodes, Indications.—He was able to confirm from personal observation, the peculiarity pointed out by Mr. Allan, that almost without exception the position of a pyrite lode is strongly indicated by the distinct dark red color of the earth immediately overlying the pyrites deposits, and he was inclined to believe that this is due to the chemical action of water percolating through the overlying *débris* to the pyrites below, decomposing the same and giving rise to fumes which colour the surface soil. However this may be, it is worth noting that the most valuable pyrites lodes are indicated in this manner.

Mining, Rio Tinto.—Some idea of the magnitude of the scale at which mining is carried on at the Rio Tinto mines may be gathered from several indications. For example it is stated that the produce of pyrites per day is 4,400 tons; that some 60 locomotives are required on the works to deal with the traffic. It may be roughly stated that some 40 miles of railway are necessary on the company's mines and works, for the purpose of transporting minerals to and from the calcination grounds, precipitation tanks, blast furnaces, refining furnaces, etc., etc. And in order further to assist in grasping the magnitude of the mining done he might mention that during his experience there, an average of 20 tons of dynamite per month was necessary to carry on the blasting work, the consumption of dynamite having occasionally reached 25 tons per month.

Opencast Mining.—The large opencast or quarry at the Rio Tinto Mines, several hundred feet in depth, is a sight not easily forgotten. A faint idea of this opencast may be obtained from Plates V. and VI., which have been reproduced from some photographs taken about three years ago from the place itself. These plates will give a better idea of the mode of working than any long verbal explanations which may be made. Here, however, it may be as well to observe that as the South Lode opencast is situated well up on the side of a steep hill, advantage is taken of this to run all the produce from the mines and opencast, out on the level, by means of five different level tunnels, these tunnels are made large enough to allow locomotives and trains of five tons wagons to enter the opencast as well as the gallery workings, thus obviating the necessity of winding engines, except from workings below the lowest tunnel level.

The necessity to remove by the present system of working increasing millions of tons of overburden, consisting of porphyry or clay slate from the sides of this opencast, will, in the course of time, compel the abandonment of this opencast as regards depths, and Mr. Allan has suggested a system of, as it were, " working broken," by a vertical longwall face.

The Rio Tinto opencast is worked entirely by means of blasting by dynamite, and at stated intervals each day a bugle sounds as a warning to engine drivers to remove all locomotives and wagons out of the way ; a second and third bugle-call clear all workmen from the benches and neighbourhood ; the fuses are then fired and for ten minutes a terrific din of blasting bursts on the ear ; at a distance of 300 or 400 yards one may look on with comparative safety, when a fine sight of falling rock, smoke, and dust may be witnessed, which, together with the roar, makes a never-to-be-forgotten impression. This opencast is provided with a set of powerful electric lights arranged around its surface edges, and by the aid of reflectors a strong ray of light can be thrown on any part of the chasm and work continued during the night.

Exploring on unknown Pyrites Lodes and the mode of carrying on underground work.—When a new lode is to be explored the usual method

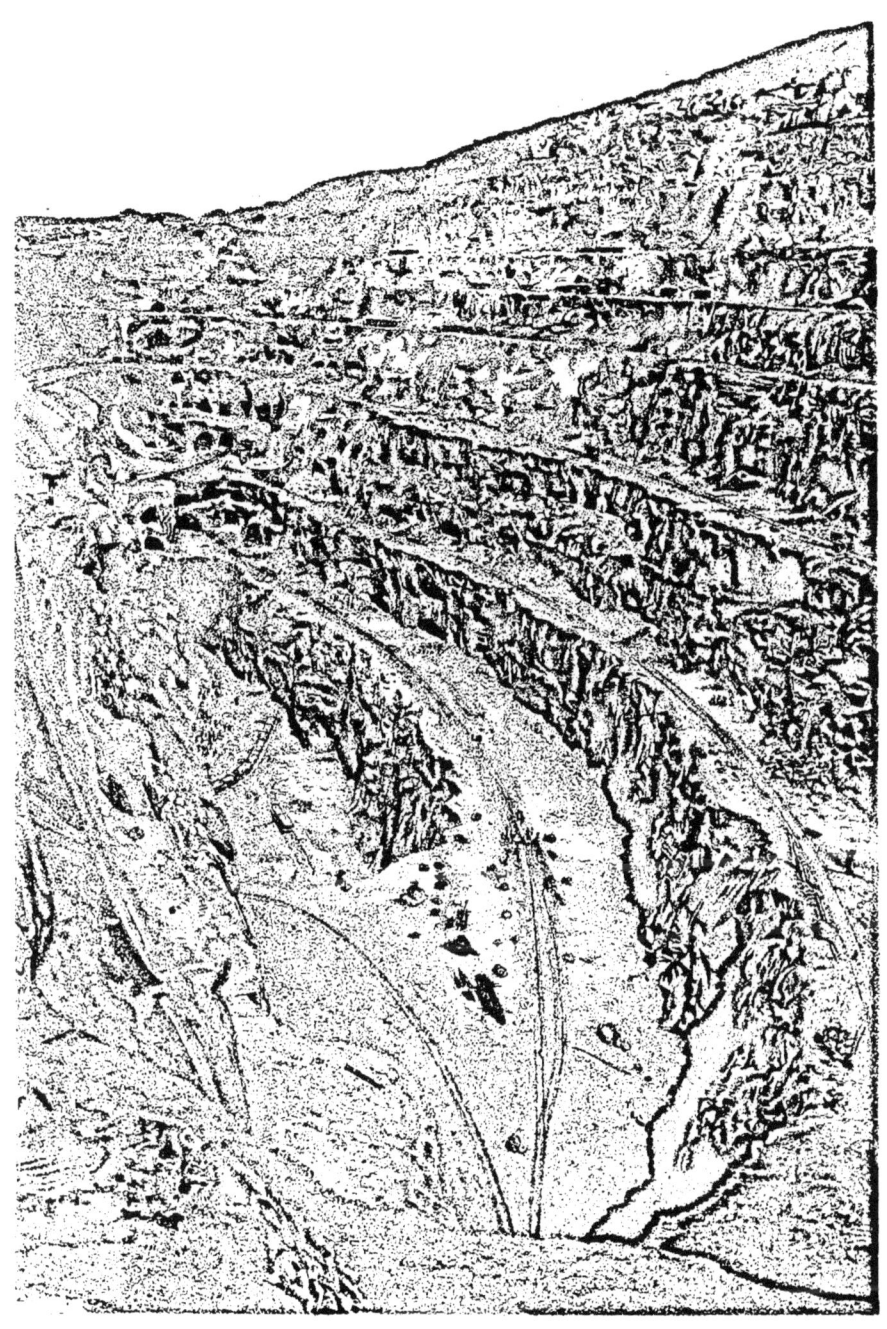

m"The Pyrites Deposits of the Province of Huelva."
de of the Rio Tinto Mines - June 1886.

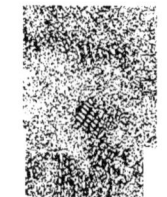

employed is to mark off on the surface at right angles to the course of the lode (which is usually roughly indicated by the red surface appearance, formerly alluded to, as also by the outcrop of clay slate), a series of small trial shafts; these are sunk through the loose *débris* which frequently covers these lodes, until the clay slate or porphyry is reached, marking the limit north or south of the line of lode : these trial shafts enable the engineer to fix the approximate line of the lode as well as the "hade" or angle at which it lies from the vertical position. When sufficient evidence is given by the trial shafts that a workable lode containing enough copper to warrant more extensive explorations has been met, a suitable low-lying valley is usually selected, often at a considerable distance, and a tunnel heading is driven forward towards the centre line of the lode; this tunnel is made sufficiently large to admit a locomotive. The work of driving these tunnels is performed by rock-drills, four drills on one frame, and compressed air at 70 lbs. per square inch, the air being taken from the compressor in 7 inch diameter flange pipes ; the electric light is used at the face of these tunnels, and the usual rate of progress in clay, slate, porphyry, and mineral, in a heading 13 to 14 feet square, varies from 80 to 110m. per month, or say, from 88 to 120 yards per month. The ventilation generally relied on is the air from the compressor which keeps the face clear, but a dense volume of dynamite fumes hangs about these tunnels till communication is made with the large ventilating shafts, which are usually sunk every few hundred yards ahead, when the lode is reached. On reaching the mineral the tunnel headings are guided in the direction they are to take by the explorations going on in the larger ventilating shafts which have been steadily sunk in advance every two or three hundred yards on the lode. From these large exploring shafts, on reaching sufficient depths to encounter the top of the mineral, are driven at every 11 yards of depth, headings at right angles to, and on line of the lode, care being taken that each shaft commences these numerous floors at the same level, so that, on communication being made, one to the other, no inequality of level is experienced.

These main exploring shafts (subsequently to be used for ventilation as the tunnel comes up) are as quickly as possible connected with each other on the line of the lode at the different floors by means of galleries 6½ feet square ; but usually only sufficient mineral is raised to keep the exploring galleries free to travel in. Immediately the low tunnel comes up near enough to the exploring shafts to take off the water from the pumps there necessarily employed, winzes or staples are sunk from each floor down to the tunnel level, a little to one side of the locomotive way, wooden shoots with trap doors are arranged over the wagonway, and the work of opening out the top galleries by pillar and stall is commenced by driving with 6½ yard walls, cross-cuts or headings 13 feet square over to both north and south walls of the lode ; the mineral so obtained is tipped into the nearest winze till it is full; a train of 30, five tons capacity wagons can by this means be filled up in a very short time.   When the number of floors become too many, and the traffic in the main tunnel congested, an intermediate locomotive way is arranged, and the mineral is tipped every two or three floors to an engine way.

In some of the old work performed at Rio Tinto mines the winzes or staples are almost as numerous as the headings, and were evidently resorted to as working places to extract mineral when short of pit room ; these staples, as a rule, were all left open a few years ago, and many deaths resulted from these traps to the unwary.   This state of affairs has been improved lately, and they are now protected.   One of the great difficulties to be encountered, and every precaution taken against, is the occurrence every here and there of old Roman shafts and headings, in some cases full of water and fine running sand, in other cases full of mud, and again charged with fine mineral-like powder.   These "Cuevas" as they are called, are usually suddenly opened out by a blast of dynamite, and a whole train of wagons has been sanded up and lost in a few seconds in a heading 14 feet high, for many days together, necessitating the removal of 3,000 to 4,000 tons of run before the work could be proceeded with.   These mine runs are not unfrequently the cause of loss of life when men do not get quickly enough out of their way.   Blasting accidents from dynamite are of frequent occurrence, and again men lose their lives in the exploring shafts and headings, which are not usually ventilated, by returning too soon after their blasts to see the effects of their shots, when they are overcome by the dynamite fumes.

The water in and about these copper pyrites mines is so charged with copper in solution, in an acid state, that a precipitate of metallic copper is immediately formed on all iron brought in contact with this water, the result is that all pumps, plungers, buckets, clacks, bolts, nuts, spear plates, etc., exposed to the direct action of this water have to be made of a special kind of bronze which resists the action of this acid.   In the case of the rising column of a set of pumps, which are unlikely to be immersed, the difficulty is got over by lining the set with an inside cleading of ¼ inch wood, this, if well done, preserves the iron from the corrosive action of the water.   Another curious fact regarding this water charged with copper is that timber after long immersion becomes in a curious way saturated with metallic copper; so that the pores of the wood which originally contained its natural sap, are filled instead with copper in a metallic state; this is immediately seen on cutting into a piece of wood that has been long under water. How long a time is required for this metamorphosis it is impossible to say, some of the specimens of Roman timber from old workings which have been unearthed were in good preservation and may possibly have been a thousand years or more in soak. It is probable that the old Roman water-wheel discovered underground at the North Lode, Rio Tinto, at a depth of 360 feet, (shown in Plate VII., taken from a photograph), owes its preservation to saturation by the copper liquor, which eventually becomes partly metallic. Iron rails submerged for a short time in the water of the mines become completely coated with an eighth of an inch of bright copper, this chemical action is of course taken advantage of on the company's works to produce large quantities of copper precipitate, called "Cascara." Every drop of water from the mines is consequently of value, and miles of laundry boxes and carefully arranged open water conduits, convey the liquor to the precipitation tanks.

It cannot be wondered at that the price of copper keeps low when it is stated that 4,400 tons of ore, averaging 3 per cent. of copper, are produced daily at Rio Tinto mines; this means that the quantity of metallic copper, if all saved, would amount to 792 tons per week, or, adding the copper precipitate produced from the water of the mines, there would be in round numbers 800 tons of copper per week from these mines only.

Professor LEBOUR said, Mr. Tyzack had mentioned the geological age of the fossil he had obtained. Mr. Allan seemed himself to have no special view as to the age of the deposits, but had simply mentioned other people's views. Mr. Allan stated that the age of the formation was indicated in the geological maps of Maestra and De Verneuil as Silurian, Roemer described it as belonging to a low horizon of the Culm measures, and J. A. Phillips as apparently of Silurian, Devonian, and Carboniferous age. It was very probable that John Arthur Phillips was right. Culm in Germany was simply a slaty or shaly deposit of the Lower Carboniferous formation which passed perfectly regularly into the Devonian shale, which was similar to that in the Carboniferous. In the south-west district of Portugal there was the same sort of thing. Scarcely a fossil was found, and the succession might run from Silurian to Carboniferous. The place where the slate shown by Mr. Tyzack was found furnished a fossil which enabled them to say what age that particular part of the series belonged to, and there was no doubt it was the Carboniferous. This was the Carboniferous Culm of the district. There were on the table a number of other specimens of exactly the same fossil, only distorted by the crushing they had been subjected to by pressure. There was no doubt this fossil was the *Posidonomya Becheri*, which was one of the most widely distributed fossils in the world. He showed specimens of the same fossil from Budle, near Bamborough, Northumberland, from the Hartz, in Germany, and another from Silesia. The rest, as Phillips, a very acute observer, said, is probably Devonian as to the part immediately below this, and possibly Silurian as to the lowest part. He (Professor Lebour) was not speaking only from the rocks of that district, but also from reading accounts of similar rocks in Portugal, which were a continuation of the same great band extending from the borders of Russia.

The CHAIRMAN said that these papers had a peculiar interest to him, inasmuch as he was well acquainted with a large portion of the district described, having some years ago visited many of the mines there, among them being Rio Tinto. The Tharsis he had not seen. Huelva was a place which had become of great commercial importance in later years, and he was glad that this Institute was now in possession of a paper of the description of Mr. Allan's. He did not think he had before met with a paper on these mines in this form. He believed Mr. Collins wrote one on the subject; but this Institute had not been in possession of anything of the kind. There were several interesting points, but he was not able to deal with them at present. One of the most remarkable things in the workings of past times was the smallness of their size. He had seen levels driven 2 feet 3 inches square in this hard rock, beautifully cut; and how any man could have so cut them was beyond comprehension. There was no evidence of powder having been used, and, in fact, powder was not known when this work was done. The Buitron deposit mentioned in the paper he knew a great deal about. There was no doubt that, after a large expenditure of money, and making a railway some 40 miles long, it was a failure; but the company were fortunate enough in getting possession of another deposit, and so the railway may have still answered their purpose. The district, as a whole, was limited, extending only a few miles north and south, and still less east to west; but it was of very great interest. He did not know that it would always hold good, but, as a rule, the mass of pyrites was very much in the shape of a ship, it narrowed downwards, and then came to a bottom and the same length-ways. They might compare the shape of the mass to the hull of a ship, or, in some cases, like a cigar. He moved a vote of thanks to Mr. Allan for his paper, and to Mr. Tyzack for his supplementary communication.

Mr. J. B. SIMPSON seconded the vote of thanks, and it was agreed to.

The CHAIRMAN said he thought the discussion of Mr. Walton Brown's paper, on "An Account of Experiments in France upon the Possible Connection between Movements of the Earth's Crust and the Issues of Gases in Mines," would stand over to a future meeting.

Mr. WALTON BROWN said, if it was allowed to stand over till the next meeting, he expected a report from the Committee appointed to make observations would then be ready.

This concluded the business of the meeting.

PROCEEDINGS.

GENERAL MEETING, SATURDAY, DECEMBER 10TH, 1887.

SIR LOWTHIAN BELL, BART., PRESIDENT, IN THE CHAIR.

The SECRETARY read the minutes of the last meeting and reported the proceedings of the Council.

The following gentlemen were elected having been previously nominated :—

HONORARY MEMBER—

Mr. WILLIAM BEATTIE SCOTT, Mines Inspector, Wolverhampton.

ORDINARY MEMBER—

Mr. WILLIAM STEPHENSON BLACKBURN, Mining Engineer, Astley House, Woodlesford, near Leeds.

ASSOCIATE MEMBERS—

Mr. BENNETT HOOPER BROUGH, Assoc. R.S.M., F.G.S., &c., Assistant to Professor of Mining at the Royal School of Mines, 5, Robert Street, Adelphi, London, W.C.

Mr. JOSEPH R. IRVINE, Hendon Ropery, Sunderland.

Mr. WILLIAM LEE, Felling Colliery, Newcastle-on-Tyne.

Mr. JOHN EMANUEL TYERS, Mechanical Engineer, Mohpani, Central Provinces, India.

Mr. JACOB WALLAU, c/o Messrs. Black, Hawthorn, & Co., Gateshead.

Mr. JOHN ANDREW YOUNG, 7, Tyne Vale Terrace, Gateshead.

The following gentlemen were nominated for election :—

HONORARY MEMBER—

Mr. ARCHIBALD EDWARD PINCHING, H.M. Inspector of Metallic Mines (Cornwall), Osborne Lodge, Stoke, Devonport.

ASSOCIATE MEMBER—

Mr. LANCELOT DOBINSON, Hebburn Colliery, Newcastle-on-Tyne.

The PRESIDENT said that in connection with the resignation of the Secretary the matter was in the hands of a committee. At the earliest period possible the committee would report to the Institute generally as to what line of conduct they would recommend to be taken.

———

The SECRETARY read the following Report of the Committee appointed to inquire into the observations of Earth Tremors :—

REPORT OF THE COMMITTEE APPOINTED TO INQUIRE INTO THE OBSERVATIONS OF EARTH TREMORS WITH THE VIEW OF DETERMINING THEIR CONNECTION (IF ANY) WITH THE ISSUE OF GAS IN MINES.

YOUR Committee beg to report as follows :—

It was agreed in the first stage of their investigations that observations of the time of the motions of the earth's crust were more important than any others, and consideration has been accordingly given to the most efficient means of obtaining such observations.

Professor Garnett was requested to advise them upon the most suitable forms of seismological instruments for this purpose and the valuable report supplied by him is hereto annexed.

After deciding upon the form of seismograph to be used, approaches were made to several makers of philosophical instruments who were unable to execute the requirements of the Committee. Finally, in 1886, arrangements were completed with the Cambridge Scientific Instrument Company for the supply of one of Professor Ewing's duplex pendulum seismographs which would record horizontal motions of the earth upon a plate of smoked glass.

In the meantime, Mr. John Daglish gave permission for the use at Marsden Colliery, on the surface, of a seismograph constructed in accordance with the drawings of Mr. Walton Brown and similar in all details to those employed by Professor Ewing in Japan.

The instrument used in these observations (see Plates VIII. and IX.) recorded the existence of tremors, tiltings, or other movements of the earth's crust and the time of their occurrence, without making any record of their extent or direction.

SEISMOGRAPHS.

The apparatus to be described is largely in use in Japan.

It consists of a pendulum (so controlled by friction as to be dead beat for small displacements) which is used as a steady point. To enlarge the motion of the earth relative to this steady point a lever is attached.

H H H H is a wooden box with a door at the lower end for inspection of the apparatus.

A lead ring R is suspended by brass wires as a pendulum from the screw S. This screw passes through a small brass plate P which can be moved horizontally over a hole in the top of the box. The motions of the point of suspension allow the pendulum to be adjusted.

Over the top of the pendulum is a wooden bar W, carrying two sliding pointers *h h* resting on a glass plate placed on the top of the pendulum. The points give the frictional resistance above referred to.

There is a brass bar across the inside of the pendulum perforated with a small hole (conical) *m*. A stiff wire passes through *m* and forms the lever I. This wire passes through a small ball *i* which rests upon the upper side of a small brass plate perforated with a conical hole and resting on the wooden bar O crossing the box.

The index I is connected with one pole of the circuit, and hangs freely in the centre of a depression in a small cup of mercury M. This depression is produced by screwing a small pin into the bottom of the. wooden cup. The mercury forms the other pole of the circuit. Should the index move more in any other direction than the vertical the circuit is immediately closed.

For oral demonstration the instrument is shown in connection with a single stroke bell which is rung by each closure of the circuit.

For obtaining permanent records of the time of the tremors an American clock was used, the hour hand being connected with a paper disc *a*. A needle is attached to a lever *p*, (Plate IX.) moved by the electro-magnets E each time the circuit is closed, making a puncture in the paper disc for each closure.

The records cover a period of nearly seven months, from October 19th, 1886, to April 30th, 1887. (See Plate X.) An interesting feature of these records is the irregular and perturbed movements which lasted from February 7th to March 12th, 1887. They appear to be connected with disturbances originating at places very distant from the observatory at Marsden.

It seems highly probable that the shocks experienced at St. Louis in the United States on February 7th were a more violent result of the motions recorded at Marsden on the same day. These motions continued until February 23rd, the date of the disturbance at Nice and adjacent district, and ceased on March 12th when that series of Italian disturbances ceased. The shocks recorded on March 14th seem to have been a reverberation of those experienced in Bohemia and Burmah.

The pulsations recorded at Marsden on April 7th and 13th, are evidently the results of the severe shocks felt at Aden and elsewhere on the 6th, and at Charlestown on the 11th.

The experiments now placed on record have been made with a somewhat rough apparatus and will be shortly extended by means of a more perfect form of seismograph made from the designs of Professor Ewing, of University College, Dundee. These continued observations will be accompanied by measurements of the percentage of gas found in the return air of the mine, made by some of the perfected apparatus which are applicable to such purposes.

The measurements of the proportions of gas have not been made up to the present time. It may however be mentioned as at least a curious coincidence, that the disturbances of December 6th to 8th were closely followed by increased issues of gas at several of the collieries in this district.

APPENDIX A.

SEISMOGRAPH RECORDS AT MARSDEN.

	Made Contact.	Out of Contact
	1886.	1886.
1.	October 25, (?)	October 29, 9 a.m.
2.	November 1, 1·30 p.m.	November 4, (?)
3.	November 4, 4·5 p.m.	November 5, (?)
4.	November 16, 12·30 noon.	November 17, between 9·30 a.m. and 1·30 p.m.
5.	November 17, 2·20 p.m.	November 19, between 9 a.m. and 12 noon.
6.	November 27, 1 p.m.	December 3, during night.
7.	December 6, 0·45 a.m.	December 7, between 5 p.m. and Dec. 8, 9 a.m.
8.	December 9, 2·30 p.m.	December 11, 10 a.m.
9.	December 18, 6·53 a m.	December 18, 10·30 a.m.
10.	December 30, 10·40 p.m.	A shock.
	1887.	1887.
11.	January 9, 12·25 noon.	January 17, between 5 p.m. and Jan. 18, 9 a.m.
12.	January 18, 9·15 a.m.	A shock.
13.	January 25, 12·30 noon	A shock.
14.	January 26, did not record.	February 5, between 5 p.m. and Feb. 7, at 9 a.m.
15.	February 7, 5 p.m.	March 12, between 12·30 noon, and
16.	March 14, between 9 a.m. and noon,	March 14, 9 a.m.
		A shock.
17.	March 16, 1·30 a.m.	March 16, between 9 a.m. and 1 noon.
18.	March 18, 11·30 p.m.	A shock.
19.	April 7, 9·20 a.m.	A shock.
20.	April 13, 2·20 p.m.	A shock.

Instrument out of use after April 30th, 1887.

APPENDIX B.

REPORT UPON SEISMOLOGICAL INSTRUMENTS.

For most of the matter on which the following report is based, the writer is indebted to the many published papers of Professor Ewing, of University College, Dundee.

The earth tremors observed in Japan have generally consisted of a succession of oscillations, each having a period somewhat less than a second or very little greater, and an amplitude in the horizontal directions varying from about $\frac{1}{250}$ inch to $\frac{1}{4}$ inch, while the amplitude in the vertical direction has been considerably less. The maximum force required to act upon a body in order to make it oscillate through $\frac{1}{250}$ inch and complete the double oscillation in one second is about $\frac{1}{5000}$th of the weight of the body. If the periodic time be increased the force required will be diminished, varying inversely as the square of the periodic time.

In all seismographs advantage is taken of the mass, or inertia, of a body in virtue of which it tends to remain stationary while the earth vibrates beneath it. From what has just been stated it appears that the mass must be suspended with great delicacy if it is to record very small earth tremors, since a force arising from friction or any other cause and equal to $\frac{1}{5000}$th of the weight of the body may be sufficient to cause it to move in precisely the same manner as the earth. Hence the friction and other resistances must be so reduced as to be small in comparison with $\frac{1}{5000}$th of the weight of the suspended body. It is also clear that the equilibrium of the suspended body must be nearly neutral or astatic so that when its supports are displaced through $\frac{1}{250}$th of an inch relative to the body, the forces tending to restore equilibrium and supposed to act at the centre of gravity of the body must be small compared with $\frac{1}{5000}$th of the weight of the body.

For the purposes of this report the instruments commonly used may be divided into the following classes :—

1.—Instruments which require to be directly observed at the time of the tremor.

2.—Instruments which record the existence of a tremor without recording its direction or amplitude.

3.—Instruments which record the details of horizontal vibrations (horizontal seismographs).

4.—Instruments which record the details of vertical vibrations (vertical seismographs).

5.—Clocks or other mechanism for moving the paper, glass, or other material on which the record is made.

6.—Apparatus for recording the instant at which some particular phase of the tremor occurs.

1.—Among the first class may be mentioned the pendulums of Bertelli and Rossi, the indices of which are observed directly by a microscope or, after reflection, in a glass prism, in which case the motion in all azimuths can be observed with one fixed microscope. In the same class are the different forms of instruments in which the microphone and telephone indicate the existence of vibrations. As it is not possible to keep an observer permanently stationed at each instrument we may dismiss this class without further notice.

2.—The second class of instruments may be employed when it is required only to record the existence of a tremor and the time of its occurrence, without making any record of its extent, direction, or duration ; but these instruments are most useful in a subsidiary capacity ; i.e., to set in motion an instrument of class 5 in order to start the motion of a plate or strip of paper on which a full record is made. The apparatus may consist of a very delicately-balanced body which falls over, or otherwise moves through a considerable distance, and in doing so releases a detent, starts a train of mechanism, and either stops a clock or, by pressing a disc of paper against the hands of a clock, which have been furnished with ink pads, records the position of the hands and immediately withdraws the paper without disturbing the going of the clock. But the most sensitive as well as the simplest instrument of this class is a light pendulum carrying a piece of platinum wire below the bob. Underneath the pendulum is placed a small wooden cup of mercury into the centre of which an iron pin is fixed. The mercury rises through surface tension in a convex ring around the pin, while the end of the platinum wire lies within the dimple. When a horizontal displacement occurs in any direction, the platinum wire comes in contact with the mercury and completes a circuit through which a current then passes, excites a magnet, and either starts a recording apparatus of class 5 or a simple time register such as that above alluded to. Instead of the platinum wire being fixed directly to the pendulum, it may be attached to a pointer movable about a point, fixed relative to the frame of the apparatus a little below the bob of the pendulum. The upper end of the pointer is moved by the pendulum ; the platinum wire at the other end moves through a much greater distance within the mercury dimple. This instrument may be called an *electric seismoscope*. (See Trans. Seismological Society of Japan, Vol. IV., Plate opposite p. 97.)

3 and 4.—It is more difficult to construct a vertical seismograph than a horizontal seismograph, and the vertical movements to be recorded are very much smaller than the horizontal movements.   Hence it seems desirable in the first instance to attempt to measure only horizontal movements.   Horizontal seismographs consist of

(a) Long suspension pendulums.

(b) Duplex (astatic) pendulums.

(c) Horizontal pendulums.

(d) Heavy plates supported (astatically) on rolling spheres or
       cylinders.

The instruments (d) possess the advantage of absolute astaticism, but it is impossible to reduce the rolling friction to one-thousandth of the weight of the plate, and hence only very severe tremors can be recorded by them.   In order to obtain sufficient astaticism in instruments of sub-class (a), it is necessary to employ a long suspension.   The record may be traced by a pointer mounted as described under class 2.   The method of making the record will be referred to under class 5.   De Rossi employs four pendulums a b c d (Plate XI.) suspended from the corners of a square, and one e from the centre having a longer period of vibration than the others so as to check the natural swinging of the system. Between the central pendulum and each of the others is suspended a light needle f f f f attached to a coil of very thin wire.   The needles are suspended by fine silk, so that the portions of the silk make an angle of 155°, and so multiply the motion of the pendulum about six times upon the needles.   Whenever the pendulums approach one another sufficiently, contact is made by the needle with mercury in a cup.   It is not easy to see the special advantage of this arrangement.

(b) The duplex pendulum consists of an ordinary pendulum, com-bined with a mass which is supported on a rod, and free to turn about a point beneath its centre of gravity.   This mass is generally introduced within the hollow bob of the suspended pendulum.   Its equilibrium being unstable and that of the surrounding bob stable, the equilibrium of the composite system may be made as nearly neutral as is desired.   A short suspension may thus be made to take the place of one of indefinite length.

The supporting rod may be continued through the hollow bob of the suspended pendulum and made to carry a tracing point, the motion of which may be several times as great as that of the earth.   (See Trans. Seismological Society of Japan, Vol. V., p. 89.)   The instrument is well adapted for giving in one tracing a complete record of the horizontal movements of the earth on a fixed plate, but it appears to be difficult of construction and, for preliminary purposes, inferior to a pair of horizontal pendulums.

(c) Horizontal pendulums record only those horizontal movements which take place in a direction perpendicular to their own plane. Hence, for a complete record of the horizontal movements of the earth, a pair of such pendulums, with their planes at right angles to each other, should be employed. The motion may be multiplied to any extent by sufficiently increasing the length of the tracing pencils.

Formerly the horizontal pendulum was mounted, as shown in Plate XIII., so as to turn about the axis $a\ b$, the line $a\ b$ being very slightly inclined to the vertical, so as to give a very small amount of stability to the instrument. The line $c\ d$ passes through the centre of percussion of the frame with respect to the axis $a\ b$. The mass M turns about the centres at c and d, and thus balances as a particle fixed at the centre of percussion of the bracket. During an earth movement perpendicular to the plane of the bracket, the line $c\ d$ remains fixed, and the end of the pointer makes an enlarged tracing of the displacement.

Instead of employing the centres $a\ b$ for the pendulum to turn on, it is simpler, and it introduces less friction, to suspend the instrument by a wire $a\ f$, and allow it to turn on a knife edge or point b, which rests in a small steel or agate cup (see Plate XIV.) let into the vertical support. The depth of the depression in which b rests determines the stability of the instrument. In the most recent form of suspension a slot is cut in the strut $b\ g$, so that the supporting pillar may pass through the slot, and the end b of the strut is then tied to the back of the pillar by a piece of watch spring held in screw clamps. In this way the flexure of the watch spring takes the place of the turning of the knife edge, and it is said that the resistance is still further diminished. A pair of horizontal pendulums of this construction appears to be the most promising instrument with which to commence investigations. (See Plate XV.)

Fig. 1, Plate XV., is an elevation showing one of two horizontal pendulums, and Fig. 2 is a sectional plan of the pair. The post p^1 firmly stuck in the earth carries two horizontal levers $l^1\ l^2$ set at right angles to each other to record the two rectangular opponents of horizontal motion. The bobs m^1 and m^2 are fixed to the rods $l^1\ l^2$, and tied to a pair of small vices at the top of the post by fine steel wire t. Fig. 2 is a plan of the post p^1 and shows the arrangement of the vices which are fixed in a manner to allow the rods two horizontal degrees of freedom of adjustment. At the back end of rod l is a fork—shown on a larger scale in Figs. 3, 4. and 5,

—which consists of two parallel cheeks of brass a a terminating in a vice b, which is clamped by the screw bolt and nut c. A split upright pin p is fixed to the post, and a short thin flat band of very flexible steel is clamped between it and b to the end of l. This is kept in tension by the thrust exerted by l, and when the horizontal pendulum swings the spring bends at, or close to, a vertical line in the centre of the pin. A sector of the pin x facing towards a Fig. 3, is cut out to give the spring room to bend about the axis of the pin. The split sides of the pin p are pressed together by the nut n and so caused to hold the spring between them.

In Fig. 5 one of the cheeks a is removed and the spring shown in vertical section, the flat spring appearing there lettered s. Only one part of the pin p is screwed, the lower part is a smooth cylinder and the cheeks a a pass just clear of it on either side, their distance apart being adjustable within certain limits by the screw bolt and nut d. Hence no horizontal translation of l can occur, and the only freedom of motion possessed by the suspending rod *en masse* is that of revolving in vertical line joining the upper end of t with the axis of the pin p. A pair of long bamboo rods r r serve to multiply the motion and record their displacement side by side on a revolving smoked glass plate g, by means of horizontal pointers y y y. These are short pieces of straw tipped with steel, and each is provided with a little balance weight w behind the hinge, which lightens the pressure of the pointer on the plate and so reduces the friction.

To bring the records parallel the rod l^2 is set at right angles to l^1, and the counterpoise w^1 serves to bring the centre of gravity of the system back into the line of l^2.

5.—The records may be written on—

(*a*) A revolving smoked glass plate.

(*b*) A smoked glass plate drawn in a straight line beneath the recording pencils by mechanism started by an electric seismoscope.

(*c*) On Morse telegraph paper drawn past the pencils by an ordinary Morse instrument.

(*a*) If a revolving plate be used it may be made to revolve continuously or may be started by an electric seismoscope and allowed to run for two or three minutes. A continuously revolving plate is useless if inspected only at long intervals, because the slow motion of the zero due to variations of temperature or other accidents causes the pencils to trace a broad band in course of time and thereby to erase their own records when the amplitude of the vibrations is small. (Plate XVI.) Hence when the instruments can be inspected only after an interval of some hours, the electric seismoscope should be employed to start the glass plate.

(*b*) The chief difficulty in this case is to employ a strip of glass of sufficient length to obtain a record extending over a sufficient length of time, say three or four minutes.

(*c*) In this case the paper may be kept in a continuous motion or started by the seismoscope. The record may be made by means of electrified ink, as in the syphon recorder; by the employment of Bain's method of preparing the paper; or by means of small sparks from an induction coil sent in rapid succession through the recording pencils. This method involves the least mechanical resistance but necessitates the constant employment of an induction coil.

6.—The time may be determined by causing a separate current controlled by a clock to make a record on the paper at the end of every minute. A double mark might be made at the hours. If a seismoscope be employed to start the recording apparatus the same current might be utilised to stop a clock, or to print the position of the clock hands on a disc of varnished paper as mentioned above.

W. GARNETT.

Plate XVII. shows a complete set of apparatus, as recommended by Professor Ewing, for automatically registering the motion of the ground during earthquakes. They were originally designed by the Professor for the Seismological Observatory of the University of Tokio, where some of them have been in use since 1880. The forms now offered contain many improvements in detail, suggested by experience of earthquake measurement in Japan.

The Horizontal Pendulum Seismograph *a*. This records two rectangular horizontal components of each successive displacement of the ground, in conjunction with the time, on a revolving plate *b* of smoked glass. Two horizontal pendulums, pivoted on the same base by sharp steel points in sapphire sockets, furnish two steady points, with respect to north-south and east-west motion respectively. The pendulums are furnished with an adjustment by which they are put in nearly neutral equilibrium, and stand at right angles to each other. The motion is multiplied and recorded by two pointers projecting from the pendulums, and set so that they trace, with exceedingly little friction, a pair of magnified records of the earth's horizontal motion, in two components, side by side on the smoked glass plate. The pendulums and pointers are shown on the right-hand side of the plate. On the left is a clock *c*, which drives the plate at a uniform speed by a projecting friction roller *d*. The speed of the clock is controlled by a balanced centrifugal governor *e*, furnished with vanes which dip into oil. The clock may be arranged to run continuously, but is more generally started by means of an electro-magnetic detent, which acts whenever an electric circuit is closed by the small Palmieri Seismoscope *f* which appears in the woodcut behind the plate. This seismoscope is a short pendulum, ending in a platinum point which hangs just over a depression in a cup of mercury, and ready, when the slightest disturbance occurs, to make contact by touching the edge of the depression. This happens during the preliminary tremors of the earthquake, and causes the plate to begin revolving before the principal motions are felt. A record is then traced in the form of two undulating lines, which may extend without confusion throughout two or three revolutions of the plate. The record is preserved by varnishing the plate and using it as a "negative" to print photographs. The plate is supported in a way which allows it to be removed and replaced without disturbing the rest of the apparatus. A spare plate is supplied with each seismograph.

A special appliance is provided for the double purpose of showing the speed of the plate during its motion, and of determining the hour and minute at which the earthquake occurred. This is a second clock, not shown in the plate, which is started into motion along with the first, and has an escapement which marks time on the plate during the *first* revolution and then ceases to mark. The clock, however, continues going, and an inspection of its dial shows the interval that has elapsed from the time of the shock.

———

The PRESIDENT asked Professor Garnett if he had anything to add?

Professor GARNETT said the main departure from the reports which had been made in regard to instruments arose from the high prices which were charged for the horizontal pendulums. It was found that a duplex pendulum, made by the Cambridge Instrument Company, could be got at a much lower cost than the pair of horizontal pendulums; if the horizontal pendulums could have been made in Newcastle, the cost would have been still lower. They thought the best thing they could do was to recommend the Council to purchase a duplex pendulum on the score of economy, and this recommendation had been adopted by the Council, and the instrument was now in the possession of the Institute.

Professor LEBOUR said that in connection with the subject of earth tremors and this Institute, he should like to mention what was stated more fully in his report as a delegate from this Institute to the meeting of the British Association, and which he thought ought to be stated here, that, after considering the work that had been done by the Earth Tremors Committee of the Institute in the North, the British Association, and especially Sections A, C, and G of the Association, thought it advisable that this work should be extended over a greater part of the country. Of course this Institute could only pretend to conduct such work in its neighbourhood; and much of the value of the work depended upon it being extended by a network of observations all over the country. At the last meeting of the British Association, at Manchester, there was appointed a very influential committee to consider the advisability and the possibility of establishing such a network of seismographical observatories over England. That committee was now beginning its work. A great many gentlemen of very well known names were on that committee, and this was a guarantee that its work would be done carefully. He might add that at Birmingham they had already started the setting up of a seismoscope of a simpler kind, and three were about to be set up within a radius of ten miles of that town; three more next year, and so on. At Birmingham they intended to do this little by little on account of the expense. Others were going to be set up in Scotland, and he heard of inquiries from various parts of the country for information about setting up these things. In this way the work which this Institute begun was already receiving recognition; and it was specially mentioned in reference to the British Association Committee that they were to consider whether the work, as done by this Institute, was to be imitated or not. Sunderland was the locality which first called attention to these earth tremors; and within the last few days Mr. Backhouse, of Sunderland, had published in the local papers an account of some very careful surveys, which had been made by a very competent surveyor, of his house and grounds, proving there had been a sinking of the land in that part of Sunderland of something like 3 feet—he forgot the exact figure—in less than thirty years, and that the sinking was now going on, and that there had been a sinking of several inches in the last few years.

The PRESIDENT—Independent of coal workings?

Professor LEBOUR—That was the question. He (Professor Lebour) would not say what the cause was. This sinking had been going on for years, and that it was still going on seemed to be absolutely proved by the earth tremors occurring in Sunderland in the last two years, and which were constantly going on; he heard of one last week. Whether they were really earth tremors, or due to underground workings, or due to simple local falls of stone—as he believed they very likely were—

would be proved by a seismological network of observations. The seismograph recording various local tremors, in various parts of the country, would be a check upon one another, and it would be seen whether the tremors were local or spread over a wide extent. As to the word seismoscope he found that Professor Ewing, an authority upon the subject, applied the word seismograph to the larger kind, and the word seismoscope to the smaller kind, which recorded where, and the time the tremors took place.

Mr. A. L. STEAVENSON said, in the report of the Committee it was stated that it might " be mentioned, as at least a curious coincidence, that the disturbances of December 6th to 8th, were closely followed by increased issues of gas at several of the collieries in this district." He would be glad if the Committee would give a little more information as to how these observations were made. They knew that gas appeared, in the first place, by the gradual or continuous exudation from the face, or it might come off at the goaf edges at a low state of the barometer; or by a man's pick point entering a cavity where gas was; but he did not see how earth tremors were likely to affect these three cases. In observing the earth tremors and gas issues they should agree upon a system by which the giving off of gas at the collieries could be equally and regularly noticed as the seismological observations. This could be done by taking two or three collieries, and have an indication of the gas given off in those places.

The PRESIDENT— Can anyone answer Mr. Steavenson's observations?

Mr. WALTON BROWN said, the Committee had carefully considered the question of measuring the gas, but they had experienced great difficulty in finding an apparatus sufficiently rapid and delicate for their purposes. In an ordinary mine the usual amount of gas in the main returns might be, say $\frac{1}{2}$ per cent., and if doubled by any unusual circumstances it would only amount to 1 per cent. Now it was very difficult to obtain an instrument to measure such small percentages of gas quickly and correctly. Of course they could certainly take it to the laboratory and have the samples analysed, but that would be a slow and expensive process.

The PRESIDENT—Yes, that could be done.

Mr. WALTON BROWN—He was corresponding with Mr. Maurice and Sir William Thomas Lewis as to their fire-damp indicator It was not ready for use yet, but he believed it would be what was wanted.

The PRESIDENT—It might be interesting to the meeting to know what is about the cost of these instruments.

Professor GARNETT thought the cost of the duplex pendulum was about £14, and the complete set of horizontal pendulums, etc., cost something like £60. This answer was subject to correction as he spoke from memory.

Professor LEBOUR—One of these simple recording apparatus cost about £2.

The PRESIDENT—This is a most interesting question, and thanks are due to these gentlemen for taking so much trouble in connection with these investigations.

————

The following paper by Mr. JAMES McKINLESS "On a Gauzeless Safety-lamp," was read:—

FRONT ELEVATION OF SEISMOMETER.

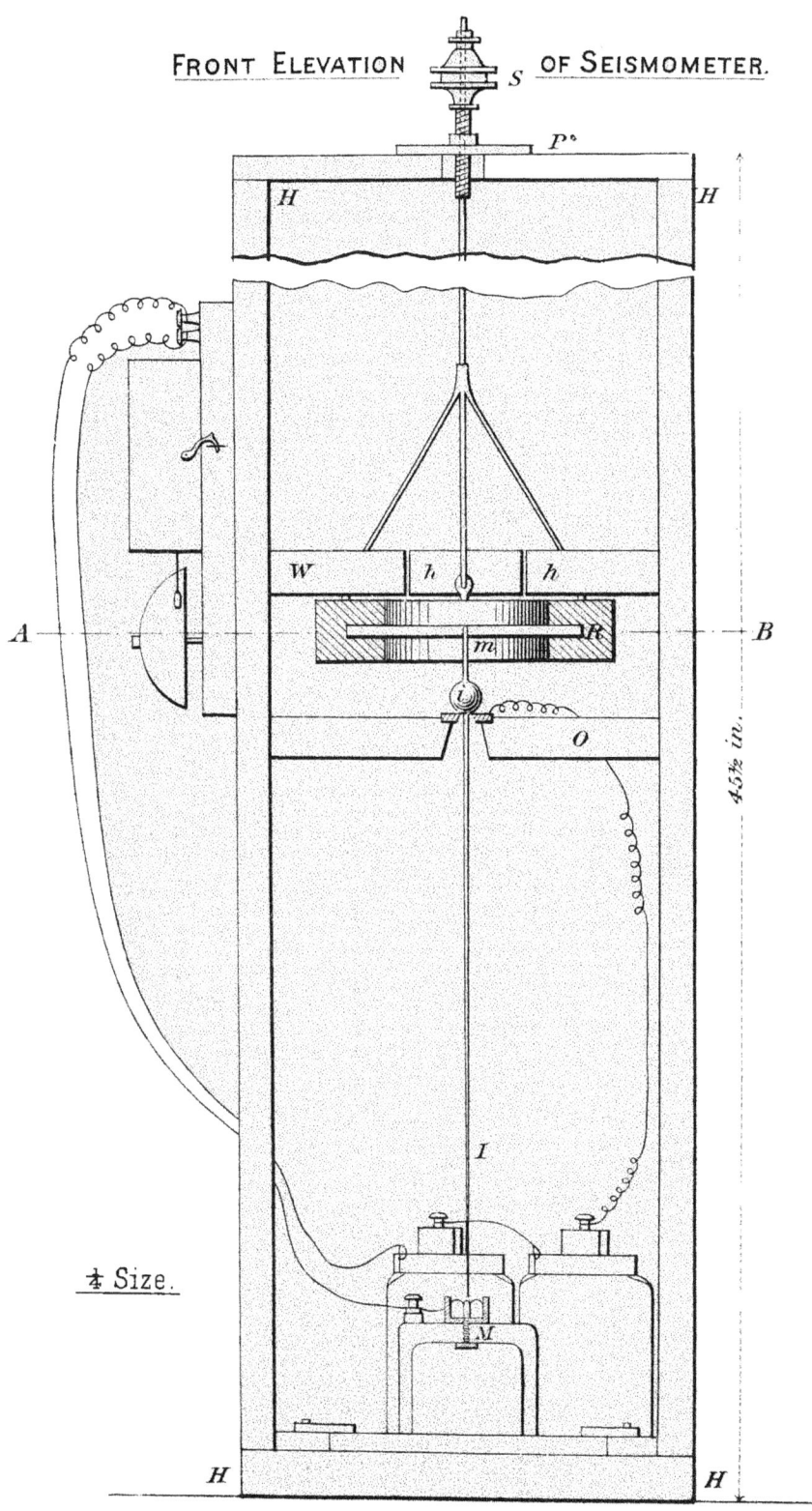

¼ Size.

INSTRUMENT USED AT MARSDEN.

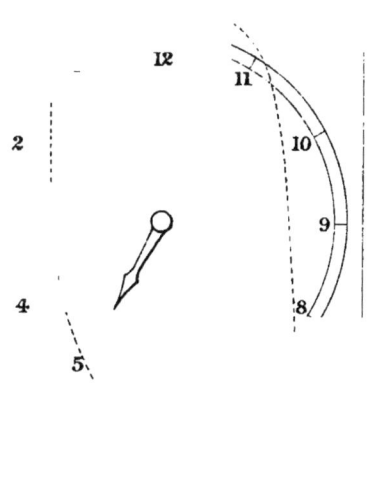

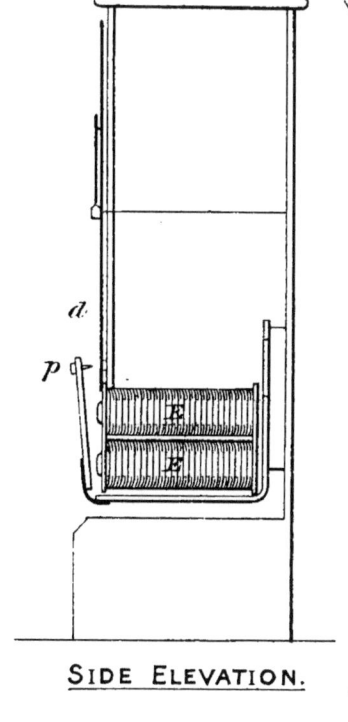

AMERICAN CLOCK.

SIDE ELEVATION.

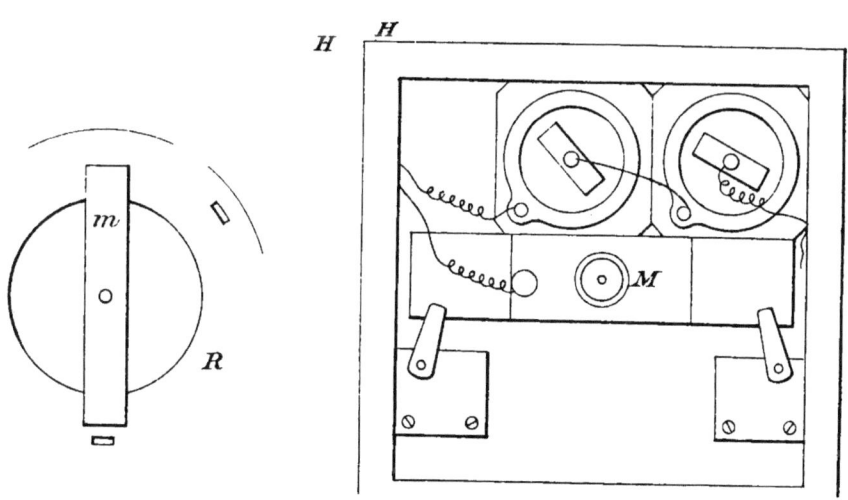

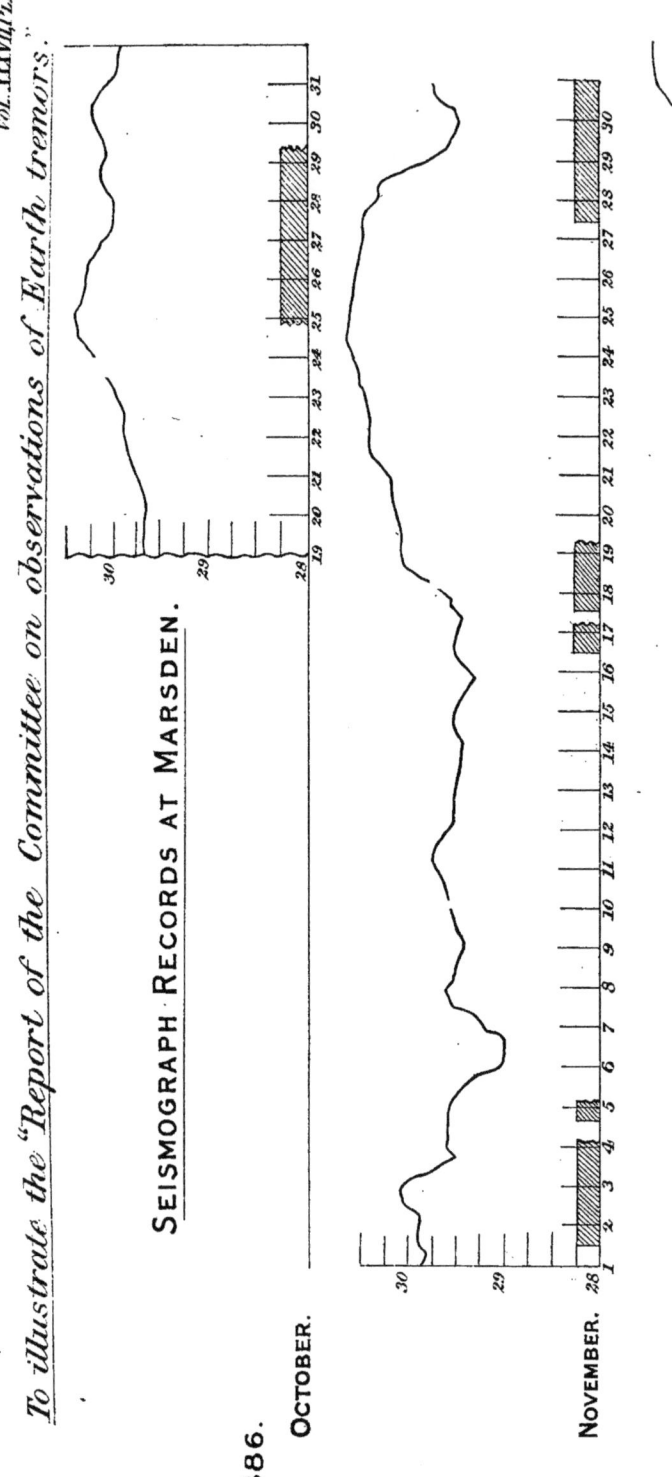

To illustrate the "Report of the Committee on observations of Earth tremors."

SEISMOGRAPH RECORDS AT MARSDEN.

1886.

OCTOBER.

NOVEMBER.

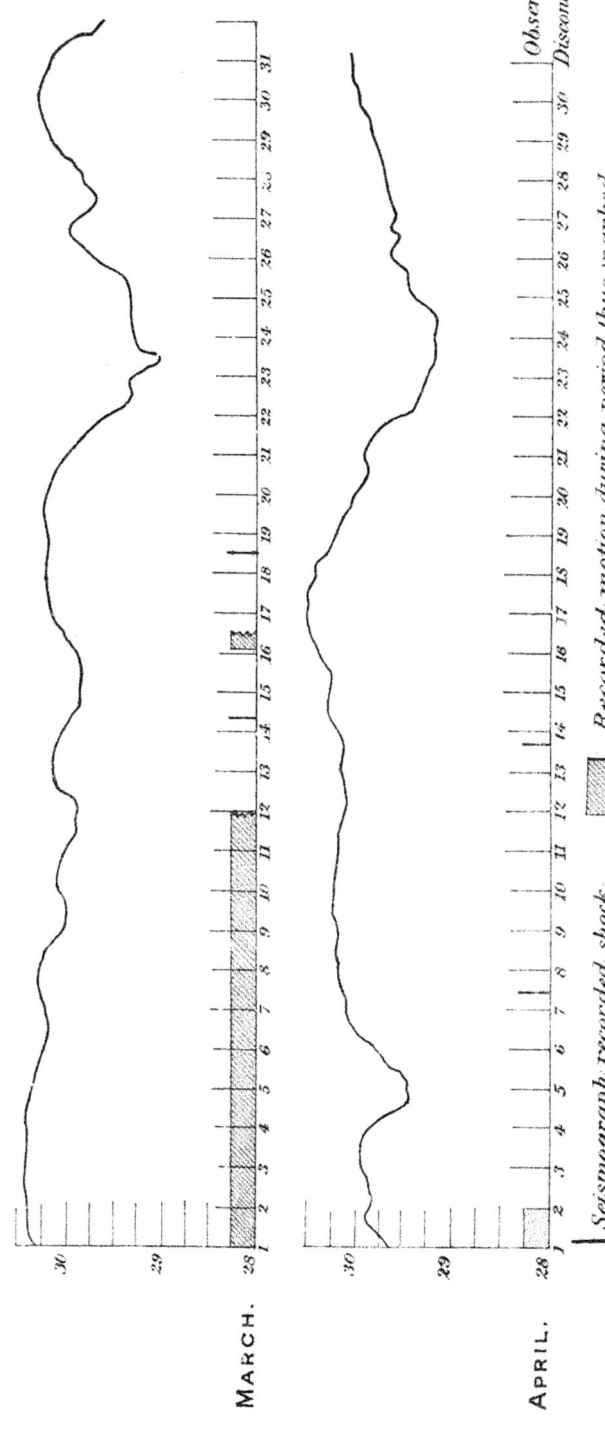

29

FEBRUARY.

MARCH.

APRIL.

| Seismograph recorded shock.

▨ Recorded motion during period thus marked.

| Observations

| Discontinued.

trate the "Report of the Committee on observations of Earth tremors."

DE ROSSI'S MACHINE.

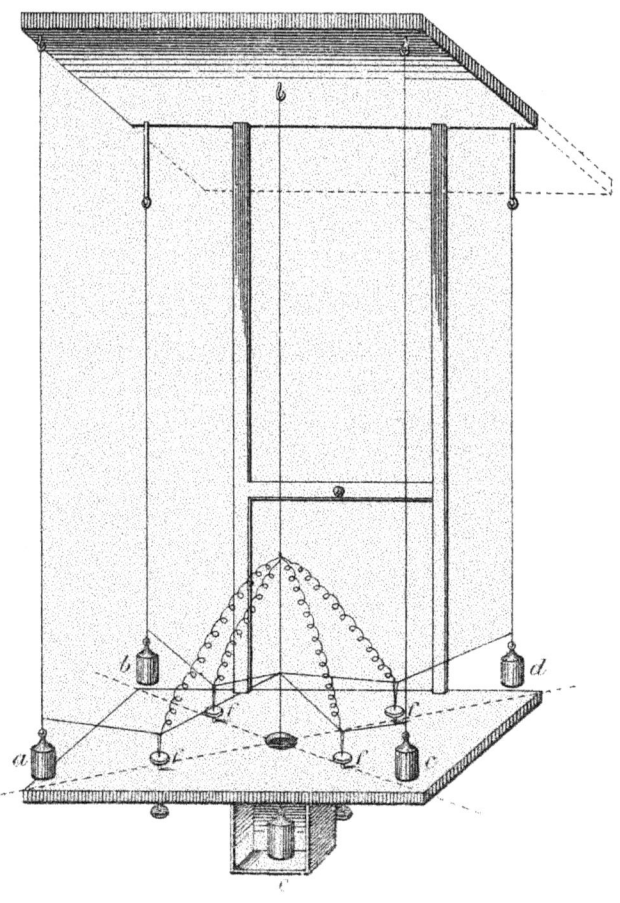

To illustrate the "Report of the Committee on observations of Earth tremors."

EWING'S DUPLEX PENDULUM SEISMOGRAPH.

FIG. I.

SECTION THRO' C. D.

Scale, Figs. 3. 4. 5. Half size.

FIG. 3.

FIG. 4.

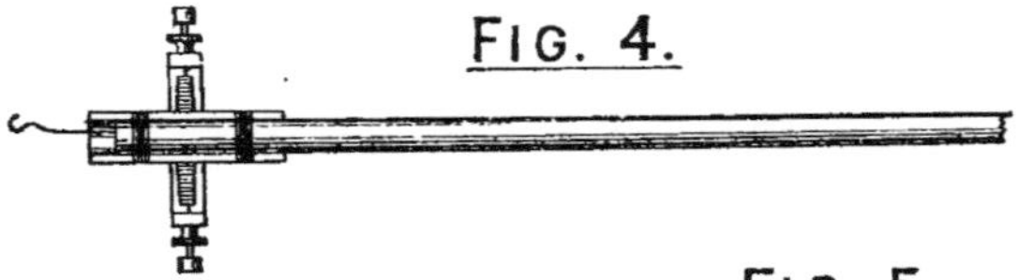

FIG. 5.

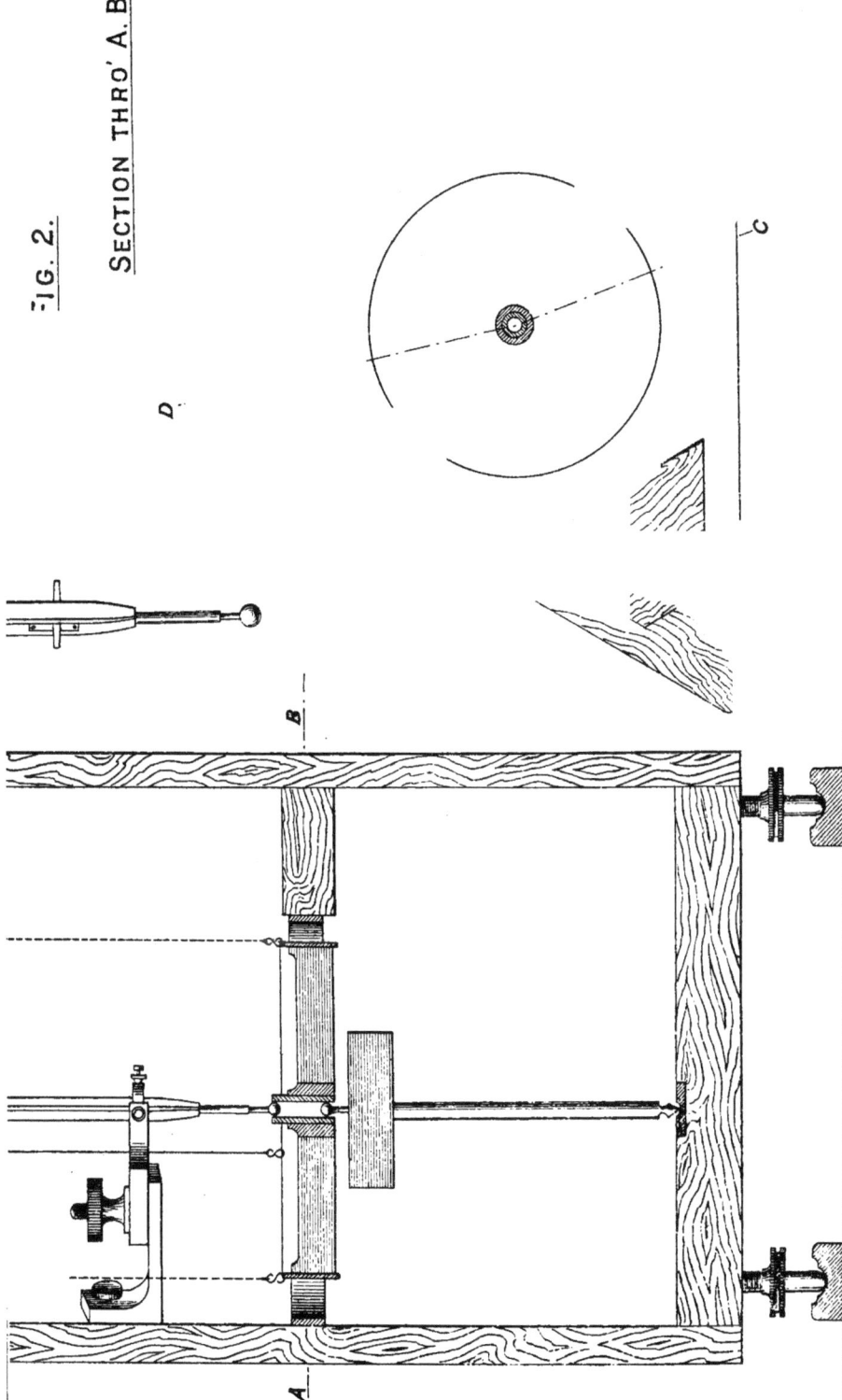

FIG. 2.

SECTION THRO' A. B.

lustrate the "Re ort of the Committee on observations of Earth tremors."

A FORM OF HORIZONTAL PENDULUM.

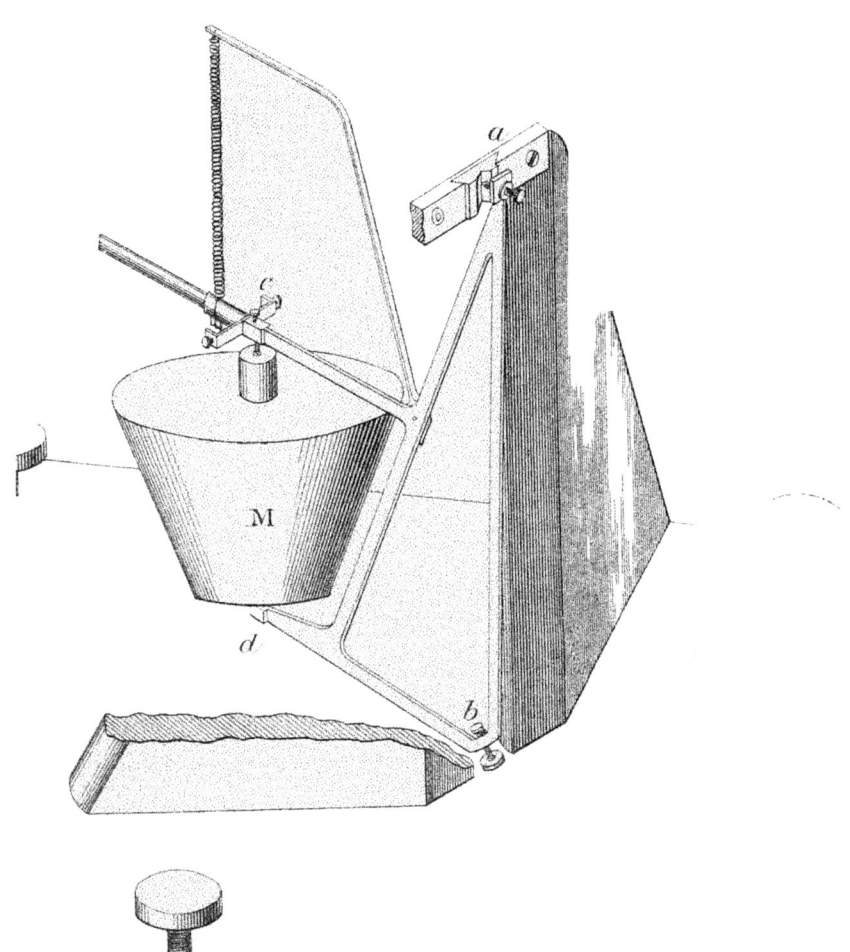

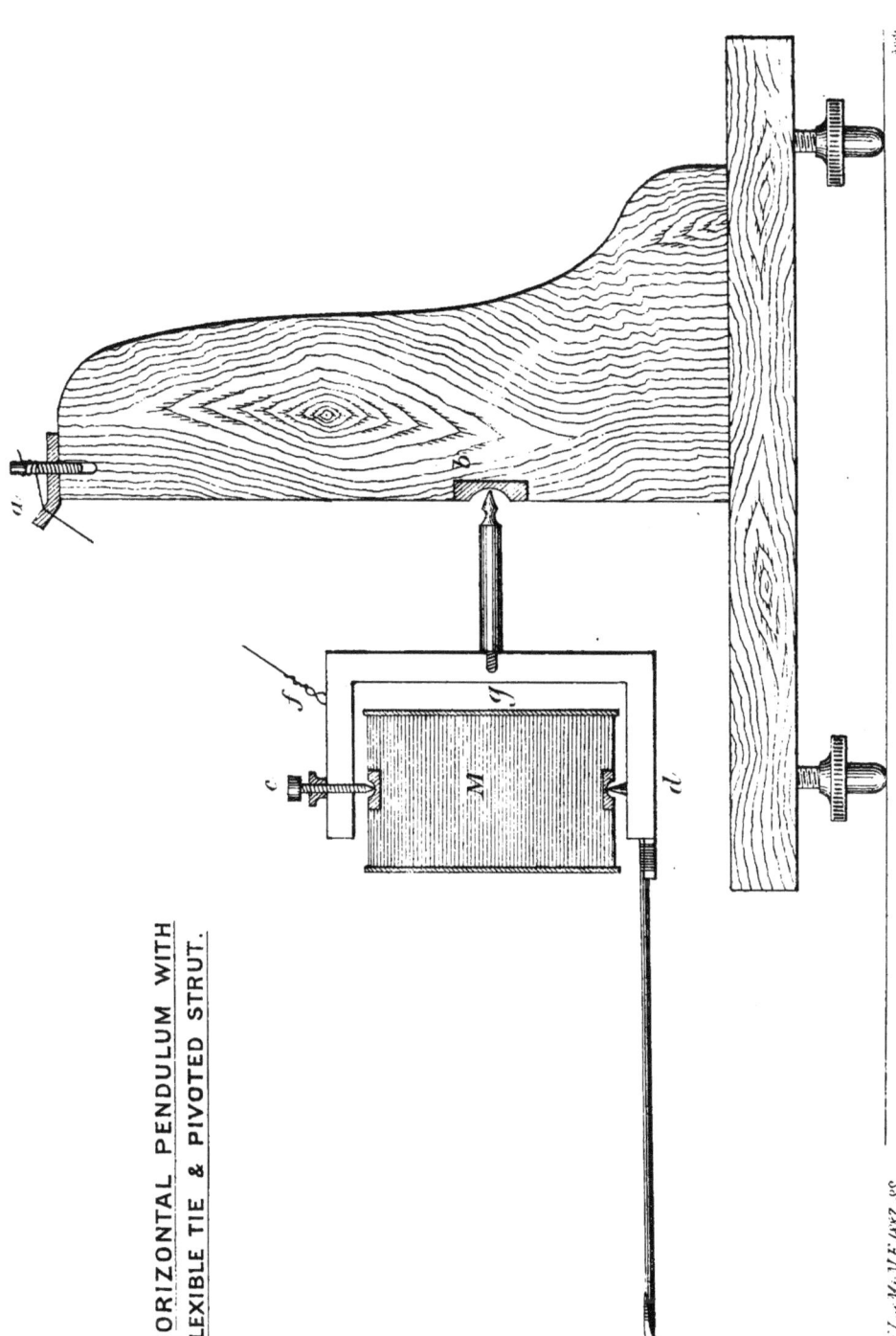

HORIZONTAL PENDULUM WITH
FLEXIBLE TIE & PIVOTED STRUT.

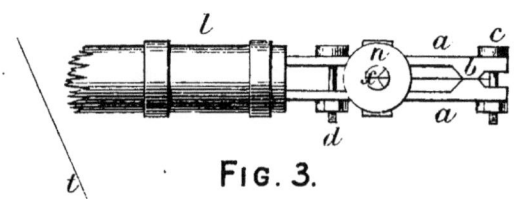

FIG. 3.

p'

p

l

l'

w'

FI

l^2

T JOINTS.

p

n

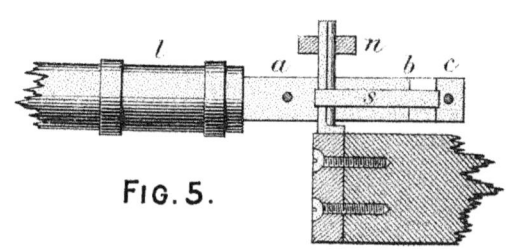

l   *a*   *n*   *b*   *c*

s

FIG. 5.

y

y   *g*

r   *y*

To illustrate the "Report of the Committee on observations of Earth tremors."

COPY OF A RECORD MADE BY THE HORIZONTAL
PENDULUM MACHINE SHOWN IN PLATE XV.

F.May 89.

To illustrate the "Report of the Committee on observations of Earth tremors."

A COMPLETE SET OF APPARATUS.

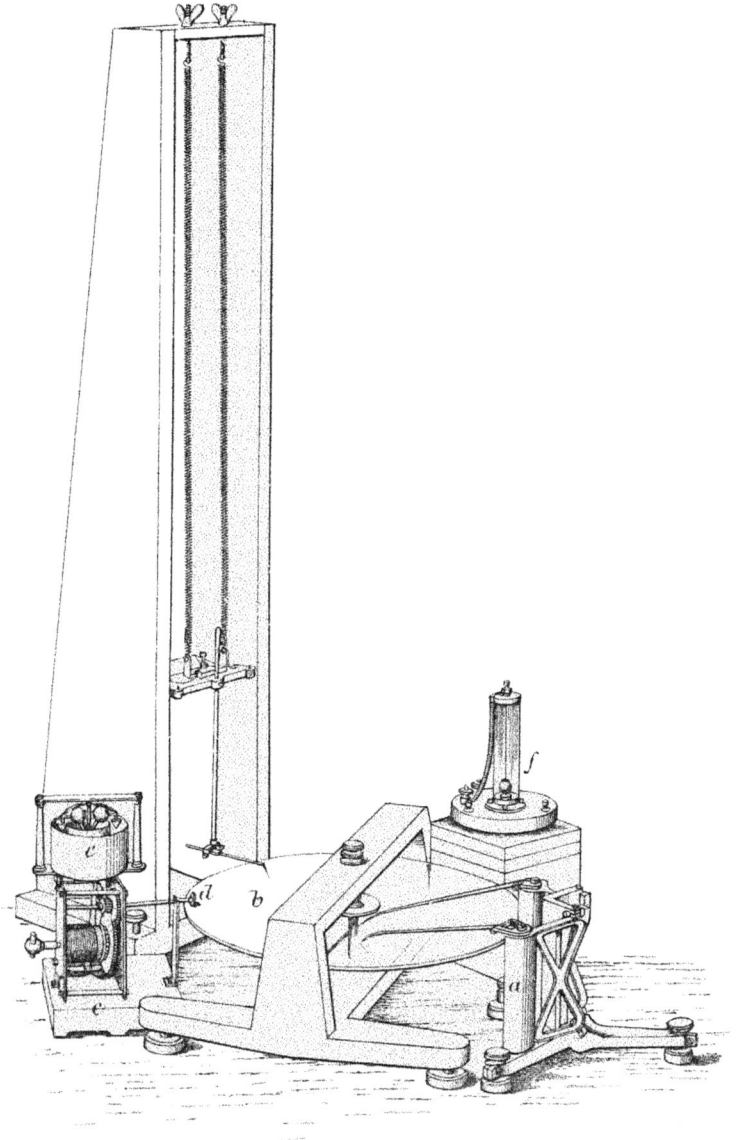

ON A GAUZELESS SAFETY-LAMP.

By JAMES McKINLESS.

THE chief feature of this lamp is that gauze is entirely discarded, the feed air is admitted through a large number of very small holes drilled in a belt or band h (Plate XVIII.) above the glass m and middle ring of the lamp f passing downwards between this belt and an inner conical chimney l into the combustion chamber k. The results of combustion pass upwards through this conical chimney l (on the top of which a cap l^1 is fixed to collect the unconsumed particles of carbon or soot), and thence through a large number of very small holes drilled in a cap j, and escape near the top of the lamp.

This outlet cap is fitted and riveted to a cylindrical tube k, the lower end of which fits into a collar of the feed air belt h, on which collar the diaphragm of the conical chimney rests. This arrangement separates the inlet and outlet currents, so that all the feed air entering the lamp passes into the combustion chamber.

There is also an upper ring d or diaphragm on a level with the top of the bonnet through which the outlet cap is visible, and this arrangement allows the results of combustion to escape without remixing with the feed air.

There are two shields a and b outside the feed air belt, having large openings to allow the free ingress of air, but so arranged as to prevent the force of a current from striking directly against the small feed air holes and disturbing the flame. The outer shield can easily be removed in case of damage, and, when put right, can as easily be refixed, being fastened by simply turning in four or five tongues t of its own metal under the top ring d. The combustion chamber k and the drilled band h are kept in their place by the ring f; g is the bottom ring that secures the whole to the oil reservoir o; m is the glass, and n a ring for holding the glass in its place.

Most modern lamps depend for their safety on the number or arrangement of bonnets or shields which protect the gauzes, and many are the devices which have been adopted in this respect, but this is a really safe lamp without any shields.   By the kindness of Mr. C. F. Clarke, it was tested at the Park Lane Apparatus, near Wigan.   The inner lamp was simply placed on the glass with an asbestos washer between the glass ring and oil vessel, and wedging it to the top of the test box to prevent tilting in this position, with the force of the current playing directly on the feed air belt n the outer gas could not be fired. It is only fair, however, to say that the highest velocity obtainable at this apparatus (with a very explosive mixture) did not exceed 17 or 18 feet per second.   What greater velocity it would stand without shields there has been no opportunity of ascertaining, but most probably it would be safe in a much stronger current.   The complete lamp, with its shields, has withstood a velocity which fired nearly all other lamps, including the Marsant, Morgan, Donald, bonneted Mueseler, Protector, and many others which are largely in use at the present time.

Probably the most severe tests to which safety-lamps have ever been submitted in this country are those which took place at the Neepsend Gas Works, Sheffield, the reports of which appeared in the Sheffield papers of the 20th and 27th June last.

It is unnecessary to occupy space by giving the full report of these tests, but one short paragraph may be of some interest.

"Now, referring to these vertical descending tests, it ought to have been stated that this lamp would not burn in the wind under the high pressures, and it was, therefore, put in the box at 1·40 water gauge, and after the gas had ignited inside the lamp the pressure was increased to 2·50 water gauge.   It then took two minutes to fire, so the lamp behaved well even under these conditions."

These tests seem to prove that the Clifford, McKinless, and Purdy No. 2 lamps are the three best in the order given, so far as safety tests go; but there are other points which are very essential to a good, practical collier's lamp.   It should be strong and durable, simple in construction, not easily put out of order or rendered unsafe.   It is admitted by many mining experts that gauze is the weak part of a safety-lamp, because it is so easily damaged, and a very small defect (which may escape notice) may be the cause of a great disaster.   It is also unreliable because from burning, oxidation, and brushing, the wire gradually becomes thinner and loses its absorbing power, and it is, therefore, difficult to determine at what point safety ends and danger begins.   This lamp is strong in every part, and instead of gauze the inlet and outlet are through holes drilled in special metal $\frac{1}{8}$ of an inch thick, every hole is $\frac{1}{26}$ inch diameter and $\frac{1}{8}$ inch long, and therefore represents a tube. It has very few parts, and these can be so easily and quickly taken to pieces and put together that a great saving should be effected in the lamp cabin. It also possesses the great desideratum (which cannot be too highly appreciated) that the collier, as well as the official who inspects, can see at a glance that the lamp is complete in all its parts, because with the outlet cap and the conical chimney in sight the lamp cannot be improperly put together. There is nothing, therefore, to examine but the lock, which is a simple arrangement for a lead plug.

By covering up the three lower rows of openings in the outer shield a with asbestos cloth or leather, the lamp feeds from the top, and is then an excellent gas trier; and the writer has been told by a manager in Lancashire that gas could be found with this lamp when a Davy failed to show any trace of it.

It gives a very good light which is not easily put out, and it will bear a considerable amount of tilting, swinging, and jerking. All lamps having chimneys are supposed to be more or less sensitive, but the chimney seems essential to the safety of the glass by preventing the flame impinging on it, and, although this lamp has a chimney similar to the Mueseler, the dimensions are different, the currents are separated, and it is much less sensitive. There are many who consider the chief merit of the Mueseler is that it compels the men to be very careful, but this can hardly be considered a very sensitive lamp.

Something may be said about its cost, and many say an ordinary lamp is good enough, and it costs much less. Some say this even of the uncovered Davy, and so said Mr. Speakman, of Bedford Leigh, until an appalling accident proved he was wrong.

It will be apparent to anyone that with 1,416 small holes to drill the lamp must be an expensive one to produce, and yet, by the aid of special machinery, they can be sold at a price not much in excess of an ordinary lamp offering the usual security; and when the saving of time in the cabin and in the inspection below is considered, and the absence of gauze which requires such frequent renewals, even on the score of economy it will compare favourably with any other lamp. Its weight is 3 lbs.

In conclusion, the writer calls attention to a paragraph from page 68 of the Report of the Royal Commission :—" If a lamp has the property that when placed in the current of an explosive mixture, the flame and the internal ignited gas are both certainly extinguished in a few seconds, such a lamp must be absolutely safe; but it does not follow that a lamp is unsafe if the gas continues to burn within the lamp."

This lamp, one of the first made, appeared to possess this very desirable property, for, at the Aldwarke Main tests, it went out entirely in all cases, but it was too sensitive, and, therefore, not practical. Afterwards the number of feed air-holes were increased, which was a great improvement; and, latterly, the feed air and outlet have been increased very considerably, with the result that the internal ignited gas does not go out always, and, in very high velocities, would probably continue to burn. This leads always to the conclusion that a lamp so nicely balanced in the intake and outlet as to secure complete extinction of flame in gas, under all circumstances, is scarcely a practical one.

The lamp could be adapted to burn spirit, but the one shown is burning "Ogilvie" oil, which is thought to be the cleanest and most satisfactory oil to be met with.

The PRESIDENT: No doubt the meeting would like to hear some of their experienced friends give their opinion upon this lamp.

Mr. STEAVENSON said he saw that the author of the paper considered gauze the weak part of a safety-lamp. His (Mr. Steavenson's) impression was that the glass was the weak part. The glass was apt to break from the heat of the flame, or a drop of water coming to it when hot. He was inclined to think that a lamp without gauze might be considered a weaker lamp than a lamp with gauze.

Mr. G. B. FORSTER said they had really got so many excellent lamps now that it was very difficult to distinguish between them. He agreed with Mr. Steavenson that a gauze inside of the glass was certainly safer, but the difficulty was that it interfered with the light so much. He had never in his own experience had an instance of a lamp which might have been rendered dangerous by the glass breaking without a blow, which might equally damage the gauze.

Mr. JOHN DAGLISH said this lamp was rather a return to the old Stephenson lamp. He remembered seeing it in some book—he believed it was by Sir Humphrey Davy—that an element of safety was absent in such lamps; that the great theoretical value and practical use of gauze was its rapid conducting power, which was absent where a large mass of metal was dealt with. He had never seen a glass broken in his own practice, but he had heard of a glass actually breaking and leaving an aperture.

Mr. J. G. WEEKS thought if lamps were always necessarily examined at bank, and only there, this lamp might be found practicable and useful; but if any value was to be attached to the examination of the lamps by the deputy-overmen and other officials at the examination stations appointed in-bye in the workings of the mine, as was the general practice in the North of England, then this lamp, from its construction, would not be suitable, as it would entail too much time and labour to take it to pieces to make the necessary examination, such as could and was easily and readily done with the the gauzes of lamps like the Marsant and bonneted Clanny.

Mr. MARKHAM asked if the lamp was as easily cleaned as the Davy?

Professor MERIVALE asked if the lamp got full of smoke and soot after a little use? Was there any difficulty in keeping it clean?

Mr. McKINLESS said, with respect to the breaking of the glass, it should be borne in mind that even if a collier struck his glass with his pick, and broke it, naturally there could be no harm accrue from that immediately, unless there happened at that very second of time to be an explosive mixture present. One second before such an accident occurred the lamp would indicate the presence of gas; and it would hardly be possible to break a glass under conditions of danger. With regard to the chimney, he looked upon it as essential to prevent the flame impinging on the glass. He looked upon the chimney as a necessity to safety-lamps depending upon glass. With regard to the examination below, it would be found on inspecting the lamp that it was impossible to put this lamp together if any part was left out. One of its strong points was, that the miner using it, as well as the official who had to inspect the lamp, could see at a glance that no part was left out. With the base of the conical chimney in view, and the cap also always in view, it was impossible to omit anything; and examination below was entirely unnecessary in any case, except that there might be a possibility of the holes being choked up, and the deputy could easily see that. With regard to choking up of the holes at the top, it had not been found to be so in practice. The lamp had been used in dusty places longer than was generally worked in this neighbourhood, and it had not been found that the holes filled up. They required cleaning certainly; they ought to be cleaned every day, the same as gauze lamps; because he had found that in a number of instances the lamp man had been in the habit of looking through the holes, and because he did not see any of the holes clogged up thought cleaning was not required. The best plan would be to clean the lamps by means of a jet of steam; an ordinary inch pipe, with a conical nozzle opening to the size of the glass, so as to fit into the top glass ring (of course removing the chimney first) would answer the purpose, and one person could clean a very large number in a short time. It was necessary, of course, to keep the holes open; but he did not see any difficulty in doing this: a jet of steam or perhaps compressed air would do it. A small amount of moisture went off with combustion, and in course of time, if not cleaned, every hole would get a coating of a sort of slimy matter, and, instead of a $\frac{1}{16}$th diameter, the holes would probably, in a short time, be found to be reduced to a $\frac{1}{32}$nd; and this would reduce the outlet by nearly half, and would not be fair to the lamp. It was necessary, therefore, that they should be kept clean. This would answer the question whether the lamps were easy to clean.

Mr. FORSTER—It is supposed that this glass may be broken if it gets very hot and a drop of water falls on it.

Mr. McKINLESS said he had tested the glass by admitting small quantities of gas so as to make the glass as hot as possible, and then plunged a brush into cold water and thrown spray against it. He had also managed to persuade one or two viewers to try this themselves, and he had not found in any instance that the glass had even cracked. At the most the cracking of the glass would be all that could result, and with a crack in the glass there was no danger whatever.

The PRESIDENT, in the name of the members, thanked Mr. McKinless for his paper, and for his attendance at the meeting.

The following paper by Mr. Emerson Bainbridge on a "Description of a Miner's Safety-lamp designed to meet the requirements of the Mines Regulation Act coming into force on January 1st, 1888," was read :—

To illustrate Mr James McKinless' paper "On a Gauzeless.
Safety Lamp."

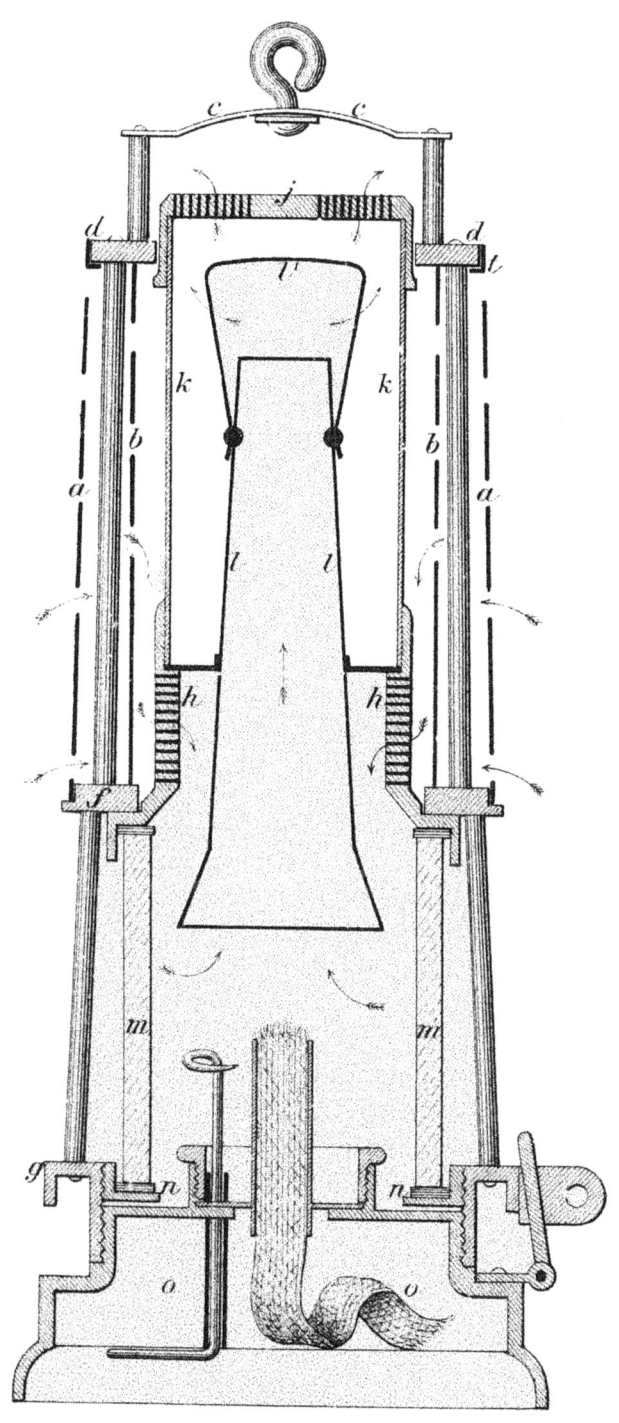

Scale ¾ full size.

Andrew Reid. Newcastle

DESCRIPTION OF A MINER'S SAFETY-LAMP DESIGNED TO MEET THE REQUIREMENTS OF THE MINES REGULATION ACT COMING INTO FORCE ON JANUARY 1st, 1888.

By EMERSON BAINBRIDGE.

Some years ago the writer was allowed the opportunity of describing a new form of safety-lamp to this Institute, and since that period the lamp has been used at various collieries, but time has shown that although it has the advantage claimed for it, it admits room for considerable improvement, and further than this, it is not, without a shield, safe in the light of the various experimental tests which have been carried on during the past few years.

The useful work of the Royal Commission on Accidents in Mines was directed for a considerable time to the important question of the actual safety of so-called safety-lamps, and their report has no doubt been the basis of the fresh regulations respecting miners' lamps imposed by the new Mines Act. That Act requires that in a pit where gas is found, a safety-lamp used underground shall be of such a description as will not fire an explosive current impinging against the lamp at a velocity the same as that which exists in the place where the lamp will be in use.

These enactments practically condemn the Davy, the Stephenson, and the Clanny lamps in all mines where safety-lamps are used at all.

Noting the elaborate character of most of the new types of safety-lamps which in recent years have been brought before the mining public, the author has for some time been turning his attention to the question of producing a lamp simple in construction, economical in first cost and in the cost of oil and repairs, easy to clean, and satisfactory in lighting. The object of this short paper is to describe a lamp which, he ventures to hope, possesses these characteristics, and which is an actual safety-lamp within the meaning of the new Mines Act.

Before describing this lamp reference may be made to some of the objectionable points relating to the majority of lamps now in use, viz. :—

1.—These for the most part use oil as an illuminant, though deodorised spirit is now largely used, and a mixture of oil and paraffin together in the proportion of 2 : 1 is also used on a small scale. From a series of careful inquiries the author finds that where oil is used the cost per annum of lamp-keeping, including everything, amounts to 15s. per lamp per annum in some cases, whereas with spirits the cost does not exceed 10s. With the lamp now before the Institute, he expects to be able to include everything in a cost of 7s. 6d. per lamp per year.

2.—Of the above cost, "repairs," of course, form an important item, and the author has come to a conclusion on this point which may surprise some of the members of this Institute, namely, that the chief repairs which a safety-lamp needs are due to the treatment it receives in the lamp cabin. The twisting due to the screwing and unscrewing of the ring which secures the glass of a lamp is probably the chief cause of the giving way of the supporting pillars, which are thus much strained, whilst the "banging" a lamp often receives when it is restored to the lamp cabin, also does much to damage it.

3.—The lamp gauze, however, is probably the item in which the most severe renewals are required, and, in cases where a double top gauze is used, it is frequently found that, whilst the upper gauze has to be changed every two months, the lower gauze, namely, that nearest the flame, has to be renewed in one month. Nor is this all, for the top of the *vertical* gauze is also so damaged (close to the gauze cap) that it has to be restored at the same time. In the lamp now referred to, this difficulty has been surmounted.

4.—In most safety-lamps hitherto in use, the gauze is exposed to the air, and in this way has not only been the means of accumulating a large quantity of coal-dust, which clogs the air holes, but has been found very liable to get damaged by a miner's pick, or by a falling body, or by careless treatment. The new lamp, in common with all bonneted lamps, provides against these objections.

5.—In the cleaning of most of the lamps in use up to this date, the lamp is separated or unscrewed into a number of parts, this alone being a fruitful cause of wear and tear; further, especially in oil lamps, there is a good deal of over-cleaning of gauzes, such cleaning by brushes contributing, probably, much more to the wear of some parts of the gauzes than the actual use of the lamps in the mine. In employing spirit, much less cleaning is necessary, and in the lamp now described it is proposed to altogether discard the use of brushes in the treatment of the gauzes, and to apply compressed air.

6.—The relative price of oil and spirit is as 2s. 4d. is to 9d. per gallon, but, as at present used, the economy of spirits is represented by not more than half the saving these figures suggest. In the new lamp, by feeding the reservoir through the wick-hole, the waste caused by the sliding lid is avoided.

7.—The ordinary small screw which locks a safety-lamp is loose, and is apt to get lost; in the new lamp the screw is so made that it cannot be detached outwards.

The author submits one of the lamps he has designed to the Institute to-day, a sectional sketch of which is shown, and on reference to which the following features of the lamp which he now summarises will be understood :—

1.—It is a lamp composed of only four movable parts (which are simple and easily put together), instead of the ten separate parts which constitute some of the modern improved safety-lamps.

2.—As the lamp bottom is, in all cases, most subject to wear and tear, the strength is specially put in that part. The inside of the brass bottom is covered with tin, which, whilst preventing corrosion, maintains the full lighting power of the illuminant used.

3.—It is suggested that, on the ground of economy and superior light, deodorised spirit should be used (the cost of which is less than one half as much as oil), the flame being adjusted by means of a ring round the wick tube; and an ordinary pricker. Although spirit is recommended, colza, seal, or other oils may be used, if preferred.

4.—The glass is about ¾ inch longer than the ordinary Clanny lamp glass, and is made of a specially clear and translucent quality of glass, polished at both ends, and properly annealed. It can be cleaned in its place, and need never be moved except in case of accident. There is hence a saving in washers. As special provision for expansion by heat is made, breakages will necessarily be few.

5.—The glass frame, or body of the lamp, is supported by four pillars, which are so placed as to allow fully three-eighths of the glass to be free from a pillar, and its consequent and objectionable shadow.

6.—The gauze can only be taken out by removing the locked bonnet. This is done by causing the top of the lamp to unscrew. This top is locked by a bolt, which is fastened by the lock which secures the lamp.

7.—The holes at the top of the bonnet are made *above* the gauze, and the cap of the gauze is of *metal* instead of gauze, the former being much more durable.

8.—By a simple and efficient shut-off arrangement, the inlet of air which supports combustion can at once be closed, if the presence of gas is discovered.

9.—This lamp has been tested in the highest explosive current, at a velocity of over 30 feet per second, and does not explode, but is found perfectly safe. It should be noted that this velocity is twice as great as is met with in any ordinary coal mine.

10.—The screw which secures the top of the lamp to the glass frame is protected from wear by means of a stop or block which brings the bolt exactly opposite the hole into which it is secured by the lamp lock.

With regard to the use of spirit for lamps there appears to be a pre-judice against its use, caused chiefly by some fires which have taken place in lamp cabins, and by small explosions which have been observed in the lamp. After careful inquiries the author has come to the conclusion that with ordinary care such mishaps could have been and can be avoided, and the fact that about 40,000 spirit lamps are now in use in the county of Yorkshire alone, indicates that it is growing in favour. There is little doubt that for cleanliness, good light, and economy, it is much superior to oil, and a spirit flame is also more sensitive in indicating the presence of small quantities of gas. A more brilliant light enables the workman to avoid danger better, and to produce the coal cleaner.

A lamp burning with spirit is more apt to go out than one using oil, but this ensures more careful treatment of the lamp.

The safety-lamp which the author has endeavoured to describe has been tested at a velocity of 30 feet per second and has been found quite safe, but at a higher velocity it will probably be found that it will not explode. The lamp does not cost more than 6s., and the author ventures, in conclusion, to express the hope that in point of first cost, economy in lighting, cleaning, and repairs, simplicity of construction, and practical safety, the lamp he has brought before the members will possess some points of interest at the present time.

The details may be thus described :—a (Plate XIX.) is the ring which does not revolve; b b, air outlets above gauze; c, cap of gauze, made of sheet copper; d d, point where shield screws into lower portion; the ring e and the lower part of the shield d are so arranged by means of a stop that the locking bolt i slips at once into its place without requiring any ad-justment of the shield; there is also a small stop and slot which allows the shield to move $\frac{3}{16}$ of an inch, and cuts off the supply of air shown by the arrow p at will; g, copper base for gauze to prevent wear; h, specially annealed transparent glass, which is only removed when broken; j, asbestos washers which allow of the expansion of the glass; k, pricker; l, screw cap for spirit reservoir, to prevent waste caused by sliding panel; m, lock which exposes bottom and top of lamp simultaneously.

Professor MERIVALE said he had got Dr. Stroud to measure the light given by this lamp, and he found that it gave $\frac{1}{3}$ of a candle-light. He (Professor Merivale) thought it gave a great deal more light, and therefore he got Dr. Stroud to measure the light.

The SECRETARY—Mr. Bainbridge writes that he is, unfortunately, unable to be here to-day, and he would feel obliged if the discussion on his lamp could be postponed for the present.

———

The following paper by Mr. George Lee on "The Endless Chain in Spain" was read :—

To illustrate Mr Bainbridge's paper on "A Safety Lamp to meet the requirements of the Mines Regulation Act of 1887."

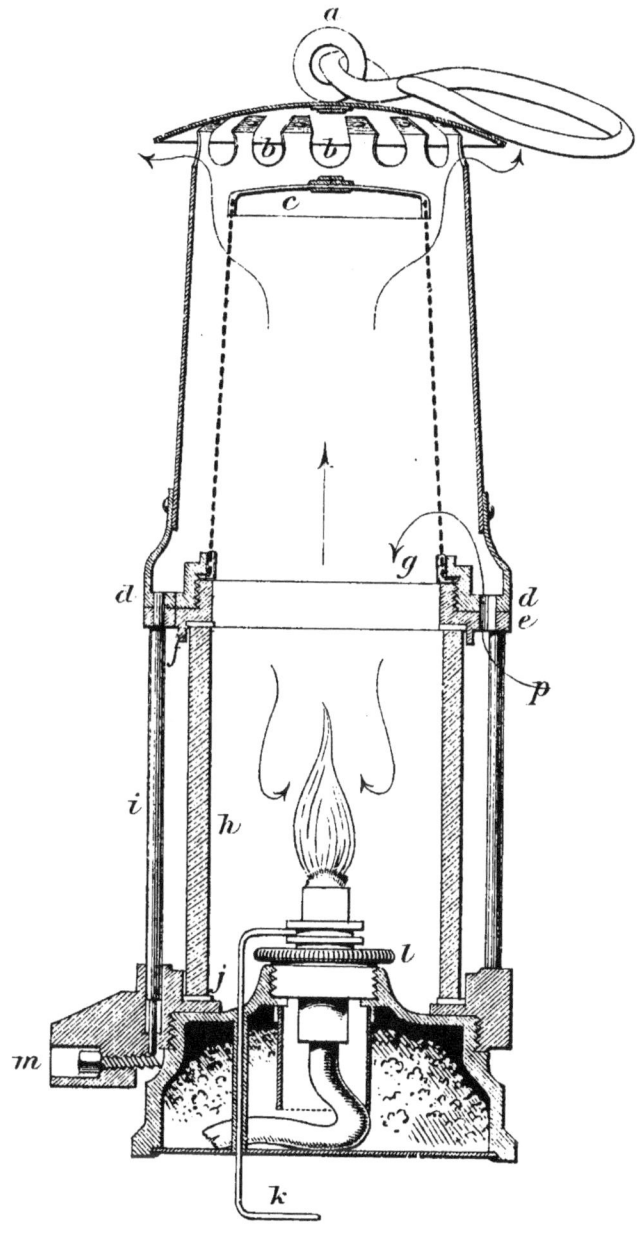

THE ENDLESS CHAIN IN SPAIN.

By GEORGE LEE.

A STATEMENT of the particulars of a railway constructed recently, together with a few remarks upon the district which it has brought into communication with the trunk line commanding a port of shipment, may not be out of place as a supplement to the previous paper on the above subject.*

The locality of the areas taken, named San Juan and Dolores, containing the deposits of iron ore, for the transportation of which the railway was designed, is situated in the highest altitudes of the Bilbao mineral zone.

The ore, which is chiefly a rich "Campanil" of a superior quality, and "rubio" is found in the limestone capping the mountains. (See Plate XX.)

To convey the produce of these mines over the mountains, down 1,640 feet to the Galdames railway, belonging to the Bilbao River and Cantabrian Railway Company, which winds round, midway up, the spurs of the mountain containing the iron ore deposit of the Somorrostro, an endless chain, whose well-known properties aided in surmounting the difficulties pertaining to the profile of the most desirable route (see Plate XX.), was deemed the best method of transport that could be adopted.

The railway is 3,408 yards long, consisting of seven lengths, determined by four angles and two stations, the particulars of which are as follows:—

* See Vol. XXXIII., p. 187.

	Wheels, Stations, or Angles.		LENGTHS.						SECTIONS.				Size of Chain.
	Designation.	Height above Datum, 15 feet below A.	Designation.	Length.	Fall.	Rise.	Inclination. Avg.	Max.	Brake.	Length.	Fall.	Average Gradient.	
		Feet.		Feet.	Feet.	Feet.	%	%		Feet.	Feet.	%	Ins.
...	H	1,489	...	...	...	...	...	...	...	...	...	...	...
1	G	1,564	GH	2,794	...	75	02·7	30·00	} G	5,944	565	9·5	11/16
2	F	1,199	FG	1,721	365	...	21·1	28·8					1
3	E	924	EF	1,426	275	...	19·3	33·0					15/16
4	D	550	DE	2,006	374	...	18·6	28·0	E	...	...	...	1 1/16
5	C	259	CD	1,010	291	...	28·8	43·5	D	...	...	...	15/16
6	B.	128	BC	754	131	...	17·2	36·0	C	...	...	...	3/4
7	A	15	AB	721	113	...	15·9	30·0	B	...	...	...	3/4
...	...	...	...	10,432	1,549	75	...	...	...	...	...	...	...

The chains, weighing altogether upwards of 63 tons, were specified to be of bar rolled net size, equal to a strain of 20 tons per square inch, made in lengths not exceeding 400 metres; out of each length. the endeavour was made to select test lengths containing the weakest link, which were submitted to the test with the following results:—

Size—Diameter in Inches	11/16	3/4	15/16	1	1 1/16
	Tons.	Tons.	Tons.	Tons.	Tons.
Specified proof strain	6·9	7·4	11·5	13·2	14·8
„ breaking strain ...	13·8	14·8	23·1	26·4	29·6
Parted after being pulled stiff at	14·5	17·6	27·5	32·7	34·6
Weight—lbs. per metre	12·86	15·77	27·30	29·21	30·80

The proof strain was not applied for the ordinary purpose of this test; but to give the open chains a set, in order to insure the kindly working of the links in the claws of the wheels.

The chief factors that affected the installation of the chain were:—

Weight of load (mineral carried by tub) ... ... 10 cwt.

„ tub ... ... 3·4 „

„ 1 1/16 in. chain, per yard ... ... 27·7 „

„ 11/16 in. „ „ ... ... 11·5 „

„ full train, with 1 1/16 in. chain, per yard ... 94 lbs.

„ empty „ „ „ ... 43·5 „

„ full train, with 11/16 in. chain „ ... 77·2 „

„ empty „ „ „ ... 27·9 „

Coefficient of resistance ... ... ·066 lbs.
Distance between centres of tubs in train ... ... 65½ feet.
Span, being the space between the tubs ... ... 61¾ ,,
Top of tub, or point of suspension above the sleepers 30¾ ,,
Height of rail ... ... 2·13 ,,
,, fork above rail ... ... 3·2 ,,

In contending with the extraordinary gradients, curves of small radius, and the exceptional weight of the trains, the chief object in the designing of the plant was to have a tub which, whilst meeting the demands of the first two, should be as light as possible. To secure this only iron and steel were used in its construction.

The system of signalling is that of the ordinary hammer rapper, one of which, combined with a lever for pulling the rapper in advance, is placed on the framework of each station. The duty of the brakesman is to so control the train that the chain shall be fed with tubs from the train that terminates at his station, at the regulation distance, during such times as the hammer remains elevated, pulling up the train gradually in twenty metres, on its fall, which liberates the lever holding the rapper in advance, thus giving the signal to stop at the whole of the stations simultaneously.

Instead of using the vertical shaft of the chain wheel, a sliding stud is provided, in separate framework at the angles and stations, for the loose pulleys of the terminating chains. In the rear of the stud is a screw by which the brakesman takes up the slack produced by the wear, or meets the variation, as affected by temperature, in the length of the chain, thus effectually placing within the power of the brakesman the prevention of the chain from trailing; also rendering unnecessary the subjection of the chain to the extra strain, far exceeding the requisite sustaining tension, from the blocks used to haul in the slack in the process of the periodical shortening where a fixed pulley is used.

To prevent the light chain of the laden train on the length G H from lifting out of the forks in the concave curves m and n, see-saw bearing-down sheaves were designed, consisting of a beam frame, sup- ported by an axle passing through its centre, at a height admitting the tubs to pass underneath. In each end of the beam is placed a sheave on the line of the chain it bears down; beneath the sheaves the sides of the beam, shod with iron, are deeper than the intervening space between, in the middle of the beam which, when in a horizontal position, is free above the top of the passing tubs. When an approaching tub comes in contact with the tapered end of the beam, the deepened sides gliding over the top of the load, the sheave is raised beyond the reach of the fork, whilst the sheave in the depressed end of the beam, in advance, bearing down, retains the chain, relieved from the pressure of the sheave under which the tub is passing, securely within the fork. In a similar manner the departing tub depresses the end of the beam in the rear, bearing down the chain in the fork whilst the other sheave is cleared.

To provide for the probability of the surplus power in the trains of the lengths C D and D E being utilised to lift mineral out of the valleys in the vicinity of the angle E, a heavier chain, capable of bearing the additional controlled load of the length C D, was put down on D E, the control of which length is, in the meantime, under the restraining force of a four-bladed fan-fly 13 feet diameter by 7 feet wide.

At the station F the two chains E F and F G are united by means of two clawed wheels secured to a shaft governed by a fan-fly, neutralising so much of the surplus power of the length E F as is over and above that required to assist the length F G in overcoming the resistance on the length G H.

The distance between the terminus of the railway at H and the mine is worked by gravitation, the empty tubs leaving the chain at a point before reaching H, sufficiently high above so as to insure their running as far as the mine's sidings, from where the laden ones gravitate to the hanging-on at H.

The rock met with in making the formation of the railway was for the first five-eighths of a mile the hard blue schist peculiar to the district and locally known as "Cayuela," which beyond yielding beds of hydraulic limestone is practically barren veins of ore, although they may have the "Cayuela" for walls to contact lodes, never piercing it. In the absence of more suitable building stone it is used, protected from the weather, under the influence of which it perishes, by a plastering of lime. On leaving the blue rock just above D, with the exception of a little cutting on approaching F, the profile of the railway skims the surface or is met by shelving the rock, as between F and G, over sandstone containing beds of freestone, especially one discovered a little up the mountain side from F, where a stone eminently fitted for architectural purposes is to be had, and available to supply a local demand which at present is met by importation.

At E was crossed one of those veins of ore easily traceable across country for miles, often by the escarpments and ridges of the brown iron ore, offering greater resistance to the denudating forces than the enclosing rocks on the official maps of the district, by the mining "takes" that dot the line of strike.

Over the limestone between G and H the chain adapts itself to the surface line of the route chosen to avoid cutting the extremely hard rock after very little preparation.

The present terminus of the railway at H is in the very heart of a group of only partially explored " takes " situated on the tops of the mountains in the eastern part of the district of Galdames. The deposits of ore are chiefly in the cavernous limestone, and although they were, compared to the situation of the mines nearer to the water-driven forges, almost inaccessible, yet there are signs of the ancients having worked the appreciated rich ore found in this limestone.

———

The PRESIDENT said, that Mr. J. T. Cackett and Mr. B. N. McLaren had prepared a very beautiful drawing of the different systems of haulage which were exhibited at the recent Jubilee Exhibition in Newcastle, and had presented the drawing to this Institute. It would be framed and hung up, so that gentlemen might have an opportunity of seeing it at their leisure. The Council had already passed a vote of thanks to these two gentlemen for their handsome gift, in which he was sure every member would join.

This concluded the business of the meeting.

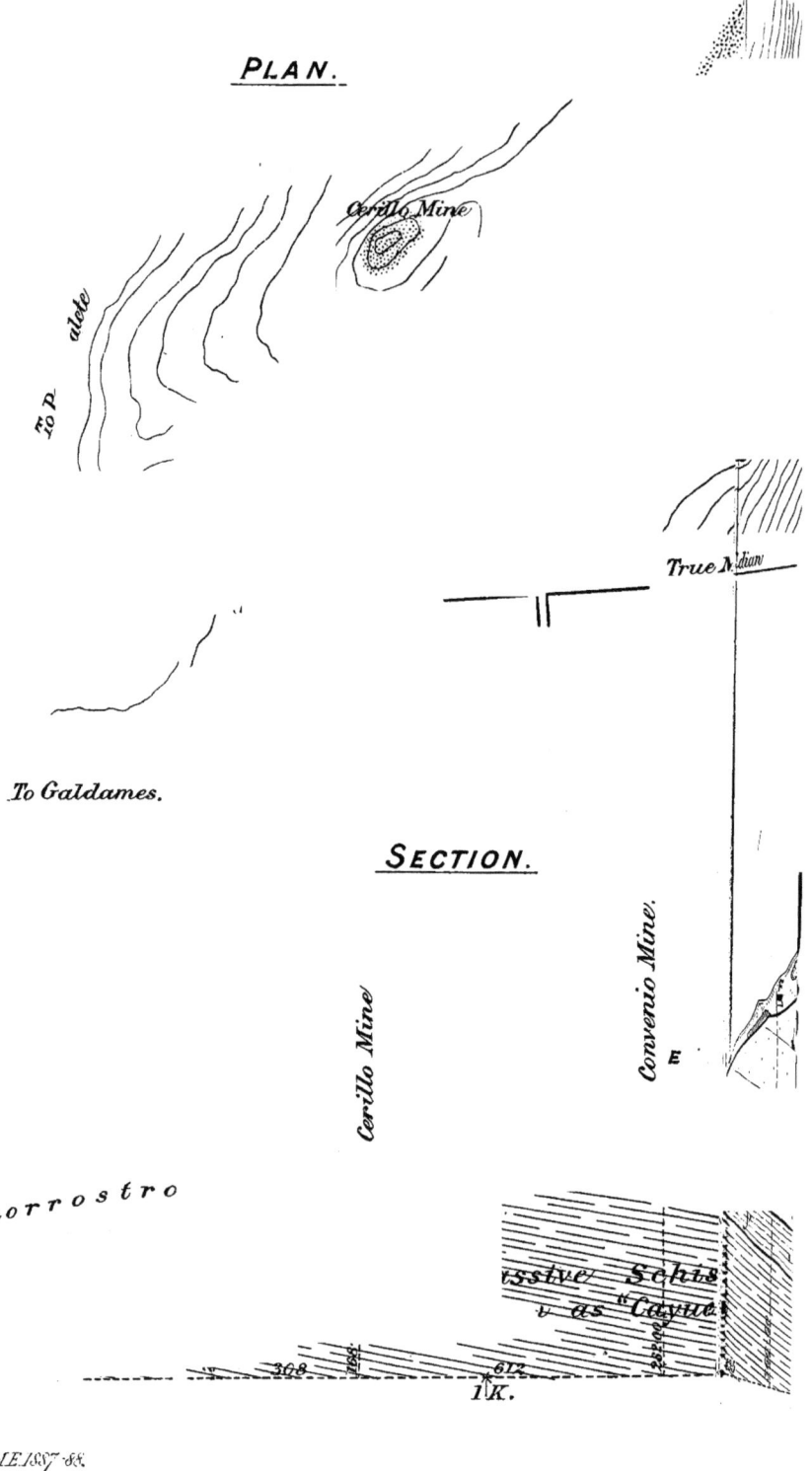

PLAN.

Cerillo Mine

Top alote

True Meridian

To Galdames.

SECTION.

Cerillo Mine

Convenio Mine.

E

morrostro

ssive Schis

v as "Cayue

308

612

1 K.

of M&ME 1887 88.

"On the Endless Chain in Spain."

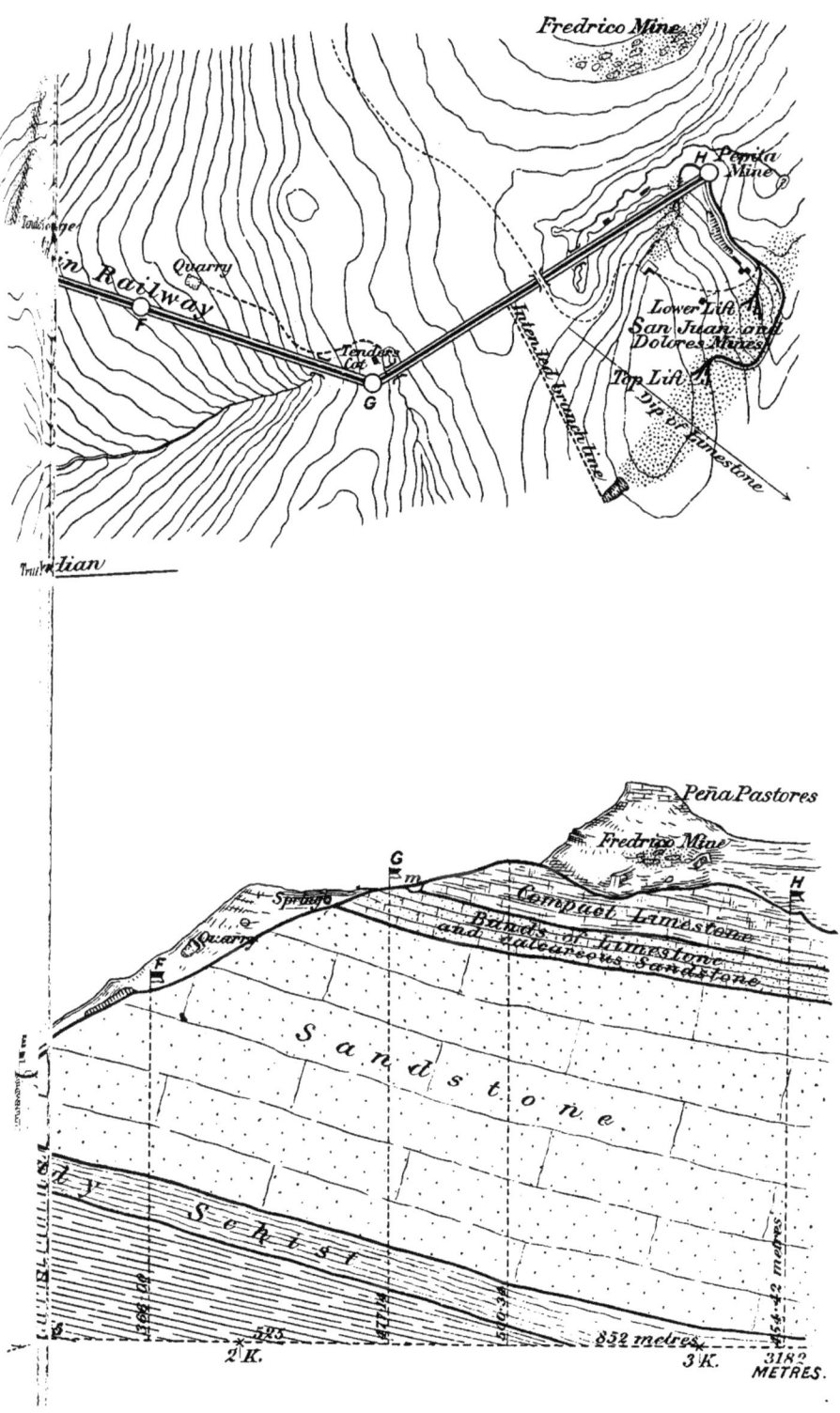

PROCEEDINGS.

GENERAL MEETING, SATURDAY, FEBRUARY 11TH, 1888, IN THE WOOD
MEMORIAL HALL, NEWCASTLE-UPON-TYNE.

J. B. SIMPSON, ESQ., IN THE CHAIR.

The SECRETARY read the minutes of the last meeting, and reported the proceedings of the Council.

The following gentleman was elected having been previously nominated :—

ASSOCIATE MEMBER—

Mr. LANCELOT DOBINSON, Hebburn Colliery, Newcastle-on-Tyne.

The following gentleman was nominated for election :—

ASSOCIATE MEMBER—

Mr. LANCELOT FLETCHER, Marsden Colliery, South Shields.

Mr. CHARLES J. MURTON read the following "Notes on the Tkiboulli Coal-Field (Caucasus)":—

NOTES ON THE TKIBOULLI COAL-FIELD (CAUCASUS).

By CHARLES J. MURTON.

In the summer of 1886 the writer had the opportunity of visiting this coal-field, and, as it is possible that it may some day enter into competition with the English coal now so much used in the Black Sea, some notes may be useful.

SITUATION.

The village of Tkiboulli is on the southern slope of the Caucasian Mountains, lying almost due east from Poti 138 versts (91 miles), and from Batoum 178 versts (118 miles), by rail. These two towns form the termini on the Black Sea for the railway which crosses over through Tiflis to Baku on the Caspian Sea.

At Rion Junction, a short line of some 7 versts (4½ miles) connects the town of Kutaïs with the main line, from which town a branch line was being constructed to Tkiboulli at the time of visit. This line of 38 versts (23 miles) is now completed, so that Tkiboulli is now in direct communication with the ports of the Black Sea and the Caucasus generally. See Plate XXII.

GEOLOGY OF DISTRICT.

Taking Tkiboulli as a centre, the Nakeral Mountains describe a semicircle, at a distance of 5½ versts (nearly 4 miles), towards the north-east. See Plate XXIII. These mountains form an abrupt termination to the valley in which Tkiboulli lies, in the shape of an escarpment, 4,000 feet above sea level. On the south, or Tkiboulli side, the fall from the crest is for some distance perpendicular; northwards the ground slopes gradually away.

This mass, except where capped by Cretaceous rocks, is of Jurassic or Oolitic age. The strata at the escarpment stand out white and distinct throughout the whole semicircle. The beds lie at a high inclination, increasing on descent to lower ground.

·Numerous streams, taking their rise from this ridge, in their quick descent have carried down boulders of limestone, some of huge size, which lie scattered in all directions, as well as in the beds of the streams themselves. These streams also have cut deep ravines, showing in their channels thick beds of limestone, sandstone, this coal-seam, and, towards the lower grounds, cutting through the many-coloured marls and shales of the Lias, a formation continuous from here to Kutaïs.

By their union these streams form the Tkiboulla River, which, flowing south for a little distance, is seemingly hemmed in by the Lagory Hills, but through which it has found or formed a natural tunnel of some two miles, joining at the other side the Kwrila, a chief tributary to the Rion, which flows into the Black Sea at Poti.

On looking at Plates XXII. and XXIII., it is seen that the district comprised between Kutaïs at one point, Kwamli at another, and Tkiboulli at another, seems to have been the scene of a vast upheaval, which has broken and exposed the strata now forming the escarpments at and between these points. These are all the same limestone formation, and traces of coal are found over an extensive line; but it is only near Tkiboulli that it is proved to exist in any great quantity.

Near Kutaïs the coal, which is in two or three thin seams (3 feet), is thought by M. Pernolet to represent a series of coaly shales, which lie above, and are separated by 500 metres from the seam at Tkiboulli, and he is of opinion that at a great depth (not less than 500 metres) the thick seam may still be found; but the probabilities are that these Kutaïs, Kwamli, and Tkiboulli coals are all the same, only in the latter case vastly thickened.

In 1872, Captain Lipgart began working the coal near Kutaïs, and many others followed suit in a small way, but the seams, thin to commence with, became so much thinner that in a few years the works were stopped.

The following is an analysis of these coals one mile from Kutaïs :—

				At Outcrop. Per cent.			50 Yards in Seam. Per cent.
Carbon	...	...	...	68	...	...	76
Vol. matter	...	...	...	10·92	...	...	15
Ash	...	...	...	20·40	...	...	12

At Coercebi, a little further off, and 50 yards from outcrop, it was :—

						Per cent.
Carbon	...	...	...	...	...	ˉ 45·50
Vol. matter	...	...	...	...	...	21·20
Ash	...	...	...	...	...	23·20
Water	..	...	...	...	...	10·00

A kind of cannel coal, called locally "Geishler," a few inches in thickness is found in places, which the natives make into ornaments, and which takes a polish like jet. There is a specimen of this on the table. Its analysis is:—

					Per cent
Carbon	...	...	...	...	46·95
Vol. matter and water		...	...	...	52·50
Ash ...	...	....	...	...	0·55

THE TKIBOULLI COAL-SEAM.

At the present time this seam only need be taken into account. In the curved basin formed by the Nakeral range, and almost at its foot, this seam crops out in a line almost parallel with the ridge itself, over a length of $3\frac{1}{2}$ miles, dipping rapidly under the mass above and under which it must soon lie at a great depth.

It has been known for many years, but hitherto little or nothing has been done so far as working it is concerned, owing, doubtless, to its isolated position and great difficulty of transport. It was mentioned incidentally, and described as a seam of coal 100 ft. thick, in the discussion on Mr. J. B. Simpson's paper on "The Coal-fields of Russia," at a meeting here in 1874.

From 1846 to 1850 work was carried on and the coals transported to the Black Sea and tried on the steamers, but was then stopped. The analysis of these is given as—

			1		2		3
Vol. matter...	...	...42·97		...	43·60	...	37·90
Carbon	...	...	...47·34	...	45·66	...	39·13
Ash	...	...	... 9·69	...	10·74	...	22·97

In 1866 and 1869 many other projects were formed, and in 1874 M. Pernolet reported exhaustively on the subject. But up to the time of the writer's visit nothing practical had been done, although many rumours of great doings were afloat.

On Plate XXIII., at point A, where the Sabourisoulis-gali cuts it, the seam is fully developed and shows two outcrops. South of this it thins out and probably soon becomes worthless. But at this spot it is 90 feet thick, including bands of shale, etc. Within 57 feet there is a thickness of 44 feet of coal, for the most part hard and of good quality, and almost all workable. It here lies at an angle of 50 degs., dipping east. A drift has been put in and cross drifts cutting through and proving the seam. At the time of visit the bank-side had been levelled, and the head of valley filled up preparatory to constructing inclines.

The following Section has been kindly supplied by Mr. F. W. North. See Plate XXIV. and page 96.

The analysis immediately above is probably of these coals.

The next outcrop is at B, Plate XXIII., which is the most notable. The stream (Naksheris-gali) has here cut completely through the seam, and on the west side the wall rises almost perpendicularly about 70 or 80 feet, the seams of coal and bands of shale, comprised between thick beds of sandstone, showing the structure very clearly. It is here about 100 feet thick, and lies at an angle of 30 degs., dipping north-east. Several short drifts have been put in on this side for experiment and analysis. On the east side, which has a slight slope on it, a drift has been put in, and cross drifts. proving the seam. Several workpeople had just been engaged to further extend the workings.

Another Section is taken from M. Pernolet's report. It shows the part containing workable coal to be about 90 feet in thickness, of which 44 feet is coal; but the workable and saleable coal itself is about 30 feet thick. See Section B, Plates XXIV. and XXVII., and page 97.

The following is the analysis given in same report, the letters, etc., referring to Section B, Plate XXIV., and again more extended in Plate XXVII.

I. Group.—Coal bad, dirty, friable, and pyritous.
II.   „   Coal good, with few bands.
III.   „   Coal fairly good.

		Vol. Matter.		Carbon.		Ash.		Calorific Power. (Pure Carbon=100.)
I.	A.	42·0	...	45·0	...	13	...	65
	B.	37·4	...	38·6	...	24·3	...	56
	C.	40·0	...	46·7	...	13·3	...	66
II.	D.	46·4	...	46·0	...	7·6	...	72
	E.	41·7	...	45·0	...	13·3	...	65
III.	F.	45·3	...	47·1	...	7·6	...	73
	G.	44·0	...	46·4	...	9·6	...	69

At the point C the stream (Kribikaris-gali) has not yet cut down through the whole seam, and so makes a waterfall of about 40 feet right over it. The coal is very hard and appears good.

Between here and point D a few slight traces are visible, but nothing distinct. It appears to take a sharper bend here which may betoken some slight dislocation, but the seam being at such an angle it will probably not be of material consequence.

At the point D it is again well exposed, this time by the Tiknaris-gali. Here it is about 60 feet thick, with 28 feet of coal, of which about 20 feet may be found workable. It lies at an angle of 40 degs., dipping north, a drift with cross drifts having been also put in here proving the seam. See Section D, Plate XXIV., and page 98, the letters and marks referring to specimens from respective beds. A more extended Section at D showing the angle of the dip is given in Plate XXV.

The following is an analysis of the middle bed, by Professor Kupfer, of Moscow, which shows well :—

Water	...	...	7·60	per cent.	After drying		
Carbon	...	...	73·11	,,			
Hyd.	...	...	4·72	,,			
Oxy.	...	...	9·81	,,	Coke	...	68·81 per cent.
Nit.	...	...	·75	,,	Vol. Mat....	31·19	,,
Sulphur ...		...	·03	,,			
Ash	...	...	11·58	,,			

Further westward, a few outcrops and traces are seen, but it is apparently now getting much thinner and dying out.

The samples on the table are the shattered remains of what were brought back from the outcrop at D. They do not at all give an adequate idea of the quality and texture of the coal, which along its entire outcrop is remarkable for its hardness, and, except at a few points, the absence of signs of weathering.

It is, generally speaking, a bright black coal, but sometimes cuts with a brownish tinge ; and it is only to be expected that on getting deeper the quality will improve. It ignites quickly and burns freely, with much flame and smoke, and leaves a large proportion of ash, but this shows little inclination to form "clinkers." It makes a coke which is easily crushed, and, as will be seen from the analyses, is better fitted for house or gas than for steam use.

The chief difficulties in the working of it will be the steep angle at which it lies, the number of bands contained in it, with consequent difficulty of keeping the produce clean, and the great depth which it must soon reach. See Plate XXVI.

On the other hand a large proportion could be worked above the levels of the various intersecting streams; but whilst this was being done to properly win and work the coal, shafts would have to be sunk at suitable places and stone drifts or galleries driven from them, cutting the seam to win the coal at a lower level, which in turn would serve whilst another still lower stretch was being prepared. Another way might answer, to make and keep good engine planes in the seam itself, and following its dip.

If the coal was worked by one company, the best surface arrangement might be to bring the produce down the hill sides by endless rope inclines of 2 to 2½ versts (nearly 1½ miles), converging to a point about 2 versts (1¼ miles) above Tkiboulli station. From this point, as the ground is fairly level and even, a connecting branch could easily be made, and it could then be screened and loaded direct into railway wagons.

LABOUR.

In starting any new industry to which the natives are unaccustomed, this must be an important point. Although the district is thinly populated and by a poor peasantry, the attraction of permanent employment would soon bring together the strongest and best of them from long distances, especially those from the hilly grounds further north in the Ratcha district. They would be found willing and quick to learn, and steady workers (with intervals, however, for their somewhat numerous holy days). Joiners could be made out of native material, but mechanics, hewers, and other skilled labour would have to be imported or taught by imported hands.

It would also be necessary to erect houses for them, although scarcely in so substantial a way as is customary hereabouts. Probably one room of wood or wood and rubble to hold ten or twelve each.

Of wood, chiefly chestnut, there is a great but diminishing abundance in close proximity to each outcrop. Good building lime can be made on the spot, and sandstone quarried.

The price of labour is somewhat as follows:—Boys and women, 5d. to 7d. per day; surface labourers of all kinds, 1s. 2d. to 1s. 8d.; underground, including hewers, 1s. 10d. to 2s. As little grain is grown in the neighbourhood, harvest time would not greatly affect these prices, and they would be lessened when engaged by the month, as would be the case.

COMMERCIAL AND FUTURE.

The fact of the Russian Government having guaranteed 6 per cent. to the promoters of the Tkiboulli branch line, and its construction being now an accomplished fact, seems a proof that it strongly believes in the value and usefulness of this coal-seam. The railway itself has solely the prospective carriage of the produce to depend upon for any remunerative income; and the expenditure must have been, and working expense will be, very heavy, it being a well constructed single line, with substantial stations, but, owing to the nature of the ground, with some steep gradients, and formidable embankments and cuttings. It passes through a poorly populated agricultural district. At the present time three trains per week are being run, the engines using coal got from outcrop at B.

To illustrate Mr Charles J. Murton's paper "On the Tkiboulli Coal-field (Caucasus)."

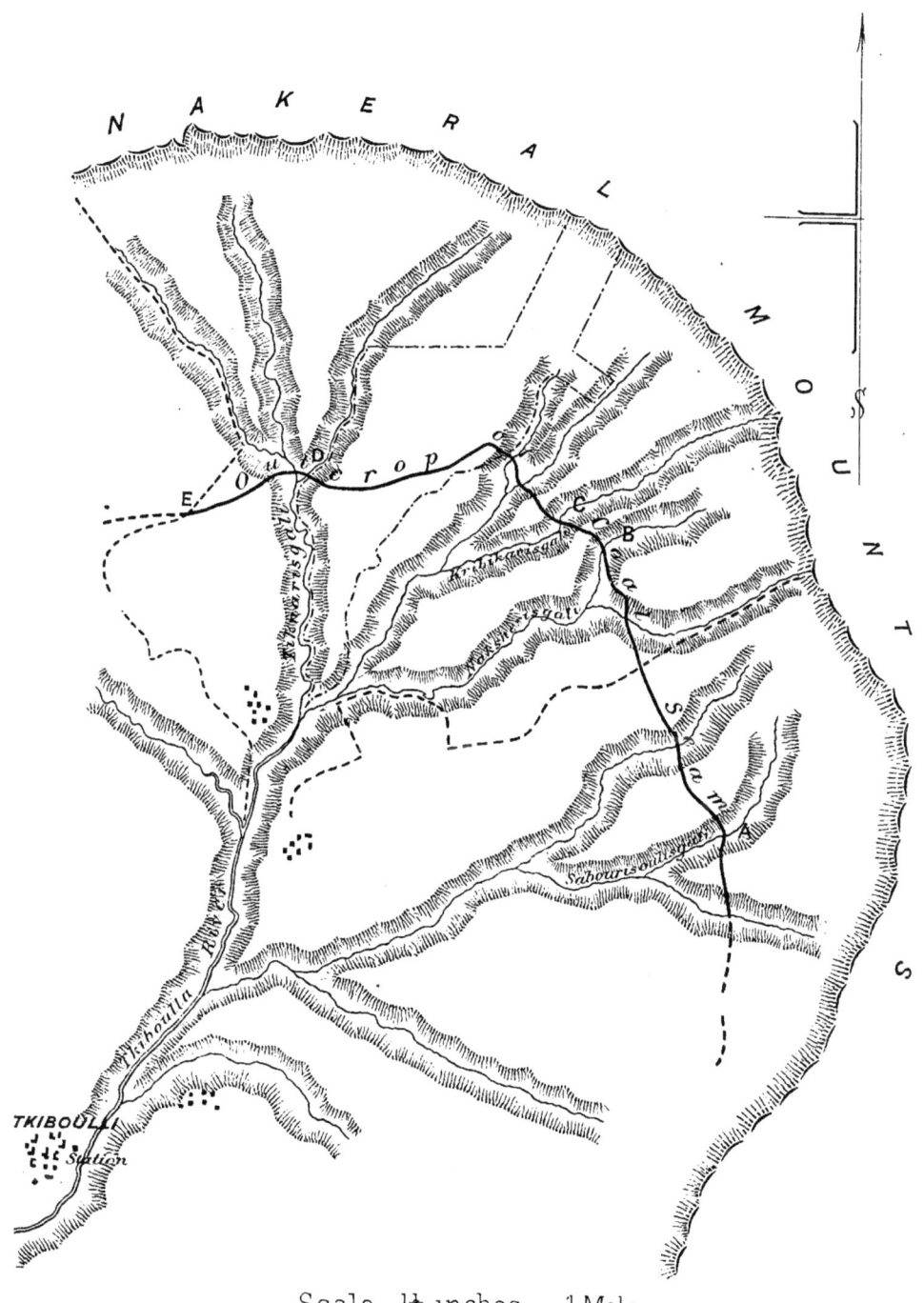

To illustrate Mr. Charles J. Murton's paper "On the Tkiboulli Coal-field (Caucasus)."

Scale, 1½ inches = 1 Mile

To illustrate M. *Charles J. Murton's paper "On the Tkibouli Coal-field (Caucasus)."*

SECTIONS AT OUTCROP A. B. D.

A   B   D

25

26

28

30

36

38

39

40

41

42

43

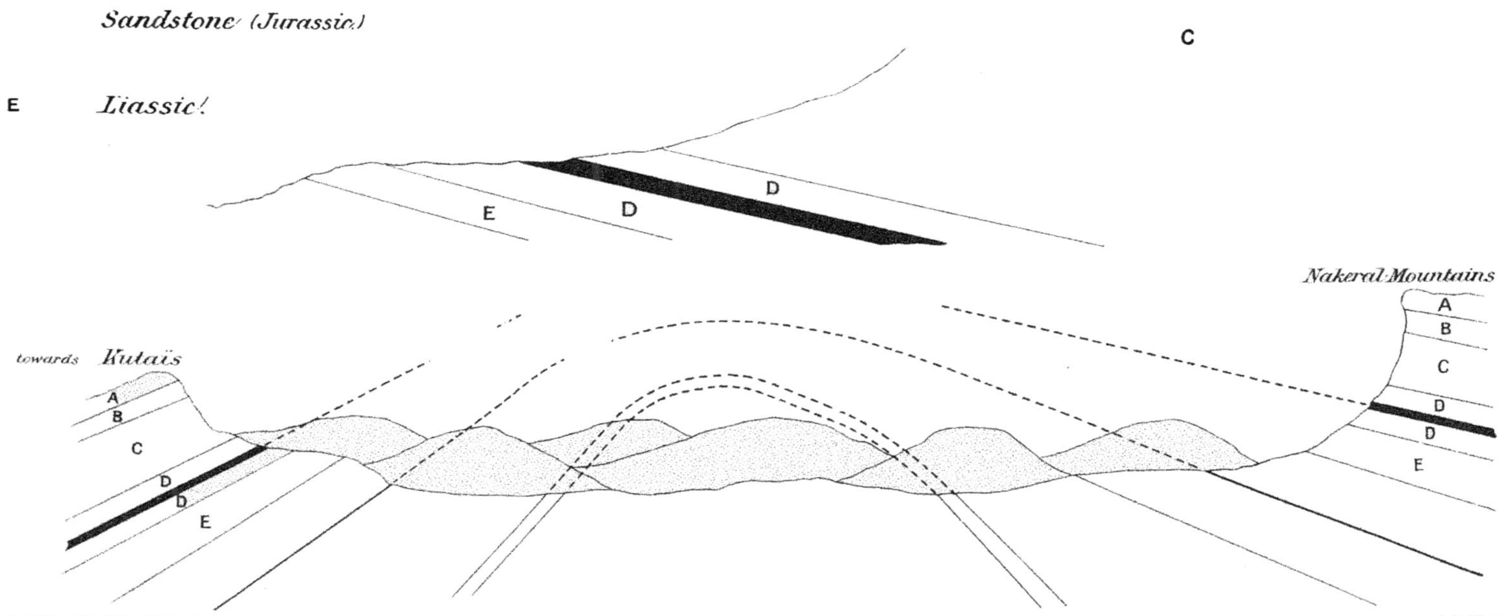

C

Sandstone (Jurassic)

E   *Liassic!*

Nakaral Mountains

A
B
C
D
D
E

D

D

E

towards *Kutaïs*

A
B
C
D
D
E

Andrew Reid, Newcastl

Angle 40°

Sandstone..

Coal (hard, coarse)..........................

Band (Clay)......................................

Coal (coarse)..................................
Band (Clay)......................................
Coal (hard).....................................

Coal..................
Black Band	0' 2"
Coal	7
Band	1
Band	2"
Coal	3
Band	1
Coal	4
Band	1

Coal (hard)....................................
Coal.....................
Band	2"
Coal	2
Band	1
Coal	4
Band	1
Coal	5
Band	2

Coal (good)...................................
Coal (soft).....................................
Band (Clay)...................................
Coal (soft)	
Band (Clay)	
Coal	
Band	0' 1"
Coal	4
Coal | Ironstone | 2 |

Sandstone (bands.).........................
Coal....................3
Band..
Coal..
Band..
Coal..

Sandstone..

Coal (coarse)....................................

Bands..

Coal.......................................
Band.......................................
Coal (coarse)...............................

Sandstone.......................................

To illustrate Mr Charles J. Murton's paper "On the Tkiboulli Coal-Field (Caucasus)."

SECTION AT B OF THE OUTCROP OF TKIBOULLI, RIGHT BANK OF RIVER

7.11.0 7.11.0	Shale Coal	} Barren			

G column, upper section (F):

Measurement	Description
10.0	Coal and Shale
2.3.6	Coal, dull, friable
7.8	Clayey Shale
1.6.8	Coal very hard
8.1.7	
7.8	Coal and Shale
1.7.7	Coal, good but friable
4.0	Coal and Shale
11.9	Coal
4.0	Shale
7.8	Coal
4.0	Shale
6.7.3	
2.5.6	Coal, hard, slightly pyritous
10.2	Shale
11.8	Coal
11.8	Coal and Shale
4.7.8	do. do.
5.1.2	
2.5.6	Coal scaly

} All this mass must be worked simultaneously and will be difficult to separate.

} Unworkable

4.8.0 4.8.0   Shale with layers of Coal

Lower section:

Measurement	Description
	Shale
	Clay
	Shale
	Leafy Limestone
15.5	Alternation of Shale and Coal in very thin beds
7.8	Shale, Clay and Sandstone
9.9	Shale impregnated with veins of Coal
	Brown calcareous shale
	Sandstone, calcareous
	Clay
	Shale
16.8	
	Limestone
	Clay
	Shale
Very thick bed	Sandstone hard

No workable Coal

Right-hand numeric columns:

47.1	45.3	7.6	.73	2.5	5.6	5.6	-	5.6	F	8.1
46.4	44.0	9.6	.69	8.8	7.5	5.0	2.5	5.0	G	16.3

Group →

By the heavy protective duties successively laid on foreign coals, and prospects of more to follow, the Government seems determined to make itself independent of outside aid. The duty, which is now about 6s. 10d. a ton, has had the effect of greatly lessening the foreign imports into Odessa, and a great increase in the output from the Donetz coal-field. English coal imports in 1884 were 309,275 tons; 1885, 193,850 tons; 1886, 117,853 tons; 1887, 47,899 tons; Azoff, 1887, 103,000 tons. Several new pits on a large scale are being sunk in the Donetz district, and new companies being formed; and although low prices still prevail, there are great expectations of better things, owing to the many schemes set forth for lessening the lighterage charges, etc., which are very excessive; besides which the Sea of Azoff is closed three months of the year by ice. One project is the making into a good port Mariopol, and this is being done rapidly. The coal is of very good quality, but scarcely equal, as it is claimed to be, to Cardiff or Newcastle.

Up to the present the principal effect of these duties has been to increase the cost to the general consumer; but the policy only needs to be rigorously carried out—as it will be, no matter at what general inconvenience and cost—to fully develop the field, and this result will as likely follow in the case of the Tkiboulli coal. Here the carriage from the mines to Poti or Batoum will not exceed 4s. 3d. per ton; freightage to Odessa, 4s. 6d.; transhipments, etc., 3s.; so that this coal could enter the market at a much lower figure than either the English at 24s. average, or Donetz at 21s. 6d. At the same time the quality is not equal to either of these, so lesser prices would necessarily have to be taken. At Batoum the present price of English coal is 32s. to 34s. per ton, and Donetz 27s. In Odessa Cardiff is now selling at 37s., Newcastle 34s., and Donetz 29s., but these prices are exceptional, owing to severe weather and scarcity of coal in port.

Besides Odessa, Constantinople, and other much smaller markets round the coast, there would be a small local consumption for railways, and a few industries, whilst others might be developed.

There is another easily accessible coal-field on the coast of the Black Sea, that of Heraclea, or Eurekli, belonging to the Turkish Government, and of which a description is given in Vol. III. of the Transactions by Mr. Longridge; but it is as yet undeveloped, although close to the coast. It is about 120 miles to the east of Constantinople.

A formidable competitor may appear in the oil from Baku. This trade is developing so rapidly that the railway, with its heavy gradients at Souram, is quite unable to transport it, together with their ordinary traffic. At present the oil is carried in tank-waggons and stored at Batoum, whence most of it is taken by steamers specially constructed and owned by the Russian Navigation Company to Odessa. A project for laying a pipe line a part or the whole of the distance (800 versts=533 miles) has long been talked of; but now the concession has been granted to the railway company, so that in a few years there will be an enormous supply of petroleum available. This will probably affect the use of coal on coasting steamers, but not ocean-going. The pipe line is to be of 8 inch pipes, with screwed joints. Starting near Baku it will go straight across country for first half of distance, then will be alongside the railway to Batoum. There will be 24 pumping stations connected by telegraph. The quantity of oil brought into Odessa in 1815 was 19,353 tons; in 1886, 40,332 tons.

In connection with the Tkiboulli Railway there is one other point perhaps worth considering. A command of the Black Sea is of vital importance to Russia, and the Black Sea fleet is being greatly augmented. In the event of war breaking out, or such like contingencies, by means of this new line there could be immediately brought into use a large supply of cheap coal in addition to the supply drawn from the Donetz field, already in extensive operation; so that, apart from any strictly commercial value, the railway and coal may be of great military value.

Other specimens on the table are limestone and chert, iron-ore and lead-ore from Ratcha, and maganese from Kurila, where it is extensively worked.

SECTION AT A.

	Ft. In.	Ft. In.		Ft. In.	Ft. In.
			Brought forward	24 10	12 4
1. COAL	0 6		19. COAL	3 0	
2. Parting		4 0	20. Parting		0 4
3. COAL	0 6		21. COAL	4 6	
4. Parting		3 4	22. Parting		0 2
5. COAL	4 6		23. COAL	2 0	
6. Parting		1 0	24. Parting		0 2
7. COAL	2 6		25. COAL	4 9	
8. Parting		1 6	26. Parting		0 6
9. COAL	1 0		27. COAL	3 0	
10. Parting		0 1	28. Parting		0 2
11. COAL	9 6		29. COAL	2 5	
12. Parting		0 3			
13. COAL	1 5		COAL	44 6	12 8
14. Parting		0 2	Spoil	12 8	
15. COAL	2 4				
16. Parting		0 9			
17. COAL	2 7				
18. Parting		0 3	Total	57 2	
Carried forward	24 10	12 4			

SECTION AT B (SIMPLIFIED FROM M. PERNOLET'S).

Group.	Analyses Letters.		Ft.	In.	Ft.	In.
		1. Sandstone				
	A.	2. Shale and *coal*			3	3
		3. **COAL**, very friable	5	2		
		4. Shale			3	3
I.		5. **COAL**, friable	4	5		
	B.	6. Shale, hard			3	1
		7. **COAL**, good	0	10		
		8. **COAL** and shale			5	7
	C.	9. **COAL**, good	2	3		
		10. Shale, very hard			6	4
		11. Clay			1	3
		12. **COAL**, hard	2	0		
	D.	13. **COAL** and shale			0	5
		14. **COAL**, very friable	2	6		
		15. Shale			0	7
		16. **COAL**, hard, with pyrites	1	0		
II.		17. Shale			1	0
		18. **COAL**, good	1	4		
		19. **COAL**, dirty	0	11		
		20. Shale			1	4
	E.	21. **COAL**, good and bright	3	3		
		22. **COAL**, rather dull	4	7		
		23. Shale			⊥	⊥
		24. **COAL**, traces of pyrites	3	3		
		25. Sandstone and shale			8	9
		26. **COAL**, dull and friable	2	3		
		27. Clayey shale			0	8
	F.	28. **COAL**, very hard	1	7		
		29. **COAL** and shale			0	7
		30. **COAL**, good but friable	1	7		
		31. **COAL** and shale			0	4
		32. **COAL**	1	0		
		33. Shale			0	3
		34. **COAL**	0	7		
III.		35. Shale			0	3
	G.	36. **COAL**, hard, slightly pyritous	2	5		
		37. Shale			0	10
		38. **COAL**	1	0		
		39. **COAL** and shale			2	6
		40. **COAL**, scaly	2	4		
		41. Shale, with *coal*			4	8
	H.	42. Shales, with thin beds of *coal*, sandstones, and clays			49	3
		43. Sandstone, hard.				

	44	3	46	0
	46	0		
Total coal-bearing	90	3		
	49	3		
Total between the thick beds of sandstone ...	139	6		

SECTION AT D.

Samples Marked			Ft. In.	Ft. In.
	1.	Saidstoie ...		
A 1.	2.	**COAL** ...	3 6	
1.	3.	Baid, clay ...		3 0
	4.	**COAL**, coarse	1 7	
	5.	Baid, clay ...		1 0
	6.	**COAL**, hard	1 6	
2.	7.	Black baid ...		0 2
	8.	**COAL** ...	0 7	
	9.	Baid ...		0 1
3.	10.	**COAL** ...	1 7	
	11.	Baid ...		0 1
	12.	**COAL** ...	0 2	
	13.	Baid ...		0 1
	14.	**COAL** ...	0 4	
	15.	Baid ...		0 1
4.	16.	**COAL**, hard	3 0	
5.	17.	**COAL**, hard	1 2	
	18.	Baid ...		0 2
	19.	**COAL** ...	0 2	
	20.	Baid ...		0 1
	21.	**COAL** ...	0 4	
	22.	Baid ...		0 1
	23.	**COAL** ...	0 5	
	24.	Baid ...		0 2
6.	25.	**COAL**, good	3 0	
X 1.	26.	**COAL**, bad...	2 8	
	27.	Baid, clay ...		0 8
		Carried forward	20 0	5 8

Samples Marked			Ft. In.	Ft. In.
		Brought forward	20 0	5 8
	28.	**COAL**, soft	0 10	
X 2.	29.	Baid clay ...		0 8
X 3.	30.	**COAL** ...	0 6	
	31.	Baid ...		0 1
	32.	**COAL** ...	0 4	
	33.	Baid, iroi-stoie ...		0 2
X 4.	34.	**COAL** ...	1 0	
	35.	Saidstoie ...		2 0
	36.	**COAL** ...	0 3	
	37.	Baid ...		2 0
	38.	**COAL** ...	0 10	
	39.	Baid ...		2 0
	40.	**COAL** ...	0 7	
	41.	Saidstoie ...		7 0
	42.	**COAL**, coarse	0 9	
	43.	Baids ...		8 0
	44.	**COAL** ...	1 0	
	45.	Baid ...		1 0
X 5.	46.	**COAL**, coarse	1 6	
	47.	Saidstoie ...		
			27 7	28 7
			28 7	
		Total	56 2	

Mr. WILLIAM COCHRANE said, he would be glad if the writer of the paper could give them an idea of the calorific value of the coal he had described in comparison with our Carboniferous coal. It was evident, if there was only 45 per cent. carbon in the Tkiboulli coal as compared with 65 to 75 per cent. in our Carboniferous coal, it would require at least one ton and a half of the coal to give out the same effect as one ton of our carboniferous coal. This would be a very serious item in connection with Mr. Murton's first statement in his paper, that the Tkiboulli coal-field "may some day enter into competition with the English coal now so much used in the Black Sea," since the use of this coal would involve a serious matter in the way of bunker room in steamers. According to Mr. Murton, the cost of taking this coal to Odessa would be 11s. 9d.; taking this coal to have one-third less calorific value than our coal, this would run the price up to 17s. 8d. for the transport alone. He (Mr. Cochrane) considered that the expenses of working the coal would render it, probably, not a very strong competitor considering the increased value of our English coal. He saw that in only one case was it stated that carbon was present to the extent of 73 per cent., which was a nearer approach to that of the Northumberland coal.   This seam would probably become later of very great commercial importance to the Black Sea district, and might be used for some purposes, but not in competition, he thought, with English coals in the world's markets—purely for local purposes.

Mr. MURTON said, there was no doubt that the Tkiboulli coal was not equal to the English coal, and could not command the same price ; but the Russian Government were showing such a firm determination to put increased duties on coal—if the present duties were not sufficient to develop their own coal-fields—that he thought that in the end our coal trade would suffer.   The Russian Government seemed to be determined to make themselves quite independent of outside help.   In the open market he thought these coals could not compete with English coals.

Professor LEBOUR said, Mr. Murton's paper was very interesting from a geological point of view.   He regretted that Mr. Murton had not exhibited some fossils by which he could have proved the age of the beds.   But fortunately there was very little doubt as to this.   He had with him a map, made for the Russian Government by General Steinmann, which showed the beds in the Caucasus; in fact, it was a part of the geological map of the country.   He (Professor Lebour) had brought down two other numbers of the same publication, in which were plates of fossils from these very deposits, which might be taken as specimens of those likely to be found at Tkiboulli; and they indicated, without any doubt at all, the age of these beds.   It was particularly interesting to find these coals in the same series as the moorland coal of Yorkshire.   He had himself, in Austria, seen several thin seams arranged pretty much as these were, and grouped in the same beds.   It would seem as if in going eastwards, the conditions became more and more terrestrial until India was reached.   This was, no doubt, a kind of link between the Oolitic coals of the east and the west.   He had taken great interest in Mr. Murton's paper, especially as Mr. Murton was an old student of the College; and he was glad to say that this was not by any means the first valuable paper which had come from old students of the College.

Mr. MURTON said, that at the time he went to the place he could not put the cost of working above 7s. a ton, so that there was a very large margin still between what it would cost at Odessa and the price of English coal.

The CHAIRMAN—Can it be worked continuously during all the year ?

Mr. MURTON—Yes.   In the winter time the working is not stopped much, but there are several holidays in the district.

The CHAIRMAN—In the Donetz coal-field work is stopped about six months in the year.

Mr. MURTON—They are stopped in the Donetz for three months, as the Sea of Azov is frozen, and everything is at a standstill. At Tkiboulli they will be able to work from the beginning of the year to the end.

The CHAIRMAN said it was very interesting to have a description of coal-fields in other parts of the world. Every one must feel that in Russia especially there were very large districts of coal which would, in time, come very largely into competition with this country. He proposed a vote of thanks to Mr. Murton for his paper.

Mr. W. COCHRANE seconded the motion ; and it was agreed to.

Professor STROUD read the following paper by Professor A. S. Herschel, " On an Improved Form of Seismoscope :"—

ON AN IMPROVED FORM OF SEISMOSCOPE.

By Professor A. S. HERSCHEL, D.C.L., F.R.S.

In the Report presented on November 26th to the Council of the Mining Institute by Mr. M. Walton Brown, on Automatic Records obtained with an ingenious form of Seismoscope at Marsden, during the months from October, 1886, to April, 1887, a defective action of the instrument in indicating, apparently, incessant vibratory motion at the station for days, weeks, and even for a whole month together, must, it would seem, be due to permanent departures of the tremor indicator from its neutral point, either produced by adhesion between the tremor-pointer and the mercury globule with which it comes in contact, or by slow heelings of the whole instrument from time to time from its normal vertical position. The latter seems to be the real explanation, since the platinum pointed indicator is not amalgamated, and has, therefore, no natural tendency to cling to the little drop of mercury in the dimple of which it oscillates when the instrument is disturbed.

All seismoscopes of the pendulum form, the neutral position of whose indicator relative to the supporting base is fixed by gravity, are liable to this defect from small unavoidable distortions of the frame and feet of the pendulum, when left for months to themselves in positions of varying atmospheric conditions; and yet no means of obtaining approximate astaticism and return to a neutral point otherwise than by neutralising gravity by various means, especially with pendulums, appears yet to have been attempted.

The object of the present contrivance is to make the neutral point of the indicator, on and relative to the supporting base, *independent of gravity for its position*, and, therefore, unexposed to change by any slow motions of flexure or yielding in the support. The first suggestion for

this purpose which presented itself was that of a shallow saucer of mercury, in which a vane, movable about a fulcrum fixed to the saucer or
support, and bearing a long slender index, should dip. Owing to the
mercury's inertia, every horizontal jolt of the saucer carrying the vane
will produce a pressure upon the vane, and move the index from some
position made normally neutral to it relative to the base, by means of a
spiral hair-spring, or other means of keeping it naturally steady in one
fixed position. Slow movements of the mercury, due to gradual heeling
of the base, will be ineffective on the vane and index, the slender
strength of the directing spring being quite able to resist the weak
pressures which slow motions of the liquid can impress upon the vane.
But mercury or liquid, in any form of shallow dish, has its own natural
oscillation period like a pendulum ; and if this should happen to coincide
with the oscillation period of an earthquake tremor, the liquid is not
more suitable to record the motion than a pendulum of the same period
would be, since it would, like a pendulum of the same period, accompany
the dish and the vane exactly in their oscillations. To make an instrument of this kind thoroughly detective, instead of being placed upon the
solid base, the mercury dish must itself be suspended astatically. But
here the usual astatic pendulum combinations are easily available, since
although their neutral position depends on the support's stability, yet as
this is only the neutral point of the mercury and dish, its slow variations
are unable to affect the place of rest of the vane and indicator, since
these are secured from such slow actions by the slender power of the
directing spring.

The simple form of astatic suspension of the mercury cup here used
occurred to the writer a year or two ago, but its properties presenting
difficulties of theoretical investigation, its excellent suitability for seis-
moscopic suspension was neglected, and it is with some little difficulty
that the mathematical rules of its construction have been now determined,
so as to be able to assign the right proportions of its rods and wires
required to produce exact neutrality, and make the suspension perfectly
astatic. It consists of two parallel horizontal bars, A B, C D, one suspended underneath the other, by means of two strings
of equal length attached to the ends of the bars, and
crossing each other in the middle between them like a
letter X. The middle point of the lower bar, when in
swinging motion describes a curve, which is either concave or convex,
upwards, or can be made nearly a straight horizontal line by varying
the lengths of the supporting strings. If a weight is hung to the middle

point of the lower bar, it will be instable or unstable or in neutral equilibrium in these three cases; and a process of mathematical treatment by approximation shows that if a is the inclination of the strings to the horizon, a and c the lengths of the upper and lower bars, the state of equilibrium in the symmetrical position is

$$\left.\begin{array}{l} \text{stable,} \\ \text{neutral, or} \\ \text{unstable,} \end{array}\right\} \text{ according as tan } a \text{ is } \gtreqless \sqrt{\dfrac{a}{c}}.$$

In the figure here given, the lower rod c is twice as long as the upper one a, and therefore $\dfrac{a}{c} = 0.5$, and $\sqrt{\dfrac{a}{c}} = 0.7$ nearly. If, therefore, $a = 35° 16'$, or tan $a = 0.7$, as in the figure, a weight suspended from the middle of the lower rod will be suspended astatically as regards motion endwise in the direction of the rod. But if C D is the upper, and A B $= \frac{1}{2}$ C D the lower rod, the position of the lower rod will now be one of neutral equilibrium, if tan $a = \sqrt{2} = 1.4$, or $a = 54° 44'$, as in the adjoining figure, longer strings being required in this case than in the former one to make the construction an astatic one.

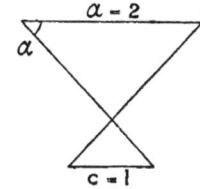

If the two rods are of exactly equal length, then $\dfrac{a}{c} = 1$, and $\sqrt{\dfrac{a}{c}} = 1$, and the lower rod will be astatically suspended if tan $a = 1$, or $a = 45°$. In this case the two rods and the two strings form the two opposite sides and the two diagonals of an exactly square arrangement. It is a preferable one to either of the other two constructions, as the range along which the middle point of the lower rod moves, nearly in a straight horizontal line, is larger than in those other cases; and this simple construction is therefore a very convenient one for obtaining suspension, as free from friction and as astatic as any that can be chosen for use, in the inexpensive class of instruments known as seismoscopes, or earthquake detectors.

Instead of straight rods, two flat boards, or long planks (of which the above figures would only show the cross sections), may be suspended one astatically from the other by means of strings crossing each other, as in the above figures, between their edges, one such pair of strings being used near each end of the two planks; and so long as an end-view of the strings and planks, looking lengthwise along the latter, presents the appearance of any of the above figures, the strings need not form pairs, each pair belonging to some one cross section of the boards, but may be all in different cross sections, or variously oblique to such cross sections. But in the oblique directions which they may have, they must together pull, in the direction of its length, oppositely upon the lower board, so as to hold it at rest lengthwise ; and they must be so distributed along the lower plank's length as to be able to carry its weight collected at its centre of gravity. Not only the board itself, but any weight suspended from it along its middle line will also, supposing the board to be weightless, and that the strings are properly distributed to support these weights, all be found to be astatically supported if they are moved together breadthwise across the plank. They will not tend—that is to say, to return very sensibly (unless the displacement is a considerable one)—towards the position of normal rest from which they are displaced. These generalisations of the construction afford useful means of applying it practically, because two crossed strings only, tending naturally to hang in one plane, rub against each other and form an *unsensitive* suspension ; but with one string crossed by two strings contact between the suspending strings can be avoided without in the least involving the astatic condition or perturbing the above very simple rules of magnitude which are needed to obtain it. Suspending strings, also, along the edges of two planks can be kept separate so as not to rub against each other ; and if they are inclined to the edges, some towards one and some towards the other ends of the planks, so as to pull lengthwise oppositely upon the lower one, that plank will be unable to oscillate lengthwise, and can only move in an astatic way parallel to itself from side to side.

From any two points in the centre line of the lower plank (or board or light frame) a pair of crossed suspending strings, carrying a rod below it parallel to the centre line, can be used, capable, therefore, of moving astatically

lengthwise, while by the freedom of the upper attachment points of the strings to move in neutral equilibrium across that direction no solicitations of gravity can make it oscillate from side to side, and it is, therefore, as neutral in its equilibrium towards one direction as towards the other. A weight suspended from its centre point will be in the same state of indifference to oscillation in any direction ; and this is the construction used in the form of seismoscope here proposed for giving the cup of mercury (or of any other liquid) a kind of suspension, which can easily be made as perfectly astatic as required.

Other modes of using the
principle, as shown in the
annexed figures, may, however, also be resorted to. In
that consisting merely of
two rods hung one below the
other, transverse to each
other, with an intervening
hooked frame between them,
the use of spreading wires

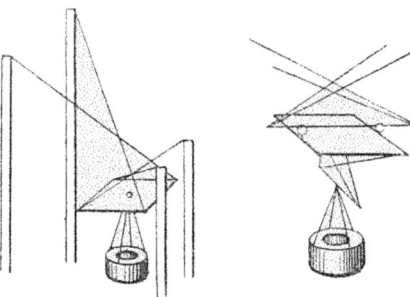

to avoid contact, and friction, between them, may be
avoided by giving single suspending wires a slight
twist in the same direction in each crossing wire, sufficient to resist their tendency to hang both in the same
plane from the points of suspension. With fine steel
crossing wires this moderate twist will preserve itself
indefinitely, and this form of the contrivance, as the
annexed figure shows, then becomes of the utmost
simplicity.

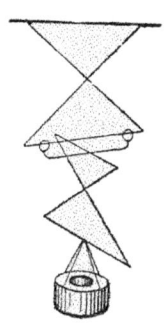

In the other construction, figured with a single
plate or frame, and suspension points *of unequal heights,* a crossed
arrangement is given to the obliquities of each spreading pair of strings,
by which freedom to oscillate laterally is given to the suspended frame;
but its equilibrium is at the same time made neutral in that new direction, and is, therefore, neutral in both transverse directions, and in every
direction in which the frame's displacement may occur. The suspension
points of one of the crossed pairs of spreading strings is made lower than
those of the other pair, in order to separate the strings in one pair from

contact with those in the other. This does
not preclude making the equilibrium in
the principal crossing direction of the two
pairs neutral as easily as when the two
pairs' suspension points are both on the
same level, for it deserves notice that for a
constant equal inclination (a) to the horizon, of two crossing strings, as $A_0 D_0$, $B_0 C_0$
in the figure, carrying a horizontal bar, C D,
at their feet of a constant length (c), the
condition that this bar's state of equilibrium
may be neutral is the same in all such modes
of suspending it as $A_0 D_0 C_0 B_0$, $A_1 D_1 C_1 B_0$.

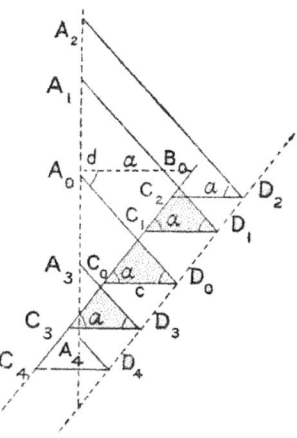

$A_2 D_2 C_2 B_0$, etc., where the points of suspension A_0, A_1, A_2, etc., are all on the same vertical line, as the requisite one above given, which determines the relations for the figure in the level position $A_0 B_0$ of the two points of suspension, viz., tan $a = \sqrt{\dfrac{a}{c}}$, if we simply observe to denote always by a the *distance in a horizontal direction* between the two points of suspension, A and B, while c is the length of the horizontally suspended bar. This *general form* of the astatic relation affords a rather wide choice of different ways of using it, among which the above examples are selected as being the most obviously convenient and simple ones.

Next to astaticism of the trough of liquid, easy mobility of the tremor-recording blade or vane which dips into the liquid, and which carries the long contact-making pointer round a fulcrum borne solidly on the frame of the apparatus, is of the greatest importance to the instrument's proper action. The blade must also receive pressure equally from the liquid's motions in whatever direction they occur, and must move with equal ease on its fulcrum towards any one of these directions. A circular hoop-like blade, quite immersed in the liquid, as shown in the accompanying drawing of the instrument, fulfils best the first of these conditions, and a pin-point bearing of the most frictionless kind, not liable to be easily unseated from its socket by earth-jolts and tremors of the latter, offers a means adopted in this instrument for satisfying the second of those two conditions. The principle used for holding the pivot-point of the light indicator steady in its socket is as follows (see Plate XXI.) :—

A helical wire spring, when strongly drawn out or compressed in length in direction of its axis, does not thereby lose sensibly any of its original easy flexibility and pliability across its axis, although exerting a rapidly increasing force along the axial direction when the extension or compression is increased. The latter force arises from combined torsion and flexure of the wire throughout its length, by which the loops of its spiral coil are all brought nearer to or removed further from each other than they were before, and it increases in strength very nearly in proportion as this torsional and flexural deformation is increased. But if in this state the spring is bent, its loops widen their distances from each other on one side, and contract them on the other ; more twisting of the wire occurring to produce increase, and an equal untwisting to produce decrease of the existing distortion in opposite halves of each spiral turn or loop of the wire simultaneously. The extreme flexibility which a helix spring exhibits when this relation is perfectly fulfilled can be easily observed by compressing and then tying together the two ends of such a spring by an inextensible string along its axis, which prevents the two spring-ends effectually from changing their distance from each other. The loops of the spiral then receive with extreme ease any relative lateral displacements, being really in unstable equilibrium with each other by reason of the slight increases of length which the axis effectively receives by any flexures which the spring's form undergoes between its ends. If the upper end of such a compressed spring is secured to the frame of the seismoscope, and the lower end presses down the bearing-pin of the indicator against its socket, being firmly attached there to the pivot, if the length and compression of the spring are at the same time not so great as to cause the spring to bulge and fly out to one side, advantage may be taken of the unstable forces of the spring's pressure which are ready to act, to overcome the spring's natural stiffness and resistance to flexure at its lower end, especially by so weakening the spring towards that end either by tapering its wire or web, or by narrowing its coil, that bendings of the coil produced by oscillations of the cap may begin there sooner than in any other part of the spring, and may so bring the bulging forces of the compression to keep that end in a nearly unstable or practically neutral state or equilibrium. No exact adjustment of any sheet or wire spring of this kind to render the pivot point which it presses and is fastened to, just neutral in its equilibrium, can very well be fashioned and produced, except by trial; but a conical wire spring having its point downwards and fastened to the cap, especially if a short part of its narrow end is nearly cylindrical, may be found without doubt to produce easily enough, when attached as described and pressed down from above, both the desired pressure on the pivot and a balance of the required delicacy round a position of rest of very slight stability. The counterpoising knob over the pivot cap allows gravity's directing force to be removed entirely before affixing the spring, two little side arms being added for laterally balancing the weight, so that the indicator always takes up, by the action of the spring alone, a constant neutral position relative to the instrument and its base. To adjust this place of rest of the contact wire to the dimple in the mercury, a rough adjustment by help of the pivot and spring carrier in two directions (with the handles c and h in the drawings) is provided, and a more delicate final one by means of the movable hair-spring k surrounding and retaining the wire in a permanent position from any small variations due to flexures in the frame and spring which may accidentally arise. It is, however, on this last spring *close to the mercury dimple* that it is intended to make the wire's place of repose depend as ex-

clusively as possible, because there is less room for warping distortions of
the structure between the guiding spring and the mercury drop m, than
between the mercury drop and the upper directing spring near the pivot-
point of the indicator. It is, therefore, the main object of this instrument's construction, after the directing power of gravity has been
removed by help of the astatic liquid trough and by counterpoising the
indicator very perfectly, to make the spring used to hold the pin-point in
its socket, which is the only other guiding force in action, as free from
directive influence upon the indicator (except such as tends as permanently and steadily as may be towards the mercury drop) as good and
fitting means can possibly be found to make it. It is owing to the
unsteadiness and easy variability of such weak directive force as it is
liable to develop, that suspension of the indicator by means of four
stretched wires, and counterpoising it round their point of attachment
to it, is for the present abandoned, but it may, perhaps, still be found
possible to adopt this plan, and in that case a great simplification would be
gained, and the use of a spring-bound pivot-point might be dispensed with.

Besides a compressed conical spring, other means of using spring
pressure to retain the light indicator freely against its pivoting point are
shown in the side view (Fig. 2, Plate
XXI.) of the instrument, in Fig. 5,
and especially in Fig. 6. That shown
in Fig. 2, and better in Fig. 6, of
producing the downward pressure by a
downward extended conical or parallel-
sided thin wire spring appears to be of
all the methods the simplest and the
best, since the upper end of the wire at
the conical point, or open top, of the
spring can be passed as a hook through

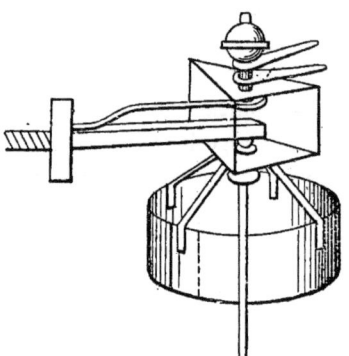

a smooth eyehole of the pointer at its base, or in an arch across a knife
edge of its stirrup there (see Fig. 6), pulling it without rigidity or friction
directly downwards towards the mercury drop, or towards any desired
point in the base of the instrument, for setting the spring to draw towards
which correctly, means of adjusting it easily at its lower point of fast-
attachment are provided.

It should be noticed that to retain the light indicator with *really
perfect* neutrality of turning in all directions against a perfectly fixed
pin-point socket, where even its counterpoised weight must purposely be
made so light that it cannot be certainly relied upon to keep it, recourse

to an astatic contrivance for applying the spring pressure, like the fine wire-connected plates shown in the adjoining figure—somewhat similar in their mode of action to those already described for suspension of the liquid vessel—is theoretically indispensable. All the modes of using a spring which have been figured (except the compressed conical one) are only approximate substitutes for such a jointed structure, because the spring's point of attachment to the pointer can never coincide actually with the pin-point centre of its motion, and must therefore always produce some oscillation of the pointer round a neutral direction. In the constructions last mentioned (Figs. 2 and 6), however, the eyehole in the base of the pointer may be brought almost as close to the pin-point in the indicator cap as we please, leaving room in the pointer's stirrup-arch for a very small-drawn socket-holder only, well fastened to the pillars of the instrument, to introduce itself between them. All needful nicety and ease of construction seems, therefore, to be very well attained by that arrangement. (See Fig. 6.)

The counterpoising knob g and cap d of the indicator may be made of brass; but the ribs d, f, to resist attacks of mercury, if it is used, must be thin steel watch-spring pinched to corrugate it, except at the ends where the pivot cap and the mica hoop-blade are riveted to them—the latter with iron rivets. The hoop form of the mica blade will give it much stiffness, though as light and thin as paper, and its ends will be bound together by the same rivets which secure one of the steel ribs to it. Its buoyancy in mercury will nearly counterpoise the indicator's weight, and will not alter, by inclination of the instrument, for any liquid used if the blade's submergence in the liquid is complete. With olive oil, or glycerine, or treacle (oil covered) also, whose viscosity will contribute as much as the weight of mercury to the blade's impressibility, no sensible changes of the neutral point are likely to be produced with the instrument's small variations from the vertical by capillary action of the liquid surface on the slender ribs where they are wetted by it.

The liquid cup may conveniently be the bottom part (cut off) of a round glass bottle, with a central hole ground through it, into which a short straight piece of wide glass tube is cemented; and four holes in the rim of the cup serve for its attachment to the hooked rider hanging by a frictionless knife-edge from a knife-edged notch in the middle of the lowest pendulum bar, as Figure 2 of the apparatus shows. The frame junctions at E and F are also knife-edged notch bearings in the meeting edges of the frames, allowing very free rotation of one frame upon the other. Small steel split rings passing through well smoothed holes in the edges of the frames form equally frictionless means of suspending the frames at those points, by their crossing wires tied to the rings. Four steel pegs with well smoothed eyeholes, carrying similar rings, can be inserted in the top-board of the case at AA′, CC′, and made to pass through it to its upper side, where they can be drawn in or out to adjust the level of the upper pendulum frame. Platinum or fine steel wire may be used for the suspensions; but the latter will probably be found preferable for its inextensibility. The four wires of the upper frame having to be all of exactly equal lengths, may be made so by forming small loop ends upon them round two round wire pegs fixed at the proper length from each other on a slab of wood, these loop ends being then slipped into the split rings.

The proper lengths for the three wires of the lower pendulum being properly determined, pegs similarly set in wood at the right distances from each other, two of the wire lengths for this lower swing carriage being equal to each other, can be formed in the same way upon one of the pairs, and the remaining unequal length upon the other pair of pegs. The bars of the frames, BD, EK, HL, should be made of thin hoop steel, with their underneath edges turned over to render them inflexible.

The four leading principles adopted and attempted to be carried out practically in *Professor Herschel's non-gravitating* electric *Seismoscope* are:

1. Tremor indication by means of the *inertia* and *viscosity of a liquid* acting on a sensitively mounted paddle blade.

2. Motionless or *astatic suspension* of the trough of liquid *by a lattice arrangement of parallel frames or bars*, suspended singly or from each other *by crossed wires*.

3. Confining the pivot-point of *an exactly counterpoised light* indicating blade and pointer to its fixed socket fulcrum *by spring pressure*, either quite neutrally or with *only the least* sensible directing force *tending constantly to any desired fixed point in the base* or pedestal of the instrument.

4. Retention of the pointer's neutral position constantly *and exactly* at the desired spot by *a flat spiral hair-spring* fixed *to a part of the instrument's base* or pedestal, *in close proximity to* the said spot.

EXPLANATION OF PLATE XXI.

AB, A′B′.—Left hand pair of crossed wires, *out of contact* with each other by suitable points of attachment.

CD, C′D′.—Right hand pair of do.; points of attachment symmetrical to those of the left hand pair.

BDD′B′ (Figs. 1, 2, 3).—Thin rectangular frame of narrow hoop steel, with lower edges bent inwards. Small smooth holes for wires or rings in the top edge at B, D, B′, D′, and two knife-edged notches in the same edge, at E, F.

To illustrate Prof. Herschel's paper on "An Improved Form of Seismoscope".

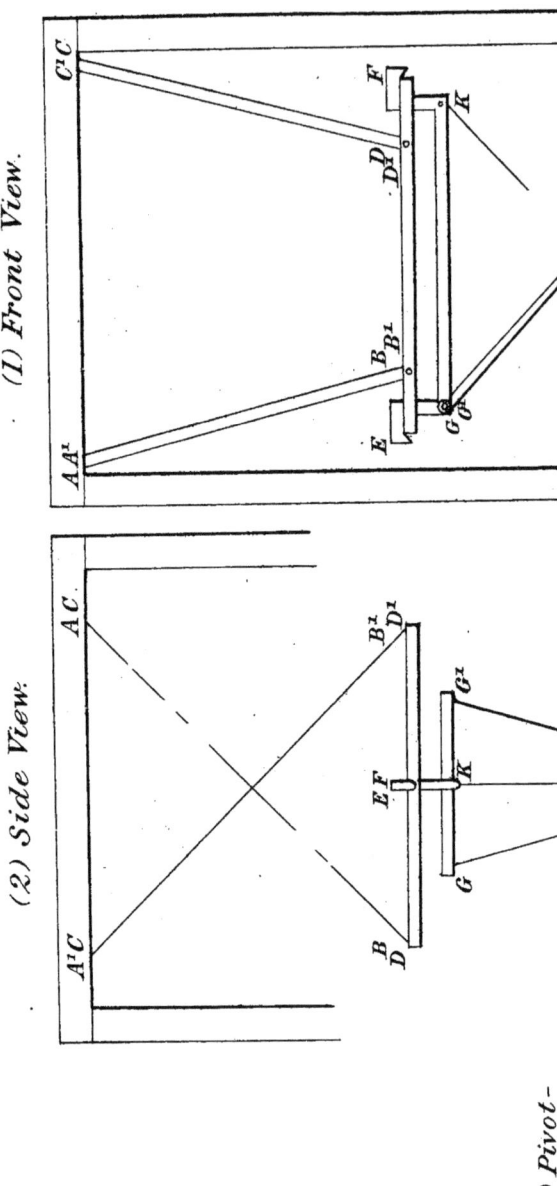

(1) Front View.

(2) Side View.

(4) Pivot -

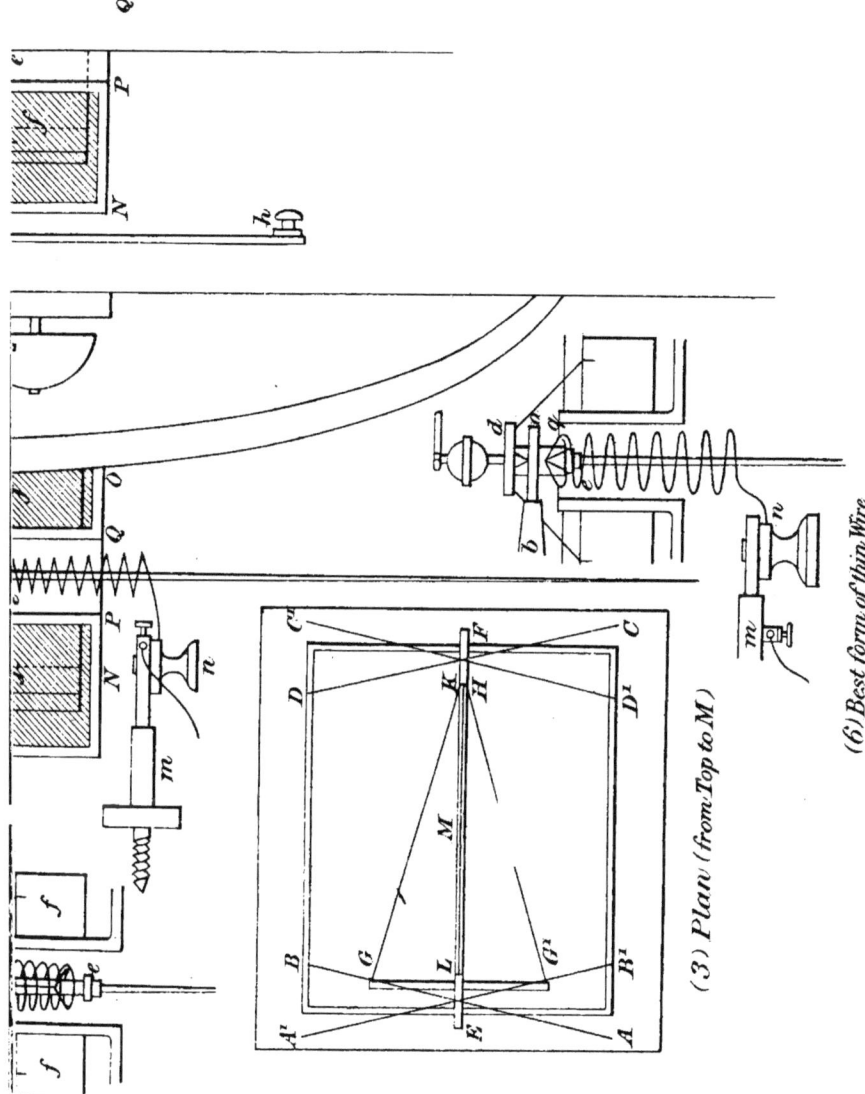

(3) Plan (from Top to M)

(6) Best Form of thin Wire

GG′K (Figs. 1, 2, 3).—Hooked steel blade with knife-edged notches at E, F, and a stiffened cross piece of angle steel at GG. Small smooth holes in the lower edges of do., at GG′K.

GḢ, G′H (Figs. 1, 2, 3).—Pair of suspending wires, crossed by the single wire KL, to carry the bar LMH. The latter is a hoop-steel blade with two small smooth holes at K, L, and a knife-edged notch at M, at its upper edge.

abc (Figs. 1 and 4).—Brass indicator support; *c*, hinged part of do., with a socket hole for pivot on top part of its hinge, and conical wire spring *g* attached by hole and peg to cap *d* of indicator, and to battery wire binding screw at *c; h*, handle to rotate the support and indicator.

defg (Figs. 1, 2, 4, 5).—Brass table cap of indicator *d*, with light brass bridge foot *de* of contact wire *eM*; four pinched watch-spring ribs *df*, riveted with soft iron to a thin mica hoop *ff; g*, a brass counterpoise knob and two small brass wing blades on a screw stem carried by the cap.

NOPQ (Figs. 1, 2).—Cylindrical glass cup with glass tube passing vertically through the central hole PQ; four holes drilled at *ll*, near its upper edge, to receive suspending wires *lM*, hanging it from a knife-edged rider hook M (Fig. 2).

mn (Fig. 2).—Brass draw-tube and turning knob for varying the lower fixture of the spiral spring *ne*, and battery connecting screw to the spiral spring; *k* (Fig. 1), adjustable spiral hair-spring to retain the index point M in the mercury dimple.

Fig. 5.—The spiral spring *ce*, fixed at the top to a hole in the handle *c*, and resting also at the top against the under side *a* of the slit brass draw-tube *ab*, which is supported by the frame, surrounds the stirrup or bridge foot *de* of the contact wire. Across its lower end is stretched a short wire, through the bridge *de*, from the middle of which a vertical wire draws up the lower end of the spring to a cross peg of the bridge or stirrup, just underneath the draw-tube *a;* the latter has on its upper side a socket plate at *a*, for the pin-point of the indicator.

Fig. 6 represents the most efficient practical mode of applying a spiral thin wire spring to keep the pivot-point *d* of the indicator, by its pressure, steadily seated and centred in the fixed socket *a*. The bridge or stirrup piece *de*, to which the pointer or contact wire is attached below, is narrow in both its horizontal dimensions, and made of thin sheet brass, so that it is easily contained within the open top of the stretched spiral spring *ne*. A small copper crupper, or smoothly rounded notch, rests, notch upwards, on the foot of the stirrup, being soldered to it to ensure electric contact. The upper end of the spiral spring forms a horizontal wire ring surrounding the stirrup. From one end of a diameter of this ring, *passing through the opening of the stirrup*, to the other, a short arch of copper wire is stretched, spanning over the crupper notch of copper on the stirrup foot, and resting on it, so as to apply the pressure of the stretched spring to it, and to make electric contact with the wire pointer. The piece *a* in which the socket-hole is made is a short slight rod of hard steel projecting from the more solid brass rod *b*, so that the copper notch and steel pivot-point *d* can be brought very near together. The spiral spring is of copper wire, to avoid heating by the current, and adjustable below, as in Fig. 2.

Mr. WALTON BROWN said he had not carefully considered the details of the seismoscope devised by Professor Herschel, and he was not competent to give an opinion upon the theoretical propositions adopted in its construction. Professor Herschel's seismoscope appears to be in form similar to the parallel motion devised by Watt, in which the weight is only able to move in an almost horizontal plane. It is possible that the numerous joints would give rise to considerable frictional resistance, which may consitute a serious form of defect in the practical working of the instrument. Any degree of stability can be given to the instrument by placing the centre of gravity of the weight or ring N O P Q more or less below the point of suspension M (Plate XXI.) The rotation of the weight round a vertical axis is prevented by the pairing of the links into V form, as shown in Fig. 4. The index seemed to be somewhat complex, and might prevent the use of the instrument by unskilled persons.

The CHAIRMAN—Do you know if Prof. Herschel has made an instrument of the kind described ?

Mr. WALTON BROWN—No, he has not made one.

Mr. W. H. WOOD—Will the seismoscope not be subject to be acted on by movements in the building in which it is placed ?

Mr. WALTON BROWN—It is fixed to the top of a post placed in the ground, and independent of any building.

Professor LEBOUR asked Professor Stroud whether this instrument was simple enough to be used by ordinary observers, who were not specially skilled ?

Professor STROUD said the instrument was simple enough. Professor Herschel had introduced a number of springs, but they were really not essential to the main portion of the instrument. It was rather difficult, reading this paper, to be able to distinguish exactly what were the novelties which Professor Herschel had introduced. The chief complications were in connection with the indicator, and particular attention had been paid to keep the end of the indicator always pointed to a fixed point. The hair spring beneath, and the spring above, were with a view to keeping the end of the indicator very exactly in the centre of the little dimple of mercury, so that in any heeling of the whole instrument the indicator should not get out of gear. In fact, Professor Herschel tried by this arrangement to keep the end of the indicator always pointing to a certain definite position by means of the hair springs. He (Professor Stroud) did not think anyone looking at the diagrams would think the instrument very complicated ; but as there were many alternative plans mentioned in the paper, they might fancy that the instrument was more complicated than it really was.

The CHAIRMAN proposed a vote of thanks to Professor Herschel for his paper, which was agreed to.

———

The SECRETARY said, Mr. Urquhart had brought a new electric miner's lamp for the inspection of members, which had not been put in the list of papers for this meeting; but knowing that gentlemen engaged in mining were always very anxious to see improvements in lamps, he had invited that gentleman to be present. The lamp was the new "Sun Light," which had been before the Royal Society in London, and was very much extolled there.

Mr. URQUHART exhibited the lamp, and said, the lamp had been brought out by a small syndicate of scientific gentlemen in London, amongst whom was Mr. Maskelyne, the well-known chemist. The lamp was a new form of secondary battery. It was new in so far as the plates which produced the current consisted entirely of the material which was required for it, and were not burdened and weighted, as in other secondary batteries, with a large amount of lead, which made the batteries very heavy, and was such a waste of material as to bring about fifty hours' light down to about five in an ordinary secondary battery; all the rest was simply waste material. In this lamp, by a new process, all the extra labour was done away with, and that was the reason why it was so much lighter. There was no lamp before them able to approach this one yet in lightness. Of course the lamp had to be made as strong as possible, and some had to have a heavy ring; but the battery itself was remarkably light. The lamp weighed 4 lbs.; sometimes a little over and sometimes a little less, according to the material used for the metal work. The lamp (exhibited) had a reflector on the top; but there was a lamp which had no reflector—with the glass placed on the top so as to give a general all-round light, which was not so powerful in one direction, but which, for some purposes in a colliery, might be found more useful than a lamp with a reflector. Another point which the designers of this lamp had attempted to meet was the risk which electric lamps ran of causing an explosion. This risk was very slight, because the glass secured the filament in such a way as not to break the filament itself. If it happened that the carbon was broken, at the moment it broke a small arc would be produced between the two ends, and a very high temperature would be produced; and under these circumstances it was just barely possible to ignite an explosive atmosphere. This lamp had a particular arrangement to protect it against this danger. The outside cover of the light was made of toughened glass, so that if this outside glass was cracked it would break, and the instant the glass was removed there was an inside spring released which put the light out immediately. It was quite impossible to break the outside glass and the inside glass without the filament going out before the air reached the filament. This was tested by Mr. Rhodes with explosive mixtures, and was always found to answer, so that he expressed himself satisfied with it. A very important question with reference to anything new in the way of a lamp was what would be the cost. An electric light lamp, giving so much more light than any existing kind of oil lamp, and at the same time giving certainly greater safety, could not be supplied at present at anything like the price of an ordinary safety-lamp even of the better class. Still it was thought that at the price of one guinea this lamp would be found to be a cheap instrument—as saving in time, in working, and facilitating work generally. He thought that such an instrument would in the end pay. The cost of maintenance was remarkably low. The horse-power required to charge one of these lamps was very small indeed, and it took only $3\frac{1}{2}$ to 4 hours to charge the lamp; other secondary lamps took 8 to 12 and even 14 hours to charge. That was a good point in favour of this lamp. They had reckoned up carefully that, including the expenditure and power, and attendance for a fair average number of lights, 3d. per week should keep each lamp going. As to the question of the special shape of the lamp, they would understand that this was really a detail. Different shapes might be preferred at different collieries, and probably some collieries would have various shapes, according to the class of workmen in whose hands the lamps would be placed. A putter or a trimmer would probably prefer a lamp to hook on to tubs, whereas a man working in the face would prefer a lamp showing a good light on the exact spot where he was working. The lamp with a reflector (such as he exhibited) gave a very good light, and could be turned on to any particular point.

The CHAIRMAN—What is the illuminating power of this light?

Mr. URQUHART—The light itself, without any artificial reflector, is $1\frac{1}{4}$ to $1\frac{1}{2}$ candles. Lamps vary a little; it is difficult to get in a small lamp exactly the same amount of illuminating power. With a reflector, taking a general measure of the light all-round, the illuminating power is $2\frac{1}{2}$ candles.

Mr. J. B. ATKINSON—How long will this light burn?

Mr. URQUHART—Nominally for 10 hours, but really 12 hours. For 10 hours it gives the light of one candle; and then it gradually dies away.

Mr. BLACKETT—Where a large number of lamps are used, say 100 or 200, a long time would be required to charge the lamps alone, unless a dynamo of a very large power was employed.

Mr. URQUHART—A dynamo of, say, 100 volts might be used, with a large number of strings, each with 12 to 20 lights across it. Each string of lights would occupy the same position on the wires that one small lamp does in ordinary electric lighting.

Mr. J. B. ATKINSON—Has any comparison been made between the cost of maintaining this lamp and the ordinary miner's lamp?

Mr. URQUHART—The amount of labour with this electric lamp is less. In reply to the question as to the number of horse-power 100 lamps would require in charging—$1\frac{1}{4}$ to $1\frac{1}{3}$ horse-power would be sufficient. Theoretically it should take one, but allowing for the waste he would say $1\frac{1}{4}$ for 4 hours.

Mr. W. H. WOOD—Would a stone falling upon the lamp break the glass?

Mr. URQUHART—If a stone broke the glass it would put the light out.

Mr. WOOD—If a stone hit the spring it would put the light out.

Mr. URQUHART—It is not the blow that does it. There is a spring kept in by the glass. The instant the glass is broken the spring comes out, and puts out the light.

A MEMBER—What would be the result if the lamp was knocked over when it was lighted?

Mr. URQUHART—A lamp could not be expected to stand upside down for 12 hours and be all right. The lamp is not fragile; it will bear knocking about; but it could not be expected to burn properly upside down. An ordinary safety-lamp goes out if it is jerked; but a jerk does not send the light in this lamp out.

Mr. J. B. ATKINSON—Is there any liquid in this lamp?

Mr. URQUHART—Yes, dilute sulphuric acid.

A MEMBER—Does the 3d. per week include the cost of keeping up the lamp?

Mr. URQUHART—Yes, it includes the renewal of the filament lamp, the acid and the plates—which last 6 months—the dynamo expenses, and wages of the attendant. There is nothing reckoned in this for interest on capital. The real novelty is in the material of which the accumulator is made.

Mr. W. COCHRANE proposed a vote of thanks to the gentleman for exhibiting the lamp, which, he said, seemed to be very effective. Whether its duration would be equal to what had been stated had yet to be proved.   In regard to portability the lamp seemed to be very satisfactory.

The vote of thanks was agreed to.

———

The following papers were taken as read, the writers being absent :—
"Bornét's Hand-Boring Machine," by E. L. Dumas ; " Ackroyd's and Best's Miner's Safety-Lamp Cleaning Machine," by William Ackroyd.

BORNÉT'S HAND-BORING MACHINE.

By E. L. DUMAS.

THE object of the writer has been to obtain a simple, strong, and cheap drill which can also be used as a substitute for the pick, or for percussive machines in mining operations.

In the greater number of cases, machines of this description, worked by hand, have been made with a view to reduce the work of drilling to a minimum, and to augment the useful effect of hand labour even when drilling hard rock.

They have been constructed of various dimensions and types according to the nature of the rocks to be drilled, but they all partake of the same common principle which characterises the system.

The following is a description of a very complete machine which, by the great power it exerts, can be employed in drilling very hard rocks, and will compete successfully with percussive drills ; each portion of the drill, with the particular object of the same, is here described :—

The machine is represented in Fig. 1, Plate XXVIII. It is composed of a cast-iron tube a, which is fixed to a supporting frame by means of the trunion b, situated somewhere about the centre of gravity of the machine, and which permits it easily to follow any deviation from the straight line the drill may make during its work.

This arrangement prevents the tool from fixing and breaking when it is penetrating substances of unequal hardness, and at the same time materially reduces the power necessary for working the machine.

Inside the cast iron tube a is placed another hollow tube c, carrying the drill, which is made to turn by means of the bevelled wheels d and the handle e. The driving motion is carried by an arm f attached to a collar which can turn freely round the tube a and which can be fixed by means of a set screw g. This method of attaching the motive power enables it to be shifted to any position which will best suit the conditions of the place in which it has to work.

The feed motion is communicated to the drill-carrier c by means of the nut h adapted to a central screw i which is relatively fixed. The collar j presses against a number of springs placed in pairs k to give a certain elasticity to the pressure of the drill upon the rock. This interposition of springs is very important, and has the result not only of preserving the drill but also of rendering its pressure proportionate to the variable hardness of the different kinds of rock which it has to pierce.

The hinder part of the screw i, which traverses the end of the apparatus, receives the lever l by the aid of which the workman can turn it either way according as he wishes to accelerate or diminish the progress of the drill. The friction of the drill against the end of the tube a is relieved by means of a number of small steel balls m, and a bolt n keeps the lever l stationary whilst the machine is in its normal condition of work.

This system, therefore, permits the workman to regulate at will, by the lever l, either continuously or from time to time, the advancement of the tool according to the hardness of the rocks which he encounters. The action of the springs k also permits the screw i to spring back slightly in hard strata, so that the lever l when it springs back escapes automatically from the restraint of the bolt n. As soon as the pressure upon the drill becomes too great, it can be easily understood that the instant the screw i ceases to be kept in its place, it is turned round by the friction exercised on the nut h of the tool-carrier, and turns with this latter without making the drill advance until the elasticity of the springs k have caused the drill to penetrate the rock sufficiently to allow of the progressive movement being continued without inconvenience.

The drill, Fig. 3, which is fitted to the extremity of the carrier c, has a characteristic form which is of great importance. It will be seen by the drawing that it is formed of a blade of steel o, bent in the form of a screw, the section of which is in the form of a lozenge, the cutting head p is of a diameter slightly in excess of the body of the drill. This special disposition of the drill has the following advantages :—

1.—The swelled portion of the blade o in the middle imparts great strength to the tool.

2.—Its cutting edges prevent the small fragments cut from the rock from jamming the tool.

3.—The increased size of the head of the drill permits it to turn with greater liberty during the whole of its course.

These three conditions together have reduced to a minimum the power necessary to work the machine.

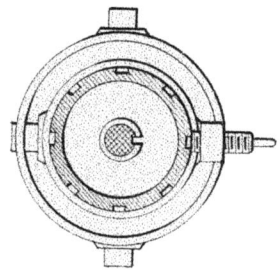

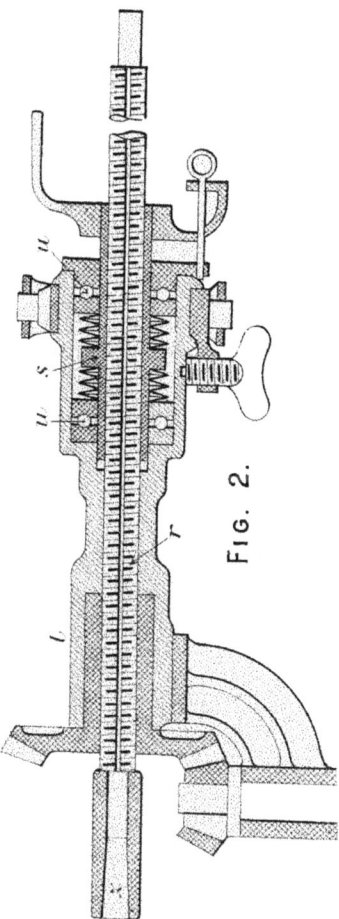

FIG. 2.

imensions of
for working
ill is applied
ve in which
rate of feed,
s, while *u u*
se shown in

the pressure

g motion of
in the same

means of a
in the most

sal joint *y*,
e drill may

ed round to
of bringing
it has been
a double set
e feed screw
rill.

in a large
r cross-cuts
rboniferous
er cent. less
e workmen
n in hewing
it requires.
l and shale
's tool, on
ch it can be
coal to be
mechanical

The feed motion is communicated to the drill-carrier c by means of the nut h adapted to a central screw i which is relatively fixed. The collar j presses against a number of springs placed in pairs k to give a certain elasticity to the pressure of the drill upon the rock. This interposition of springs is very important, and has the result not only of preserving the drill but also of rendering its pressure proportionate to the variable hardness of the different kinds of rock which it has to pierce.

The hinder part of the screw i, which traverses the end of the apparatus, receives the lever l by the aid of which the workman can turn it either way according as he wishes to accelerate or diminish the progress of the drill. The friction of the drill against the end of the tube a is relieved by means of a number of small steel balls m, and a bolt n keeps the lever l stationary whilst the machine is in its normal condition of work.

This system, therefore, permits the workman to regulate at will, by the lever l, either continuously or from time to time, the advancement of the tool according to the hardness of the rocks which he encounters. The action of the springs k also permits the screw i to spring back slightly in hard strata, so that the lever l when it springs back escapes automatically from the restraint of the bolt n. As soon as the pressure upon the drill becomes too great, it can be easily understood that the instant the screw i ceases to be kept in its place, it is turned round by the friction exercised on the nut h of the tool-carrier, and turns with this latter without making the drill advance until the elasticity of the springs k have caused the drill to penetrate the rock sufficiently to allow of the progressive movement being continued without inconvenience.

The drill, Fig. 3, which is fitted to the extremity of the carrier c, has a characteristic form which is of great importance. It will be seen by the drawing that it is formed of a blade of steel o, bent in the form of a screw, the section of which is in the form of a lozenge, the cutting head p is of a diameter slightly in excess of the body of the drill. This special disposition of the drill has the following advantages :—

1.—The swelled portion of the blade o in the middle imparts great strength to the tool.

2.—Its cutting edges prevent the small fragments cut from the rock from jamming the tool.

3.—The increased size of the head of the drill permits it to turn with greater liberty during the whole of its course.

These three conditions together have reduced to a minimum the power necessary to work the machine.

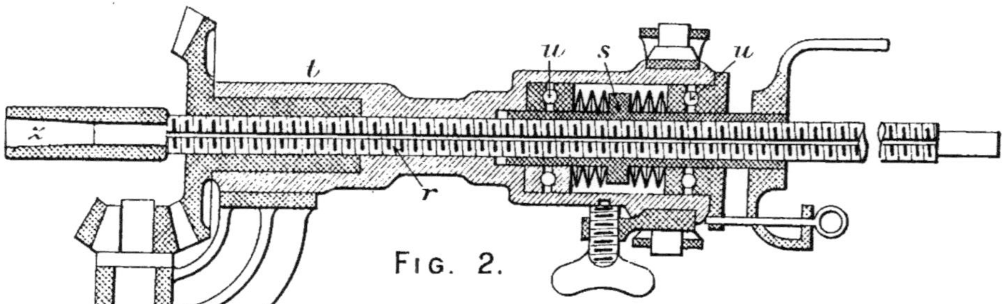

FIG. 2.

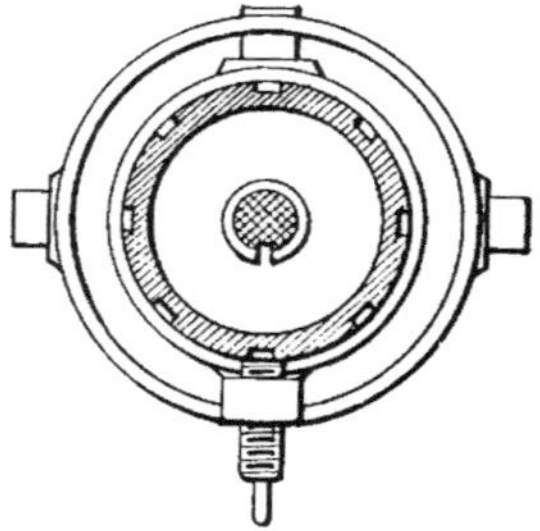

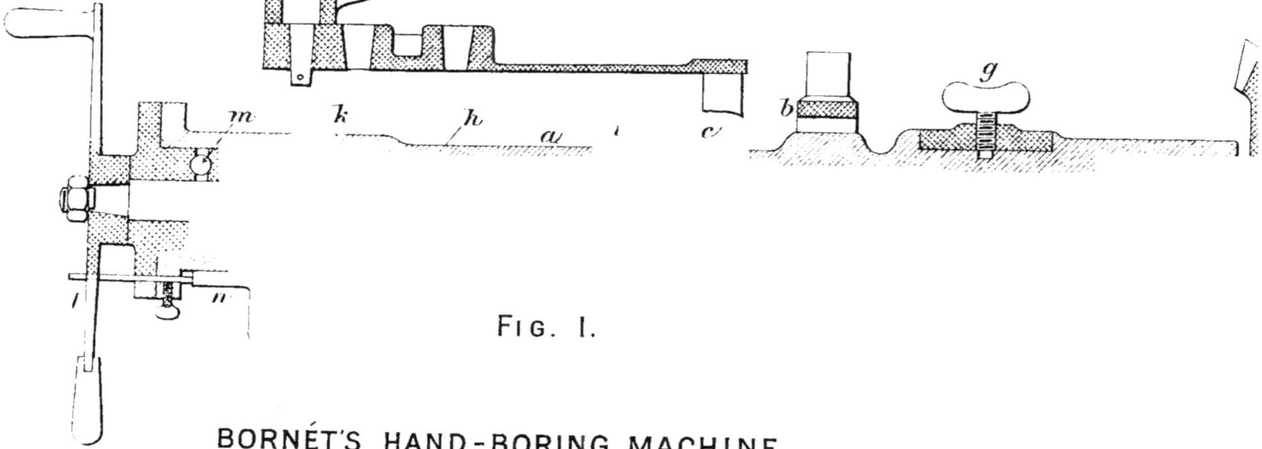

FIG. 1.

BORNÉT'S HAND-BORING MACHINE.

FIG. 3.

p

Fig. 2 represents a very simple form of machine, the dimensions of which are specially reduced so as to make them applicable for working in tender strata, such as coal. In this type of machine the drill is applied direct to the central screw r, which carries a longitudinal groove in which slides a driving key, and the nut s, which determines the rate of feed, is relatively fixed to the inside of the body t of the apparatus, while u u are collars and slut balls answering the same purpose as those shown in Fig. 1, only in duplicate to take the pressure each way.

This type of machine has, like the preceding one—

1.—The employment of springs u to give an elasticity to the pressure of the drill.

2.—The employment of a screw lever v, and the rotating motion of the drill is carried by means of small steel balls in the same manner as the one just described.

3.—The machine can turn round in any direction by means of a suspending collar, so that it can be placed at will in the most favourable position for work.

4.—The mode of suspension, by means of the universal joint y, permits the machine to follow any deviation the drill may make during its working.

This arrangement of the machine also allows it to be turned round to work in either direction, which does away with the necessity of bringing back the feed screw after each operation. To do this, it has been necessary to arrange the nut so that it can be acted upon by a double set of springs and friction balls and to make both ends of the feed screw square, so as to carry the socket z at each end to receive the drill.

The drill and the coal-getter are constantly used in a large number of collieries and quarries. The drill is employed for cross-cuts and exploring galleries driven in such rocks or shale and Carboniferous sandstone. Its economy may be stated as from 20 to 30 per cent. less than hand-drilling, and it is important to remark that the workmen themselves are anxious to use the apparatus, even at a reduction in hewing price. This proves its economy and the small amount of work it requires.

The coal-worker is arranged especially for working in coal and shale of average hardness. It may be specially called the miner's tool, on account of its lightness, its power, and the facility with which it can be worked. It makes perfectly cylindrical holes, which enables coal to be easily acted on, either by gunpowder or by the employment of mechanical wedges, where it is necessary to avoid danger from explosion.

NOTES AS TO THE WORKING OF THESE MACHINES.

DRILL No. 1.

Weight of machine	101·2 lbs.
Weight of a support suitable to seams of 6½ feet	105·6 „
Greatest pressure exerted by the tool	2,200 „
Average speed of drill per minute	2 inches.
Force necessary to work it	$\frac{1}{10}$ horse-power.
Price of the machine with its support and all accessories ...	£22.
Price of the steel blade section, as before described, cutting boring surface, in the form of a heart, from 1 to 2⅓ inches in diameter	1s. 2d. per lb.

COAL-GETTER No. 2.

Weight of machine	35·2 lbs.
„ „ support	48·4 „
Greatest pressure exerted by the drill	1,650 „
Average speed of drill while being driven by the screw direct	7 to 8¼ inches per minute.
Average speed of drill when driven by means of gear ...	3½ to 4¼ „ „
Power necessary to work it	$\frac{1}{15}$ horse-power.
Price of the machine with its support	£8 10s.

ACKROYD'S AND BEST'S MINER'S SAFETY-LAMP CLEANING MACHINE.

PATENT 3089—1879.

By WILLIAM ACKROYD.

THIS machine has been invented for the purpose of cleaning safety-lamps and their gauzes by machinery, and superseding the usually tedious process of cleaning them by hand.

Plate XXIX., Figs. 1 to 5, shows the arrangement of the machine.

Fig. 1 is a front elevation; Fig. 2, a side elevation; Fig. 3, a cross section through A, B, Fig. 1; Fig. 4, a plan looking at the top; Fig. 5, a modification in the construction of the machine.

a is the framework, by preference of metal, fixed on suitable bed or table *b*; on this framework *a* is mounted in a suitable adjustable bearing *c* a shaft *d*, carrying on it a large circular brush *e* and smaller circular brush *f*, the larger one *e* being for the purpose of cleaning the exterior of lamp gauzes, and the smaller one *f* for removing foreign matter from the interior thereof. For the purpose of cleaning the exterior any required number of spindles *g* are provided round the brush *e*; on these the gauzes are placed and retained in contact with the brush *e*. A rotary motion is imparted to the larger brush *e*, and simultaneously to the spindles *g*, in order to bring the whole external surface of the gauze in contact with the brush *e*. In order to cleanse the interior of the gauze the smaller brush *f* is introduced into the interior, the gauze being held in the hand of the attendant whilst the small brush *f* is rotating in the interior. In order to assist in cleansing, a horizontal reciprocating motion is imparted to the spindle *d*, producing thereby two motions of the brushes *e* and *f* in different directions simultaneously. The reciprocating motion is obtained in one direction by the rotary cam *h*, which is secured to the spindle *d*, being brought in contact with the fixed cam *i* on the framework *a* during its rotary motion, and in the other direction by the spring *k*; or the arrangement shown at

Fig. 5 or its equivalent may be employed, in which 1 is the spindle, 2 the crank or eccentric, from which motion is transmitted thereto through connecting rod 3 and projecting collar 4. The required rotary motion is given to the crank or eccentric 2 through cone pulley 5, bevel gearing 6, and shafts 7 and 8.

On the framework a are other spindles l and m, by preference in adjustable bearings, on which are fixed discs n, having on their faces brushes o of suitable form; these are caused to revolve at any convenient speed, and are used to clean and polish the body and other parts of the lamp; these brushes o may be used either wet or dry, in the former state for cleaning the dirt off and in the latter for polishing.

On the ends of the spindles l and m conical forms p are provided for screwing on and off the metal ring known as the "glass ring."

In combination with this machine stationary projecting pieces q are employed, whereby the attendant is enabled to slacken the metal ring known as the "glass ring" or tighten it in its position.

Motion is transmitted to the various spindles and brushes through pulleys r from any suitable or convenient motive power.

Every mining engineer knows how destructive to the gauzes the present mode of lamp cleaning is; of course when the miners take them home to clean this destruction is intensified, but even when done in the best regulated lamp cabin it is very great. There is no doubt also that by keeping the brass work clean and bright more light from each description of lamp is obtained. It is claimed that by the use of this machine not only can the lamps be more rapidly taken to pieces and put together, but that the gauzes can be better and more carefully cleaned while the brass work is being polished, and the whole process of cleansing and preparing the lamps for a large colliery be carried out by much fewer hands.

————

The discussion on Mr. M. Walton Brown's paper on "A further attempt for the Correlation of the Coal-Seams of the Carboniferous Formation of the North of England; with some Notes on the probable duration of the Coal-Field," was taken :—

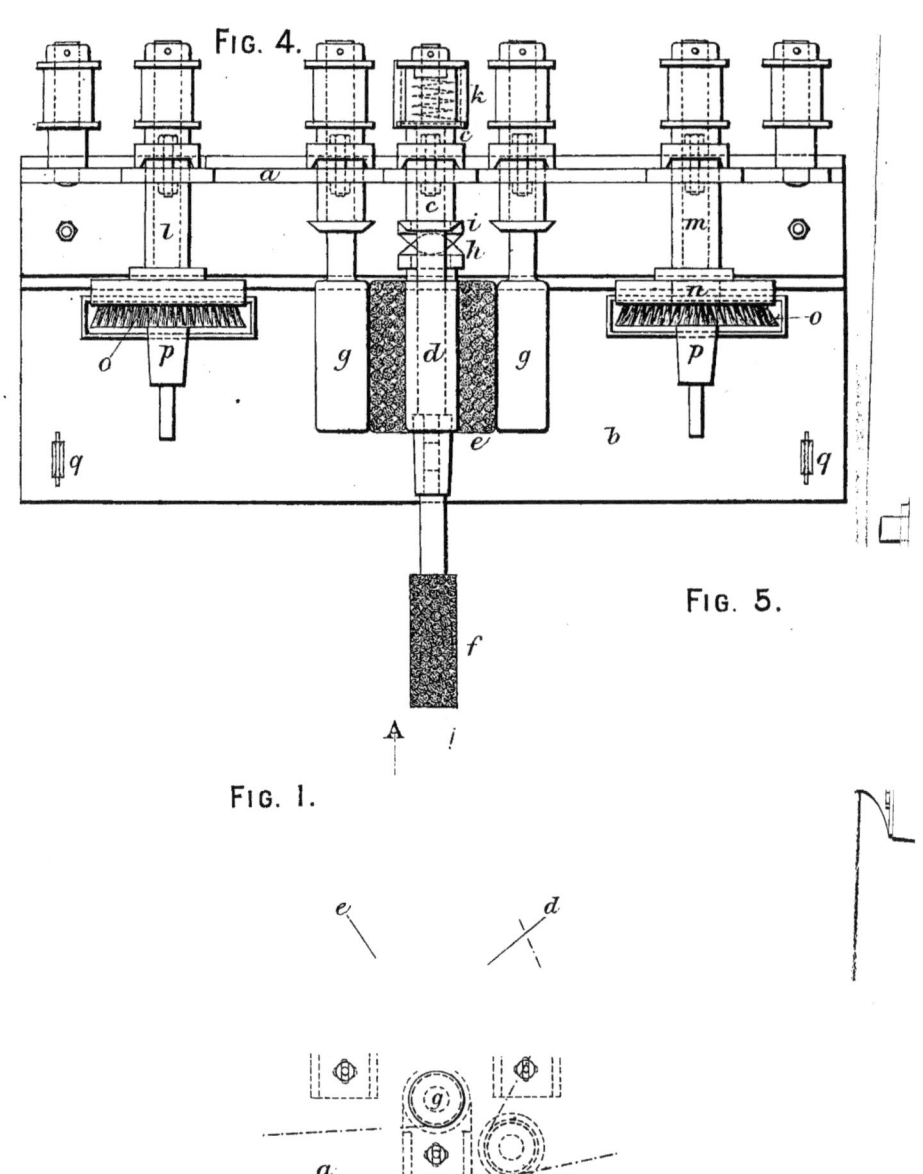

FIG. 4.

FIG. 5.

FIG. 1.

iners Safety Lamp Cleaning Machine."

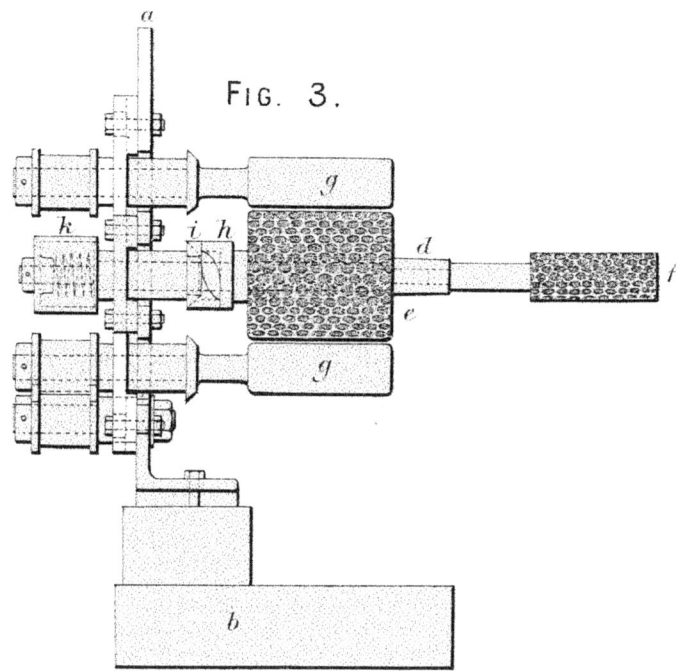

FIG. 3.

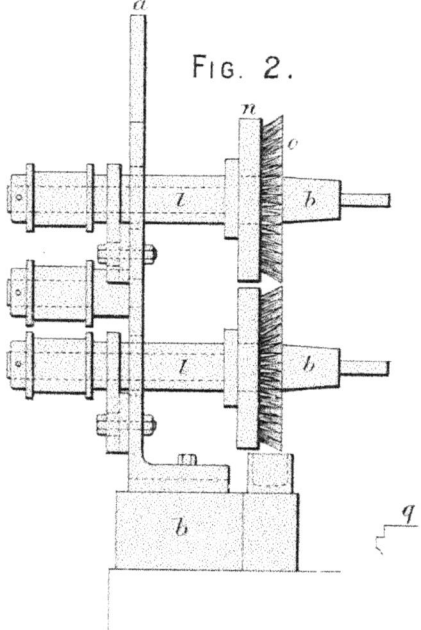

FIG. 2.

DISCUSSION ON MR. WALTON BROWN'S PAPER ON A FURTHER ATTEMPT FOR THE CORRELATION OF THE COAL-SEAMS OF THE CARBONIFEROUS FORMATION OF THE NORTH OF ENGLAND ; WITH SOME NOTES ON THE PROBABLE DURATION OF THE COAL-FIELD.

The SECRETARY read the following letter from Mr. Gresley:—

OVERSEAL, ASHBY-DE-LA-ZOUCH,
1st February, 1888.

THEO. W. BUNNING, ESQ., Secretary.

DEAR SIR,—With reference to Mr. M. Walton Brown's paper on coal-seams, I should like to contribute a few words in the discussion when it comes on; and as I cannot get to the meeting perhaps you would have the kindness to read my remarks which I make as follows :—

When trying to puzzle out some problems concerning the probable mode of formation of coal-seams a few years ago, it struck me that if a growth *in situ* theory be accepted (though I myself am no upholder of the idea that coal-beds are the remains of forests of *trees* which grew on the spot) at no time during the continuance of the deposition of the Coal-Measures was the living vegetation of which coal is composed wholly placed under water, or otherwise killed off from every part of the earth's surface undergoing the accumulation of such deposits, because such a destruction of plant-life would I imagine have necessitated the creation of a new set of plants for each seam of coal, which seems altogether irrational to suppose for a moment. And it therefore seems to me to follow that each bed of coal, be it thick or thin, where sunk or passed through at any particular point represents only an *offshoot*, so to speak, or one particular period of time (a pause in the deposition of sediment over that area) out of the vast ages during which I conceive the coal-plants had a continuous existence either over one area or another. Now, Mr. Brown's theory for coal-seams would seem to imply what I have just said, and I am pleased that he has brought out the idea before this Institution. In attempting correlations of coal-seams we must, of course, keep in view the idea that there can be very little doubt (if any at all) that our various coal-fields are nothing but the *remnants*, so to say, of the original enormous coal-field which probably not only overspread by far the greater part of the area now occupied by the British Isles, but also large areas of the Continent (Germany, Belgium, France, etc.) and portions of what are now seas and oceans ; so that we shall never really be able to set this interesting question at rest by *proof*.   The most remarkable instance I know of of the splitting-up of a coal-seam into numerous smaller ones, is in the South Staffordshire district where the "ten yard" or "thick" coal near Dudley when followed in a northerly direction becomes divided up into about a dozen distinct beds interstratified with some 400 feet of shales, etc.   A somewhat similar splitting-up of this "thick" coal to the south, west, and east of Dudley occurs.   Also in the Warwickshire coal-field we find the coal-seams all close together in the south near Coventry, but widely separated as followed towards the northern end of the district. Can Mr. Brown inform us whether any attempt at a correlation of the coal-seams of county Durham with those of Yorkshire has yet been made, or of those of the Cumberland with the North of England ?   It would be also interesting to outsiders like myself to learn from Mr. Brown where and in which coals the *anthracite* occurs, or has been met with, in the northern coal-field.   Does it occur in lenticular patches, *i.e.*, over very limited areas in particular bands of particular seams, or is the full thickness of the coal converted into anthracite ?   Does it seem to occur only near to *faults* or to *whin dykes ?*   Is *cone-in-cone* coal (crystallised coal, as it is sometimes termed in South Wales) met with in the neighbourhood of Newcastle or in county Durham and associated with the anthracite ?

I am, Sir,

Faithfully yours,

W. S. GRESLEY, F.G.S.

Mr. M. WALTON BROWN—There would be great difficulty in carrying out Mr. Gresley's suggestion regarding the correlation of the seams within this district with those in Yorkshire and Cumberland.

The CHAIRMAN—There is no seam of anthracite in this district.

Mr. J. B. ATKINSON—There is a little seam of coal worked in the neighbourhood of Alston of anthracitic nature.

Mr. MARLEY said, he was sorry he was not present at the meeting when Mr. Brown read his paper, and he would take this, the first opportunity, of calling attention to what he thought was an unfortunate omission in connection with the correlation of these coal-seams, and that was the omission of all reference to the work of their ex-President, Mr. G. C. Greenwell. Mr. Greenwell followed Mr. Buddle with a synopsis in 1855. Mr. Brown has only referred to Mr. Buddle's synopsis of 1830 in the Transactions of the Institute, and which he revised for the "Wise Week" in Newcastle in 1838. But afterwards Mr. Greenwell prepared a synopsis, and he (Mr. Marley) called attention to it now, as it was their duty to put this on record as well; and then the work of the late Mr. N. Wood, the late Mr. J. Taylor, with himself (Mr. Marley) came third. In connection with a geological division which Mr. Brown had made, he joined issue seriously with Mr. Brown in introducing the names of the Gannister beds and Millstone Grit series. In his paper, Mr. Brown considered or took the Brockwell coal-seam as the line of demarcation, and as the bottom of the whole of our coal formation proper. He did not know where Mr. Brown had got his reason for making such an assertion, because, close to the Gannister beds, the sandstones and other strata were entirely in harmony with the rest of the coal formation; and he differed from Mr. Brown very materially on the point. He hoped, before the paper was printed and published, that Mr. Brown would add some addenda, because the Victoria and the Marshall Green, in the Auckland West district, lie far above the Millstone Grit strata. He (Mr. Marley) considered these seams, and a third lower seam, should be coupled in and form part of the coal formation proper. The synopsis publication of the Chairman was referred to by Mr. Brown, and he (Mr. Marley) was glad that the Chairman followed the original line, and carried the coal formation proper down to the Millstone Grit. This was an important question, and should be put right. If Mr. Brown would refer to the 133rd sheet of the Geological Map, he would find that it supported him (Mr. Marley) in these views.

Mr. W. H. HEDLEY said, on looking over Mr. Brown's synopsis of the coal-seams yesterday evening, he found that he could not altogether agree with some portions. With regard to the synopsis of the coal-seams in Table I., that, as to the county of Durham, appeared to him to be in the main correct, so far as the correlation of the seams, apart from their nomenclature, was concerned. But for the names of some of the seams, those in the Consett district in particular, Mr. Brown had apparently gone back to ancient history; and, it struck him that in a new synopsis, which might be regarded as in a measure corrective of those that had gone before, it might have been better to abandon the older names in favour of those now generally recognised. Thus, in the Consett district, the seam termed the Pasture Drift, had been worked and known for the past twenty years as the Three-Quarter Seam ; that termed the No. 1 had been known for an equally long time as the Townley ; the Stone Coal as the Tilley; the Busty Bank as the Busty ; the Five-Quarter or Splint Coal as the Five-Quarter. All these more modern names, as he might term them, were used in the table prepared for the Royal Commission on Coal eighteen years ago. Again, at Garesfield and Prudhoe, the seam termed the Barlow Fell had been for long known and spoken of only as the Townley ; whilst, although at Prudhoe the names Five-Quarter, Six-Quarter, and Yard were those recognised as applying to the seams as placed in the table, the same seams at Garesfield were now respectively known only as the Stone Coal, Five-Quarter, and Three-Quarter. In the Cockfield district also, nobody for the last thirty years had, he should imagine, thought of speaking of the Five-Quarter Seam there as the Crow Coal; and it was news to him that the thin coal, occurring between the Five-Quarter and Main Coal Seams, had anywhere in that neighbourhood been deemed of sufficient importance to be called by a special name, although it would, of course, occupy a corresponding position with the Three-Quarter Seam of Consett, Garesfield and Ryton.

Mr. T. E. Forster asked Mr. Brown where he got the name "Glebe" in the Blyth district ? It was always known as the High Main Seam.

The Chairman—Is there not a seam there called the Moorland Seam?

Mr. T. E. Forster—That is another seam.

The Chairman—Is it not the High Main Seam?

Mr. T. E. Forster thought not.

The Chairman—The Glebe and the Moorland are the High Main.

Mr. Walton Brown—They are two seams, the Black Close or Moorland, corresponding to the Three-Quarter Seam ; and the Glebe, to the High Main Seam.

Mr. T. E. Forster—The High Main is the Glebe in the neighbourhood of Bedlington.

Mr. Simpson, jun., said that in the small section which Mr. Brown gave from Murton to Pontop he showed the Low Main and the Maudlin, at Ryhope, to be separate. This was exactly opposite to the recognised position. The Low Main Seam at Murton was some ten fathoms below the Maudlin, it rose to the level of the Maudlin Seam north of Seaham shaft, and continued close on the Maudlin over the Ryhope district.

Mr. White—At Murton Colliery the Low Main Seam is within a fathom of the Bensham, about half a mile from the shaft, and they have been known to have almost come together, although at the shaft they are about 15 fathoms apart.

Mr. MARLEY said the remarks just made reminded him that there was a great desideratum in connection with Mr. Brown's paper. Mr. Brown had not attempted to give the thickness of each seam, or the distance between each seam, and therefore it was difficult to compare correctly and trace how far each correlation was correct. For instance, Mr. Brown spoke rather indefinitely as to the seam of coal found 12 fathoms below the Brockwell Seam, and of a second seam 15 fathoms further down, making 27 fathoms. This was the only instance he gave of the depth or thickness. But actually at Marshall Green, the first seam, the Hargill Hill, or what was called the Victoria Seam, was found at $10\frac{3}{4}$ fathoms ; the next seam below was about $8\frac{1}{4}$, and the next seam worked again below that at Marshall Green, was about $8\frac{1}{4}$ fathoms, making $27\frac{1}{4}$ fathoms. There were three seams there which had been worked and coal band 8 fathoms lower, making $35\frac{1}{4}$ fathoms of Coal-Measures below the Brockwell Seam. But without the depth being given it was difficult to make any comparison and trace them. Perhaps it was now too late to do it; but it would be of great additional value to the synopsis if Mr. Brown could give such information. He was perfectly well aware that what he asked for involved a great deal of labour and trouble being spent.

Professor LEBOUR said he would like to say one or two words as to what Mr. Marley had said respecting the Gannister beds. He agreed with Mr. Marley that below the Gannisters were measures just like the Coal-Measures above the Brockwell.

Mr. MARLEY—The Gannister beds are much below coal beds at Marshall Green.

Professor LEBOUR—Yes ; they occupied a position between the Mill-stone Grit, and that part of the Upper Carboniferous series where the good, thick, workable seams were found. In this district it had been usual to use the Brockwell Seam as the lowest of the so-called middle Coal-Measures. He did not attach importance to this. It was all Coal-Measures from the highest seam known to the top of the Millstone Grit. The name Gannister was not a very good one ; because there were not only Gannister beds there, but also in other parts of the Carboniferous series. He had used "Gannister beds" himself, for this reason, that he thought the alternative name, "Lower Coal-Measures," was an exceedingly deceptive one. In Scotland they had the habit of giving the name "Lower Coal-Measures" altogether to the limestone coal series there. When a Scotchman referred to the "Lower Measures," he meant beds different from those which Englishmen called "Lower Coal-Measures." This was one reason why, in this district, we did not like to use the term "Lower Coal-Measures," because, being so near to Scotland, we were sure to confuse ourselves when dealing with the Scotch sub-divisions. He agreed with Mr. Marley as to the position of the Gannister beds, and, in fact, that the Brockwell Seam was a mere arbitrary line of division, of no value at all except for convenience. As a matter of convenience it was a good one, for it was advisable to have a distinct datum to start from.

Mr. MARLEY said, that in the geological maps they called the seams below the Brockwell Seam the "Lower Coal-Measures," and then began with the Millstone Grit. In this district his friend (Mr. Brown) said it was preferable to call all coals above the Brockwell the "true Coal-Measures;" whereas he (Mr. Marley) contended that the Marshall Green and Victoria seams belonged to the "true Coal-Measures," and that it was unfortunate to introduce the word "Gannister."

Professor LEBOUR said, there was no doubt Mr. Marley was right. This was the first time he had seen the word "Gannister beds" on the geological maps as in the "Lower Coal-Measures," instead of synonymous with them.

The CHAIRMAN—There was a deep boring put down at Witton, and they reached the Millstone Grit at a considerable depth.

Mr. MARLEY—At a depth of about 80 fathoms below the Brockwell Seam.

The CHAIRMAN—In this neighbourhood there is a deep boring below the Brockwell Seam of 50 fathoms, and the Millstone Grit was not reached at that depth.

Mr. MARLEY moved that the discussion should be adjourned.

The CHAIRMAN said, that at the last meeting he stated that he would offer some remarks upon Mr. Brown's paper when it came up for discussion ; but, unfortunately, he had not noticed that the paper was to be brought up for discussion to-day, and, therefore, he had not come prepared to give his remarks. He would like the discussion to be adjourned, and so give other members an opportunity of coming forward with other facts respecting this extremely interesting matter. He seconded Mr. Marley's motion for adjournment.

The motion for adjournment of the debate was agreed to.

The meeting concluded.

PROCEEDINGS.

GENERAL MEETING, APRIL 14th, 1888, IN THE WOOD MEMORIAL HALL, NEWCASTLE-UPON-TYNE.

Sir LOWTHIAN BELL, President, in the Chair.

The Secretary (Mr. Lebour) read the minutes of the previous General Meeting, and they were confirmed.

The Secretary reported the proceedings of the Council.

The following gentleman was elected, having been previously nominated :—

Associate Member—

Mr. Lancelot Fletcher, Marsden Colliery, South Shields.

The following gentlemen were nominated for election :—

Associate Members—

Mr. William G. Wears, 28 and 29, St. Swithin's Lane, London, E.C.

Mr. William Rich, Minas de Rio Tinto, Provincia de Huelva, Spain.

PRESENTATION TO MR. T. W. BUNNING.

The President—Gentlemen, before we commence the usual business of the meeting, there is a duty which I have undertaken, and which I will endeavour to perform to the best of my ability. I am afraid the manner in which this duty will be discharged will not be equal to the occasion, because, as you will hear, I am suffering from a very bad cold, and not only is my voice weak, but I can only speak with considerable difficulty, and with some pain to myself. The duty to which I refer is that of assuring our friend, Mr. Bunning, of the entire satisfaction of the members with the way he has discharged his duties as Secretary of this Institute for the last twenty-one years. You are aware that some time ago Mr. Bunning, on account of his health, was obliged to tender his resignation. It is a matter of regret to us that Mr. Bunning has been compelled to resign his appointment, and we still more regret the cause; and I hope and trust—and you will all join with me in the hope—that with his retirement from active life he may find his health sufficiently restored to enable him to have many years to pass with advantage to his family, and to those who have the pleasure of his acquaintance. With respect to the manner in which Mr. Bunning has performed his duties, I would refer the members to the records of the Institute itself. As I have already stated, Mr. Bunning has been in our service for the last one-and-twenty years. He succeeded a gentleman well known in the North of England, namely, Mr. Doubleday. Mr. Doubleday was a man of high literary power, and very great reading; but, unfortunately, he had arrived at that condition which we shall all reach, I suppose, if we live long enough, when his health unfitted him to meet the constant claims upon his attention, and in consequence—I believe I am speaking what is strictly correct—the fortunes of the Mining Institute somewhat languished under the later period of his superintendence. Mr. Doubleday's resignation having been accepted by the Council of the day, Mr. Bunning was appointed in his room; and I believe any gentleman who will give himself the trouble to ascertain the change which took place in the position of the Mining Institute after Mr. Bunning assumed the direction of its fortunes, cannot fail to recognise a very material and very important alteration in the prosperity and general usefulness of the Institute itself. Under the circumstances I think the Council would have merited your displeasure if they had allowed Mr. Bunning to sever his connection with the Institute without giving him some formal assurance in recognition of the high appreciation in which his services have been held. We thought we were justified in voting him 100 guineas, and also in having prepared a memorial which Mr. Bunning can take away with him as a reminiscence of the value in which his services have been regarded by the Council and the members of the Institute generally. That we have acted rightly has, I believe, received ample confirmation in the conduct of the other bodies which are in the habit of meeting in this hall—I mean the different sections of the coal trade. Each of these bodies has, I understand, voted even a larger sum than we have done; and, of course, it is very gratifying to us that the example that we set to our more wealthy—I hope I may say more wealthy—colleagues, who are in the habit of assembling here, has been followed, and that the step we have taken has thus been confirmed. Perhaps, as I am addressing you on the subject of the secretarial work, I may be allowed to intimate to you that Mr. Lebour has been appointed Secretary in succession to Mr. Bunning. Mr. Lebour, as you all know, is a professor of a science closely allied to your own profession, namely, that of geology. It would be idle for me, especially in his presence, to dwell at any great length upon Mr. Lebour's merits. I believe I am speaking the strict truth when I say that his merits as a scientific geologist extend far beyond the confines of the city where, happily for the college, his fortunes are at present cast. Mr. Lebour will attend at the office from 11·30 a.m. to 1·30 p.m. regularly every day. You are aware that we consider that the office of Secretary of our Institute does not require that constant attendance which Mr. Bunning was in the habit of giving, owing to his having, as you all know, many other duties to perform in this building, and which will not have to be discharged by Mr. Lebour. Inasmuch, however, as members coming from the country might require to see the Secretary, we have thought it desirable that there shall, at all events, be two stated hours in which members can come here and count on finding Mr. Lebour. These hours have been fixed for the present at 11·30 a.m. to 1·30 p.m., but when we have had some experience of the system, it may be that it will be found necessary to have some modification introduced for the convenience of Mr. Lebour, or for the convenience of the members, or of both. Coming back to the more immediate business, I must say I cannot allow Mr. Bunning to retire from our service without tendering to him my own personal thanks, as President for nearly two years, for the unfailing assiduity and attention he has paid to my slightest wish in connection with my official duties. No trouble appeared too great or too unimportant to secure his attention; and I take this opportunity of tendering him my personal thanks for the very efficient way he has carried out the various instructions I may, on your behalf, have given him from time to time. I have much pleasure in placing in Mr. Bunning's hands this cheque for 100 guineas, and I will ask the Secretary, as my voice is gradually growing less audible, to read the memorial.

The SECRETARY (Mr. Lebour) read the following address, which is neatly engrossed, and surrounded by an ornamental border :—

"At a meeting of the Council of the North of England Institute of Mining and Mechanical Engineers, held on the 31st March, 1888, it was unanimously resolved to express to Theophilus Wood Bunning the deep regret of the members of the Institute that the state of his health has compelled him to retire from the office of Secretary, which he has held to their entire satisfaction for 21 years; and, as a mark of their respect, it was further resolved to place at his disposal the sum of 100 guineas, accompanied by the sincere wish that his health may be restored.—LOWTHIAN BELL, President."

Mr. T. W. Bunning—Gentlemen, I regret that my voice is much in the same state as that of our distinguished President. I can with very great difficulty speak, am almost totally deaf, and afraid that many of the very kind remarks, which no doubt have been made, have escaped my hearing. I think the President did say something about my having paid some attention to your wants and wishes; but how could I have done otherwise, seeing how kind you all have been to me? I am sure that at all times it has been a pleasure to me to work with you, and I only wish that my health would have spared me to work longer; but really, with the infirmities that now cling to me, I find it utterly impossible for me to go on satisfactorily performing those services which I know ought to be carried out by persons in the position in which you were kind enough to place me. I can only say that I thank you sincerely for your kindness, and assure you that I will ever have a pleasant recollection of the time spent in your service. I also thank you, Sir Lowthian, very much for your great personal kindness to me at all times.

THE LATE MR. T. E. HARRISON.

The President—There is one more matter of a still more painful kind to which I must call your attention—the death of a gentleman who for many years has been a member of this body. I allude to my friend of fifty years and more standing—Mr. Thomas Harrison, engineer to the North-Eastern Railway Co. It is usual, I believe, in our Transactions to record the death of any member of the Institute; and I think you will agree with me that in the death of Mr. Harrison not only we, but engineering science generally, and society at large in the North of England, as well as elsewhere, have suffered a very great loss in his death. I will not trespass at any length upon your time in enumerating the great services which Mr. Harrison rendered to railway engineering science during his long life; I say his long life, because if he had lived until the 4th of this month he would have completed his 80th year. The last time I saw him he told me he had summoned all his more immediate relatives to assemble at his home in order to celebrate the completion of his 80th year. Unfortunately, as you all know, the assembly of his friends took place under circumstances of a very different and of a most melancholy character. Before the 4th of April arrived Mr. Harrison was laid in his grave in the village where he had lived so many years of his life. What I invite you to do is to empower the Council to assure Mrs. Harrison, in a far more perfect way than I am at present capable of doing, of our feelings of sincere and profound regret at the death of one who was loved and respected not only by his neighbours in the North of England, but wherever the science of railway engineering was known. If you are good enough to confirm this appeal to you, the Council will take care that the sentiments of our body generally will be forwarded to Mrs. Harrison and her family.

The proposal of the President was agreed to.

Mr. W. J. BIRD read the following paper, by G. Meyer and W. J. Bird, "On the use of Iron Supports in the Main Roads of Mines instead of Masonry or Timbering":—

THE USE OF IRON SUPPORTS IN THE MAIN ROADS OF MINES INSTEAD OF MASONRY OR TIMBERING.

By G. MEYER and W. J. BIRD.

THE employment of iron supports in the main roads of coal and other mines has of late years greatly increased on the Continent. The economical and efficient form in which it can now be applied has induced the writers to give some illustrative particulars respecting the cost of the various forms of iron supports at present in use in Continental mines.

Rails, either new or old, of wrought iron or steel, are bent or arched into the shapes required by the dimensions of the roads for which they are intended. These rails are often bought as scrap iron from the railways, and when used new are of the I, T, and U sections, besides the flat-bottomed rail.

At Creuzot, in France.—Plain horizontal bars are here used, supported by wooden props. Sometimes the bars are supported on two side walls of masonry. In the latter case, when the superincumbent pressure is great, the bar is arched into a shape such as that shown in Fig. 1, Plate XXX. No space must be left between the ends of the bars and the side walls, so that yielding in that direction is impossible. The lining above the rails or bars along the roof is made with oak planking, put close together. The same style of support is often utilised for stables, shaft bottoms, engine houses, etc., and then, instead of oak lining, brickwork is built from bar to bar, the distance between which is from 3 feet to 3 feet 6 inches.

The greatest employment of iron supports is in gateways, main roads, engine planes, etc. For this purpose the rails or bars are bent into arches, or parts of an arch connected by fish-plates. Without attempting to enumerate the many different styles of arches employed, some typical examples may be given.

In the Prussian Government Lead Mines, in the Hartz Mountains.—
One of the writers (G. Meyer) has been occupied here with this kind of
work. The rails used are flat-bottomed and 14½ lbs. per yard section.
They are bent "at bank" (with the heads inwards) into a shape such as
that shown in Fig. 2, Plate XXX. The ends of the iron arch are
lodged in holes drilled in large stones set in the bottom and fastened by
wooden plugs or cement. Between these stones stone blocks are inserted
to keep them apart and thus ensure the stability of the whole. These
side stones or blocks also serve instead of sleepers for the tramway line,
the bolts being driven into oak plugs wedged into holes drilled 3 inches
deep. The lining outside the arches is done with the same kind of rails,
each 19 feet 8 inches long, arranged longitudinally, and the flat bottoms
being inside in contact with the base of the arch rails. The space
between these longitudinal rails is lined with flagstones obtained from
neighbouring quarries.

The following comparative estimates of cost of the different methods
of support are extracted from the Mines Inspector's Reports of the Hartz
Mining district.

I.—Iron supports with flat-bottom rails (cost £7 per ton) in main
roads. For length of 19 feet 8 inches :—

	£	s.	d.	= per yard. £	s.	d.
6 arch rails (14½ lbs. per yard)	1	14	3		5	2
21 longitudinal rails do.	6	6	0		19	3
Basement stones		3	4			6
Flagstones for lining		18	0		2	9
Transport of material		1	6			3
Rail-bending at bank		1	0			2
Labour (cost of)	1	13	0		5	0
	£10	17	1	£1	13	2

II.—Walling with quarry stones (masonry). For same length :—

	£	s.	d.	= per yard. £	s.	d.
Quarry stones	8	10	0	1	5	11
Mortar		14	6		2	2
Transport of material		14	8		2	3
Centres, laths, etc.		12	0		1	10
Wages	5	10	0		16	10
		16	1	2	2	9	0
If temporary timbering is necessary and this timber be lost, add cost of timber... ...	2	0	0		6	1
	£18	1	2	£2	15	1

III.—Cost of timbering (props and planks) same length :—

		£	s.	d.	= per yard. £	s	d.
Props and planks	... ...	3	18	0		11	10
Lining timber ...	... ...		12	0		1	10
Working holes for prop bottoms	... ...		15	0		2	3
Removing *débris* and transport of material	...		15	0		2	3
Wages ...	... ...	1	4	0		3	10
		£7	4	0	£1	2	0

It will be seen that these estimates relate to first cost only.

Iron Supports in the Altenwald Coal Mine (near Saarbrücken).—In the gateway of a seam where an underlying seam had been worked out, a good deal of settling and shifting was always observed. To obviate this difficulty, it was determined to put in iron supports in the form of elliptic arches, as shown in Fig. 3, Plate XXXI. To prevent the arches from longitudinal shifting, horizontal props were inserted from arch to arch at the highest points. The lining behind the arches is done with oak planking. This method of maintenance has proved very satisfactory, the only repairs sometimes necessary being the renewal of the oak planking, which is easily effected. It is important that the planks should not be placed edge to edge, but slightly overlapping, as shown in Fig. 3, Plate XXXI. so as to allow some play to the planking when under heavy pressures. Great importance is attached to the perfect vertical position of the arches. The cost of this kind of support is as follows :—

For length of 6 feet 10⅜ inches—

		£	s	d.	= per yard. £	s.	d.
3 elliptic **T** rails, with fish-plates and bolts (cost £10 per ton)	... ...	2	17	0	1	4	9
Oak timber (cost 9d. per cubic foot)	... ...	1	1	0		9	2
Working out full space and putting in sup- ports ...	... ...	1	10	0		13	0
Preparing rails ...	... ...	1	4	6		10	8
Preparing planks	... ...		1	3			7
		£6	13	9	£2	18	2

Cost of timbering for the same length—

						£	s	d.
Wages	... ...					1	2	2
Material	... ...					2	12	0
						£3	14	2

Brickwork or stonework arching was not applicable in this road, owing to the continuous settling of the floor.

In a gateway where the lateral pressure was very heavy, so that it would have been necessary to put the iron arches very close together, thus involving a considerable increase in cost, a combined system of walling and iron supports was employed. This system greatly resembles that applied at Creuzot, as shown in Fig. 1. The ends of the bars are laid on sheet iron, thus distributing the pressure over a greater area of wall surface. A complete arching of masonry would have involved a great deal of blasting to make the necessary room; the dangerous loosening of the roof under such circumstances was thus avoided by the adoption of this combined system.

A combination of walling and iron supports was also adopted to secure the engine plane, near the shaft bottom, in the same mine. As the width of the plane was there 22½ feet, masonry arching would have required a great deal of extra blasting, and no doubt have involved a considerable loss of timber, and the work would have occupied a considerable time. The cost of this system, as compared with masonry arching, is shown as follows, for a length of 32¼ feet :—

Combined system—

	£	s.	d.	£	s.	d.
				= per yard.		
Making room	54	12	0	5	1	7
Masons' wages	12	12	0	1	3	5
Quarry stones, 211 cubic feet	14	0	0	1	6	0
Mortar, 883 do.	11	19	0	1	2	3
Said, 1,059 do.	4	12	6		8	7
Oak planking	2	15	0		5	1
Wages—preparing planking		6	6			7
14 rails	34	17	8		4	11
28 pieces boiler plate put under ends of rails	1	0	0		1	10
Preparing rails		11	8		1	1
Preparing boiler plates	1	12	2		3	0
Transport of material	6	0	0		11	2
	£144	18	6	£13	9	6

Masonry arching complete, same length—

	£	s.	d.	£	s.	d.
				= per yard.		
Making room	42	0	0	3	18	1
Masons' wages	37	16	0	3	10	4
Quarry stones, 6,282 cubic feet	41	10	8	3	17	3
Mortar, 2,825 do.	38	15	0	3	12	1
Said, 3,178 do.	13	10	0	1	5	1
Wood (lost timber)		12	0		1	1
Do. centres	2	2	0		3	11
Do. planks		12	0		1	1
	£176	17	8	£16	8	11

In another gateway, the iron supports were made circular in shape, as shown in Fig. 4, Plate XXXI. Wrought iron plates (costing £5 per ton as scrap iron) were used for lining instead of oak planking. The following account shows the cost of this method for each length of 6 feet 10¾ inches :—

	£	s.	d.	£	s.	d.
					— per yard.	
Rails, fish-plates, and bolts	2	6	0	1	0	0
Wrought iron lining plates (17 cwts. 0 qrs. 25 lbs.)	4	7	6	1	18	1
Preparing rails and fish-plates		9	0		3	11
Preparing lining plates		19	2		8	4
Making room and putting in supports ...	2	18	2	1	5	3
	£10	19	10	£4	15	7

In a return air-way in the same mine, where the sides stand firm but a strong pressure from the roof was observed, a system of iron supports was adopted, as shown in Fig. 5, Plate XXXI. In this place the warm and damp return air was very destructive to timbering. The cost of this method is shown as follows for each length of 6 feet 10¾ inches :—

	£	s.	d	£	s.	d.
					= per yard.	
Rails, fish-plates, bolts	2	7	0	1	0	5
Wrought iron lining plates (14 cwts. 1 qr. 2 lbs.)	3	12	6	1	11	6
Preparing rails, fish-plates, etc.		8	0		3	6
Preparing lining plates, making room, and putting in supports	5	5	9	2	6	0
	£11	13	3	£5	1	5

The cheapest method of iron supports adopted is shown in Fig. 6, Plate XXXII., the cost being as follows per length of 6 feet 10¾ inches :—

	£	s.	d.		s.	d.
					= per yard.	
Rails, fish-plates, bolts	2	1	0		17	10
Oak planking		16	8		7	3
Preparing rails, etc.		12	0		5	3
Preparing planking		2	6		1	1
Making room and putting in supports ...	1	1	0		9	1
	£4	13	2	£2	0	6

The average cost of labour for this kind of work was from 2s. to 2s. 4d., clear of deductions. When the work previously described was done in the Altenwald mine, the whole system of iron supports was in an elementary stage of development. No workmen skilled in this special kind of work were available. Doubtless, in the future, when more experience has been gained, the cost of the iron supports is likely to be considerably diminished.

Segen Gottes Pit, Austria.—In this mine the shape of iron support used is shown in Fig. 7, *a*, Plate XXXII. A double roadway has here to be maintained. The lower ends of the arch rails are connected by cross rails, to which the tramway rails are fastened. The timber supports for the same roadway are shown in Fig. 7, *b*, Plate XXXII. The following figures show the comparative costs of the two methods :—

First cost of $\dfrac{\text{iron supports}}{\text{timbering}} = \dfrac{13}{10\cdot76} = \dfrac{1\cdot208}{1}$, thus iron supports cost 20·8 per cent. more than timbering.

After nine years' use, cost of $\dfrac{\text{iron supports}}{\text{timbering}} = \dfrac{21\cdot96}{24} = \dfrac{\cdot915}{1}$. Here iron supports cost 8½ per cent. less than timbering.

After eighteen years' use, cost of $\dfrac{\text{iron supports}}{\text{timbering}} = \dfrac{36\cdot25}{64\cdot54} = \dfrac{\cdot562}{1}$. Here iron supports cost 43·8 per cent. less than timbering.

No account is taken in this comparison of the value of the old rails when taken out as scrap iron. In the same pit, the iron supports are sometimes painted, the cost of which is considered about equivalent to that of cleaning the fungus (*Xylophagus*) from the timber. A gateway is mentioned where strong timber was crushed within three months by the heavy pressure. Iron supports, of elliptic section, were put in, and lasted fully five years without the slightest evidence of injury. At the end of that time, the district of this gateway was worked out, and the iron supports were taken away and set up in another place.

The writers might have given a number of further cases from Continental mining practice, but those cited above are sufficient to establish the efficiency and economy of the system of iron supports in main roads. In the United Kingdom, as compared with the Continent, iron is cheaper and timber dearer, which would show a still greater comparative advantage in this country. The-matter is well worth the consideration of mine managers. It may be objected that several of the methods previously described involve the use of oak lining planks at considerable cost. To meet this objection, a method of iron supports in conjunction with corrugated iron lining (Meyer's patent) has been suggested, and is likely to be experimentally introduced at an early date in some of the collieries in this country.

The PRESIDENT suggested that the cost per ton should be given.

Mr. BIRD said he would take an opportunity of adding that information.

Mr. STEAVENSON thought they had learned from this paper a process which was very useful in mines in the circumstances under which the iron was applied, and they seemed to be all exceptional circumstances. Iron was used at Altenwald "in the gateway of a seam where an underlying seam had been worked out;" and they were told that "brick or stone-work arching was not applicable in this road, owing to the continuous settling of the floor." Again, they were told that, "in a return air-way, where the sides stand firm but a strong pressure from the roof was observed, a system of iron supports was adopted. In this place the warm and damp return air was very destructive to timbering." Further on in the paper it was stated that "a gateway is mentioned where strong timber was crushed within three months by the heavy pressure." These were all exceptional circumstances, and he had no doubt, under such exceptional circumstances, the system described by Messrs. Meyer and Bird was exceedingly useful; but, under general circumstances, in coal mines his own impression was that there was hardly room for steel or iron, so far as he was able to judge. In 1885 Mr. Hugh Bell asked him to look into the question of using steel for the main gateways in Cleveland, where they used heavy timber, and he did so, and exhibited a section of rail-girder, which had been used for three years with great advantage, supplied by the Darlington Steel and Iron Co. Before going far into the question he made experiments, and he had accumulated a pile of documents which he would not deal with that day, but would prepare in a form to be read at a future meeting. In making the experiments, he selected a part of the mine where he should be disturbed as little as possible by the passing traffic, and where he would disturb the passing traffic as little as possible. If they referred to Barlow, "On the Strength of Materials," they would see he experimented on small pieces—many pieces not larger than an ordinary desk ruler. He (Mr. Steavenson) doubted whether it was fair to judge of the strength of a large balk of timber by testing only a very small piece. [He illustrated on the blackboard how the tests were made.] He said that one of the steel girders carried 13 tons, and then it simply bent down without the smallest sign of fracture, and it could be straightened and used again. When he put in the timber, the best balk came away with 4 tons, and, of course, was of no further use. He got channel iron, of 2, 3, or 4 feet, and put it on the top of the girders, and it made an efficient support for the roof. He

used channel iron, and he did not know how far this was an infringe-
ment of Mr. Meyer's patent, or whether Mr. Meyer's was an infringe-
ment of the Cleveland system.   In the case of timber, they were subject
to damp roof, and some had been three times renewed in two years, but
with steel girders this was entirely overcome.   He had gone fully into
the cost, and would give details in his paper.   He was further induced
to go into the relative strengths of material, in consequence of the ver-
dict of a coroner's jury in June, 1885.   An inquest was held at Brotton,
in Cleveland, and although the jury decided that everything possible
had been done, they suggested that larch was a better material than
Norwegian timber, and that the latter was liable to snap suddenly.   He,
himself, was under the impression that Norwegian and Riga timber was
more liable to snap as compared with larch; but, on making experiments,
he found that the strength of larch was represented by 395, Norwegian
454, and Riga 475, so that the jury and himself were mistaken.   He did
not quite learn from the paper what Mr. Meyer's patent was.   He should
like to know what was embodied in the patent, and the length of time
it had been in operation. .

Mr. BIRD—So far as Mr. Meyer's patent was concerned, he only
heard of it about three weeks ago.   It consisted in the placing of corrugated iron longitudinally from arch to arch of the road.

The PRESIDENT did not think this material to the business in hand.
They were not an authority upon patent rights.

Mr. LAWRENCE said, it struck him that the cost of labour in the two
methods, for the length of 19 feet 8 inches, was not quite correct.   The
labour for the supports was put down at £1 13s., and wages for timbering £1 4s.   If the wages for putting in timber props—which, they knew,
were very readily put in—cost £1 4s., then he could not conceive that
for the iron work, which would require a lot of smithing, and be difficult
to get into place, £1 13s. for labour could be correct.   With regard to
the application of corrugated iron in mines, so far as his experience went,
he should think it would not last more than two years.   Corrugated iron
went very rapidly; and he thought it would break out into holes, as they
frequently saw underneath railway bridges where the corrugated iron
was exposed to water.   He had seen holes in such cases in two or three
years.   He thought this would be dangerous for roofing, inasmuch as
there would be a lot of *débris* on the top, and the fact of its rotting
would not be seen until it gave way.

Mr. W. H. HEDLEY said he noticed it was put forward, as an item
in favour of the use of iron work, that less blasting was required with

it than in putting up a masonry arch; and yet the cost of making room for the "combined system" was shown to be greater than for the "masonry arch complete."

Mr. BIRD, in reply to Mr. Lawrence's criticism of the comparative cost of labour, said he was not in a position to add to the figures given in the paper. The figures were taken from the mine inspectors' reports. It was possible, as Mr. Lawrence said, that the figures had not been accurately given; but without further information he could not answer Mr. Lawrence. As to the comparative cost of making room, mentioned by Mr. Hedley, his (Mr. Bird's) idea was that the cost of making room on the combined system was the greater, although the room required was less, because it was required to make it exactly and not to have space to spare; and no doubt it would be done by pick work without blasting, while for masonry work the greater part of the making room would be done by blasting, and consequently would be cheaper.

The PRESIDENT tendered Mr. Bird the thanks of the members for the paper. In regard to the use of iron it had always appeared to him that the disinclination which existed in this country against the use of iron or steel, instead of wood, was most anomalous. He believed he was within the mark when he said that there was no country in the world where iron was more cheaply or better made than in England; and he further believed he was equally correct when he said there was no country in the world where timber was dearer than in England. And yet, with the single exception he believed of ship builders, there seemed to have been a studied disinclination on the part of engineers to use iron instead of wood. He had travelled over the railways of Europe, and he always made it his particular study to examine the wagons and sleepers on the railways over which he travelled; and he found that both the sides of wagons and the coverings of wagons were made of iron; and yet it was only within—he was telling no secret—the last few years that the North-Eastern Railway Company have consented to build 500 wagons constructed exclusively of steel and iron. Not using iron he had felt to be a great injury to the great iron trade of the district, as well, he thought, as a loss to the Railway Company.

———

Mr. T. E. FORSTER read the following paper "On Coal Nodules from the Bore-hole Seam at Newcastle, New South Wales":—

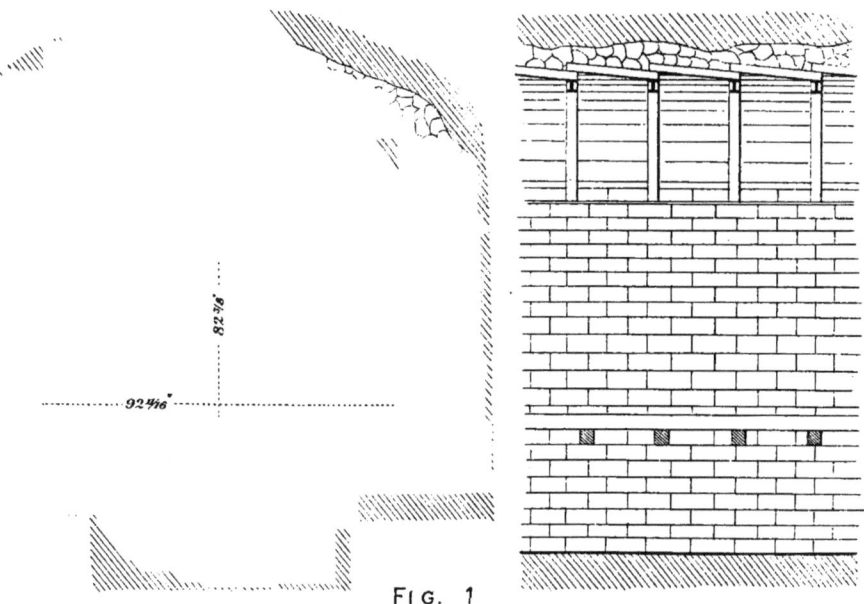

FIG. 1

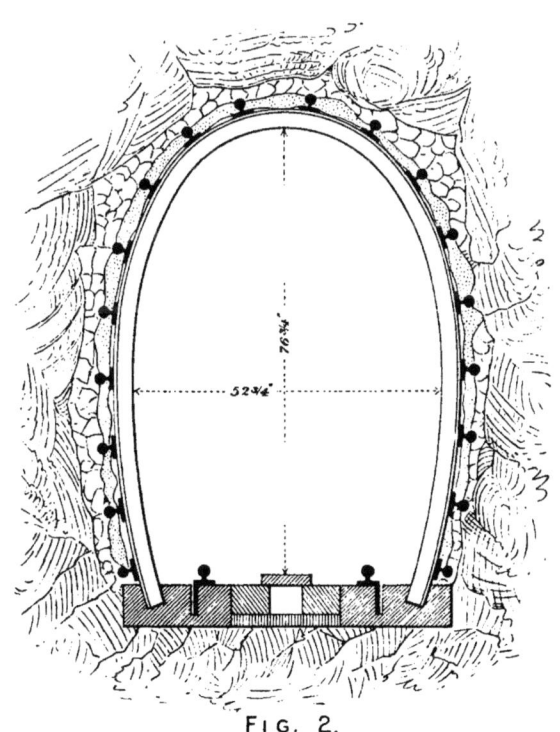

FIG. 2

To illustrate Mess.ʳˢ G. Meyer and W. J. Bird's paper on "The Use of Iron Supports in the Main Roads of Mines instead of Masonry or Timber."

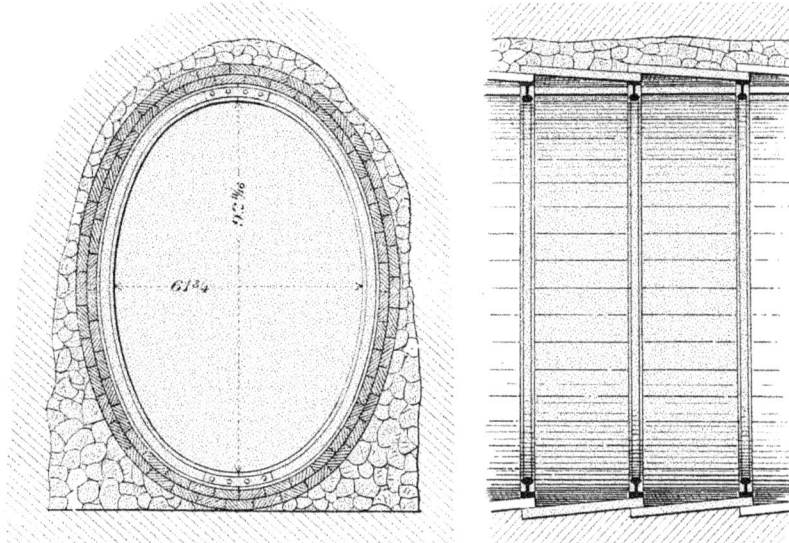

FIG. 3.

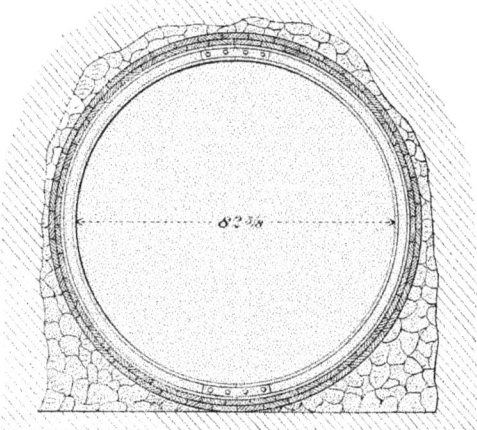

FIG. 4.

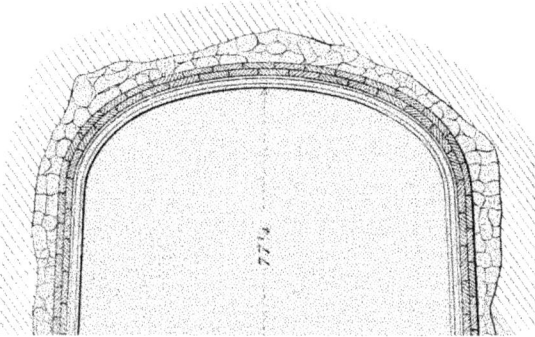

To illustrate Mess.ʳˢ G. Meyer and W.J. Bird's paper on: "The Use of Iron Supports in the Main Roads of Mines instead of Masonry or Timber."

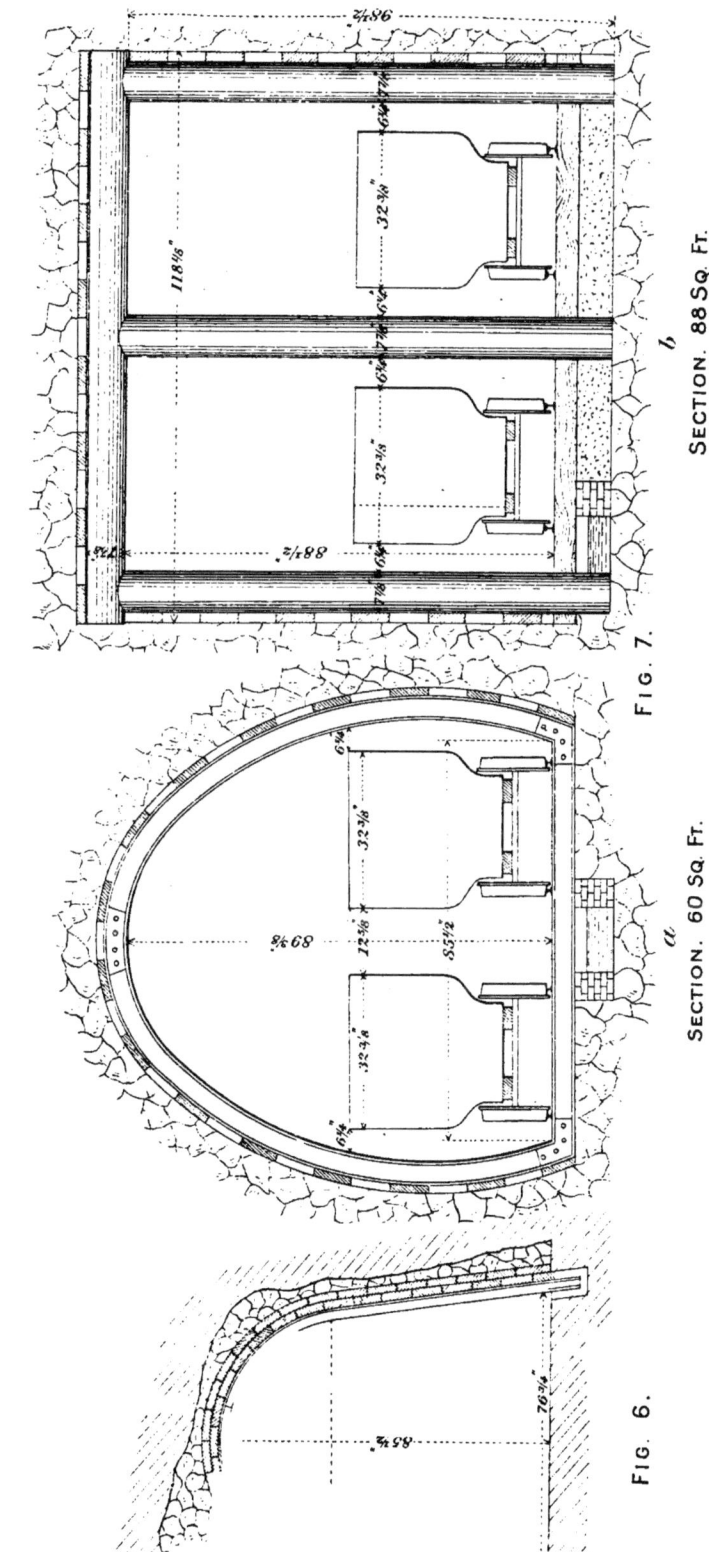

SECTION. 88 SQ. FT.

b

FIG. 7.

SECTION. 60 SQ. FT.

a

FIG. 6.

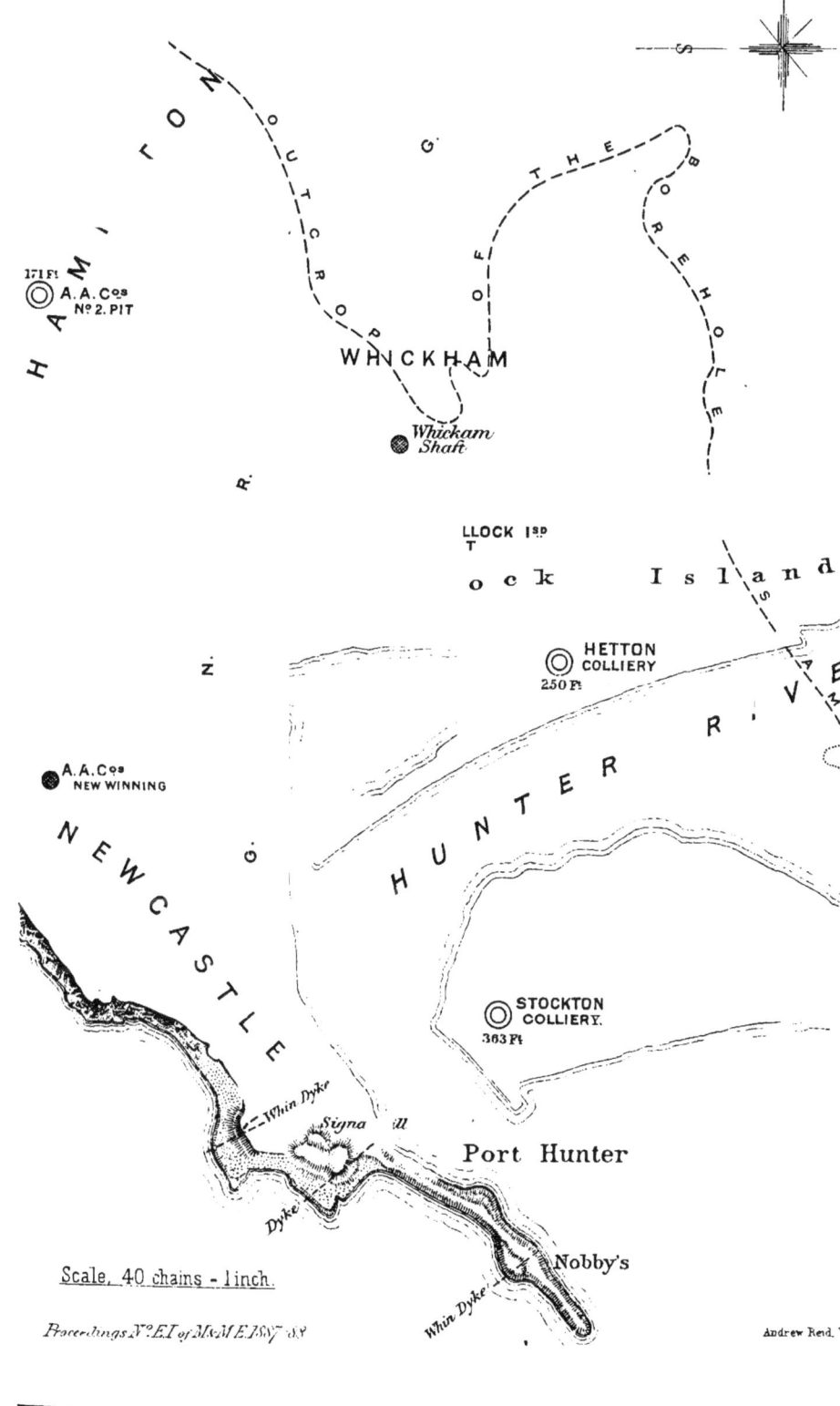

Scale, 40 chains - 1 inch.

Proceedings N.º EI of N. of M.E. 1887-88

Andrew Reid.

To illustrate Mr. T.E. Forster's paper on "Coal Nodules from th Borehole Seam at Newcastle, New South Wales".

SECTIONS OF THE BOREHOLE SEAM.

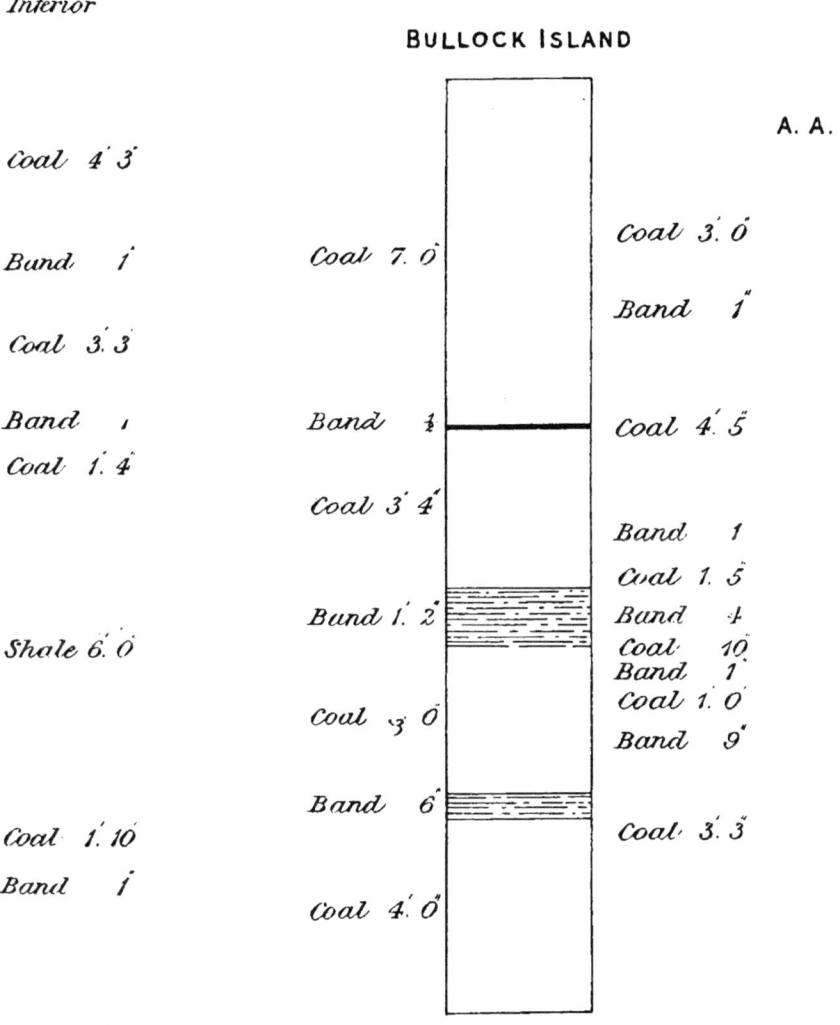

To illustrate Mr. T. E. Forster's Paper on "Coal Nodules from the Bore-hole Seam at Newcastle, New South Wales."

3/4 NAT. SIZE.

3/4 NAT. SIZE.

3/4 NAT. SIZE.

3/4 NAT. SIZE.

3/4 NAT. SIZE.

SLIGHTLY REDUCED.

(THIS SPECIMEN SHOWS FRACTURED INTERIOR.)

COAL NODULES FROM THE BORE-HOLE SEAM AT NEWCASTLE, NEW SOUTH WALES.

BY T. E. FORSTER, M.A.

THE specimens shown are taken from the Bullock Island, Stockton and Australian Agricultural Company's Collieries, at Newcastle, New South Wales, and are all from the " Bore-hole " or Main Seam of the district. This seam is extensively worked at Newcastle and in the immediate neighbourhood, almost the entire output of this coal-field, which in 1886 amounted to 2,178,000 tons, being drawn from it. It lies at a depth of about 300 feet under the town of Newcastle, a short distance to the west of which it is found to crop out, the line of outcrop running in a westerly direction past the Lambton and Wallsend Collieries, which work by means of adit levels, to the north of Lake Macquarie. The general dip of the strata is in a southerly direction towards Sydney, which lies some 70 miles further south, and is generally regarded as representing the centre or deepest part of the basin. Under the hilly ground in the neighbourhood of the lake the seam is overlaid by a series of grits and conglomerates of a peculiarly hard nature, as well as by several higher seams, the identification of which is as yet a matter of doubt.

The presence of the nodules or balls of coal is of frequent occurrence, more especially in the above-named collieries, and the seam is still further remarkable for the tendency of the lines of cleavage to run more or less in a curved form, often causing the sides and corners of the blocks of coal to present the partially rounded appearance peculiar to and characteristic of the Newcastle coal.

The positions of the collieries named may be noticed on the accompanying sketch, Plate XXXIII., on which also the depths in feet to the bottom of the seam at the different points are given.

The surface is here part of the Delta of the Hunter River, and consists of beds of sand, clay, and gravel, of considerable and varying thickness, deposited after the denudation of the Coal-Measures, when the estuary of the river was of wider extent, owing to the lower relative level of the adjacent land. These deposits have been proved to have a thickness of no less than 160 feet at Bullock Island, where the presence of quicksands as well as at Stockton rendered the employment of metal cylinders necessary during the sinking operations. The solid measures overlying the coal consist of alternations of shales and sandstones, while the seam itself rests on the thick bed of grey sandstone, which has been quarried near Waratah.

In this immediate vicinity the seam has a gentle inclination towards the harbour, *i.e.*, in a northerly direction, the greatly increased depth of the coal at Stockton being, in all probability, due to a fault which, it is surmised, passes under the river bed, and the existence of which is perhaps rendered more probable by the fact that the dip on the Stockton side is in the contrary direction.

The following section of the seam is taken in the workings of the Australian Agricultural Company's No. 2 Pit, about a mile east of the shaft and in the direction of Newcastle :—

	Ft.	In.
COAL, *Top*	3	0
Stone band		1
COAL—*Big Tops*	4	5
Stone band		1
COAL—*Big Tops*	1	5
Band—*Morgan*		4
COAL—*Four-Inch*		10
Stone band		1
COAL—*Little Tops*	1	0
Band—*Jerry*		9
COAL, bottom	3	3
	15	3
COAL	13	11
Bands	1	4

At Bullock Island the section sunk through was :—

	Ft.	In.
COAL	7	0
Band		0½
COAL	3	4
Band—*Morgan*	1	2
COAL	3	0
Band—*Jerry*		6
COAL	4	0
	19	0½
COAL	17	4
Bands	1	8½

Only the lower portion of the seam is being worked here as yet; the dip is slight and towards the harbour, while the seam is practically free from faults or other disturbances.

At Stockton Colliery, sunk on the north side of the river, the seam in the immediate neighbourhood of the shaft is in a very disturbed condition owing to a mass of intrusive dolerite, appparently an overflow from some adjacent dyke as yet unproven, having forced its way into the seam, which it has charred and destroyed in almost every direction, spreading horizontally in sheets and tongues to very considerable distances. It is probably connected with one or other of the dykes which are visible at the surface in close proximity to the Signal Hill, a third dyke, presenting an unusually fine section through the cliff at Nobby's Head, where also the peculiar "chert" rock described in Mr. H. Plew's paper* is exposed to view. (See Plate XXXIV.)

The seam is here divided into two distinct beds, between which, at the shaft, 6 feet of shale is interstratified, caused most probably by the local thickening of the band known as the Morgan.

The difficulties which have been met with in the shape of dykes and faults have rendered the laying out of the workings of this colliery in a regular and orthodox manner an impossibility, and consequently the upper and lower portions of the seam appear to have been worked indiscriminately as the nature of the ground and the exigencies of the moment demanded. The following is a full section of the seam :—

	Ft.	In.	Ft.	In.			Ft.	In.
COAL, inferior ...	3	6						
COAL ...	4	3						
Band ...		1						
COAL ...	3	3						
Band ...		1						
COAL ...	1	4						
			9	0	{	COAL ...	8	10
						Bands ...		2
Shale ...			6	0				
COAL ...	1	10						
Band ...		1						
COAL ...	6	0						
Band ...		1						
COAL ...		11						
			8	11	{	COAL ...	8	9
						Bands ...		2
			23	11				

Speaking generally, the coal under the estuary of the Hunter River is in appearance particularly bright and clean, and of a very superior quality. It is shipped for steam and gas purposes, for the last-named of which it is perhaps more especially fitted, giving high results both in productive and lighting power.

It may here be noted that the exceedingly open nature of the " backs" or facings in the seam is a further characteristic of this coal. These facings, which are exceedingly well defined and open, often present, when bared, a smooth and polished surface, so much so that the course of the headings is frequently determined by that of the backs, between two of which the places are driven.

In the workings of the Stockton Colliery, where the coal is in an unusually disturbed condition, the nodules occur with the greatest frequency, and may be easily separated from the surrounding coal, which splits away from them with the greatest readiness. The seam at this colliery is remarkable, not only for its splendid proportions, but also for its extremely bright and rich appearance, which gives the working places a strong resemblance to those in the Pennsylvanian anthracite mines, the glitter of the coal and the height of the workings combining to form an impressive sight. The cleavage of the coal is in some parts of the pit noticeable for a series of small impressions or concavities about the size of and resembling roughly in shape a mussel shell raised vertically on its longest edge. The coal at Stockton is more tender than that worked either at Bullock Island or at the Australian Agricultural Company's Collieries, and therefore less fitted for transport and rough handling.

The question of the formation of these nodules is one on which, so far as is known, no explanations or suggestions have been offered, and the solution of which can only be a matter of vague surmise. To some extent the nodules appear to have a more or less concretionary structure, and to be, roughly speaking, composed of several concentric layers, through which the ordinary cleavage of the coal passes, while in others thin layers may be observed, resembling the coats of an onion. (See Plates XXXV. and XXXVI.)

On first consideration it might be assumed that the peculiarly disturbed state of the ground in which the workings are being prosecuted at Stockton, where the nodules occur with the greatest frequency, is to some extent accountable for or connected with their existence. Looking, however, to the fact that the nodules are so closely imbedded in the seam, being in appearance and quality of the same nature and one and the same with the surrounding coal, and also to the disturbance having been almost entirely due to the basaltic overthrow of later date, which has altered and cindered the seam in its immediate vicinity only, it would hardly seem probable that their occurrence is anything further than a coincidence.

In a paper on the composition of New South Wales coals, read before the Royal Society of New South Wales by Professor Liversidge of Sydney, an account of a coal nodule from the Waratah Colliery is given.

Professor Liversidge describes this nodule as being an anthracitic coal, and apparently of a concretionary nature. On being struck with a hammer, the mass flew to pieces as if it had been in a state of strain or tension, the fragments being small, and showing conchoidal fracture surfaces.

The analysis of this nodule which is given differs from that of the ordinary sample of the seam from the same colliery, containing a slightly larger percentage of carbon and less ash, the analysis being as follows:—

			Sample of Seam.	Nodule
Carbon	...	...	81·06	83·828
Hydrogen	...	...	5·81	5·437
Oxygen	...	...	6·52	8·236
Sulphur	...	...	1·14	·190
Nitrogen	...	...	1·23	·530
Ash ...	...	...	4·24	1·779
Specific gravity	...	...	1·303	1·294

PROXIMATE ANALYSIS.

			Seam.	Nodule.
Moisture	...	...	2·21	3·32
Volatile hydrocarbons	...		36·70	32·41
Fixed carbon	...	...	55·82	62·35
Ash ...	...	...	4·15	1·72
Sulphur	...	...	1·12	·19

It may, however, be mentioned, that the first analysis is that of a sample of the whole of the seam, and the fact that the composition of the several beds comprised in it are distinctly different may perhaps explain the discrepancy.

The numerous open facings and the well-defined cleavage of the coal, with its singular tendency to curvature, seem to be unusual conditions which may probably open the way to a more feasible explanation, and it is possible that their formation is due to this alone.

The PRESIDENT said, he should be very glad if any gentleman would give an explanation of how these nodules came to be formed.

Mr. STEAVENSON said, he did not rise to make any comment upon the paper, but would like to take this opportunity of congratulating the Institute upon the fact that the writer of this paper was the third generation of the family he had had the pleasure of hearing here. He did not know whether there was gas in the seams or not?

Mr. T. E. FORSTER said, the seam was generally free from gas. In the pits there gas was almost unknown. They lay very near to the surface.

The PRESIDENT said, it was his duty to thank Mr. Forster for his very excellent paper, which was a most interesting and instructive one.

————

Mr. HUGH BRAMWELL read "Notes on the Horizon of the Low Main Seam in a portion of the Durham Coal-field," as follows:—

NOTES ON THE HORIZON OF THE LOW MAIN SEAM IN A PORTION OF THE DURHAM COAL-FIELD.

By HUGH BRAMWELL.

In the following notes the writer would draw attention to the change of horizon of the Low Main Seam, in that portion of the Durham coal-field which has the town of Sunderland as a centre, and which extends as far as the Tyne, Pensher, and Seaham, to the north, west, and south. In so doing, it is also necessary to mention one or two of the other peculiarities exhibited in the shaft sections referred to. Although well known in the immediate neighbourhood, the detailed correlation of the seams in this area does not appear to have been previously recorded, and it is hoped that the information collected may form an appendix to the general synopsis of the coal-seams recently compiled by Mr. Walton Brown.*

In order to clearly trace the position of the various seams to which attention is directed, three sets of detailed shaft sections are appended (Plates XXXVIII., XXXIX., and XL.), whilst the conclusions to be drawn from an examination of these are shown on Plate XXXVII.

NOTES ON THE SHAFT SECTIONS I., II., AND III.

PLATE XXXVIII.—The correlation of the seams between the Maudlin and the Hutton, in the St. Hilda and Wearmouth sections, must be regarded as doubtful ; it, however, appears to be probable that the seam called the Six-Quarter in the former is represented at the latter point by several thin seams lying immediately below the Maudlin.

At Wearmouth the Maudlin appears with a distinct " bottom coal," below which, as stated, are two other thin seams, all within a foot or two of each other.

At Ryhope the Maudlin " bottom coal " is 4 feet 10 inches thick and separated from the seam itself by a band only 2 inches thick. (In the Ryhope sinking account the seam lying 9 fathoms below the Maudlin is called the " Low Main.")

* See Transactions, Vol. XXXVII., page 3.

The thick well-defined " bottom coal " of the Maudlin continues in that position to within a short distance of the Seaham Shaft, when it rapidly descends, and in the shaft section exists as a separate seam—the Low Main—some 10 fathoms below the Maudlin. Tracing this Low Main Seam to Murton and South Hetton, there appears to be no doubt as to its identity with the well-known seam of that name in the southern portion of the coal-field.

PLATE XXXIX.—A staple sunk about two miles west of the Seaham Shaft proves the same change of horizon, the Low Main Seam again forming the bottom coal of the Maudlin at the staple.

At Eppleton the seams are, however, separate. The Eppleton section is easily correlated with those of Rainton and Lumley.

Attention is here drawn to the position of the Brass Thill Seam at Lumley and Rainton, and its probable representatives at Eppleton.

PLATE XL.—At Houghton the Low Main Seam exhibits a tendency to split, and at Newbottle it appears to be represented by three distinct seams.

The division is carried still further at Pensher, whilst at Boldon and St. Hilda its horizon is occupied by two or more seams, one of which at the last-mentioned place is the Six-Quarter of the Tyne district.

Attention is also drawn to the probable identity of the seam called the Low Main at Ryhope, with that called the Five-Quarter at Boldon and St. Hilda, also to its position as compared with the representatives of the Brass Thill Seam at Eppleton.

CONCLUSIONS DRAWN.

From the foregoing sections it is submitted that there is sufficient evidence to warrant the following conclusions :—

1.—That the Low Main Seam of South Durham forms the " bottom coal " of the Maudlin, over the area shaded horizontally on the accompanying plan, Plate XXXVII.

2.—That to the north and north-west of this area it leaves the Maudlin by successive splits, the area shaded vertically on the plan representing that portion of the coal-field in which a part of this seam still remains as Maudlin " bottom coal."

3.—That it is finally represented either in whole or part by the Six-Quarter Seam of the Tyne.

4.—That the seam called the "Low Main" in the Ryhope section is the Five-Quarter Seam of the Tyne, and is possibly
identical in whole or part with the Brass Thill Seam of
the Wear.

In a correlation of coal-seams it is sometimes stated that "thin seams"
are of little value as guides, on account of their liability to thin out
altogether.   They are, however, just as liable to thicken ; hence it was
thought necessary, in the foregoing sections, to give as complete details
as were available.

———

The PRESIDENT—The discussion on Mr. Bramwell's paper, and that
on Mr. Walton Brown's paper "On a further attempt for the Correlation of the Coal Seams of the Carboniferous Formation of the North
of England; with some Notes on the Probable Duration of the Coal-
field," were adjourned at Mr. Marley's request.

———

The following papers were open for discussion; but none took
place:—
   "On the Coal-field of Tkiboulli (Caucasus)," by Charles J. Murton;
   "On Bornét's Hand Boring Machine," by E. L. Dumas;
 . "On Ackroyd and Best's Patent Safety-Lamp Cleaning Machine,"
      by William Ackroyd;
   "On an improved form of Seismoscope," by Prof. A. S. Herschel,
      F.R.S., etc.

The meeting concluded.

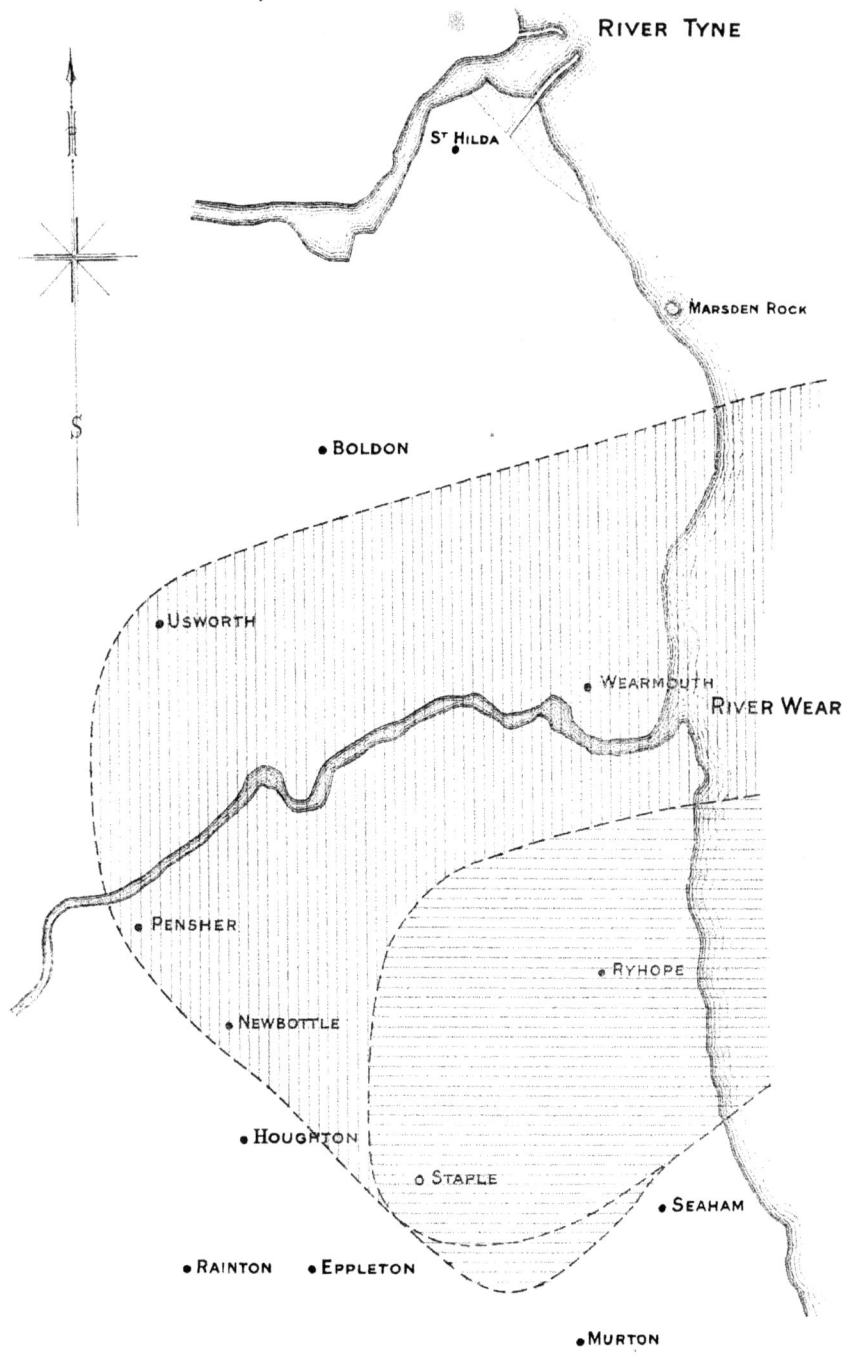

illustrate Mr Hugh Bramwell's paper "Notes on the Horizon of the Low Main Seam in a portion of the Durham Coalfield."

RIVER TYNE

Sᵀ HILDA

MARSDEN ROCK

S

BOLDON

USWORTH

WEARMOUTH

RIVER WEAR

PENSHER

RYHOPE

NEWBOTTLE

HOUGHTON

o STAPLE

SEAHAM

RAINTON   EPPLETON

MURTON

Scale, ½ inch = 1 Mile.

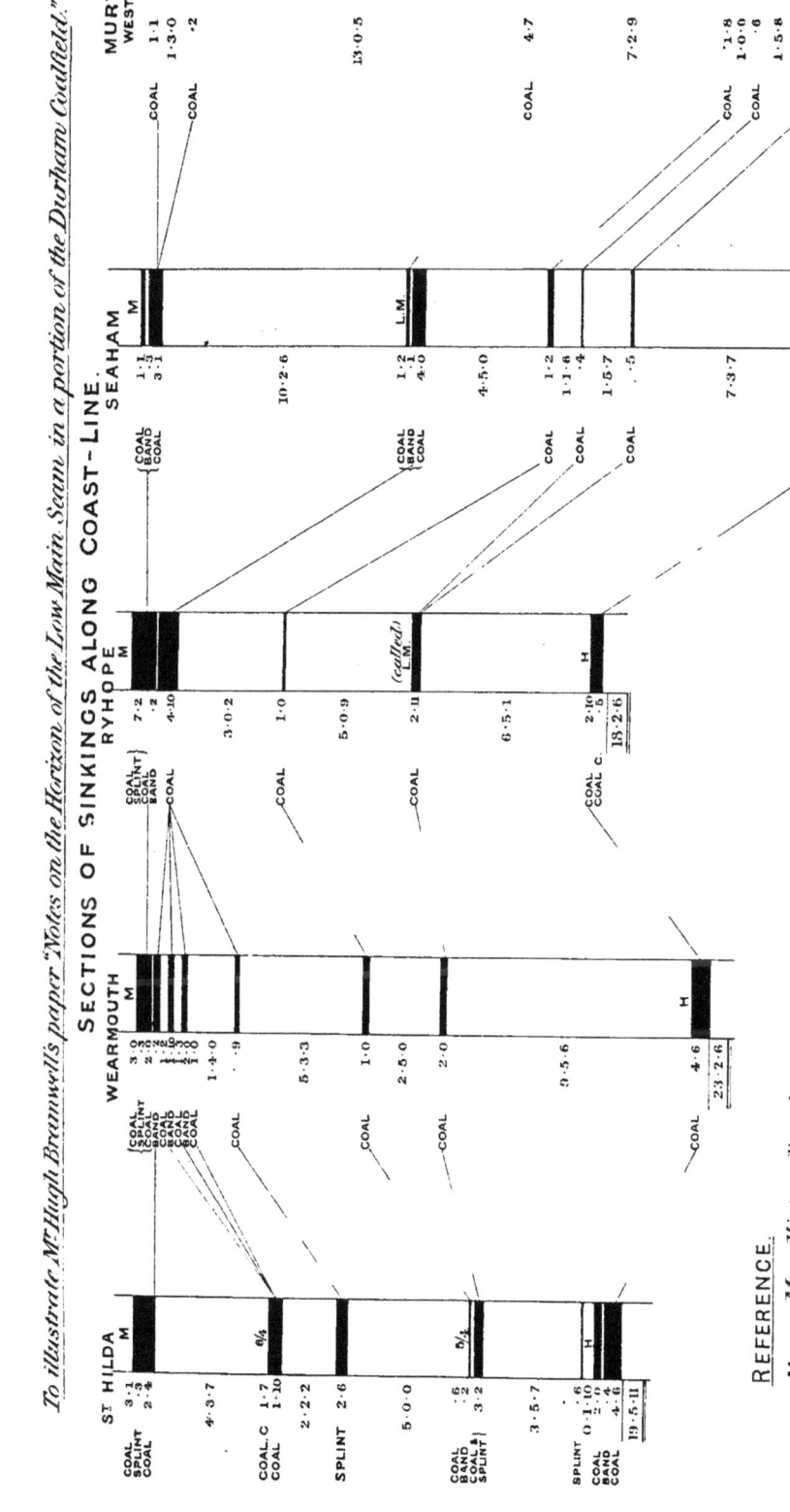

To illustrate Mr Hugh Bramwell's paper "Notes on the Horizon of the Low Main Seam in a portion of the Durham Coalfield."

SECTIONS OF SINKINGS ALONG COAST-LINE.

REFERENCE.

M – Maudlin or Bensham.
LM – Low Main.
6/4 – Six Quarter.

To illustrate Mr Hugh Bramwell's paper "Notes on the Horizon of the Low Main Seam in a portion of the Durham Coalfield."

SECTIONS OF SINKINGS ALONG SOUTH BOUNDARY.

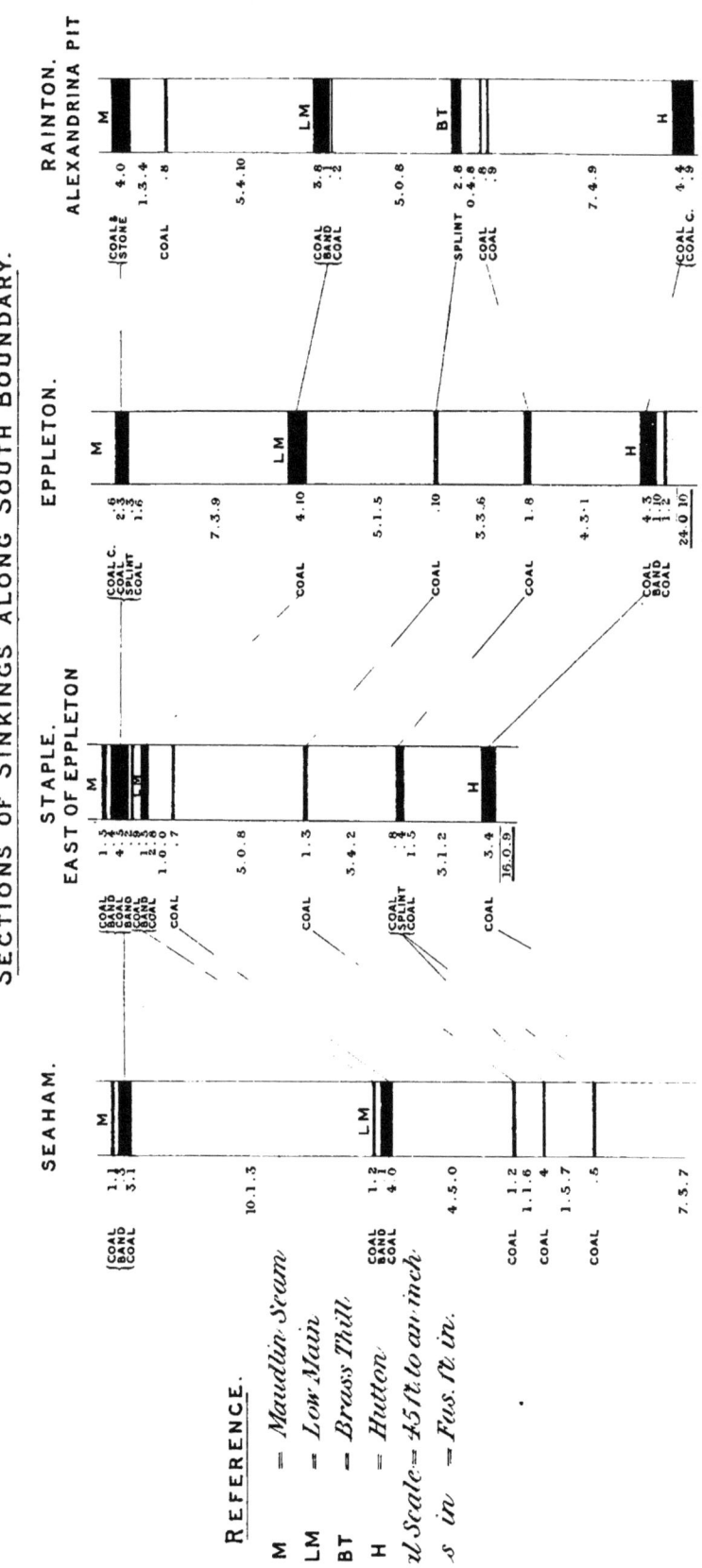

REFERENCE.

M = Maudlin Seam
LM = Low Main
BT = Brass Thill
H = Hutton
Vertical Scale = 451 ft. to an inch.
s in = Fms. ft. in.

SECTIONS OF SINKING

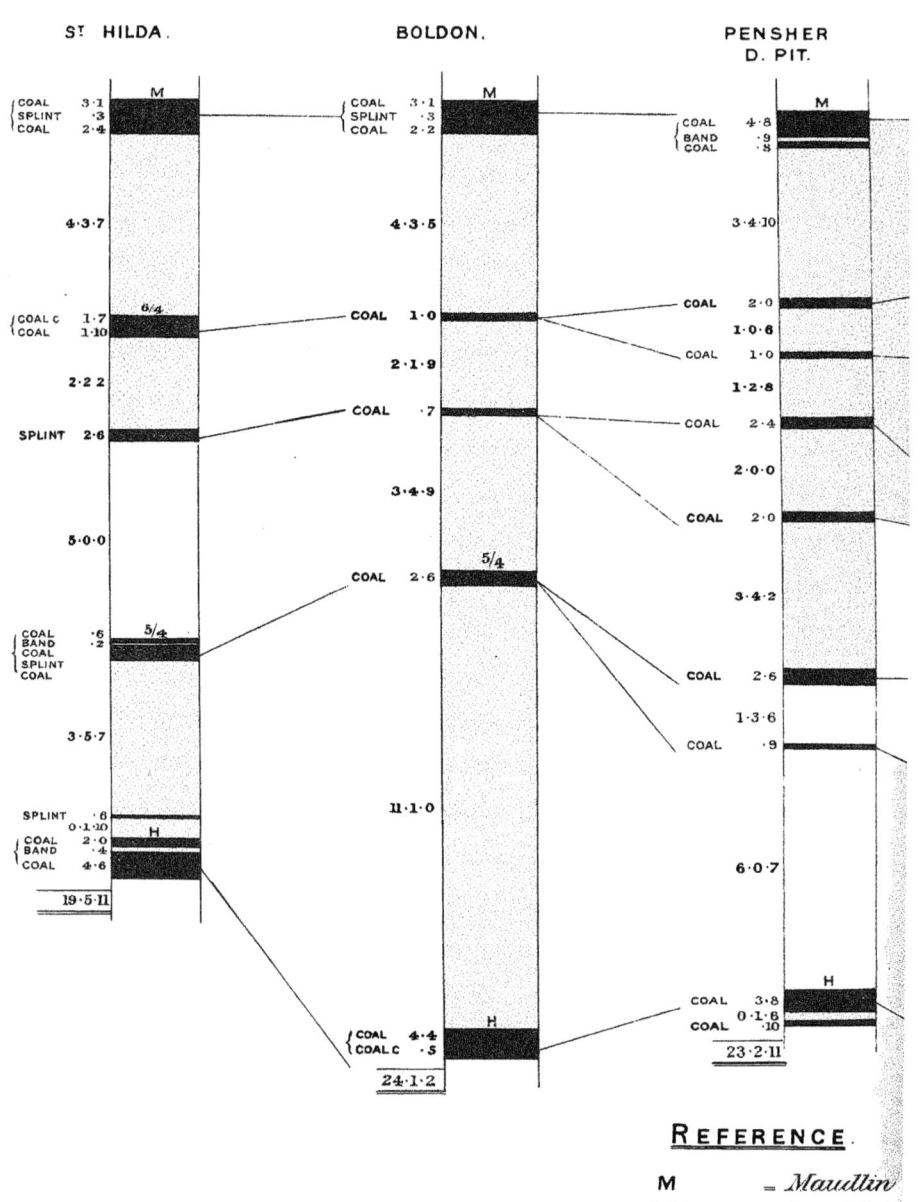

ST HILDA.

COAL	3·1
SPLINT	·3
COAL	2·4

M

4·3·7

| COAL C | 1·7 |
| COAL | 1·10 |

6/4

2·2·2

SPLINT 2·6

5·0·0

COAL	·6
BAND	
COAL	·2
SPLINT	
COAL	

5/4

3·5·7

SPLINT	·6
	0·1·10
COAL	2·0
BAND	·4
COAL	4·6

H

19·5·11

BOLDON.

COAL	3·1
SPLINT	·3
COAL	2·2

M

4·3·5

COAL 1·0

2·1·9

COAL ·7

3·4·9

COAL 2·6

5/4

11·1·0

H

| COAL | 4·4 |
| COAL C | ·5 |

24·1·2

PENSHER
D. PIT.

COAL	4·8
BAND	·9
COAL	·8

M

3·4·10

COAL 2·0

1·0·6

COAL 1·0

1·2·8

COAL 2·4

2·0·0

COAL 2·0

3·4·2

COAL 2·6

1·3·6

COAL ·9

6·0·7

H

COAL	3·8
	0·1·6
COAL	·10

23·2·11

REFERENCE.

M	= *Maudlin*
L M	= *Low Mair*
6/4	= *Six Quar*
5/4	= *Five Quar*

ALONG WEST BOUNDARY.

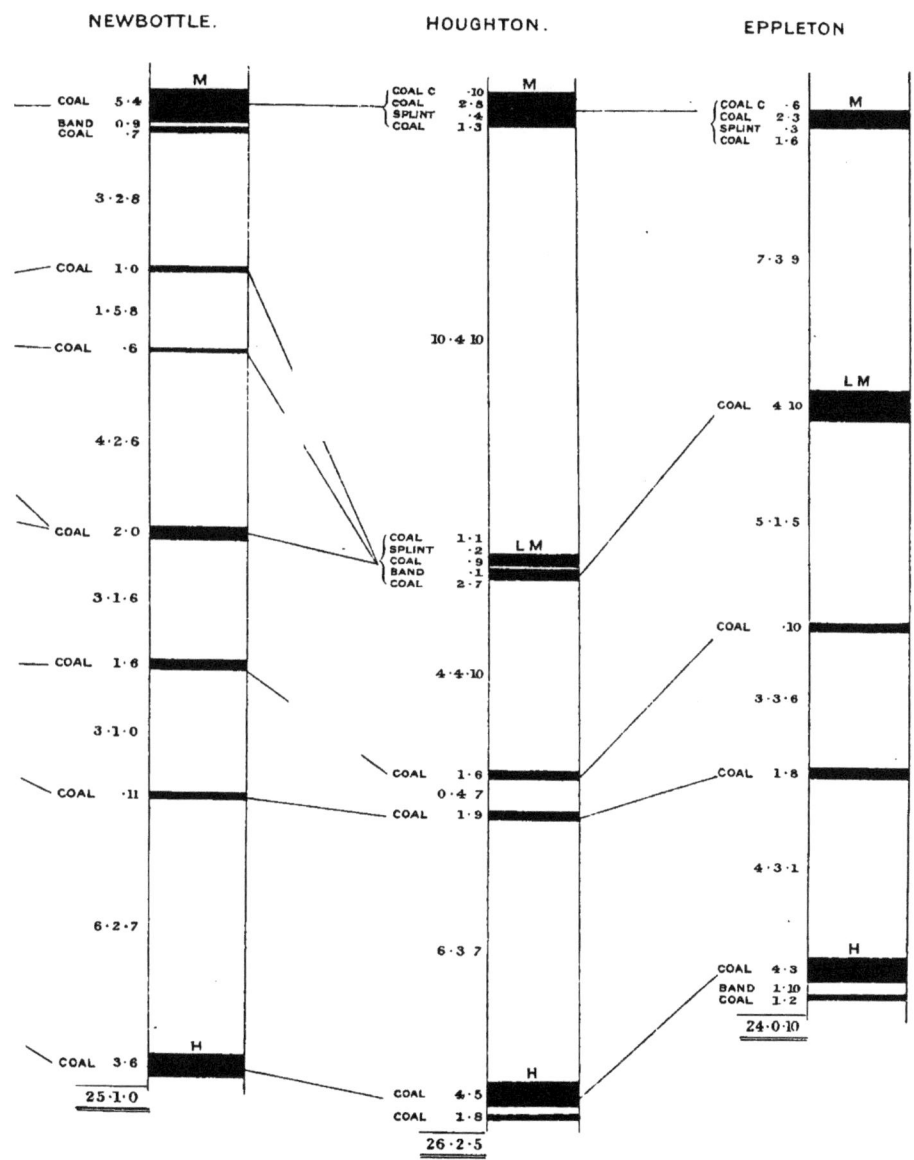

PROCEEDINGS.

WEDNESDAY, JUNE 6TH, 1888.
IN THE COUNCIL CHAMBER OF THE INSTITUTION OF CIVIL
ENGINEERS, 25, GREAT GEORGE STREET, LONDON.

Sir LOWTHIAN BELL, Bart., in the Chair.

FEDERATION OF MINING INSTITUTES.

PRESENT.—Sir Lowthian Bell, Bart., Messrs. J. Marley, A. L. Steavenson, M. Walton Brown, W. Cochrane, T. J. Bewick, W. Armstrong, Jun., J. Daglish, and T. Forster Brown (North of England Institute); W. H. Howard, J. Jackson, and M. H. Mills (Chesterfield); Professor Benton and Mr. Alex. Smith (South Staffordshire); Messrs. G. B. Walker, Jos. Mitchell, T. W. H. Mitchell, and A. M. Chambers (Midland); R. Haines and J. Lucas (North Staffordshire); and Professor G. A. Lebour (Secretary).

Mr. T. FORSTER BROWN begged to propose that Sir Lowthian Bell take the chair. .

Mr. W. COCHRANE seconded the resolution.

The resolution was carried unanimously.

The CHAIRMAN said, he would not detain them at any great length, because he presumed that they had all made themselves acquainted with the business upon which they had met that morning. It was to discuss a project which, although set on foot mainly by a paper which had been read before the North of England Institute a few months ago, yet he believed the credit of originating the scheme itself was due to his friend and predecessor——

Mr. J. DAGLISH—It goes further back than that, Sir Lowthian. It was in the time of Mr. Forster's presidency.

The CHAIRMAN said, it seemed to be buried in the mists of a remote antiquity; it had lived during all that time, and had never been called into operation. They had met there that morning in order to hear their views upon the subject, and to ascertain whether it would be favourably received, so as to justify the North of England Institute of Mining and Mechanical Engineers proceeding in their attempts to carry it into execution.

So far as he was personally concerned, he might say at once that he was in favour of the scheme, and, if for no other reason than this, that by the co-operation of all the mining engineers in the country they must, he thought, expect that they would get by a more direct road to the truth, in connection with those enquiries which it was their business to originate and discuss, than they could do single-handed. In the first place, the enquiries themselves must more or less take the colour and direction of the peculiar coal-field in which they originate, and, to correct this when necessary, it could not but be of very great advantage that the experience of one district should be compared with the experience of other districts. In addition to these, there were many questions which were almost beyond the means (he meant beyond the financial means) of a single body to investigate, but which might quite easily be brought within the powers of the union of Mining Institutes like their own. The prospectus placed in their hands very properly pointed out the example of other bodies of a cognate character, which they might themselves follow with advantage. They had the Society of Chemical Industry. Now, he believed the North of England had the credit of being one of the first, if not the first, to originate a Society of Chemical Industry. The London Society soon followed; and it saw the desirability of gathering within its fold, as it were, the Societies of Chemical Industry from the provinces which led to its establishment. The Iron and Steel Institute, of which he (the Chairman) was one of the early promoters, began at once as an Institute embracing the iron trade of every part of the country. If an instance were wanted to point out the desirability of such a mode of procedure, it was that afforded by the Iron and Steel Institute. He believed that there was no industrial Institute in the country which rose so rapidly to a position of eminence and usefulness. In speaking of the desirability of co-operation, he might mention one case in the history of that body which seemed worthy of notice, where it was thought desirable to investigate the so-called mechanical puddler, an American invention, and in order to do that in the most satisfactory way, the Iron and Steel Institute appointed four gentlemen—four, if he remembered rightly, " commissioners" as they called them—who were deputed to visit, and did visit, the United States, in order to examine the nature and success attending the use of mechanical puddling—a matter at that time of very great importance, because, as they all knew, the labour in puddling was an extremely severe one, which it was desired to alleviate. These gentlemen went over there, and reported fully upon the mechanical puddler; and although, practically, it died a natural death in this country, it must not be inferred on that account that they disagreed with the report of the commissioners, but because the introduction of steel had in a great measure superseded puddled iron.

He would just mention another matter which was germane to their own particular profession : he meant the question of coking coal. Now there had been various kinds of coking ovens recommended, and different kinds had been tried without a proper consideration of the quality of coal to be treated.   In consequence considerable sums of money had been wasted which might have been saved had the subject been examined with the care its importance deserved. He could not but think that if mining engineers in other parts had heard of the experience of the North country coke manufacturers possibly large sums of money would have been saved.

Now, might he venture, in a company of mining engineers, to say a few words in regard to the Davy lamp? The Davy lamp was an invention made fifty years ago or more.   He believed it was only very recently dis-covered that the presence of fine coal-dust in the interior of a Davy lamp might constitute a source of danger. Then, more recently, they had been told how dangerous the presence of coal-dust in the workings might be in promoting explosions, or at all events intensifying the effects of explo-sions. Having regard to the very small quantity of coal-dust which might convert atmospheric air into a highly explosive mixture, he thought the importance of a proper investigation, which might be undertaken by the united colliery districts, could not very well be over-rated.

These were a few of the ideas which had led him to give his ready and very hearty willingness to co-operate with the other Mining Institutes of the country in securing the combination that he had endeavoured to bring before them.

He would now call upon their Secretary, Professor Lebour, to let them know what progress the movement had made in other quarters than the North of England, and then they would be better able to judge, he thought, of the probability of their carrying to a successful issue the establishment of a federation of the chief coal Mining Institutes of this country.

Professor Lebour (Secretary) then read abstracts of the replies which had been received from the various Mining Institutes to the question which was addressed to them generally—"Is or is not such an arrangement as that outlined in Mr. Bunning's paper desirable ?" The Chesterfield and Midland Institute, in a letter of December 12th, 1887, answered the question in the affirmative, and appointed Messrs. J.

Jackson and M. H. Mills as representatives. They had confirmed this answer by sending their representatives there that day. The Midland Institute, by their letter of January 16th, 1888, considered the scheme desirable, and named Messrs. T. W. Embleton, A. M. Chambers, W. Carrington, J. H. Walker, and Joseph Mitchell as representatives. They also had sent their representatives, and so far, therefore, confirmed their previous letter. The North Staffordshire Institute had appointed Messrs. J. Lucas, W. Y. Craig, and Richard Haines as representatives ; but they did not say whether they agreed to the scheme or not. The South Staffordshire Institute considered the scheme desirable, and had appointed Messrs. W. B. Scott, J. Hughes, and Alex. Smith as representatives. Their representatives had also come to that meeting. The Mining Institute of Scotland, from their letter of December 30th, 1887, were generally of opinion that a federation for the purposes contemplated in the paper was inexpedient and unnecessary, inasmuch as the Mining Association of Great Britain already occupied the position proposed to be established. He should add, however, that in the letter in which the Secretary stated this, he added that his personal opinion was that the Mining Association of Great Britain had nothing to do with the matter at all. There was, however, an expression of opinion that an arrangement might be made for the first publication of Transactions somewhat similar to that proposed in the paper ; but no representatives had been appointed to attend this meeting. That had been confirmed by a letter received quite recently—they still wished to have nothing to do with the scheme. The South Wales Institute were not disposed at present to appoint a committee as suggested. That was in a letter dated October 27th, 1887, and that action had been confirmed. The Mining Institute of Cornwall were of opinion that the time had not yet arrived for the Society to join the proposed federation. That also had been confirmed. The Manchester Geological Society, although its expression of opinion was informal only, was adverse to the proposal. They had also declined to send representatives to the meeting.

The CHAIRMAN said, in the meantime, he would be very glad to hear the views of any gentleman present upon the subject.

Mr. JACKSON, as representing the Chesterfield Mining Institute, the first on the paper before them, said that their council was quite of the opinion that a federation of this sort would be desirable, and of great advantage to the mining community of Great Britain. But as an Institute they did not desire to lose their individuality; they would be glad to be, as it were, a branch of a federation; but the difficulties they saw were, that they had so many members amongst them that were under-viewers or students—young men who were learning mining engineering, and also workmen in the pits—a class of men who at home were in a position to take part in discussions, and to take an active interest in the welfare of their Institute. They felt that if they became extinct, and attempted to embody themselves in one large federation, they would be doing an injury and an injustice to a large class who ably supported them; but as to the general principle of the thing they were very strongly in favour of it. They also advocated the idea of Mr. Bunning, namely, with regard to papers, that if there was a general federation they would in time be able to send the papers that were written, first, to a central committee, which would have the power of saying whether they were worthy to be read before the General Institute or not, and then confining to themselves the right of reading those papers and discussing them at home. But the better ones would go to the general committee, and, if worth anything, would be received in the Institute. Then another question came before them, namely, that of expense. They felt that if they still had to continue the same subscription, and contribute a guinea for the privilege he had mentioned, that would be detrimental to their interests. Mr. Mills and himself had come there to express these views, and to do what they could to further the objects of the meeting, so long as it was not going to place them at a disadvantage, or to make them extinct at Chesterfield.

Mr. BEWICK said, he would merely suggest that each gentleman, as he addressed the meeting, should say for what Institute he appeared.

Mr. MILLS said, as the other member for the Chesterfield Institute, he simply wished to say that he endorsed what Mr. Jackson had already said. There was a strong feeling in their Institution that such an Institution as the late Mr. Bunning had proposed should be established. As to the exact details of that Institution, of course neither they nor their Institute were able to say anything at present; but he was sure of this, that they had the strongest feeling that some sort of Institution such as the late Mr. Bunning had proposed should be established. He hoped that that meeting would arrange some of the details necessary.

Mr. WALKER, on behalf of the Midland Mining Institute, said he could hardly add anything to what had already been said by Mr. Jackson. They felt, as he did, that their Institute was not so important, nor composed of men (in great part, he meant) of such high status, as the North of England Mining Institute; and recognizing as they did, that theirs was a weaker Institution, and that it had a particular work to do amongst the class of men Mr. Jackson had mentioned, namely, to a large extent amongst under-managers, under-viewers, deputies, and mining students, they could not help feeling that anything that tended in any way to make their Institution less suitable to their needs would be a loss to them, and therefore, although they desired very thoroughly to support the idea of a federation of Mining Institutes, they did feel that in drawing up a scheme for that purpose the council ought to bear in mind the nature of Institutions like theirs, whose status, perhaps, was somewhat more humble than that of the North of England Institute, from which the proposal originally came. The council of the Midland Institute very thoroughly agreed in the general idea of the late Mr. Bunning's scheme; but still, it lacked precision. It was put forward merely in the first instance as a feeler to elicit, he presumed, the views of the different mining engineers throughout the country; and that being so, it remained, as it were, to formulate at that meeting, if the idea was gone on with, something of a more definite kind which the different councils of the different Institutions might consider. They did not gather from the paper very clearly to what extent the subscription to their own Institution would have to be increased. Their subscription was only a guinea, and that guinea was a good deal to men of the class he had alluded to—such as under-managers and so on; and they could not recommend anything that very largely increased the amount of the subscription. Then again, they would like to know what was to be done in the case of those papers which were perhaps somewhat old to the profession as a whole, but which it was very desirable should be brought before local Institutions? There were certain papers which had appeared in the old numbers of the North of England Mining Institute Transactions, in connection with the special features which had been dealt with in particular collieries. He remembered a very curious question on ventilation—he thought it was at Pontop Colliery—which was brought forward by the late Mr. Atkinson, Government Inspector of Mines, and which he treated in such a way as brought out very clearly the principles of natural ventilation, and the influence of the different sizes of shafts, and so on. Those papers, to a certain extent, were buried in the old numbers of the Transactions of the North of England Institute; and something of a similar character might be brought forward by some one in another district and re-treated. Well, the feeling of the profession might be—" This is very old; this is something we do not very much care to be told again;" and yet, for the kind of people they had in their minds, it might be exceedingly useful. No one could read the correspondence in the weekly newspapers, such as the *Colliery Guardian* and the *Iron and Coal Trades Review*, without seeing that these papers were being constantly made the medium of correspondence of interest. Therefore, one thing he should very much like the present meeting to consider was—in how far they could permit to the different Institutions some freedom as to the selection of papers, and not impose too rigid a rule as to those which should be submitted for acceptance. Had he known that Mr. Chambers was there, he would have much preferred that that gentleman should have said what was to be said on behalf of their Institution. Later on, perhaps, Mr. Chambers might give them their views more clearly.

'Mr. LUCAS said, speaking for the North Staffordshire Mining Association and Mechanical Engineers, they found themselves in a similar position to that of the former speakers, and something beyond that. Their Institution was composed partly of mechanical engineers, and it would affect those members if one part of the whole, the mechanical element, which used to combine with them, refused to subscribe. He (Mr. Lucas) had very little to say about it, because he thought the former speakers had said everything. So far as he was concerned he would simply add that personally he was strongly in favour of federation, because he saw its necessity, not only as a federation to promote their common well-being, and to exchange views and that sort of thing, but it would be a very powerful instrument in the hands of a Mining Association as regards resolutions from such a federation, for example, if it went before the Home Secretary on matters of mining interest. In fact, on the late Mines Regulation Bill, he knew the difficulties that they had to contend against there, so that personally he was strongly in favour of such a federation being formed, but he was at a loss to define the means, and he would advocate in their Institution the formation of such a federation.

Mr. HAINES said, he also represented North Staffordshire Association, and he could only endorse what Mr. Lucas, and also the representatives of the Chesterfield and the Midland Institution, had already said. The mining engineers of North Staffordshire, he might say, most heartily supported the scheme, but the colliery managers and the other members of their body, who were not mining engineers pure and simple, did not see their way as a body to join it. He was sure it would have the hearty support of all those who practice as mining engineers, and he believed that advantage might accrue from it, but, as a body, he thought they had expressed very nearly the same views as those of the representatives from the Chesterfield and Midland Institute.

Mr. SMITH said, as representing the South Staffordshire Institute of Mining Engineers, he would simply express the opinion given by the council of their Institution when this matter came before them, and that was that they thought there could scarcely be two opinions upon the subject, and certainly that such a scheme as that suggested by Mr. Bunning was very desirable. It was rather surprising to see some of the objections and answers given on the paper that they had before them. No one but the secretary of an Institution could so fully appreciate some of the troubles and difficulties set forth by Mr. Bunning in his paper. They found very often that inventors and suggesters of schemes for the improvement of mining science brought their papers to them and made them a sort of advertisement; and although, in the rules of almost every Institution, it was laid down distinctly that papers were the copyright of the Institution, still they constantly found that the same papers were being read throughout the country. Then again, there was no doubt whatever that if they could have Transactions of a Central Institution as suggested, they would be a very valuable addition to the mining literature of the age. Another thing was that, having a Central Institution where all the great questions would be thoroughly investigated, they would not have the difficulties they sometimes met with in the local districts, where they sometimes, he might almost say, floundered upon some questions because they were not fully conversant with the whole of the ideas involved, or where they had not the advantage of getting the pick and selection of the mining science of the day. All that would be at an end if the questions were considered by a Central Institution. A great many of the objections—well, not exactly objections, but doubts—expressed by previous speakers, were really met in Mr. Bunning's paper. In regard to the subscriptions, for instance, it was not a *sine quâ non* in accordance with the paper, as he read it, if the Institution adopted the idea, that the whole of the members should join—that the Institution should come over as a body— they might be federated, but they might have a considerable number of their members (it was distinctly stated there) who might not be members of the Central Institution. That was very well dealt with in Mr. Bunning's paper, and although he stated distinctly that they should not lose their individuality, every Institution, as the gentlemen before have expressed it, would object strongly to the scheme if there was an idea that they would lose themselves, as it were, in the Central Association. Mr. Bunning pretty clearly put it, that such a thing was not desirable, because of the varying requirements of the different districts, and he (Mr. Smith) did not see that really he contemplated such a thing in his paper

Mr. CHAMBERS said, he did not think he had anything to add to what had been already said. The Midland Institute thoroughly approved of the scheme generally, though criticising some of the details. They were particularly anxious, having regard to the great number of members, as Mr. Walker had already said, that their subscriptions should not be increased. He did not entirely endorse that himself, because, of course, they were going to get additional benefits, and he thought they ought to be willing to pay something for them, and he was quite sure a scheme of that kind would be a great benefit to the whole of the mining districts of the country. They approved of it generally, and should be happy to co-operate as regards the details of the scheme.

Mr. HOWARD—Mr. Bewick suggested that each speaker should say for what Institute he appeared, but as he had not been delegated formally by the Institute with which he was immediately connected (the Chesterfield Institution) he did not know whether his summons there came from the North or Midland Institution. He felt a little more at liberty perhaps than he otherwise would, in consequence of not having been delegated by the Chesterfield Institution, to express the opinions formed in his mind, namely, that it was scarcely federation that was practicable in their case. It was more like affiliation to a Central Institution. He thought, however they were to go about it, that would be the result, and he thought it was well worthy of the consideration of the gentlemen then present; and his hope was that, before the meeting separated, something would be formulated of that character that the delegates could take back with them to their several councils and put before them; and also that means should be taken to ascertain what strength there was and what probability there was of establishing a Central Institution, with the object of assisting and furthering the views and objects of the local societies, and doing all the good that a Central Institution could do to them; not draining them, but really helping them on, and charging them no more than need be for anything that it might do for them. There would have to be something, he thought, in the way of contributions from the societies for anything that was done for them, but it would have to take that form rather than that which had been suggested in the paper. He saw great difficulties with regard to the class of members that they and other Institutions of the same kind had; and there was no doubt the Central Institution would be composed of what they would term the cream of the profession. The subscription itself would, no doubt, do that to a great extent, and, he thought that that being so, the thing would be worked out best upon those lines.

Mr. FORSTER BROWN said, he was there, not as representing the South Wales Institute of Engineers, of which he happened to be a member, but as a member of the North of England Institute. He had long held that the mining interests of the country, with which the mining profession particularly have so much to do, were of sufficient importance to justify a Central Institute, which would add weight to the particular mining profession, both for legislative and for other purposes ; and from that point of view he had gone so far as to hold that the parent Institute (the North of England Institute) ought to take the thing up, and, whatever the consequences were, to promulgate a proper scheme. But the effect of Mr. Bunning's paper, and the opinions that had been expressed that day, showed that five-eighths of the whole of the Mining Institutions of the country were in favour of such a scheme ; and it seemed to him that the next step to take was that a committee should be appointed, and that the gentlemen representing all those Institutions which were in favour of federation should be members of that committee, with a view to propounding some scheme of an Institution in London, leaving and still maintaining the local Institutions, but which would ultimately become the Mining Institute of England. And he had not the slightest doubt that if such an Institution was started on a sound basis all those dissentients would join in due time.

Mr. A. L. STEAVENSON said, the great doubt in his mind was whether the thing was financially possible. That it would be advisable in the interests of the profession of mining engineers there could be no doubt ; but, looking at Institutions of a similar character, so far as he could read the matter, the Mechanical Engineers spent £4,000 a year in doing what they practically proposed to do with the Imperial Institute; and he did not see how it was possible with an additional subscription of one guinea to meet all the contemplated expense. If it would not do so, it was possible that members such as they had now in the various Institutes would be able to contribute to the funds of the council. Again, he would refer to the possibility of there being drawn away from the local Institutes a good many of the members of their Institute belonging to South Wales and Yorkshire. He thought those members would entirely give up their connection with the Institute at Newcastle. That course might tend to spoil their local Institutions, although it might be beneficial to the Central Institution. There was just one other point, and that was as to the decisions upon the papers which were to be printed by the Central or Imperial Institute. Mr. Bunning, he thought, in his paper suggested that a meeting of the council should be held at certain times to select the papers that were to be printed; but he considered it was very essential that they should be decided upon at the time they were in print; because if the type had to be taken down and renewed it would add materially to the cost of producing the Transactions; so that that was a difficulty which would be met, he thought, by a very competent secretary. The central secretary in London would be better able to select the papers that he thought suitable, and would do it better than a committee; he would sit more as an impartial arbitrator in the matter, and decide what he thought best for the Institute. He could then at once give orders for the type and the plates to be put in hand and printed at a much less expense than if first done by the local centres, or if done for the local centres and afterwards reprinted for the Central Institute.

Mr. CHAMBERS said, he did not know whether it was taken for granted that because Mr. Bunning had suggested that the Central Institute should be formed in London, those who approved of the scheme generally agreed to that part of it.

The CHAIRMAN—Certainly not.

Mr. CHAMBERS thought that was assumed by one or two speakers.

Professor BENTON said, he had nothing to add to what they had already heard. He saw in Mr. Bunning's paper the germs of an excellent scheme, and he was waiting, personally, with great impatience to see its full development.

Mr. COCHRANE said, there was one remark he should like to make. It was thrown out by one of the members that the federation should be a body dealing with parliamentary matters and resolutions being passed in that direction had been mentioned. He took it that their impression would be that such a federation would be for scientific purposes only. If those Institutions were to federate with any idea whatever of troubling themselves about the outlying matters which were indicated in the speech which fell from one gentleman, he thought it would be a great mistake. That was his impression at the present moment. He was quite capable of being impressed otherwise afterwards, but he thought the object of their Institute—at any rate, in the North of England—had been so strictly confined to the scientific and purely mining part of engineering, that he should be sorry to see any Association connected with that body that was otherwise intended. He also saw the great difficulty of dealing with an entire Institute and saying that that Institute was to pass over to the federation as an Institute, thereby forcing all its members to become chargeable therewith; and he was certain that Mr. Steavenson's prophecy would turn out to be accurate, namely, a considerably increased expense; nor did he quite agree with other speakers who said the cream of the profession would go up there—by which, he was certain, they meant not the cream intellectually, but the cream so far as their pockets were concerned.   It would be a great mistake if the federation aimed at that, or thought that that was the way in which a federation was going to be supported.   The scheme certainly ought to encourage not members as a body but each individual of his own voluntary action to come out from the existing Institutes and join the federation under arrangements made by each local Institute as regarded each member, having due regard to their status in that body, whether under-viewers, mining engineers, rich men or poor men, and it should be left voluntarily to each local society to make all such arrangements as to distribution of Proceedings (which, after all, was the great thing) and the attendance upon the joint meetings as they liked; but to force any member of a local Institute into any higher subscription than what he now pays, would, to his mind, be a very great mistake.   One other thing was, the subject of the use of papers which already were almost taken as text books, and particularly in the North of England Institute of Mining Engineers.   That was a very important point.   He thought it was Mr. Jackson who mentioned that.   If the affiliation was to be perfect and thorough, one of the items they would have to consider would be a retrospective affiliation as well as a forward one.   They had, as every other Institute had, a very strong idea upon the subject of the copyright and value of their Proceedings.   Many of the Proceedings of the North of England Institute were at that moment absolutely out of print.   He did not say that that Institute would be prepared to give up the question of copyright; but, certainly, he thought, if the affiliation was to be perfect, that it would be a very desirable thing to consider the question of allowing an absolute reprint of those papers in other local Transactions; therefore, if they really intended to marry, they must go in for better or for worse, and they would have to make such a consideration as that retrospective as well as general in the future.

Mr. MARLEY said, he should have been glad if their ex-President (Mr. Daglish) had given them his views first, as being to a certain extent, prior to Mr. Bunning, the father of the idea which was contained in the paper; but as Mr. Daglish had not done so, he would state very shortly some of his views upon the question.   He might say, to begin with, that it was desirable that a federation of some kind should be carried out. Mr. Bunning in his paper might have pitched the case a little too high, although it was probably well to aim high so as to get a medium result.

There was no doubt that the question of finance would be a very important one, and as was suggested to Mr. Daglish four years ago, when they first entered upon the subject, equally important would it be that the local Institutes should not lose their individuality. These were some of the points, after hearing the speakers that morning, to be agreed upon. The other matters would, to a very great extent, become details; for instance, it seemed that they were sufficiently unanimous so far, and that there was a sufficient adhesion to show that joint publication should be carried out; but that the members of a whole Institute should be transferred to the federated one did not seem desirable. The respective societies would probably federate for the purpose of publication, leaving it to their members' option, at some small individual increased fee, to become members of the federated society or not; and that would to a very great extent probably facilitate the question of finance—that was to say, they would save money in all probability by joint publication of selected parts of their respective Proceedings; and then the other expense would be met by special fees at the option of the individual members. As regarded the parliamentary question he was glad that Mr. Cochrane had touched upon that, because, although their Scotch friends said that the Mining Association of Great Britain would meet all that was required, that was not the case. Parliamentary matters should certainly be outside the scope of such an Institute, and the federation and the respective Institutions should not meddle with anything but what was scientific and for the saving of life—in fact that which was laid down by their first President as the principles upon which the North of England Institute was formed. These were the principal points which, subject to details, he thought they were in a position to carry out.

Mr. STEAVENSON said, there was one point which might be attended to at once, whether they went on with the Institute or not; that was to provide the members of every Association (there were nine of them) with a copy of the index of their annual Proceedings, so that they might know exactly where to find any paper which had been published during the last year in any one of the societies' Transactions. It would at once keep them up to the mark as to what had been done all over the country. That might be forwarded to every member of every Institute.

Mr. DAGLISH, in answer to the Chairman, said, he had really very little to say because, he thought, every point that could bear upon the question had already been touched upon. He made a few notes when they commenced, and, with their permission, he would just draw attention to the scheme which seemed to suggest itself to him, and several others with whom he had been in communication, as being the most practical, and that was to confine the federation very greatly to simply publishing joint Transactions, and, if that were so, it must be more economical than the present system. It could not lead to a greater expenditure, but would lead to a less expenditure. It would not be necessary to publish the whole of the nine volumes which were now published by the various Mining Institutes, for the reason which had already been given by several gentlemen, namely, that many of the papers were already duplicated, others were papers of only temporary interest, others of a purely local interest which it would not be necessary to publish in the more expensive form. In the French Mining Association (La Société de l'Industrie Minérale) they publish two sets of Transactions. They publish very elaborate and beautifully got up Transactions of their more important papers; but they also publish, in a very cheap pamphlet form, the papers of a more temporary or unimportant character. If such a scheme could be adopted, it would suggest itself that there should be a committee of selection, and that possibly each Institute would do that for itself. There might be some control over that by a joint publishing committee; and it would be only selected papers that would be published in the larger form, leaving out, for the time being, the consideration of publishing the papers of mere temporary interest in a cheaper form. In that way, he thought, they would get all the important and interesting papers read at each Institute which at present they did not see. In addition to the joint publication of Transactions might be added the privilege of membership of each other's Association, so that if a paper was read at any particular Association any member could attend, if he had a special interest in that subject, and take part in the discussion, not as a matter of favour, but as a matter of right. A second question of very great importance, which had already been touched upon, was that of investigations for special objects. Many of them had been repeated at nearly every Institute. Almost every Institute had had committees upon Safety-Lamps, and upon Fans, and now upon Explosives. The North of England Institute had just appointed a committee upon Explosives, and he thought the Midland Institute had recently conducted investigations on the same subject. They had now formed amongst three of the Institutes (and they would meet again that day on the matter) a joint committee for Fan investigation, and, he ventured to think, it would be attended with very excellent results. Touching on a matter mentioned by Mr. Cochrane, and supported, he thought, by Mr. Marley, he would venture to say, in reference to parliamentary questions, that many of them bore upon scientific subjects, which were not dealt with in the least by the Mining Associations. In the last Mines Act, the question of the safety-lamp was imported into the Act; and also the important question of the use of the fire-damp indicator, which was a purely scientific question, and one which ought to be specially worked out by a scientific Institute; but at present the Government, if they wished for reliable information upon scientific questions bearing on mining, had not any one body to whom they could refer for information. If they sought this from any one of the Mining Institutes, there would probably, and naturally, be a certain jealousy on the part of the others, besides which each district had not exactly the same peculiarities of condition. The circumstances and requirements of each district were rather different, and it would hardly do for any one Institute, in any one district, to take by itself any leading action; but if there was a general body, with whom the Government might communicate upon scientific subjects, he thought it would be a benefit, not only to the coal trade at large, but especially to mining engineers, whose character and responsibilities were so greatly affected by those legislative Acts. There was mention made by Mr. Jackson of the importance of not interfering with the individuality of each Institute at present. He thought that seemed to be the unanimous opinion. That was more needed in Mining Institutes than in any of the engineering Institutes, because many of those gentlemen referred to who were members of the former could not go up to meetings of a central body. They could not attend any meetings unless held in their own locality; and it was of the first importance that those gentlemen should retain their membership. He thought that it would be the opinion of almost everyone who had considered the subject, that it would be fatal to in any way interfere with the individuality of the present Institutes. If each Institute, however, did not join as a body but individually, he did not say that the scheme could not be worked out, but there would be some difficulty in supplying each member with a copy of the Transactions if they were published by a central body to which they did not belong. He would just add that the points which seemed to him to be of chief importance were—1st, combination for publishing only; 2nd, combination for experimental research; 3rd, the importance of a united body to whom the Government might apply in case of requiring reliable information on scientific questions affecting mines, and obtain the opinion of those who had given time and attention to the study of those subjects as a body, rather than as individuals.

Mr. BEWICK said, in answer to the Chairman, that he did not think he had anything to add to what had been so well said by others.

He certainly should very much like to see the project carried out ; but he could not but also see that there were very grave difficulties in the way of it.   Probably if a committee were appointed that day to consider the whole bearings of the case, all those difficulties might be overcome.   It was only as they appeared, perhaps, on the first blush of the thing, and they might be got over.   He must say, taking the scheme as a whole, he was quite in favour of it.   He did not know whether there was anyone there from those who had said " nay " to the project, such as the Mining Institute of Scotland, or the Cornish Institute, or the Geological Society of Manchester ; if so, perhaps they would express their opinion.   They were the people they would like to hear from.

Mr. COCHRANE asked if there was anybody present who had made a calculation of, or had the slightest idea as to the cost?   Was there any impression on anybody's mind on that subject—the cost of the federation ?

Mr. STEAVENSON said his impression was that it would cost £3 3s. each member, and that it would have a bad effect on local Institutes.

Mr. DAGLISH thought 10s. a member would be sufficient merely for publishing, and the present system of publishing cost more than that. Therefore combination for that purpose would be attended with no extra cost whatever.

Mr. COCHRANE—Would they excuse him rising again ?   With regard to the question of publishing which Mr. Daglish had thrown out, which was no doubt true, it seemed to him that considerable expense would be saved in that direction, and that they should still have every paper—not merely the chief, but the temporary ones spoken of—which would be a great advantage, from all their societies.   Suppose they became affiliated in a small manner, and with a secretary there, for the object of being a body that could be referred to, as had been indicated, each society might guarantee to take from each other society its Transactions to the full extent of the membership of the society.   It seemed to him that 10s. each —there were 2,200 members at that moment—would be sufficient. Those Transactions could certainly be supplied at 10s. each ; that was £1,000 or £1,100.   That was simply a trifle as compared with what was now spent, and that would furnish as many copies as would be required. He agreed with Mr. Steavenson that something like £3,000 or £4,000 would be about the cost of the society.

Mr. DAGLISH—If they established a centre in London ?

Mr. COCHRANE—Yes ; £3,000 or £4,000 per annum.   The other societies being affiliated might say, for instance, "We will undertake to take 120 copies of Transactions of each one of the other societies," and so on. If they were to do that, the cost of the printing would be much cheapened, one's productions would be open to everybody, and everybody would have a copy of everybody else's Proceedings.

Mr. DAGLISH said that was still making the Mining Institute publish nine volumes every year; whereas the Mechanical Engineers, with all their large numbers of able members, only published one. It was clear they were all publishing papers that ought not to be published, at least in such an expensive form.

Mr. WALKER said, that with regard to that idea, the difficulty now arising with respect to printing seemed to be that they had to print at a number of different places. He supposed that much could be arranged, that one printer should be appointed to print for all the Institutes, and that the size of the Transactions should at any rate be uniform. Then, in any case, the whole of the papers that were read before any Institute, and the discussions upon those papers, would be put into type and printed, but the actual number printed would depend upon the demand for them. For instance, supposing that a certain person read a paper before the North of England Institute, and that Institute required a thousand copies for its own members—he said that simply to take round numbers—a thousand copies at least of that paper would be printed, with the discussion upon it, to go to all the North of England members; but, in addition to that, either the paper might be considered of such general interest that it might be decided at once that it was worth sending out to all the other Institutions, or any Institution which desired could have copies of that paper, and the discussion might be supplied with it on certain terms to be arranged. Then he thought it would be very well if their Transactions could be sold by a publisher or by the secretary of the General Institution at a certain price fixed by the council or the author of the paper. Nothing had been said yet about discussions. Now, he thought it would be a great pity if they were to confine themselves to the publication of the papers without the discussions, or at least abbreviated discussions. Sometimes very erroneous ideas were promulgated in papers, and if such a paper went forth with the *imprimatur* of the Institute, it might seem as if they were to some extent giving their sanction to ideas which were perhaps erroneous, premature, or badly digested. Now, would they just take an illustration which had been alluded to at that meeting? They had been having papers at the Midland Mining Institute on new flameless explosives. The discussions had been, he thought, very much more valuable than the original papers—at any rate quite equally so. Different gentlemen have brought before the Midland Institute the result of their individual trials and their individual experiences under very wide and different circumstances, and the net result had been a very valuable amount of information brought together, which certainly would not appear in the original papers. Then, he thought that the discussions, so far as those discussions appeared to be pertinent (and the general secretary might strike his pen through anything that was said that did not appear to be pertinent) should be published in smaller type with the papers. As to Mr. Daglish's remarks about their not knowing what papers were being read in different Institutes, he presumed that a general circular would be sent out each month saying what papers were going to be read before each of the affiliated Institutes, and the question of time was rather important. There must not be too much delay before they got their papers. In the Midland Institute they made it a rule to always have papers month by month. Thus they were received before the meeting, and members went to the meeting prepared to discuss the papers. If any great amount of delay were to take place before members got their papers, then the result would be that a very great deal of interest in the subject would be lost, and perhaps time wasted which it was desirable to save. Then, it was proposed to call that Institute the Imperial Institute. If it was really meant to be an Imperial Institute, was there any idea of extending its operations to the Colonies? Some of them were becoming more and more interested in mining in other parts of the world, and if it was to be an Imperial Institute it certainly seemed that if its scope could be extended to Australia and Canada it might be a great help to those of them who had any dealings with coal-fields in those parts of the world. For instance, the coal-fields in New South Wales appeared to be very different in geological structure and conditions from those in this country, and the publications which had come under his own notice with regard to those matters were very scanty, so that if their Proceedings could be opened to reliable reports by engineers on the spot as to the structure of the Colonial coal-fields and so on, it seemed to him that it might be exceedingly valuable to so extend the scope of the Institution. It would be rather a mistake to call it "Imperial," unless they intended to embrace in their scope something more than the four corners of the United Kingdom.

The CHAIRMAN said, he believed Mr. Mitchell, the Secretary of the Midland Institute, had just come into the meeting, and they would be very glad to hear if he had any opinion to offer upon the subject.

Mr. MITCHELL said, he quite agreed with Mr. Cochrane, and he could only endorse his remarks so far as he was personally concerned.

Mr. SMITH, with regard to the question of the Colonies, asked if they were not rather taking it that the whole of the Transactions would have to be published, and the expense borne by the membership fees of the Central Institution? That was not the idea of the paper; because, in addition to the fund derived from the membership fees, he took it that each federated Institution would have to contribute to the Central Institution for the publication of the Transactions, and they could well do that if they were spared publishing their own, because it would be a material saving to them.

Mr. MILLS said, he would like to ask whether they could not do something that day to bring the matter to an issue? They must have funds. It had been proposed by the late Mr. Bunning that the fees, in the first instance, should be a call upon the funds of the Institution. He did not think that would be at all an advisable step. He thought that if they could raise a little money in the first instance to have a secretary for their committee, they might afterwards ascertain the extent of support the new Institution was likely to receive. He had looked into the question of subscription and finance to some extent, and he would propose that there be several classes of members. In the first instance, there might be the first class, called "fellows," or anything of that description—that was to say four guineas; then members at two guineas; non-resident members, such as Mr. Walker proposed, in the Colonies and different places, one guinea; also associates, who would be the under-viewers and people who could not attend, under-viewers in the Colonies, and students as well. He thought if they could come to some idea as to the subscription the Institution would require, and then ask each one of their own members of their own Institutions whether they would join the Institution at the subscription, it would be a great thing.

Mr. CHAMBERS saw very great difficulty in the suggestion which Mr. Mills had made. He thought they wanted a scheme first, and thought the proper thing to do would be to appoint a small committee to draw up one to meet the views of all the gentlemen; then they could have some idea of what the conditions would be, and what amount of subscription would be required. He thought that, taking the moderate scheme which Mr. Daglish had sketched out as a basis, it was quite possible to draw up one which would unite them to a certain extent, though not to the full extent which the late Mr. Bunning desired, nor to the full extent which some of them hoped it might develop into by and by; but still he thought a scheme might be drawn up which would only necessitate a moderate expense, and which would be the first stage in uniting the Mining Institutes of the country.

The CHAIRMAN said, Mr. Daglish had placed in his hand a motion, which he would leave him to bring forward afterwards. He (the Chairman) reminded them that in the first place they must prepare a scheme; but not until they received a preliminary and conditional assent of every one concerned. They had had the matter now before them, he ventured to say, in a fairly complete form; they had heard the opinions, with many of which he entirely agreed—some he was perhaps not inclined to adopt without modifications. Their observations would all be printed in a condensed form, and they would have an opportunity of considering the general views of those who had taken a part in that morning's proceedings, which might, as far as possible, be embodied in some scheme that ought to be drawn out; and it ought to be drawn out, as had been suggested by two or three of the last speakers, by a small committee. He thought a copy or copies of that scheme might be sent to the different Institutes, to ascertain how far they would feel inclined to join in the federation.

Mr. DAGLISH said, the only feeling in his mind about appointing a committee was, that it would only result in a scheme emanating from those gentlemen, individually, who formed the committee, because they had not really any scheme as yet before them, although they were met there that day as a committee appointed by the several Mining Institutes favourable to federation for the purpose of presenting some tangible scheme to their respective Institutes. They had a suggested scheme for a central confederated body for all purposes, with a special subscription, and another suggested scheme purely confined to publishing; and these were two entirely different things.

The CHAIRMAN said, he left it to the discretion of the committee to propound a scheme either on the lines spoken of in the one case or on the other as they thought fit.

Mr. DAGLISH said, if they would allow him, the only object he had in suggesting that a definite resolution should be proposed was, that it was easier to speak for or against and to a resolution, rather than to deal broadly with various facts. He was going to suggest a resolution for discussion which, if agreed to, the delegates present that day could submit to their councils as something definite.

The CHAIRMAN—They would see that if that resolution was carried they rather confined the committee to draw the scheme upon the lines of that resolution, whereas he wished to leave them entirely free.

Mr. DAGLISH said, he did not propose to limit it to that, but simply to commence with that as something definite.

The CHAIRMAN—Supposing the first step were to annul it ?

Mr. DAGLISH—They could suggest something else.

The CHAIRMAN—It would be rather an awkward thing for a committee which had been appointed by a particular resolution if their first step would be to cancel that resolution.

Mr. SMITH said, he thought the Chairman was quite right. They were not as delegates in a position to pledge their Institutes to any course. As the Chairman has said there was no actual scheme before them.

The CHAIRMAN—No; but he gathered that on the whole they were all favourable to federation.

Mr. SMITH—To the principle.

Mr. COCHRANE said, he would propose something if they would allow him which, he dared say, would commend itself to the meeting. It was—"That this meeting recommends that each society appoint its secretary to form a committee, and that each society bear the expense of its own secretary, in order to formulate a scheme to submit to a future meeting of this committee." The object in doing that would be—each secretary would go back again and have an opportunity of consulting in his own way his own council at the minimum of expense and at the maximum of convenience to the members, and also with the power of learning in the best way what his particular Institution wished. Those secretaries ought to meet, and they were best capable to formulate the matter. Let each society bear its own expenses. He proposed that the meeting appoint each secretary as a member of a committee to formulate a scheme, and then, having got their report, they might go back again to their societies simply with that as a recommendation ; and as to the meeting place, he forgot to suggest that.

Mr. MARLEY—Say, the presidents and the secretaries.

Mr. COCHRANE—No; he should only ask for the secretaries—it was practically their duty to formulate the scheme. They could collect the best opinion from their own councils ; and he proposed that the meeting place be Derby.

Some conversation as to the place of meeting then took place.

Mr. COCHRANE—Let that be a subsequent matter. If the idea be that the secretaries should do that, they being the responsible people for formulating a scheme, and each responsible to his own council, it met the proposition that far.

Mr. JACKSON said, he would be very glad to second the motion, because he thought it was a very practical way of bringing the matter to a start.

Mr. FORSTER BROWN said, although he quite agreed with Mr. Cochrane that the secretaries of the different Institutions were the gentlemen who certainly should be members of the committee to formulate the scheme, he did not think that it should be limited to the secretaries of the Institutions, because they might have certain views, whilst the members of their particular Associations might have different opinions, and he thought the committee ought to be very much wider, because, after all, it all hinged upon the report of that committee as to whether the matter was to be carried out, and whether it was to be carried out on sound lines ; therefore, he thought, the first step was to have a representative committee of all the Institutions who were favourable and who would promulgate their scheme for the consideration of the committee, and he begged to propose that.

Mr. CHAMBERS said, he quite agreed with Mr. Forster Brown that the committee consisting of the secretaries only was not wide enough. He was going to suggest a comparatively limited committee himself, and was afraid that it would not be as some gentlemen would possibly desire it, but he thought it would be sufficient for the purpose—namely, two members from the North of England Institute, and one from each other Institute, in addition to the secretaries. That would be a small workable committee. If they got a very large committee the members were liable to leave the work to others ; then, if they all attended, there would be a very great difficulty in getting through the business.

Mr. COCHRANE said, the expense was an important item. At the present moment nobody was subscribing any money. He did not want to interfere with the conditions.

Mr. MARLEY moved that a committee of three or six be appointed by each respective society to consider the necessary details for carrying the matter out, and then that a general meeting of the whole be held to formulate the result of what they in their opinion thought was best to promote a federation. It was, practically, Mr. Forster Brown's suggestion.

The CHAIRMAN said, that would make about ten more members than the whole meeting then present. He asked if Mr. Marley did not think that he was suggesting somewhat too largely ?

Mr. MARLEY thought he might say one word more, just by way of explanation. In proposing from three to six in each society, he proposed that each society should then send a deputation out of that six to meet together ; therefore, practically, it was, say two out of the number.

Mr. DAGLISH said, that each society should see that two of its members did attend the committee meeting.

Mr. MARLEY—Yes.

Mr. FORSTER BROWN said, he had no objection to that.

Mr. MARLEY said, each society would appoint its own member.

Mr. SMITH said, practically, if Mr. Cochrane's resolution was accepted in the amended form, appointing other representatives besides the secretary, the societies themselves would do that. He thought they should leave it to the Institutions themselves. As Mr. Marley suggested, they would meet and instruct their representatives how to convey their ideas to the committee. He did not think that from that committee they ought to stipulate that the several Institutions should appoint committees.

Mr. COCHRANE, addressing the President, said the secretaries were the people to formulate—not in any way to determine. After that they wanted some tangible resolutions.

Mr. DAGLISH—They must have a meeting.

Mr. MARLEY—They must have a meeting.

Mr. COCHRANE—Of the secretaries ?

Mr. DAGLISH said they thought there should be some other members present as well as the secretaries.

Mr. COCHRANE—Yes, at each council. What he proposed was—that that meeting should recommend each society to appoint its secretary, then those secretaries would be instructed by the council of each society fully. Then they would meet, they would formulate a scheme in writing—resolved this, that, and all the rest of it. He might say it was peculiarly their province to do that.

Mr. DAGLISH—The only further condition was, that in addition to the secretaries there should be one or two members of the council.

Mr. COCHRANE said, he did not object to it, except on the ground that they were putting each society to very much more expense than what was necessary. In the one case they simply sent the secretaries to that meeting, and in the other case also they would not get the attendance of those members at the meetings, whereas the secretary's official position was such that he could always go to those meetings, and the members they might appoint might not be capable of doing so.

Mr. DAGLISH said, he thought Mr. Marley and the Chairman had put that right. It was proposed that each society should see that two of their members attended by arrangement.

Mr. MARLEY said, he thought they were all pretty well agreed.

Mr. HAINES—What was proposed had been really done, and they had it before them. The representation had already been done. It was not only the representatives, but what they might or might not do.

The CHAIRMAN asked if he meant in point of numbers?

Mr. HAINES—Yes; he thought they had provisionally three names from each Institute before them.

The CHAIRMAN said, it did not follow that those gentlemen would continue their attendance. They could see that what they wanted was to have the sanction of that meeting.

Mr. HAINES said, he was merely speaking of their Institute; their representation was prospective rather, he should say.

Mr. WALKER said, it would be necessary for each council to have their secretary, he thought, so that on each point that might arise they could discuss such questions amongst themselves as there might be a little doubt about.

Mr. FORSTER BROWN—Simply two representatives.

Mr. MARLEY—Neither name nor office.

The CHAIRMAN asked Mr. Cochrane if his motion was that the secretaries be appointed?

Mr. COCHRANE—Yes; and the secretaries should be empowered to prepare the draft.

The CHAIRMAN said, the motion before the meeting was that by Mr. Cochrane, namely, that the secretaries be appointed in order to form a committee to draw up a scheme. To that an amendment had been moved that each Institute send two of its members, neither of the two being of necessity the secretary, to attend the meeting in order to draw up a scheme; and, as usual upon such occasions, he would put the amendment first.

The amendment was then put to the meeting, when 10 voted for it and 3 against it, and the Chairman declared the amendment carried.

The CHAIRMAN said, that seemed to him to complete the business; and he thought, before they left the room, they ought to allow him to convey the thanks of the meeting to the Institution of Civil Engineers for the use of their room.

After further conversation as to the proposed place of meeting,

Professor BENTON begged to move, as a representative of South Staffordshire, that Sheffield be the place of meeting.

Mr. MARLEY seconded that.

The CHAIRMAN said, the motion was that Sheffield be the place of meeting.

The resolution was put to the meeting, and carried unanimously.

Mr. DAGLISH—As Sir Lowthian Bell had been kind enough to act as Chairman on that occasion, should they ask him to be so good as to convene the meeting?

The CHAIRMAN said, he should be very glad to do that, if they wished it.

Mr. DAGLISH said he would propose that.

Mr. FORSTER BROWN said he would second it.

The resolution was put to the meeting by Mr. Daglish, and carried unanimously.

Mr. FORSTER BROWN begged to propose a vote of thanks to the Chairman for fulfilling the duties of the chair.

The resolution was put to the meeting, and carried unanimously.

The CHAIRMAN said, he had had very much pleasure in occupying the post that had been assigned to him.

Mr. DAGLISH said, he thought it should be known that the Chairman had come up specially that day from his place in Yorkshire to attend that meeting.

JOINT COMMITTEE OF THE NORTH OF ENGLAND INSTI-
TUTE OF MINING AND MECHANICAL ENGINEERS,
MIDLAND INSTITUTE OF MINING, CIVIL, AND
MECHANICAL ENGINEERS, AND THE SOUTH WALES
INSTITUTE OF ENGINEERS "ON MECHANICAL VENTI-
LATORS, 1888."

———

MEETING HELD IN THE COUNCIL CHAMBER OF THE INSTITUTION
OF CIVIL ENGINEERS, 25, GREAT GEORGE STREET, WEST-
MINSTER, WEDNESDAY, 6th JUNE, 1888.

———

Mr. A. L. STEAVENSON proposed that Mr. Daglish take the chair.

Mr. W. COCHRANE seconded the proposition.

The resolution was put and carried.

The CHAIRMAN said, they all knew why they had met, and he did not
think he need make any preliminary remarks. Mr. Walton Brown had
been in communication with the secretaries of the Midland and South
Wales Institutes, and he would commence the proceedings by asking
Mr. Brown to state exactly the position in which they stood just now.

Mr. M. WALTON BROWN said, he had corresponded, on behalf of
the North of England Institute, with the secretaries of the Midland and
South Wales Institutes, and the preliminary details were all thoroughly
understood and arranged on behalf of the three co-operating Institutes.
As to the expense of carrying on the experiments, each of the three
co-operating Institutes had agreed to subscribe not more than £100, and
it was agreed that the Report, when completed, should be the joint property of the three Institutes. Three engineers would be appointed, one
by each Institute.

Mr. HORT HUXHAM—That is right.

The CHAIRMAN—How many engineers is that then?

Mr. M. WALTON BROWN said, three experimenting engineers would
be required. Each of the three Institutes would appoint one, and each
pay the expenses of their respective engineer; that would be the simplest
way of proportioning the cost.

Mr. THOS. EVENS asked if it was understood that each engineer would be paid for his time, as well as his hotel and travelling expenses, or were they supposed to give their services?

The CHAIRMAN said, so far as the North of England Institute was concerned, up to the present time they had never paid anyone. They had always been able to obtain the services of suitable gentlemen who were good enough and able enough to undertake the duties; but it was possible they might not be able to do so, and he thought that it was left to each Institute to do as it liked; each paying their own engineer.

Mr. HORT HUXHAM—Not out of the £100 or £300?

The CHAIRMAN—No; I think not.

Mr. HORT HUXHAM—That is just the point. We are a little doubtful about it.

Mr. CHAMBERS said, he understood the Institutes joining in these experiments were three—the North of England, the Midland, and the South Wales, and no others.

Mr. M. WALTON BROWN—Yes; no others have been asked.

The CHAIRMAN asked if there was any scheme formulated?

Mr. M. WALTON BROWN said, he had drawn up a programme of observations to be made, and instructions to the engineers, which had been sent to the committees of the co-operating Institutes.

The CHAIRMAN asked Mr. Brown if he proposed that they should go through the programme *seriatim*?

Mr. M. WALTON BROWN said, he would suggest that it should be gone through *seriatim*, and ascertain whether they approved of it or not. He would read it through.

The CHAIRMAN asked if it was the general wish of the gentlemen present that they went through the programme *seriatim*?

Mr. G. B. WALKER asked if the general principles were agreed to as to what should guide the investigation? These were instructions, as he took it, to the engineers. The Midland Institute had been under the impression that they intended, in these experiments, to confine themselves very much to two things—first, to places where there were two ventilators of different descriptions working on the same mine, so that their results could be very accurately compared; and, second, to fans, which had not hitherto been dealt with. They felt that the experiments which had previously been published (for instance, those which the North of England Institute published some four or five years ago) were not quite final, and that it was very desirable to take into account all the ventilators which were now in successful operation, and that a sufficient number of each type should be experimented upon, in order to arrive at some reliable data as to their general characteristics, effects, and advantages. The Midland Institute thought that the conditions of the mines where these ventilators were at work should be clearly stated, as there were many sources of error, which have probably crept into former experiments through sufficient information not having been given when the results were published, in order to enable any one to decide what were the conditions under which the fans were worked, and under which the results were obtained. The idea of the Midland Institute was, therefore, to rather extend the scope of the investigation beyond what he understood was the idea of the North of England Institute..

The CHAIRMAN said, would not the three great fans be the basis of the tests—the Schiele, the Waddel, and the Guibal? Did not that cover nearly the whole ground of previous experiments? They would not require to test another Schiele, and another Waddel, or another Guibal in any way, because they would have tried these in the first sets of the experiments they made.

Mr. WALKER—Take for instance the Waddel at Celynen. He understood (he said it with all due reserve) that it was by far the best Waddel that had ever been erected. The results of that particular Waddel were the only ones which were given in the report of the North of England experiments.

The CHAIRMAN—Yes; but you see the Waddel will now be tested as against another fan, under exactly the same circumstances.

Mr. WALKER said, there was a second set of considerations—that was durability. A fan which, after five or six years' wear, was considerably shaken, was not so valuable a fan as one which had run for a very much longer period without any perceptible deterioration.

The CHAIRMAN asked if Mr. Walker proposed at present to move a resolution to extend this, or would he bring it forward afterwards as the work went on?

Mr. WALKER said, he simply made those remarks because he thought they were going on to instructions to experimenters before they finally decided what the scope of the investigation was to be.

The CHAIRMAN said, Mr. Walker had mentioned that the scope of the enquiry included two things: that was to say, the fans of different kinds upon the same mine, and the new fans that had never been tested. Did he propose to add a third to that, namely, to go over some of the other fans?

Mr. WALKER—That was the idea of the Council of the Midland Institute.

The CHAIRMAN—Are we not then going into a very large question, and possibly a question that will give rise to some degree of dispute and squabbling if we are going to test individual fans again?

Mr. WALKER said, he simply mentioned it because they had already appointed a Committee to experiment with a certain number of fans in the South Yorkshire district, and they suspended those experiments in consequence of the invitation received from the North of England Institute to co-operate with them, but, at the time they agreed to co-operate, it was very clearly mentioned by his Council that the enquiry ought not to be of too restricted a character. They thought that there were not before the world as yet any reliable statistics respecting the principal types of fans.

The CHAIRMAN asked if he (Mr. Walker) did not think that they, or rather most of them, had come to the conclusion that they could not place much confidence in those isolated experiments on account of the very fact of the circumstances differing so much; and unless they could get different fans under exactly the same circumstances, that really these experiments were of no value? He only mentioned that; he did not wish to put a stop to the investigation any further than it might be their wish. Perhaps Mr. Walker would test it by moving a resolution at once.

Mr. WALKER said he would rather do so, if there was time, after a little more expression of opinion. If they liked he would move the general resolution, "That the object of the investigation be to ascertain as thoroughly as possible the relative efficiency and value of the various kinds of fans now in operation." That was a somewhat comprehensive resolution.

The CHAIRMAN—Yes; that would carry it, certainly. Would any gentleman second that resolution?

Mr. ARMSTRONG, Jun., seconded it.

The CHAIRMAN—It being understood that at present the investigation is limited to two different fans on the same pit, and to fans that have not hitherto been experimented upon. It is now proposed to extend this investigation further. Those who are in favour of that, please signify the same by holding up their hands.

Mr. HORT HUXHAM—Before putting that resolution, he should like to ask what the words "experimented upon" refer to?

The CHAIRMAN—Published, I suppose.

Mr. HORT HUXHAM—Published you mean?

The CHAIRMAN—Yes.

Mr. HORT HUXHAM—Published in the Transactions of any particular Institute, or not?

The CHAIRMAN—The resolution covers everything. It is as wide as possible. There is no limit to it.

Mr. HORT HUXHAM said, what was passing in his mind was simply this, he apprehended every fan had been more or less experimented upon.

The CHAIRMAN—No ; every *system* of fan, not every fan.

Mr. HORT HUXHAM—Every system of fan ; and those experiments had been more or less published.

Mr. ARMSTRONG, Jun.—"Recorded" would be perhaps a better word.

Mr. HORT HUXHAM—Recorded in some particular Proceedings or Transactions ?

Mr. ARMSTRONG, Jun.—Yes ; in the Transactions. That is better than "published."

Mr. WALKER—The inference is, that the three Institutes combining to make these experiments would probably publish the results.

Mr. FORSTER BROWN—They would be the joint property of the three.

Mr. ARMSTRONG—Recorded in the Transactions of the three Institutes.

Mr. HORT HUXHAM—Quite so.

Mr. FORSTER BROWN said, he was going to suggest this : Would they not obtain all the practical objects they sought by testing different fans, where there are duplicates on particular pits ? By that they would get definite results as regarded those particular fans. But inasmuch as those particular fans probably comprised the principal fans which had otherwise been experimented upon, they would obtain all the objects required without going into an unlimited enquiry.

The CHAIRMAN said, probably, like every other gentleman present, he had made a number of experiments, and he found that the conditions were so utterly different that they could not compare two fans on different pits. They had engines underground with steam only; they had engines underground with a boiler. They could not tell how much of the ventilation was due to these actions; therefore, to take an experiment with a fan upon a pit was no indication of its relative value as compared with another fan on another pit.

Mr. A. L. STEAVENSON said, his impression of the origin of this Committee was that it was merely to test the fans where there were two fans of different descriptions on the same shaft, to satisfy the want that had been felt by mining engineers, and to prove whether two fans, worked under exactly the same conditions, gave different results. For his part, by a mere calculation alone, he thought they had satisfactorily solved the question; but then it would be much more satisfactory if different kinds of fans were tested on the same pit. He rather thought that they should first give attention to that point. If they began to test different kinds of fans they would get into a very extended range of examination, for there were a large number now of different kinds; but that should be considered before they started. He should like to suggest before they went to any very great extension of their work, they should decide as to how the cost was to be divided.

The CHAIRMAN—That was arranged not to exceed £100 for each Institute.

Mr. CHAMBERS said, he could not help thinking that Mr. Walker was leading them a little further than even the Midland Institute intended to go by his comprehensive resolution; and he was bound to say, after hearing what other gentlemen had said, that he thought it would be almost better for the Committee at present to confine itself to the scheme which the North of England Institute proposed. He should like also to suggest that they should make some experiments with the original fan, now forty or fifty years old—that was, the Biram fan; the first fan, he believed, put up in the country, and which had been running from the day it was put up to the present day. It would be very interesting to know what that fan was doing. He thought they could get permission to have it tested.

Mr. COCHRANE—They would find the whole of the experiments upon the Elsecar fan in the Transactions of the North of England Mining Institute, made by the late Mr. J. J. Atkinson and himself, before they adopted the Guibal type of fan.

Mr. CHAMBERS said, he was not aware of it; therefore he withdrew his suggestion.

The CHAIRMAN asked Mr. Walker if it would meet with his views to let this matter rest for the present? If the Committee had energy left, after they had finished the objects for which they were started——

Mr. COCHRANE—And money.

The CHAIRMAN—And money; or can get more. He quite agreed with him it would be very advisable not to let it drop. It would be a pity; but he thought they should limit themselves to the very large undertaking they had in front of them at present, otherwise they would never get to the report stage.

Mr. WALKER said, he should like to add some limited proposal to the effect that the Committee might experiment with fans which present certain novel features in their adaptation. Mr. Garforth, of the West Riding Colliery, had a Schiele fan at the top of a very small shaft, whose friction would be entirely abnormal, and the results of that fan should be very instructive.

The CHAIRMAN—Yes.

Mr. WALKER—And if any fan was, in the opinion of the Committee, so placed that it presented new features, he thought it would be a pity to neglect to get the particulars of the working of such a fan.

The CHAIRMAN—You will move no resolution then?

Mr. WALKER—No.

Mr. GARFORTH said, as the money that was voted was very limited, he thought it would be better to go step by step, and take it in two stages—first, the fans in duplicate at each pit, and make that the scope of the Committee's investigation at first; then, if the money ran to it, they might go into the other. The same Committee would continue the experiments.

The CHAIRMAN asked if Mr. Garforth would kindly move that resolution, and state, as the expression of the opinion of the meeting, that the Committee should report as soon as they had completed the experiments of the duplicate fans?

Mr. GARFORTH said, he should be very happy to do so.

Mr. COCHRANE—With the present money?

Mr. GARFORTH said, he should be very happy to move "That the operations of the Committee be confined to those cases where two fans of different constructions were erected on the same mine, but the Joint Committee at the same time express their hope that they will be able to extend their operations to newly invented fans after this series of experiments are completed."

The resolution was put from the chair and carried.

Mr. M. WALTON BROWN then put in a schedule of observations to be made, and instructions to the engineers, which, after discussion and amendment, were adopted by the meeting. (See Appendix p. 189.)

The CHAIRMAN said, in their case they had appointed Mr. M. Walton Brown, and the Joint Committee had asked Mr. Brown to act throughout as General Secretary also. Who would he communicate with on behalf of the other Institutes?

Mr. M. WALTON BROWN said he had communicated with Mr. Mitchell and Mr. Huxham.

The CHAIRMAN—Quite right; so long as that was understood.

Mr. HORT HUXHAM asked if it was understood that the Secretaries should accompany the experimental engineers?

The CHAIRMAN—Not unless they like; but it was expected that some of the Committee would be present always at these experiments.

Mr. THOS. EVENS—I think so.

The CHAIRMAN hoped that they would be present, both to assist and to see that the thing was carried out properly.

Mr. M. WALTON BROWN—Three engineers were to be appointed, one from each Institute.

The CHAIRMAN asked if it was the pleasure of that meeting that each Institute appoint one engineer?

Mr. GARFORTH—Yes.

The CHAIRMAN said, the next thing was where were the experiments to be made?

Mr. HORT HUXHAM—Is it left to the engineers to decide where they commence their experiments first?

Mr. M. WALTON BROWN asked if that could not be done by correspondence, so as to avoid any further meetings of the Joint Committee until the experiments were completed?

Mr. COCHRANE—The Secretaries could prepare lists of fans proposed to be tried in each district for approval by each Committee.

The CHAIRMAN—Yes; there was no need to call a meeting for that purpose.

Mr. COCHRANE—No; it would be entirely done by correspondence.

The CHAIRMAN—Yes; so that each Institute might agree.

Mr. GARFORTH moved that the best thanks of the meeting be given to Mr. Daglish for his kindness in presiding there that day.

Mr. THOS. EVENS had much pleasure in seconding that.

The resolution having been put and carried,

The CHAIRMAN said he was much obliged to them. He thought they had done a very good day's work.

Mr. STEAVENSON moved a vote of thanks to the Institution of Civil Engineers for granting them the privilege of meeting in their rooms.

The resolution was unanimously carried and the meeting separated.

APPENDIX.

OBSERVATIONS TO BE MADE, AND INSTRUCTIONS TO THE ENGINEERS.

Six separate experiments shall be made upon each ventilator, in which the friction of the mine is varied, as follows :—

(a) The return to the ventilator closed.

(b) The return to the ventilator closed, with the exception of an opening of 3 square feet.

(c) The opening doubled in area.

(d) The mine under ordinary working conditions, with all machinery at rest (hauling, winding engines, etc.).

(e) The entrance of air facilitated by opening some doors.

(f) Air admitted as freely as possible from the atmosphere.

In each trial the six experiments shall be made in the above-named order, and as nearly as possible at the normal speed of periphery, subject of course to the ability of the engines to drive the fans at the required speed when passing large volumes of air.

Two more experiments shall also be made with the mine under ordinary conditions, and the fan running at higher and lower speeds.

The normal speed of periphery shall be taken at 6,000 feet per minute.

In each experiment observations shall be made of—

(a) The number of revolutions per minute of the fan and engines.

(b) The volume of air.

(c) The water gauge.

(d) The indicated horse-power.

(e) The height of barometer.

(f) The temperature.

NOTES.

(a) *The revolutions* of the fan and engines shall be counted by an ordinary engine counter, and, if possible, two independent observers shall undertake this duty.

(*b*) *The Volume of Air.*—A Casella air meter or Biram's anemometer shall be employed, provided with some simple form of stopping and starting gear, say, started by the tension of a string and stopped by the reaction of a spring; that is to say, the revolutions would be recorded so long as the string was pulled tight.

The measurements shall be made at the same point in (1) the return air-way and in proximity to the inlet of the fan, and also at (2) the inlet (or inlets) and in (3) the shaft.

If possible a length of arching shall be taken, and the place of measurement must be of some regular geometrical form.

If all parts are not accessible to the observer, the place of measurement must be reduced in size by a rectangular wood frame or doorway.

The area of the place of measurement must be divided into 16 equal areas, and a reading of the anemometer taken in each at its centre of gravity. The division shall be made by means of horizontal and vertical strings, thus—

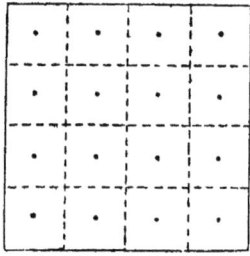 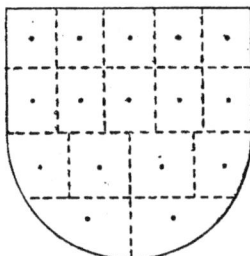

The anemometer shall be held for 30 seconds in each position, without intervals between the readings. Two observers shall attend to this, one to handle the anemometer (standing at one side) and the other to observe the seconds' watch and book the results.

When the resistance of the mine is varied, the position of the fan shutter, or other appliance for modifying the useful effect of the fan, shall be tested, if practicable, to ascertain the position which yields the highest water gauge at the normal speed.

The anemometers shall be tested at intervals with the same efficient machine.

(c) *The water gauge* readings shall be made at the centre of the drift (where the air is measured) with the end of the tube pointing to the fan, and at the same time as the anemometer readings.

Simultaneous readings must also be taken at the centre of the inlet to the fan.

The end of the tube shall be protected by a flannel cap from the effects of velocity.

The readings shall be made every 30 seconds.

The water gauge used in the experiments shall be of the ordinary form. Distilled water shall be used in the water gauge.

Flexible rubber tubing will be required to connect the instruments with the points of observation.

(d) *The indicated horse-power* shall be obtained by means of a Richard's indicator, made by Negretti & Zambra, and costing about £7 10s.

Both ends of the cylinders shall be connected thus, with a three-way

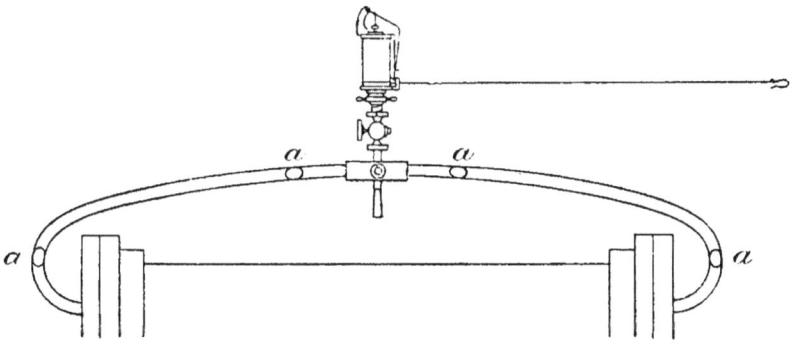

cock at the point of union, and above which the indicator shall be placed.

If there are a pair of cylinders, two indicators shall be simultaneously employed. By this means both cylinders and both ends of each cylinder will be indicated almost simultaneously.

Three sets of diagrams shall be taken during each experiment, at the beginning, middle, and end.

Experiments shall also be made to determine the friction of the engine without any air passing through the fan, or by detaching the fan from the engine where possible.

The indicators shall be tested by weights in the ordinary manner.

(e) and (f) *The readings of the barometer and thermometer* in the open air, and that of the thermometer alone in the drift, shall be registered. The hygrometric conditions of the inner and outer airs shall also be recorded.

———

Generally, all the experiments shall be made under similar conditions either when pits are idle or otherwise. All time observations shall be made with a centre-second watch, costing about 13s. 6d.

Additional information shall be obtained as under:—

(1) Depth and diameter of downcast and upcast shaft.

(2) Obstructions (if any) in shafts, with sketches.

(3) Distance apart of the shafts, with working sketch of seams.

(4) Difference in surface level of shafts.

(5) Temperature at tops of upcast and downcast shafts.

　　Temperature at middle of　do.　　　　do.

　　Temperature at bottom of　do.　　　　do.

　　If boilers, etc., are in use underground, the temperatures should also be observed (where possible) at the point where the smoke is delivered into upcast, with sketch and dimensions of the smoke drift and volume of air passing through it.

(6) Dimensions of fan, distance from pit, and dimensions of fan drift (with plans).

(7) Dimensions of engines.

(8) A record of the steam pressure at the time of taking the indicator diagrams.

(9) A record of the water gauge at the bottom of the pit, where possible.

(10) The date of erection of the fan and engines.

(11) The original or estimated cost (and date) of fan, engine, boilers, building, etc.

(12) The cost of maintenance, being the actual cost of stores and repairs of fans and engines.

(13) Particulars of all accidents, and duration of stoppages of fan since erection.

Instruments required:—

2 water gauges.

100 feet india-rubber tubing with wire core.

2 flannel caps for tubing.

2 thermometers, wet and dry bulb.

3 anemometers.

2 Richard's indicators.

1 set of reducing gear.

2 Bourdon steam gauges, 60 and 150 lbs.

2 centre split-second watches.

1 aneroid barometer.

2 Harding's counters.

2 three-way cocks.

> Tool chest, ratchet brace and 4 drills, screw spanner, pipe tongs, pincers, pipe cutter, callipers, stock, dies, taps and key, oil tin, short lengths of steam pipe of various diameters.

PROCEEDINGS.

GENERAL MEETING, SATURDAY, JUNE 9TH, 1888, IN THE WOOD
MEMORIAL HALL, NEWCASTLE-UPON-TYNE.

SIR LOWTHIAN BELL, BART., PRESIDENT, IN THE CHAIR.

The SECRETARY read the minutes of the last meeting.

The SECRETARY reported the proceedings of the Council; and said it had been arranged for the Summer Meeting in Scotland to take place, if possible, on Tuesday, July 24th; notice would be sent to all members as soon as the arrangements were completed. The Federation Meeting took place in London on Wednesday, and was fully attended, not only by members of their own Committee and Council, but also by representative members from the various Mining Institutes in the country; and the Committee on "Fan Ventilation" was also largely attended at the same place and on the same day. Both meetings might be regarded as quite successful.

The SECRETARY read the balloting list, as settled by the Council, to be sent out to the members.

The following members were elected:—

ASSOCIATE MEMBERS—

Mr. WILLIAM RICH, Minas de Rio Tinto, Provincia de Huelva, Spain.
Mr. WILLIAM G. WEARS, Mining Engineer, 28 and 29, St. Swithin's Lane, London, E.C.

Mr. M. WALTON BROWN read the following Report:—

REPORT OF THE COMMITTEE APPOINTED TO ENQUIRE INTO THE EXPLOSION OF AN AIR RECEIVER AT RYHOPE COLLIERY,

AND CONSISTING OF

MR. W. F. HALL, PROFESSOR BEDSON, MESSRS. J. DAGLISH, J. T. DUNN, G. B. FORSTER, H. LAWRENCE, W. LISHMAN, L. WOOD, AND M. WALTON BROWN (Secretary).

DRAWN UP BY
M. WALTON BROWN, WITH AN APPENDIX BY PROFESSOR BEDSON.

GENERAL ARRANGEMENTS.

The compressing engines and No. 1 air receiver are situated (Plate XLII.) on the surface. The two steam cylinders are 32 inches, and the two air compressing cylinders are 33 inches diameter, and 5 feet stroke. The air compressing cylinders have water jackets; the water being admitted at the sides, and, passing upwards, escapes by an overflow pipe. It is supplied from a small reservoir, which furnishes a flow of water through the jacket. The ends are also water jacketed in the same way as the sides of the cylinders, the water being admitted on the lower side, and rises to the top, where it overflows. There are two inlet and one outlet valves, 8 inches diameter, at each end of the air cylinders.

The lubrication of the air cylinders was at first effected by screw inlet valves $1\frac{1}{2}$ inches diameter at each end, connected by pipes with the bottom of the receiver, whence sufficient of the lubricant was drawn in at each stroke to lubricate the piston, valves, etc., any excess being returned to the receiver along with the air. The lubricant used at that time was—

Soft soap	2 pounds.
Colza oil	$\frac{1}{4}$ gallon.
Water	$\frac{1}{4}$ „

This means of lubricating was discarded during the year 1881, when about one gallon of fluid was poured into the suction valves of the two cylinders every two hours, consisting of—

Soft soap	2 pounds.
Mineral oil	$\frac{1}{2}$ gallon.
Water	$\frac{1}{4}$ „

The compressed air is delivered by the engine into pipes 10 inches diameter (formerly 8 inches), through which it passes into the No. 1 air receiver (Plate XLI.), consisting of an egg-ended boiler 6 feet in diameter and 29 feet long (the contents being about 770 cubic feet), and built of ⅜ inch iron plates arranged in rings with hemispherical ends. There were two safety-valves, each 4 inches diameter, and loaded to 50 lbs. per square inch. The two delivery pipes passed into the top of the No. 1 air receiver, and thence one cast iron delivery pipe 10 inches diameter led the air to the top of the shaft.

The air is taken down the shaft, a distance of about 250 fathoms, by means of malleable iron pipes 9 inches internal diameter, ⅜ inch thick, in 12 feet lengths. The flanges are ¾ inch thick, welded on and fastened together by eight ¾ inch bolts. They have plain joints made with cement and packing. There is a stand pipe, provided with stuffing box to act as an expansion joint, placed in the shaft at 127 fathoms from the surface.

The pipe from the shaft is connected with the No. 2 air receiver, which is placed near the bottom of the shaft; it is 4 feet diameter by 12 feet long, and made of ⅜ inch iron plate   The air is taken (Plate XLII. Fig. 2) from No. 2 receiver in metal pipes, decreasing from 8 inches to 6 inches diameter, to three hauling engines placed at various distances from the shaft. These hauling engines have each two cylinders 14 inches diameter and 18 inches stroke. They are situated at 660 yards, 990 yards, and 1,950 yards from the receiver at the bottom of the shaft. They run about 120 revolutions per minute, the air being cut off at three-quarter stroke.

THE EXPLOSION.

The engineman at the air compressor said that the three air engines had all been running, two were stopped, and he had just finished oiling the air compressing engines. He had lubricated the air cylinders, and when near the door of the house he was knocked down by the violence of an explosion. The third engine was easing up when this explosion occurred, at 10·40 p.m. on March 1st, 1883.

The pressure of air which he had observed when lubricating the air cylinders was 57 lbs. per square inch. The compressing engine had been running at twenty-three or twenty-four revolutions per minute, and had slowed down, when the two underground engines stopped, and the air was blowing off at the receiver. He further stated that he was rendered unconscious, having been struck on the head with some flying *debris*. As soon as he recovered (he could not say how long he had been unconscious) he found the engine racing away, and stopped it. He then observed a fierce fire burning, in the No. 1 air receiver, like a furnace, owing to the blast of air playing upon it.

The evidence of other persons showed that the explosion of the No. 1 air receiver was accompanied with bright flames from 20 to 30 feet high, which immediately subsided, leaving a fierce fire burning brightly within the receiver.

Other persons arrived on the site of the explosion within ten minutes of the occurrence, and means were at once adopted for the extinction of the fire, which was easily accomplished by the aid of water.

The engines usually make from twenty-three or twenty-four revolutions per minute; and as the air engines are stopped the speed slackens —steam of 35 lbs. per square inch not being sufficient to overcome a pressure of more than 58 lbs. in the No. 1 air receiver.

The No. 1 air receiver is arranged so that the air blows off at 50 lbs.; but should none of the air engines be running, the pressure actually rises to 58 lbs., and the speed of the engine is reduced to fifteen or sixteen revolutions. At this speed the air blows off at the valves as quickly as it is pumped into the air receiver.

None of the attached thermometers were in use at the time of the explosion. They are stated to have usually indicated about 180° F., although it was said that the temperature of the air occasionally rose as high as 230° F. The second engineman said the temperature of the air had been as high as 300° F.

The water was up in the jacket, and the engineman had seen that there was water in the cistern. Both the cylinders were as bright as glass, and, as well as the valves, showed no signs of over-heating. The packing of the piston was not singed, nor did it show signs of abnormal heating.

THE DEPOSIT.

The pipes leading from the air cylinders to the No. 1 air receiver used to be cleaned every three or four months when the old lubricant was used in the air cylinders, but with the use of the new lubricant the deposit had not taken place so rapidly.

On one occasion, when the pipes from the air cylinders were 8 inches diameter, they were found almost completely closed, there not being room to push in the hand. (See Fig. 1, Plate XLVI.)

This led to 10 inch pipes being put in instead of the smaller ones.

The new lubricant had been used for about twelve months before the explosion, during which time the pipes had not been cleaned. They were taken off after the accident and examined, when the deposit was found not to exceed $1\frac{1}{2}$ inches in thickness, laying on the lower side, and thinning out towards the vertical edges. (Fig. 2, Plate XLVI.)

The deposit has been found in the gauge pipe which is attached to these pipes, and the gauge, on more than one occasion, when taken off was found to be choked up. The gauge which was in use at the time of the accident was examined, and the pipe was found to be coated internally with a slight deposit about $\frac{1}{16}$ inch thick.

The deposit extended into the No. 1 air receiver, where it was found all over the bottom ; on the last occasion when the No. 1 air receiver was cleaned out, the deposit was found to be 9 inches thick at the bottom, thinning out towards the vertical sides.

The use of the new lubricant was introduced after this cleaning, and at the time of the explosion, judging from the remains then found, it is considered that the deposit may have been 2 inches thick.

The deposit extended down the pipes into the pit, and was found in No. 2 and No. 3 air receivers ; but it was not so abundant as in the pipes and air receiver at bank, and was of a different nature, being somewhat oily, and of a rusty red colour.

The deposit found in the pipes and the No. 1 air receiver was evidently a mixture of coal-dust and the lubricant carried over from the air cylinders. It may therefore be said to consist of coal-dust, mineral oil, soft soap, and water.

DAMAGES.

The receiver (Plates XLIII., XLIV., and XLV.) was found blown into four pieces, nearly detached from each other, a cylindrical rip passed through one inlet and the outlet pipe, and almost completely round the boiler ; another rip extended along the top of the boiler, through the man-hole doors, safety-valve seats, and other old holes, as far as the spherical end. This end was almost blown off, and received two distinct fractures, which laid it flat out.

There is abundant evidence to show that after the explosion the receiver was, in part, at a white heat. This tends to prove that some heat must have accumulated in the plates previous to its bursting. The plates themselves clearly evidence the fact of their having been very hot ; the heating was, however, local, and appears to have been greatest at the side next the pit, below the point where the air entered, and thence extended along the side of the unfractured end.   The fractured end of the receiver did not appear to have been subjected to any great temperature.   The metal pipes leading to the pit were not damaged.

Two of the cast iron pillars, forming a portion of the supports of the heapstead, were broken in several places, and the pieces thrown to some distance.   A cabin upon the heapstead was considerably damaged, and the windows of houses in the neighbourhood were more or less shaken.   A pulley wheel standing near had the rim broken, and four spokes were bent by the side of the boiler being blown forcibly against it.   (See Fig. 3, Plate XLVI.)

The air pipes are placed in the pit, which is bratticed, one side being used as an upcast, the other side was intended to be a downcast, but the transmission of heat is so great that it is simply used as a dumb pit. The brattice is built of bricks set with cement, upon iron girders or buntons.   These girders are formed of two side plates 6 inches deep and $\frac{7}{8}$ inch thick, and $19\frac{1}{2}$ feet long, bolted together, and studded at intervals, as shown in Fig. 5, Plate XLVI.

The slide buntons and slides are attached at one end to the buntons carrying the brattice.

Serious damage occurred in the shaft.   At 127 fathoms one segment of the cast iron tubbing was found blown about $\frac{1}{2}$ an inch outwards from the pit, and showed the sheeting ; a stuffing joint and stand pipe were also found damaged.   At a depth of 180 fathoms several pipes were partially split open.   From 200 to 212 fathoms the pipes, brick brattice, side buntons, etc., had all fallen down the pit.   At 215 fathoms a piece was blown out the side of one of the pipes.   (See Fig. 6, Plate XLVI.) From 220 to 231 fathoms about 11 fathoms of brick brattice were damaged, and the buntons and brick brattice blown into the upcast side of the pit.   (See Fig. 4, Plate XLVI.)   One of the pipes at 225 fath ms was burst open into an almost flat sheet of iron.   (See Fig. 7, Plate XLVI.)

The air receiver, pipes, guides, etc., at a depth of about 250 fathoms were displaced, and had fallen to the bottom of the pit ; there was no brattice at this point, and the pipes were found damaged 30 yards inbye. One of the workmen at the bottom of the shaft saw a flash of flame in the shaft at the time of the explosion, and was thrown about 20 yards. A tub which he was pushing was overturned and thrown about 15 yards. The effects of the explosion were felt at considerable distances from the shafts; thus an air crossing was "lifted" about 1,000 yards from the shaft, and waste doors were "moved" at a distance of 3,500 yards inbye.

There was a great similarity in the bends and fractures of the buntons carrying the brattice at the two points where the damage was most serious. The fracturing forces appear to have acted at about the same point upon all the beams, with the difference that some were only bent at the points where others were broken, as shewn by the * * * in Fig. 4, and by the arrows in Fig. 5, Plate XLVI.

PROBABLE CAUSES OF THE EXPLOSION.

It appears improbable for the No. 1 air receiver to have been exploded by the normal pressure of the air, as the engine would have stopped before the pressure became too high, in addition to which a pressure of about 220 lbs. per square inch would have been required to burst the air receiver, and to burst the air pipes a very much greater pressure would have been necessary.

The bursting of the No. 1 air receiver must, therefore, have been produced by the explosion of some substance inside. The temperature in the No. 1 air receiver must have been somewhat high, and must have heated the plates to a high temperature before the explosion, otherwise they could not have been raised to a white heat in the short time that the fire was allowed to burn after the explosion.

The bursting of the air pipes in the shaft and the damage done in the shaft itself appear to show that an explosion occurred in them, and that it took place subsequent to the explosion which (it is supposed) took place in the No. 1 air receiver. This opinion is supported by the fact that the metal pipes near the receiver leading into the engine-house and to the pit were unbroken. If an explosion had only occurred in the air receiver, it is very improbable that the bursting of the pipes in the shaft, and the serious damage in the shaft itself, could have resulted from the compressive wave or "momentum" given to the air in the pipes by an explosion in the No. 1 air receiver. It would not explain the considerable effects found for some distance inbye, where the pipes themselves were not damaged.

This second explosion would probably ensue from the explosive gases, generated in the air receiver and passed to a certain distance down the pipes in the shaft, being ignited by a flame produced by the explosion in the air receiver.

It is possible that the flashing point of the mineral oil, forming part of the lubricant used in the air compressing cylinders, might have been lower than the temperature of the compressed air, in which case the receiver might have been charged with an explosive mixture of air and oil vapour, which would be readily ignited by flame or heat. The oil was tested, and it was found at 230° F. that about 1 per cent. was volatilized daily, and at 500° F. about 9 per cent. was evaporated daily, and the flashing point was 365° F. It is difficult to understand, therefore, how an explosive mixture of air and oil vapour could be formed in the air receiver, owing to the considerable excess of air which would be present, or how it could become ignited.

It has been suggested by others that the gas given off by the oil at 300° F. might be ignited by the spontaneous ignition of a piece of cotton waste left in the receiver; but this appears to be incapable of producing the effects observed.

Professor Bedson, of the Durham College of Science, Newcastle-upon-Tyne, has made a valuable series of experiments to ascertain the temperatures at which coal-dust alone and the deposit found in the No. 1 air receiver were ignited, when heated in a current of air at ordinary and higher pressures. It was found impossible to produce combustion of the deposit, although the temperature of the air at atmospheric pressure was raised to 450° F.; but with coal-dust alone ignition took place at a temperature of 291° F., and the mass finally glowed with a dull red heat.

Further experiments were made with air at a pressure of 60 pounds per square inch, the results of which showed that the temperature of ignition of coal-dust was the same as with air at ordinary pressures, but when ignition occurred a more intense heat was produced, as shown by the coking of the dust.

It appears to be highly probable that under a pressure of 60 pounds per square inch the deposit, consisting of a mixture of coal-dust, mineral oil, soft soap, and moisture, might have been ignited in the same manner as in the experiments with the coal-dust, although no such phenomenon was produced in the experiments.

This supposition receives some support from the evidence of heating in the various parts of No. 1 air receiver. It seems possible, therefore, assuming that the deposit was in a state of ignition, for destructive distillation of a portion of the deposit to take place, forming considerable volumes of combustible gases, which would be exploded as soon as the mixture consisting of the proper proportions of air and the gas (from the distillation of the oil and coal-dust contained in the receiver) was ignited by the burning deposit.

CONCLUSIONS.

The cause of the explosion of the air receiver cannot, unfortunately, be exactly determined. It would, however, appear desirable that:

The air used for compressing purposes should be taken as pure as possible, care being taken in selecting the site for the machinery to avoid the vicinity of gas works, burning heaps of waste material, or any works such as lime kilns producing stythe or smoke, or any other gases (inflammable or otherwise) which might be injurious or dangerous to life, if carried into the mine.

The use of illuminating gas in the engine house should be avoided, as a careless workman might leave a tap open, and the escaping gas might be drawn into the compressing cylinders.

It would also be advisable to keep the machinery at some distance from the screens so as to avoid the drawing of the fine coal-dust into the cylinders and delivery pipes.

Attention should be directed to the necessity of a constant and sufficient supply of water to the jackets of the air cylinders, so as to prevent the danger of spontaneous combustion of deposits formed in the pipes or vessels.

The danger of using mineral oils of a low flashing point, for lubricating air-compressing cylinders, is sufficiently obvious.

M. WALTON BROWN.

APPENDIX.

In seeking for an explanation of the explosion of the Ryhope air receiver, there are two things to be borne in mind :—

1. The particular conditions which would produce a temperature sufficiently high to ignite an explosive mixture of air and a combustible gas.

2. The production of the explosive mixture.

These are two necessary conditions, as the existence of one without the other would not lead to results such as those detailed in the report on the explosion.

At the outset, attention was directed to the discovery of some means of obtaining a temperature sufficiently high to ignite an explosive mixture of air and combustible gas. These were sought in the conditions under which the air-compressor was working before the explosion took place, and of these the following appeared to be the most worthy of attention :—

1. The temperature of the compressed air, brought about by the heat generated in the compression of air to 58 lbs. pressure.

From the report itself, and the evidence of those examined by the Committee, no knowledge of a very definite or satisfactory character is to be obtained, and one must look to the theoretical possibilities and make them the basis of any conclusions. The heat produced by compression is estimated to be sufficient to raise the temperature of the air from 60° to 369·4° F. This is, of course, a much higher temperature than would be obtained in practice, as a considerable proportion of the heat would be lost. Still, there is sufficient indication of a high temperature having been reached to leave a very considerable margin.

2. The accumulation of coal-dust and lubricant in the pipes leading from the cylinder to the air receiver, as also in the air receiver.

The existence of this accumulation shows that in working over a considerable period a large amount of dust must have been drawn with the air into the cylinder.

In the constituents of this accumulation it would appear most probable that the heat-producer necessary for an explanation of the explosion is to be found. The constituents were :—

1. A paraffin oil, soap, and water, forming together the lubricant; and

2. The coal-dust which, when mixed with the lubricant, formed the pitch-like lining of the receiver.

There is nothing in these materials themselves which could in any way give rise to a production of heat; but this is rather to be sought for in the chemical changes which might possibly be produced in them by the constant passage over and through them of air at a temperature such as that to which the air would be raised by compressing it. In considering the possible changes which might take place, the oil, which is described as a mineral or paraffin oil (as it consists of a mixture of compounds of carbon and hydrogen, which are chemically inert), did not appear likely to act as a large contributor to any heat production. Under the conditions named soap may also be neglected, and the influence of water at low temperatures may be disregarded; whilst in the carbonaceous material (representing the coal-dust) is to be found a substance capable of being rapidly oxidised by the action of heated air, and of having its temperature speedily raised to a point sufficiently high to start ignition.

At this point it would be well to briefly recapitulate the main phenomena of combustion.

Combustion is the term used to describe those cases of chemical union taking place between the constituents of a combustible body and the oxygen of the air, resulting in the formation of new bodies—the products of combustion—and also in the evolution of heat, and sometimes light. Such a definition would embrace all phenomena of burning, but, in addition to such, there must be considered those closely allied, and only separated from them by differences arising from the absence of any visible effect, namely, the phenomena of oxidation or slow combustion.

The essentials of the two classes of phenomena are well illustrated in the differences observed between the burning of a piece of iron in the air and in the slow rusting of iron, and also in the burning of a combustible liquid like alcohol and its slow oxidation. In the one case you have heat produced sufficient to be evident to the most casual observation, in the other it is only on closer examination that heat production is made evident. The difference is essentially one of degree and intensity or rapidity. In order that a combustible body may burn its temperature must be raised to a given point, which is specific for each substance, and is known as the temperature of ignition. In the temperatures of ignition of combustible bodies there are wide differences—some substances taking fire when brought into contact with the air, whilst others again require to be heated beyond a red heat before combustion takes place.

Easily oxidisable substances, such as phosphorus, iron, alcohol, etc., can and do unite with the oxygen of the air at temperatures much below their temperatures of ignition, and in so doing heat is produced, which, if not dissipated, may serve to gradually raise them to that temperature at which combustion as ordinarily understood takes place. Facts such as these enable one to give a satisfactory explanation of the spontaneous ignition of coal cargoes, of heaps of oily waste, etc.

All conditions which bring about an intimate contact between an easily oxidisable substance and the air will in promoting the oxidation of that substance lead to a production of heat, which, if allowed to accumulate, may in course of time cause the inflammation or ignition of the substance or its rapid combustion. The physical state of the combustible has a powerful influence in such cases, as shown by the simple experiment of pouring a solution of phosphorus in carbon disulphide over filter paper; the solvent evaporating leaves finely divided particles of phosphorus disseminated throughout the pores of the paper, and so exposed to the air that in a few moments the phosphorus is ignited. The same fact is demonstrated by the spontaneous ignition of pyrophoric iron or lead, *i.e.*, the finely divided iron or lead resulting from the decomposition of iron oxalate or lead tartrate when brought into contact with the air.

It would, therefore, appear that for the spontaneous ignition of a combustible body there is required, primarily, its slow oxidation. The heat thus produced must not be allowed to be dissipated or conducted away; and, further, all conditions favourable to increasing the rapidity of this oxidation—such as fine state of division or the application of heat to either the combustible body or the heating of the air—will tend to shorten the time in which its temperature of ignition will be reached.

The application of these facts to the Ryhope explosion would appear to be justified, if by their means it would be possible in any way to explain the production of the two sets of conditions which have been stated as necessary to account for the phenomena detailed in the report. In the conditions of working, attention has been fastened on the coal-dust drawn into the air receiver, and for this reason, that it, of all the constituents of the mud found in the receiver and elsewhere, constitutes not only the greater proportion, but is also the most easily oxidisable of these constituents.

The spontaneous ignition of coal heaps, of coal cargoes, etc., the weathering of coal, all show with what readiness and ease coal undergoes change when in contact with air. The explanations of these phenomena have from time to time been made the subject of investigation, and it is to the experiments of Richters *(Dingl. Poly. Journ.*, 190, 193, and 195) and others, that one must look for the facts upon which to base a satisfactory explanation.   Richters has shown that when coal is heated at temperatures from 356° to 374° F. an increase in weight is observed; this increase in weight continues until a maximum is reached, when further continued exposure to this heat brings about a slight loss of weight.   The analysis of the coal after heating shows that the increase of weight is due to an absorption of oxygen.   This oxygen is not mechanically absorbed, but rather acts chemically upon constituents of the coal, part of the hydrogen being converted into water, and a portion of the carbon into carbon dioxide, and a coal is produced altered in some of its physical properties, and containing a larger proportion of oxygen than does the original coal.   Inasmuch as this absorption of the oxygen of the air by coal takes place at ordinary temperatures, and when special precautions are taken against a loss of heat, the action of the oxygen has been demonstrated to be attended by a considerable heating of the coal, it is evident that this phenomenon accounts not only for the weathering of the coal but also its spontaneous ignition.

With a view of testing the application of these facts to the case under consideration, a series of experiments were instituted, in which it was sought to demonstrate the oxidation of coal-dust by passing air over it at various temperatures under the ordinary pressure.

In this series of experiments the coal-dust was heated in a glass tube, through which a current of air, dried and freed from carbon dioxide, was forced.   After going over the coal the air was passed through a wash bottle containing lime-water, so that the temperature at which oxidation first began would be indicated by the milkiness of the lime-water.   The tube was placed in an air-bath, and outside the bath a thermometer was fitted into it.   (See Plate XLVII.)

The following are the results of the experiments with this form of apparatus :—

		a.	*b.*	*c.*	*d.*	*e.*
			Deg. F.	Deg. F.	Min.	
I.	...	35 min. ...	338	392	95	No ignition.
II.	...	30 „ ...	329	396	75	Do.
III.	...	30 „ ...	336	396	100	Do.
IV.	...	30 „ ...	379	478	63	Do.
V.	...	30 „ ...	351	457	90	Do.
VI.	...	No turbidity...	—	243	450	Do.
VII.	...	25 min. ...	354	401	40	Ignition.
VIII.	...	70 „ ...	302	351	81	Do.

(*a*) Time in minutes from the moment when the heating began to the time at which the first turbidity of the lime-water was noticed.

(*b*) Temperature of the bath when the lime-water first began to be turbid.

(*c*) Maximum temperature to which the bath was ultimately raised.

(*d*) Time the experiment occupied.

(*e*) Result.

In experiment V. the mixture of coal and oil from the Ryhope air receiver was substitued for the coal-dust.

It will be seen from these experiments that oxidation commences at as low a temperature as 302° F., as shown by the production of carbon dioxide. Further, in two instances the ignition of the coal-dust took place when the temperature of the bath had been raised to 351° F. in one case and 401° F. in the second.

Under the conditions of these first experiments but a very small mass of coal-dust ($\frac{1}{3}$ ounce) could be used, and exposing very limited surface to the action of the air. With a view of increasing the mass of coal-dust experimented on, and also enlarging the surface, further experiments were conducted. In these the coal-dust was (instead of placing it in a tube) spread in layers about half an inch deep over a small sheet of asbestos, supported in the air-bath on two glass rods, and a current of air was made to play over the surface of the coal. Ignition took place at the following temperatures :—

I.—437° to 446° F.	VI.—410° F.
II.—446° F.	VII.—392° „
III.—419° „	VIII.—392° „
IV.—417° „	IX.—374° „
V.—410° „	X.—356° „

In the case of ten experiments the coal fired at temperatures varying from 356° F. to 446° F., these being the extremes.

These experiments showed that ignition of the coal-dust could be more easily produced when working under these conditions than when tubes were used, as in the former experiments; and it was thought that by a further increase of the quantity of coal-dust operated on, and also of the surface of the dust exposed to the air, a still further reduction of the temperature might be effected. To test this a second series of experiments were made with a larger air-bath (see Plate XLVIII.), the coal-dust being placed on a sheet of asbestos about 9 inches square. Underneath it was a similar sheet to distribute the heat evenly, and above it a tube bent in the form of a flat spiral and pierced with fine holes, air being forced through this tube by means of a water blower. The temperature of the bath was carefully regulated by a thermostat. The following are the results of the experiments with this apparatus :—

		a.	*b.*	*c.*	*d.*
		Deg. F.	Deg. F.	Min.	
I.	...	266	275	330	No ignition.
II.	...	266	274	2,700	Do.
III.	...	271	289	615	Do.
IV.	...	293	302	300	Do.
V.	...	293	302	405	Do.
VI.	...	302	320	80	Ignition.
VII.	...	302	316	70	Do.
VIII.	...	280	293	75	Do.
IX.	...	293	293	60	Do.
X.	...	266	291	75	Do.

(*a*) Temperature to which the bath was heated before the coal was put in.

(*b*) Highest temperature of bath.

(*c*) Time the experiment occupied.

(*d*) Result.

In experiments IV. and V. half of the asbestos was covered with the mixture of coal and oil from the receiver, the other half with coal-dust.

In five out of ten experiments the coal ignited, with the temperature of the bath varying from 291° F. to 320° F. In none of these cases did the time from which the dust was placed in the bath to the moment of ignition exceed one hour.

In making these experiments it was observed that in every case in which the ignition of the dust took place, that immediately after the thermometer immersed in the dust began to rise, a peculiar odour was noted, soon followed by a rapid rise of temperature as shown by the thermometer, and a commencement of the burning of the coal as observed by the experimenter. This remarkable odour in time became of great value, serving as a more valuable indication of the progress of oxidation than the thermometer itself. The burning of the dust always commenced in different parts and appeared to have commenced below the surface, and when once started it would continue to glow until the coal was apparently burnt away. A section through the coal shows that the central portion was alone completely burnt, the upper and lower surfaces still retaining a coal-like aspect.

In the experiments in which the ignition of the coal took place the thermometer immersed in the coal, after reaching the same point as that of the bath, for example, 284° F., would rise slowly to 356° F. and then rapidly shoot up beyond temperatures registered by a mercurial thermometer. An experiment was made with an air thermometer, the bulb of which was immersed in the coal-dust; and starting with the bath at the ordinary temperature and gradually raising it to 298° F., the indications of the thermometer in the coal-dust showed that the dust took about half an hour to reach the temperature of the bath, then the temperature of the coal rapidly rose until, with the bath at 298° F., the coal-dust ignited.

The readings of the air thermometer show satisfactorily that after the coal-dust reached the temperature of 356° F. the oxidation is then greatly accelerated and the thermometer rapidly rises to 392° F., after which it was no longer able to keep pace with the rise in temperature in the coal, now proceeding with such rapidity that in a few minutes—some two or three—the coal took fire.

It was next thought desirable to make some experiments with air under pressure. With this object a strong iron tube A with a collar A[1] (Plate XLIX.) was made, provided with three smaller tubes B with brass collars c at the end of each; the ends of the tube were closed by thick glass plates D, kept in position by screw caps E fitting on to asbestos washers and screwed on to the collars. The interior of the tube was illuminated by a gas burner F placed opposite one of the ends, the light from which reflected by a mirror G opposite the other end, thus enabling the experimenter to observe the changes taking place in the interior of the tube. Coal-dust was spread over the lower half of the tube—a thermometer H with bulb immersed in the dust was fixed in the central aperture. Of the two remaining apertures one was connected with the air receiver containing air under 60 lbs. pressure, the other provided with a small safety-valve furnished with a stop-cock partially closed, so as to allow for a circulation of the air through the tube. The tube was heated to the desired temperature by placing it in an air bath carefully heated. A preliminary experiment was made in the laboratory with a cylinder of compressed air, but the working in this manner proving unsatisfactory, permission was obtained from Mr. H. Lawrence, of the Grange Iron Works, to make some trials with air supplied by one of their air-compressing engines. Two experiments were made, with the result that ignition of the dust was observed to take place when the coal-dust had been heated to a temperature of 293° F. It was further noted that in both cases the combustion was vivid at the part at which air entered from the compressor, and that a higher temperature must have been reached than when experimenting with air under ordinary pressure was shown by the fact that particles of coked dust were found amongst the coal after each experiment, a result never obtained in experiments made at ordinary pressures.

The bearing of these observations on the question under discussion would appear to be :—(1) It is shown that with an easily oxidisable material, such as coal-dust, it is only necessary to accelerate the oxidation by heating the dust in contact with air to a temperature within the limits of those likely to be reached in compressing air to 58 lbs., in order to bring about in the course of time the ignition of the dust. (2) These results have been possible even when experimenting with a comparatively small amount of the material; an influence of importance, as is shown by the results obtained in the two different series of experiments made, the conditions of which differed essentially only in the amount of coal operated upon. In the second series, with 1 oz. a temperature of 356° F. was required for ignition; whilst in the third series, with 10 ozs., the temperature required was reduced to 291° F.

Despite the fact that the one or two experiments made with the deposit found in the Ryhope air receiver, made under conditions similar to those in which the coal-dust was successfully ignited, were not in a like manner successful; still, as there can be no doubt that this material is of a nature somewhat akin to the coal-dust entering so largely into its composition, and, like it, capable of being oxidised by the oxygen of the air, the constant exposure of this substance to the action of air heated by compression must result in the acceleration of the oxidation, and, as with the coal-dust, in a rise in temperature until in some portion of the mass local ignition commences. Combustion once begun at any point in the mass, the conditions are most favourable to its rapid extension. Further, the heat generated by a combustion of a portion of the mass might serve to destructively distil neighbouring portions, and thus give rise to the formation of combustible gases, which, mixing with the air, would form an explosive mixture. An explosive mixture formed under these conditions would be readily ignited by contact with the already burning material.

In conclusion, I have to express my indebtedness to Mr. Saville Shaw, of the Durham College of Science, for the valuable assistance given in conducting these experiments.

P. PHILLIPS BEDSON.

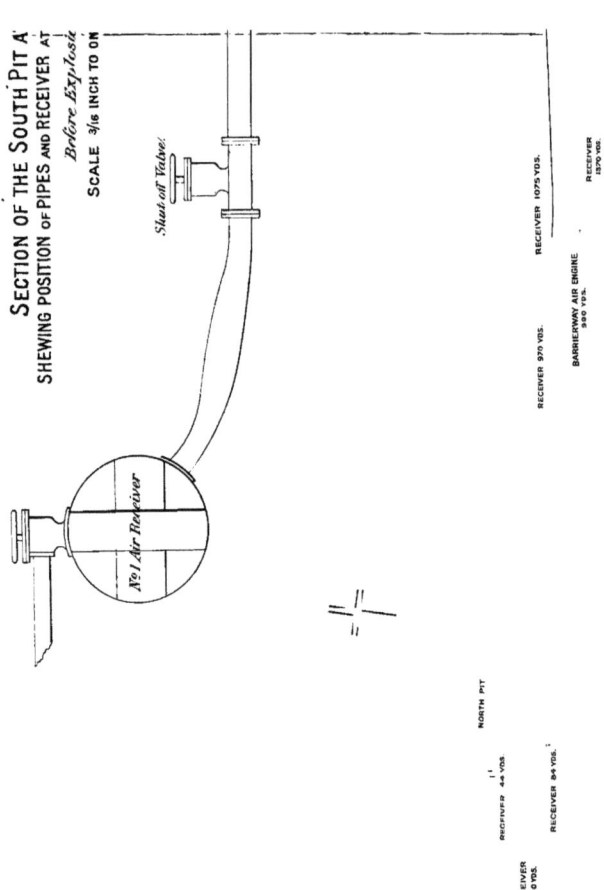

To illustrate the "Report of the Committee appointed to inq

SECTION OF THE SOUTH PIT A
SHEWING POSITION OF PIPES AND RECEIVER AT

Before Explosi

SCALE 3/16 INCH TO ON

Shut off Valve

Nº 1 Air Receiver

NORTH PIT

RECEIVER 44 YDS.

RECEIVER 84 YDS.

RECEIVER
240 YDS.

RECEIVER 970 YDS.

RECEIVER 1075 YDS.

BARRIERWAY AIR ENGINE
990 YDS.

RECEIVER
1370 YDS.

RECEIVER
650 YDS.

HIGH MAUDLIN
AIR ENGINE
650 YDS.

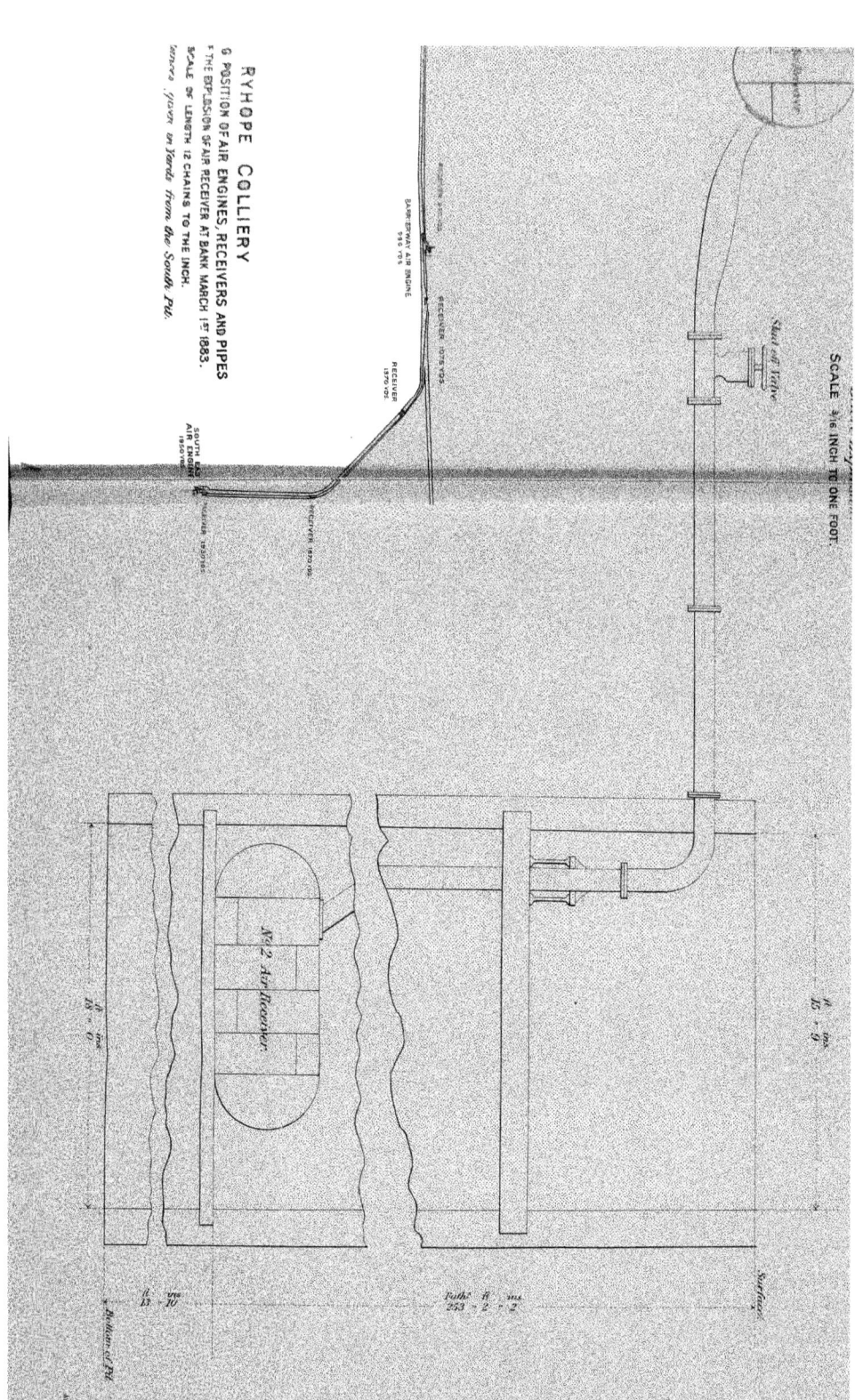

RYHOPE COLLIERY

G POSITION OF AIR ENGINES, RECEIVERS AND PIPES

x THE EXPLOSION OF AIR RECEIVER AT BANK MARCH 1ST 1883.

SCALE OF LENGTH 12 CHAINS TO THE INCH.

tures given in Yards from the South Pit.

SCALE 3/16 INCH TO ONE FOOT.

BARKERWAY AIR ENGINE
9·80 YDS

RECEIVER 1570·05L

SOUTH 63
AIR ENGINE
1860 YDS

No 2 Air Receiver.

Mr. A. L. STEAVENSON said, there could be no doubt that Professor Bedson had traced to its source the cause of the accident; and they were very much obliged to him for the trouble he had taken. The information which Professor Bedson had given might be of some assistance in tracing the effect of coal-dust in colliery explosions. There was no doubt when they once raised a certain portion of coal-dust to a given temperature that produced gas, the explosion went from one part of the pit to another; and in the same way as suggested here, the high temperature had produced gas, and the explosion had gone forward. But whilst they had ascertained the source of the explosion at Ryhope, he thought there were further means of preventing a similar explosion that might be taken beyond those suggested in the report. It was suggested that the bath should be kept full of water, and so the temperature be kept low. But this was not half sufficient. What was wanted also to be done was to allow water into the cylinders during compression.

Mr. LAWRENCE—That has been done.

Mr. STEAVENSON—But it was not stated in the report. In the report it was stated that "attention should be directed to the necessity of a constant and sufficient supply of water to the jackets of the air cylinders, so as to prevent the danger of spontaneous combustion of deposits formed in the pipes or vessels." It was not for the sake only of preventing explosions, but also for the sake of economy that water should be used. If the temperature rose it was not a question of how rapidly that temperature could be reduced. The effect of throwing water in, instead of compressing about 4 feet, was that they compressed about 3 feet; and they reduced the compression by one-third. If, instead of having the cylinder surrounded by water, they also threw water into the cylinders in the form of spray, it would not only prevent an explosion, but also have a beneficial effect in the way of economy.

Mr. G. B. FORSTER said he knew one case in which what Mr. Steavenson described was done. The water was injected into the cylinders so as to fill up and make a cushion for each stroke; and he thought the effects were very beneficial. They were very much indebted to Professor Bedson for his paper; and he hoped it would do something towards the prevention of such accidents as occurred at Ryhope.

Mr. H. LAWRENCE said when the compressor was first erected at Ryhope colliery the mode of lubricating was as described in one portion of the report. A certain portion of the lubricant was put into the cylinders by supplementary valves, which were fitted at each end of the cylinders; and the quantity to be taken in at each stroke could be regulated. It was not only put in as a means of lubricating the cylinders, and assisting to keep the air cool, but also to fill up the spaces at the end of the pistons. That mode was very shortly laid aside; and the means of lubricating was then by putting soft soap, oil, and other substances in; and this went on for years successfully, until there came the mineral oil; and after the mineral oil was used the explosion took place. As to the remarks of Mr. Steavenson, he might mention that many years previous to the explosion at Ryhope they made some compressing cylinders for Mr. Daglish, who was the means of their applying the practice used very much in France, which was exactly what Mr. Steavenson had described—that was that there was a pump fitted to the air-compressing cylinders, the stroke of which was the same, and on the compressing side was enforced cold water in the shape of very fine spray, meeting the compressor piston and assisting in the compression. This they had fitted in all the compressor cylinders at Mr. Daglish's establishments; and after the Ryhope explosion the pump was fitted to both their air-compressors. This being done there was no fear of an explosion taking place—that was if the explosion referred to took place from combustible gas in the receiver. Although it was now a long time since the explosion took place he, as one of the Committee, might say that if there was no other result of the delay than the valuable experiments made by Professor Bedson, it had certainly brought a good deal of information to the members of this Institute.

Mr. J. A. G. Ross said, he could bear out the remarks which had been made as to the economical result of applying water with each stroke. Some years ago he had something to do with the screw-hopper barges of the Tyne Improvement Commission. The machinery on board the barges was worked by hydraulic power, which was kept at a good pressure in the air accumulator, working as high as 700 lbs. per square inch; and it would be impossible to prevent pipes getting to a red or white heat sometimes. In that case they found efficacy of each stroke drawing in by suction a very small portion of water. That small portion of water, injected, or rather taken in by suction, into the cylinders, not only kept the air cool, but also filled up the air spaces, in which case it was found absolutely unnecessary to use any lubricant at all. And the same applied to low pressure. In the case of the Swing Bridge over the Tyne, during its erection, in sinking cylinders the same appliance was used with the most perfect success, without any difficulty, and without any need for a lubricant when the machinery was arranged in a proper manner. As to the explosion at Ryhope, he accompanied Mr. E. B.

Martin, a member of this Institute, and chief engineer to the Midland Steam Boiler Association, to the site of the explosion; and he could quite corroborate the statement about oxidation. Oxidation seemed to have been carried on for a very long time, the scaled portions indicating that combustion had extended over a very great period; but it was quite local, the indication that a fire had been generated in the receiver from some combustible material. There was a large quantity of coal-dust soaked in oil. He asked if Professor Bedson had a specimen of the plate ?

Mr. BROWN—Part of the deposit is here.

Mr. ROSS—There was great scale on the plate, indicating that combustion had been going on for some time.

Principal GARNETT said, there was only one point which he thought it worth while for him to mention, and that was the increasing intensity of the explosion as it went down the shaft. The phenomena observed in the shaft afforded a good illustration of what was known as "detonation" as distinguished from explosions. The explosion in the receiver started a wave of compression, which produced a sufficiently high temperature to cause the generation of combustible matter in the pipe. This combustion increased the temperature and the pressure, and thus the wave of compression went on increasing in intensity until it was capable of producing the stupendous effects observed at a distance of 200 fathoms from the receiver. This subject had been carefully investigated by Professor Dixon, of Manchester.

Mr. JOHN DAGLISH said, there were two systems of supplying water to air-compressors; one was by suction in the manner mentioned by Mr. Ross, and the other was by driving it into the cylinders. This was the mode urged by French engineers. Water was injected as a very fine spray, against a knife edge if possible, so that it might be dissipated in the form of vapour, and there was no increase in heat. This raised the interesting question, which was discussed at this Institute, and in which Professor Herschel took some interest, upon the effect on compressed air by the temperature being reduced. It was a very complicated question, and was treated very ably by Professor Herschel in the Transactions.

Mr. WALTON BROWN said, that as to the explosion in the pipe, caused, as Professor Garnett said, by the transmission of force, he thought the whole of the Committee were satisfied that the second explosion had taken place either in the receiver at the bottom of the shaft or in the pipes, because there was evidence of the flame having gone in that direction.

Mr. DAGLISH—Why should not Professor Garnett's explanation account for this ? It is very interesting. Why should the flame have gone down from the surface this way ?

Mr. WALTON BROWN—The Committee accept the explanation that the flame went down, but the pipe explosion at the bottom was caused, they thought, by the ignition of gas.

Mr. DAGLISH—Is Professor Garnett's explanation a new one ?

Principal GARNETT said, it was not altogether new. He repeated his explanation of the wave of compression increasing in intensity as it went down the pipe. The compression produced a sufficient explosion to further raise the temperature, and this went on and produced stupendous effects.

Mr. DAGLISH thought the question brought before them by Professor Garnett was of interest, because when they examined a pit where there had been what they conceived to be a coal-dust explosion, they noticed in a striking manner that the force of the explosion was very much greater at a long distance in than at the origin of the explosion. This was almost invariably the case.

The PRESIDENT said, it was now his duty to move a vote of thanks to the writer of the report, and to Professor Bedson for the admirable researches which he had set forth in the appendix. He must congratulate the city of Newcastle on the great assistance which the Durham College of Science was able to afford to the mining engineer and to the city generally. By the inauguration of the College in Newcastle they had been able to draw into their midst men competent to undertake the investigation of those natural laws which lie at the foundation of all such phenomena as had been described that day. In connection with the college were gentlemen not only able but also willing to devote their time and talents to tracing out the cause of explosions such as they had heard about that day; and the members must permit him to thank those gentlemen connected with the College for the useful information which they afforded the members. He did not remember whether Professor Bedson had taken any means to ascertain whether the temperature caused by the heating of coal-dust, or the material inflamed under the chimney that afternoon, would produce an explosion of the gas they had to deal with in mines. It was not merely an elevation of the temperature that was wanted, but they required—if he remembered rightly his reading of Sir H. Davy's experiments—a flame to cause the explosion of a mixture of marsh gas and air. He did not know whether Professor Bedson had ascertained whether the temperature of incandescent coal-dust would be sufficient to generate such a gas in the receiver itself.

Professor BEDSON said, they had not made any experiments on this point.   In working with coal-dust alone and air at ordinary pressure the mass after ignition became simply incandescent, but with compressed air the temperature in parts would be much higher, the intensity of the action being greater.   In working with oil in the last experiments, the continued heating of the mixture resulted in a gradual expulsion of a large portion of the oil, and, after the ignition of the coal itself, the vapours produced by the distillation of the oil were ignited.

The PRESIDENT—Am I right that Sir Humphrey Davy found that a dull red heat was not sufficient to ignite a mixture of marsh gas and air ?

Professor BEDSON—That is so.

Mr. JOHN MARLEY seconded the vote of thanks, and it was agreed to.

Owing to the lateness of the hour, it was resolved to postpone till the next meeting the reading of Mr. A. L. Steavenson's paper, on "Timber v. Steel in Mining."

ENGINE AND SHEWING POSITION

Safety Valve

Safety Valve

No 1 Air Receiver

Shut on

ft. ins.
3 . 6

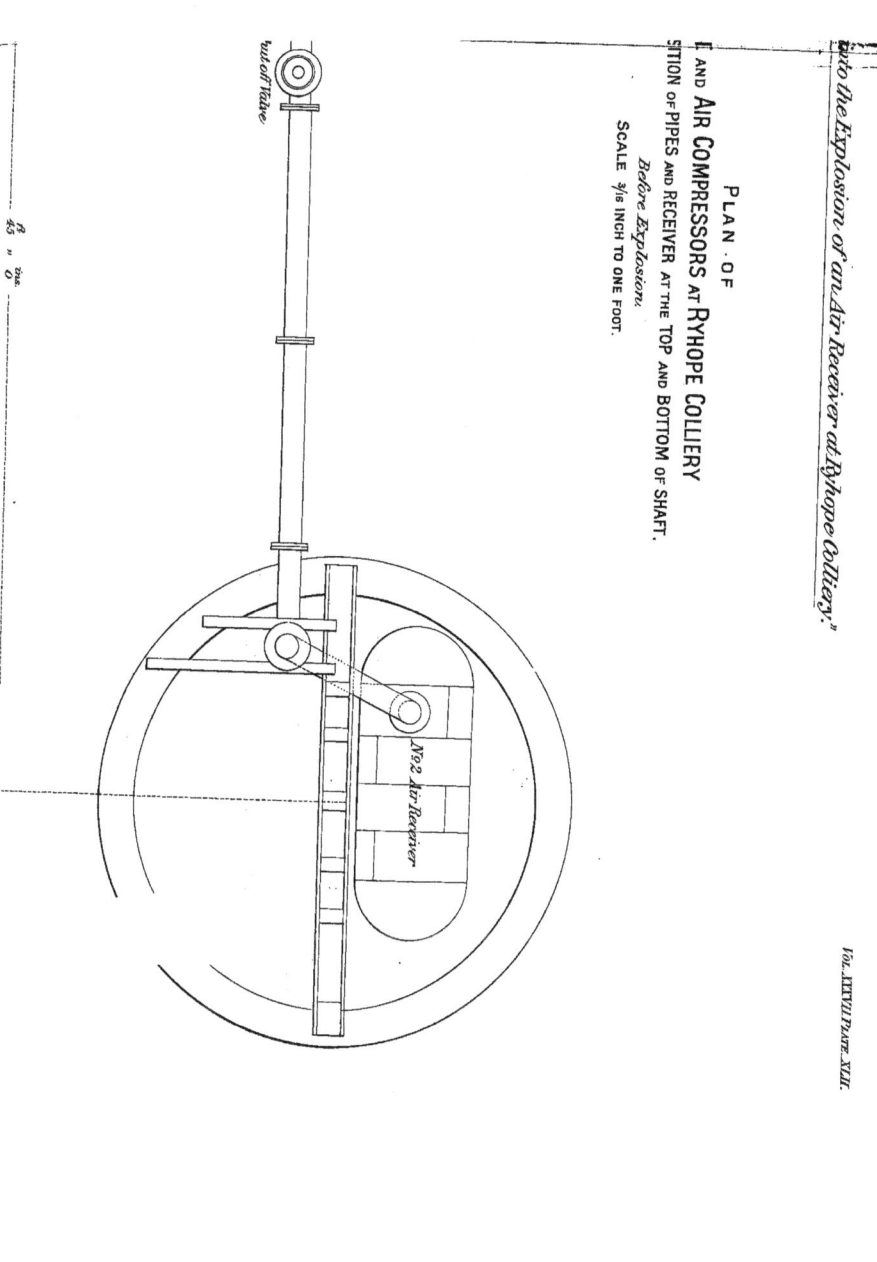

PLAN · OF

E AND AIR COMPRESSORS AT RYHOPE COLLIERY

SITION OF PIPES AND RECEIVER AT THE TOP AND BOTTOM OF SHAFT.

Before Explosion.

SCALE 3/16 INCH TO ONE FOOT.

No 2 Air Receiver

ut off Valve

To illustrate the Report of the Committee on "The Explosion of the Ryhope Air Receiver."

To illustrate the "Report of the Committee appointed to inquire into the Explosion of an Air Receiver at Ryhope Colliery."

SKETCH PLAN OF THE Nº1 AIR RECEIVER.

INSIDE VIEW OF INJURED PORTIONS. ›

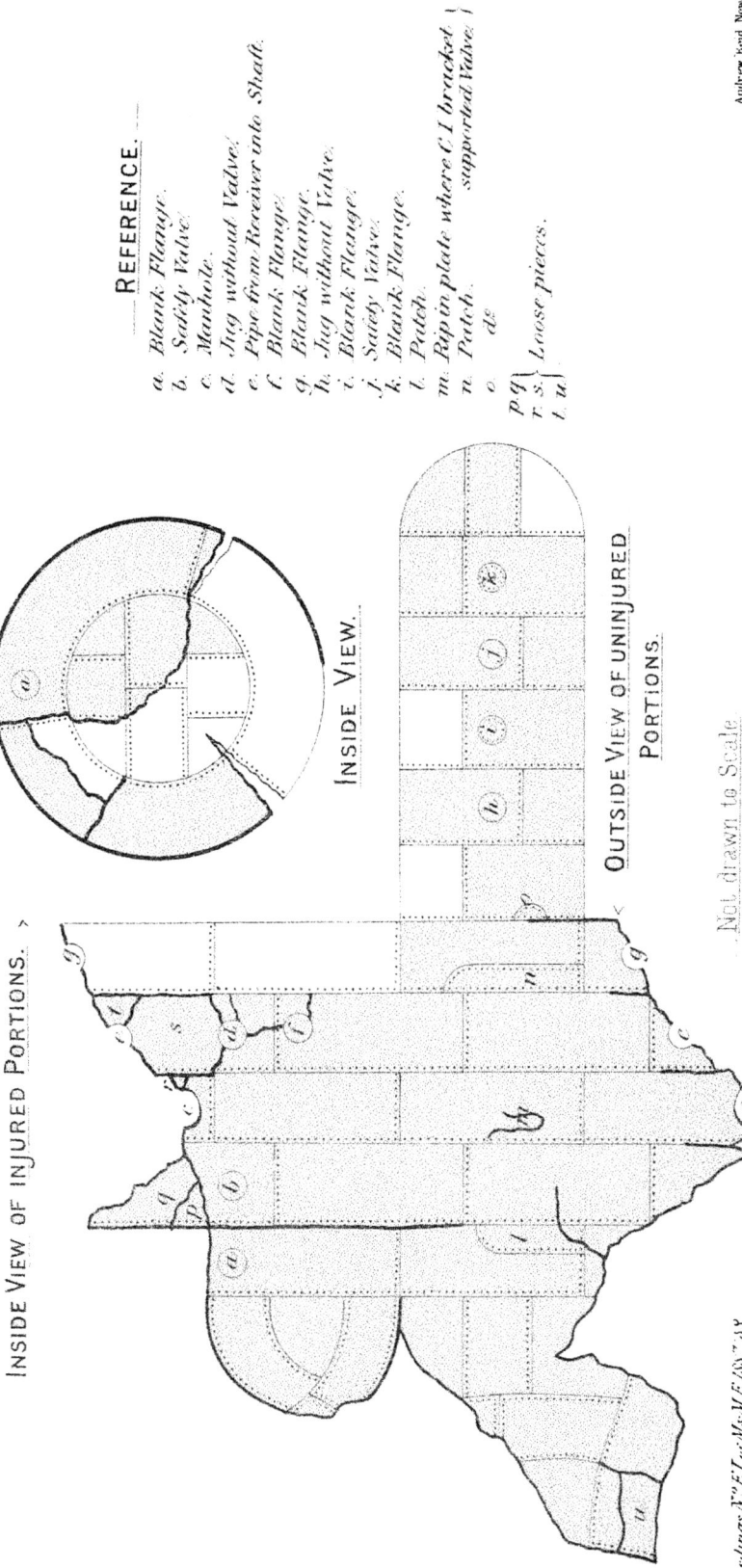

INSIDE VIEW.

OUTSIDE VIEW OF UNINJURED PORTIONS.

Not drawn to Scale

REFERENCE.

a. *Blank Flange.*
b. *Safety Valve.*
c. *Manhole.*
d. *Jag without Valve.*
e. *Pipe from Receiver into Shaft.*
f. *Blank Flange.*
g. *Blank Flange.*
h. *Jag without Valve.*
i. *Blank Flange.*
j. *Safety Valve.*
k. *Blank Flange.*
l. *Patch.*
m. *Rip in plate where C I bracket* ⎫
n. *Patch.*   *supported Valve.* ⎬
o. *d.*   ⎭
p. q. ⎫
r. s. ⎬ *Loose pieces.*
t. u. ⎭

Proc. Nº ET of M. M. E. By° 'AY.

Andrew Reid, Newca. dc

FIG. 1.

DELIVERY PIPE.

FIG. 2.

DELIVERY PIPE.

FIG. 3.

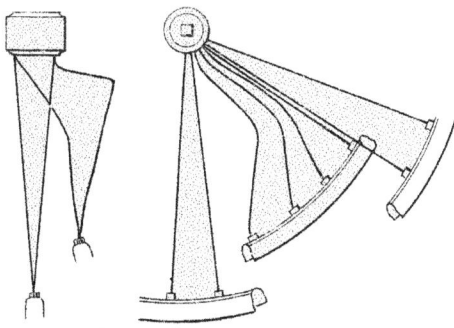

PULLEY WHEEL.

FIG. 4.

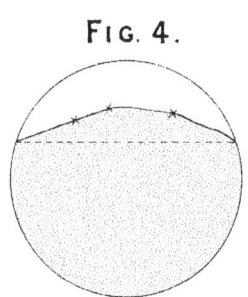

FIG. 5.

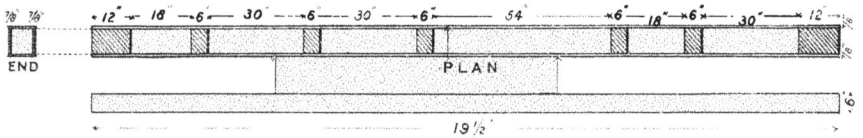

END

PLAN

19 ½

SIDE VIEW

GIRDERS IN SHAFT.

FIG. 6.

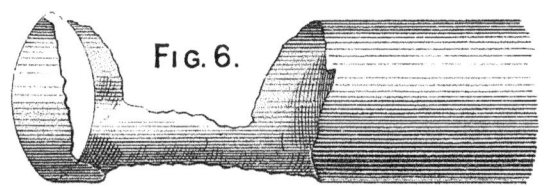

PIPE WITH FLANGE BLOWN OFF
AND PIECE BLOWN OUT OF SIDE

FIG. 7.

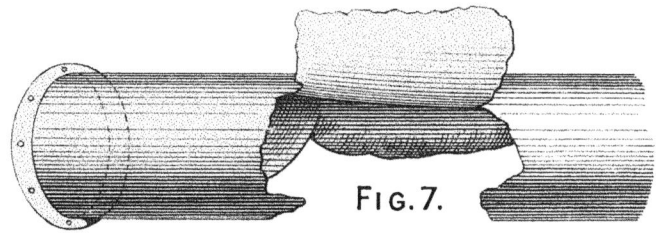

PIPE WITH SIDE BLOWN OPEN.

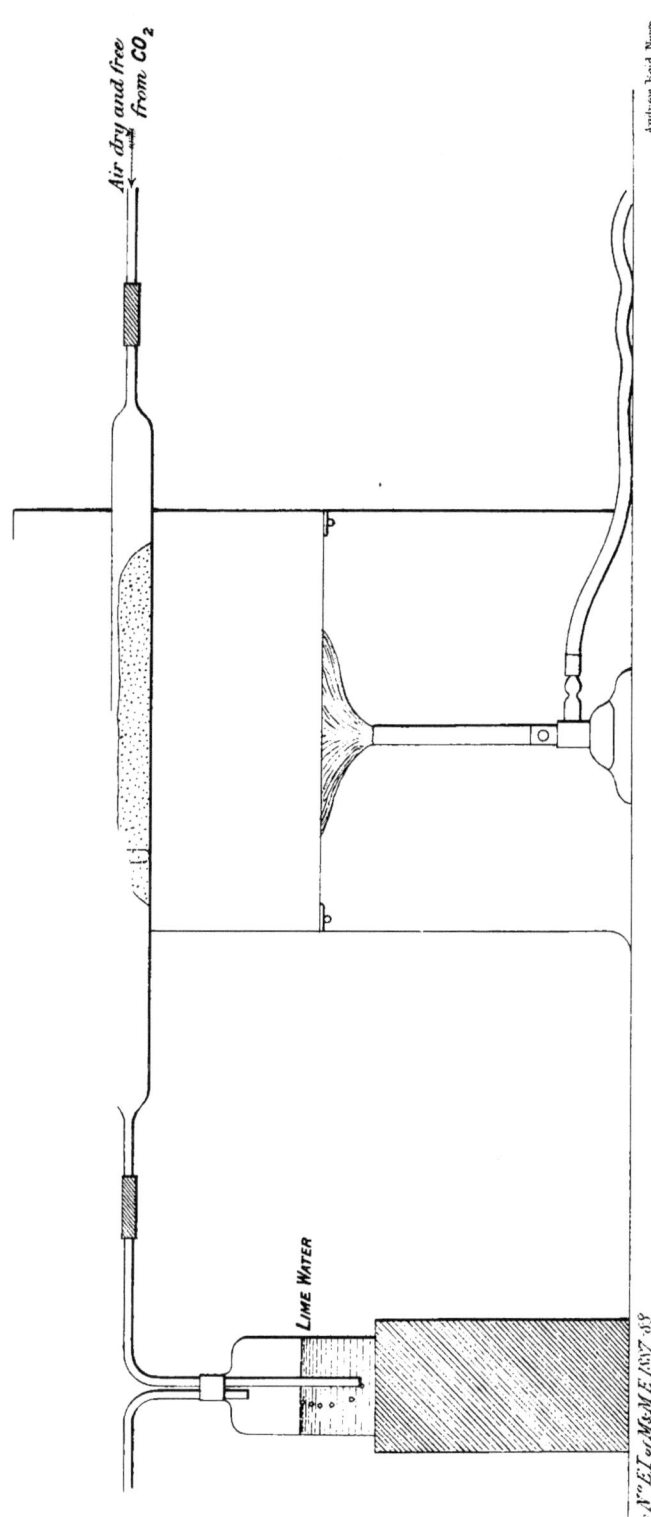

Air dry and free from CO_2

LIME WATER

Fig. N° E1 of Mʳˢ M⁻E. RN 7 68

Andrew Reid, Newc.

To illustrate the Report of the Committee on "The Explosion of the Ryhope Air Receiver.

FORM OF APPARATUS USED IN THE SECOND EXPERIMENTS.

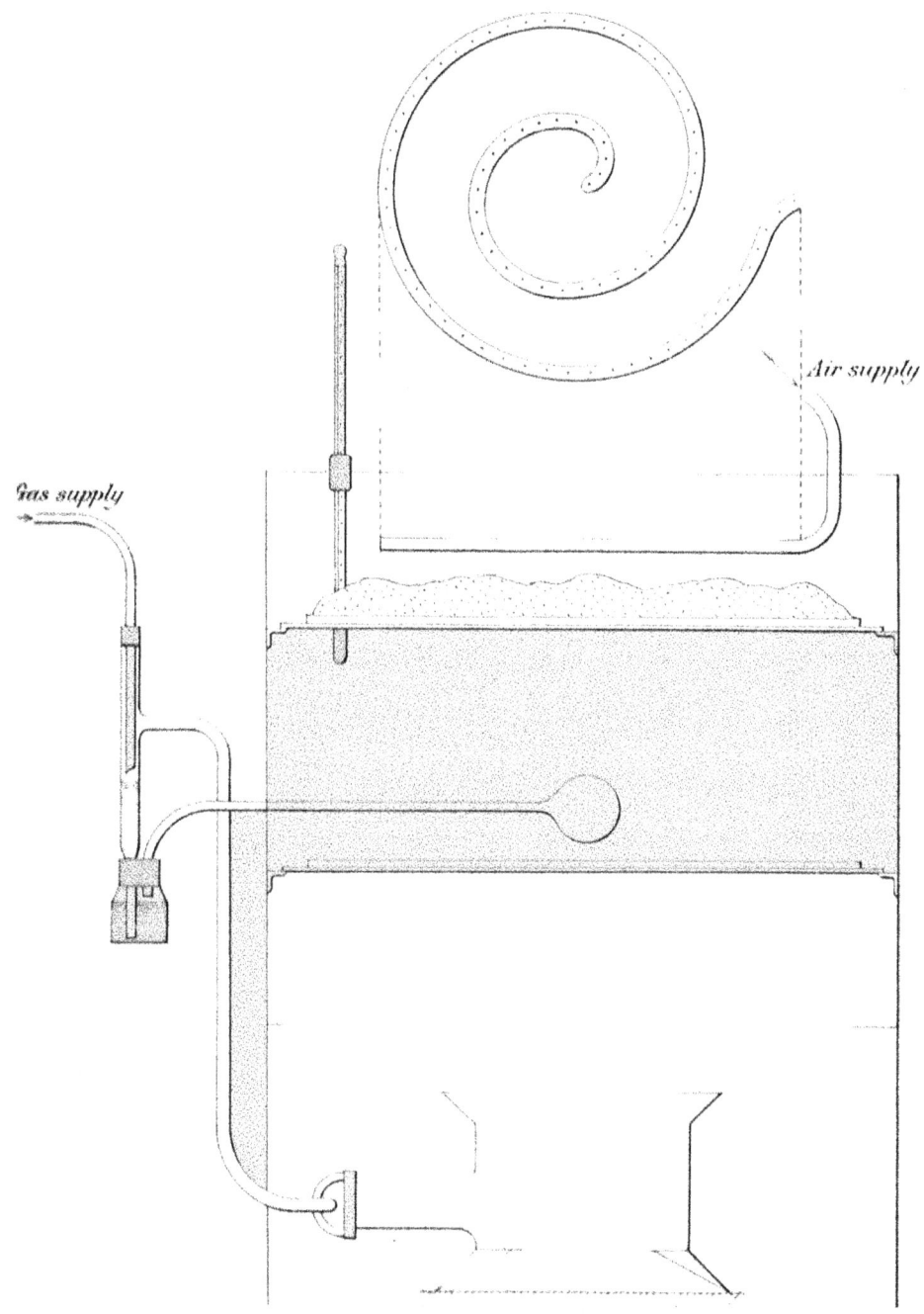

Air supply

Gas supply

To illustrate the Report of the Committee on "The Explosion of the Ryhope Air Receiver."

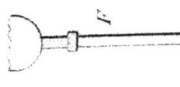

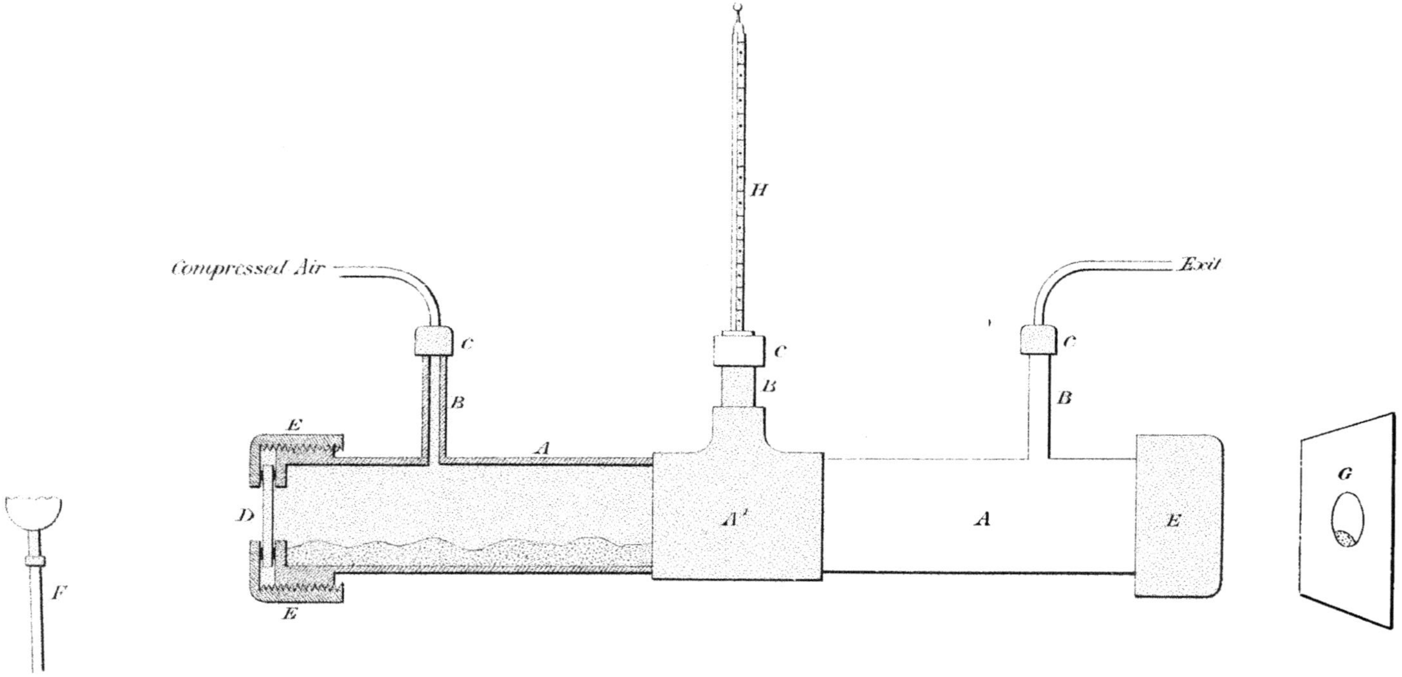

Exit

Compressed Air

G

E

A

C

B

A´

H

C

B

A

C

B

E

D

E

F

PROCEEDINGS.

ANNUAL GENERAL MEETING, SATURDAY, AUGUST 4TH, 1888.

SIR LOWTHIAN BELL, BART., RETIRING PRESIDENT, IN THE CHAIR.

The SECRETARY read the minutes of the previous meeting and reported the proceedings of the Council.

The following gentlemen were elected :—

ORDINARY MEMBER—

W. THOMPSON, Mining Engineer, Connaught Mansions, Victoria Street, Westminster, London.

ASSOCIATE MEMBER—

REGINALD GUTHRIE, Secretary, Coal Trade Associations, Newcastle-on-Tyne.

The SECRETARY read the Annual Report.

The PRESIDENT said there was nothing in the report requiring any observations from him, and he, therefore, would content himself with simply moving its adoption.

Mr. GEO. B. FORSTER seconded the motion, and it was agreed to.

Messrs. W. H. Hedley, Henry Jepson, and J. R. Irvine, were appointed scrutineers for the election of officers for the ensuing year, and they handed in the following as the list of officers :—

PRESIDENT.

JOHN MARLEY, Esq.

VICE-PRESIDENTS.

CUTHBERT BERKLEY, Esq.	THOMAS DOUGLAS, Esq.
T. J. BEWICK, Esq.	A. L. STEAVENSON, Esq.
WM. COCHRANE, Esq.	JAMES WILLIS, Esq.

COUNCIL.

WM. ARMSTRONG, Jun., Esq.	WM. LISHMAN, Esq.
J. B. ATKINSON, Esq.	G. MAY, Esq.
T. W. BENSON, Esq.	Prof. J. H. MERIVALE.
R. F. BOYD, Esq.	R. ROBINSON, Esq.
M. WALTON BROWN, Esq.	J. B. SIMPSON, Esq.
S. C. CRONE, Esq.	T. H. M. STRATTON, Esq.
W. H. HEDLEY, Esq.	J. G. WEEKS, Esq.
H. LAWRENCE, Esq.	W. H. WOOD, Esq.
T. LISHMAN, Jun., Esq.	W. O. WOOD, Esq.

The PRESIDENT—As the President-elect is present, perhaps you will allow me to congratulate him upon the choice of the members.

Mr. JOHN MARLEY—Mr. President, I feel honoured by the election that has just been made. I will do my best to imitate the ten Presidents that have gone before me, the last, and I believe, not the least of them, being yourself.

Mr. A. L. STEAVENSON read the following paper "On the Introduction of Steel Supports for the Maintenance of Main Roads in the Mines of Cleveland."

ON THE INTRODUCTION OF STEEL SUPPORTS FOR THE MAINTENANCE OF MAIN ROADS IN THE MINES OF CLEVELAND.

By A. L. STEAVENSON.

THE paper which was read at the meeting in April on Iron Supports has led the writer to contribute his share to the common fund of information on the subject.

In February, 1885, he was asked to consider and give his views upon the suitability of steel for mining purposes, and having visited the steel works at Darlington, and selected such sections as seemed suitable, he reported in favour of a trial, but suggested—"timber will bend and give notice before breaking ; will steel do the same ? "

The result was, 3 tons of girders, 69 lbs. per yard, and 2 tons for packing, known as channel, 16 lbs. per yard, of various lengths, to be used as cross pieces or packing, were obtained. See Plate L.

Most of the members probably know timber is a heavy item in the cost of ironstone at some mines ; very large balks are required, viz., from 12 to 16 feet long, and 7 to $8\frac{3}{4}$ inches in diameter at the small end, weighing from 1 to 3 cwts., and, owing to the damp atmosphere, their average life is not much over two years.

Larch is preferably used peeled, and Riga and Norway, and yet good as it always is in quality, the resident manager reports—"We have some timber crossings that have been put in three times during the last four years ;" this, of course, implies a large amount of labour as well as material.

The annual timber bill at the mines of Messrs. Bell Bros., Ltd., in Cleveland, now, does not fall far short of £10,000, although trade is slack and short time is worked.

The variation in cost of timber at different mines, is very striking, viz., from 0·10d. up to 5d. per ton, according to the conditions of roof, and the proportion of whole and broken mine being worked.

The roof is, of course, the Upper Lias or alum shale, 200 feet in thickness, but in some places there is a few inches, occasionally amounting to 2 feet, of dogger ironstone, which helps to make a roof, so that in such mines no timber is used in the whole mine; but in the frequent absence of sufficient dogger, or when removing the pillars additional weight is incurred, much of this 200 feet has to be borne by the timber.

In a large majority of mines, after the whole mine is worked and before removal of pillars, this shale gradually falls and fills up the "broken" ground.

Before deciding finally to adopt steel for the main roads, it seemed very desirable that an actual test of the strength of the full-sized girder should be made, rather than trust to any mere calculation, based upon the reputed tensile strength of steel, especially with a view to proving, in case of any future contingencies, that the change had not been made without full consideration of such questions, and in view of the some-what conflicting data given by the authorities as to the relative strength of good wrought iron and steel.

Molesworth gives the latter a greater tenacity of 39 per cent. Mr. Adamson, in his address to the Iron and Steel Institute, says about 30 per cent. A committee of the Institution of Civil Engineers puts the tensile breaking weight in tons per square inch of Yorkshire iron, 23·70; Bessemer steel, 31·92; or *30 per cent.* in favour of the steel; whilst De Bèrgne & Co., of Manchester, in some special tests of Bessemer steel, prove it be 40 to 50 per cent. stronger than iron for structural purposes.

These tests the writer proceeded to carry out upon a few steel girders and full-sized timber balks, and he also, as a part of the investigation, made a number of experiments on a small scale upon timber.

For girders and large balks a suitable place was selected in the mines, near where water pipes for other purposes had been provided.

Recesses were cut in which to place the girders near the roof, as in ordinary use, and by means of a lever, made from a 75 lb. rail (L), an empty iron tub (T), Plate L., was suspended, capable of being slowly filled with water, until it brought down the specimen under examination—the weight of water required, tub, and leverage being easily got.

The large balks of timber were in the same way put to an actual test, just as they are used, except that the load was at the centre instead of distributed.

In all questions of strength of timber, Barlow may be said to be "the authority" in this country. His first essay on the subject appeared in 1817, and a sixth edition in 1867. He recites experiments by Buffon on pieces 4 to 8 inches square, and he adds—"These are, it is presumed, all that are historically deserving of any particular notice in this place."

He also mentions experiments by Colonel Beaufoy in the dockyard at Deptford, on specimens 5 feet long and 2 inches square; also by Messrs. Peake and Barrallier, 2 inches square; and by Mr. Conch, on triangular prisms, the sides being 3 inches.

Barlow's tests were all on small sections of from 1 to 2 inches square.

The objection to testing such small pieces is recognised by Gregory in his chapter on the "Strength of Materials," when he says—" If the material is of cast metal it is found that the exterior hard crust is different to that of the interior . . . . and in the case of fibrous material or timber . . . . . in cutting the bar to the required dimensions many of the exterior fibres will be cut transversely, and will not, therefore, be capable of so great a proportionate strength as the similar fibres within the more central portion of the bar."

Another writer on the same subject, in a paper read to the Society of Engineers, says:—" It may be remarked, however, that the greater part of these (experiments on timber) also are open to the objection before referred to of having been made upon exceedingly small pieces."

Numerous and very careful researches have also been made into the qualities of Colonial timbers; but without attempting any serious investigation of the subject, the writer has made what may be called a few "every-day" tests sufficient to satisfy himself in a practical way as to what the mines timber will actually carry, how far the steel girders may with safety be used instead, together with a little insight into the question of relative cost. Of course, timber in mines has its load distributed, whereas in results now got the weight was suspended from the centre.

The beams were also to some extent fixed, increasing to one and a half times the breaking weight when freely supported; but as in daily use they are fixed, it was the best arrangement, and, in fact, necessary, to prevent the girder canting under its load.

The girders finally adopted, and which have given every satisfaction, are 50 lbs. per yard.

One of these, it will be seen, carried 9·36 tons, when it overcame the supports and canted to one side, after carrying at least double the breaking load of timber as shown in Table C; the next, a 66 lb. girder, sunk under a load of 12·62 tons without fracture, and was afterwards straightened and put into use. These were considered sufficient to show the capacity, so long as the quality was maintained; and it is satisfactory to be able to add, that out of nearly 200 tons now in operation, there has been only one girder a failure, i.e., broken short.

The writer does not propose to use these results as evidence sufficient to base any general assumption of tensile or transverse strength. They are too few, and show too great a difference to permit of it; in fact, the figures obtained and given by Barlow, hardly warrant any general conclusion to be drawn.

Take, for instance, his larch results: with similar specimens, 6 feet by 2 inches square, he gets a breaking weight in lbs. of 300 in one case, and 552 in another, and his resistance of a rod an inch square, where

$$S = \frac{l \times W}{4ad^2},$$

varies from 853 up to 1,149, the only safe manner in which to treat the question, is to deal as with a chain, where it is said that its strength is that of its weakest link.

On this point Box, in his work upon the strength of materials, says:—" Experiments have shown that there is great variableness in the strength of all materials, even when apparently of the same strength and quality. The mean strength, as found by numerous experiments, is usually taken, and it becomes a matter of considerable practical importance that the engineer should know within what limits the strength may probably vary, and particularly that the probable minimum should be known." He also shows that larch, our hitherto favourite mines timber, is very variable.

An excellent article on this point appeared in *The Engineer*, 22nd February, 1861. The writer quotes a Table of Experiments sent to the Iron Commission by Mr. Robert Stephenson, made to determine the best iron for the construction of the High Level Bridge, in which 1 inch bars on 3 feet supports gave results varying from 518 to 1,072 lbs.

But we have by the few tests made on a large scale demonstrated with sufficient clearness that the timber, as daily used, is much inferior to the steel girders in point of strength, of much importance where the roof is heavy. Fewer pieces are required, and a much better and neater arrangement is produced; it also reduces the number of props, and makes a clearer road for men and horses.

Other testimony may be found to the same effect in the *Transactions of the Midland Institute*, Vol. X. We have an account, by Mr. Smith, of the importance of additional strength in goaf roads. He says:—"These steel girders have given much greater satisfaction than the usual timbering, with which it was very difficult to maintain a road at all. So far, out of some thousands in use, only nine have broken."

Of the reduction of obstruction in the roads by the use of steel instead of timber, we find in the *Bulletin de la Société de l'Industrie Minérale*, Vol. XV.:—

	Cross Section of Roadway		Proportion.
	Gross square feet.	Clear square feet.	Per cent.
Iron frames	33·8	27·4	81
Timbering	78·6	40·9	52
Brick arching	84·5	31·2	37

which is a satisfactory confirmation. For every day use in the working places the question of length of life does not arise, since timber in such cases is not worn out by age, but by fixing, and removing frequently from place to place.

The economy to be effected is not, of course, in first cost. In fact, as the following figures show, it is considerably more; but whilst the timber often averages a life of not more than eighteen months, the steel, so far as we see, with three and a half years' experience, appears to be a permanent job.

The cost of what in mining language we call "board end crossings" seems to offer the best opportunity for comparison of steel with timber, and details are given in Tables A and B, also sketches. These do not show the packing, which in bad roofs is an important item; but it will be readily seen what a much lighter and neater arrangement is the result.

The headways 12 feet and the board ends 14 feet in width, taking six of each kind as a sample, we find that, including the packing material and all labour, the average cost of steel is £5 4s. 1d., against timber £3 16s. 6d., or an increase of *36 per cent.* as the cost of permanency and greater efficiency.

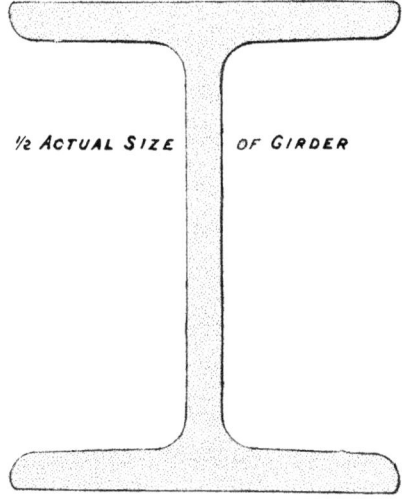

½ *ACTUAL SIZE* OF *GIRDER*

TABLE A.—STEEL.

Cost of Six Steel Board End Crossings.

No. of Balks.	Length of Balks.	No. of Pieces of Packing.	Weight of Balks and Packing. S.B. 16·66 lbs. per ft. S.P. 7¼ lbs. „	Rates.	No. of Men to put in each Crossing.	Cost.	Total Cost of each Crossing.	
	Ft. In.		Cwts. Qrs. Lbs.	£ s. d.		£ s. d.	£ s. d.	
2	14 9	...	4 1 25	5 5 0	...	1 3 4		
2	13 6	...	4 0 2	,,	...	1 1 2		
4	13 0	...	7 2 26	,,	...	2 0 6		A.
		66	4 1 2	,,	...	1 2 5		
					3	0 13 0		
							6 0 5	
3	15 9	...	7 0 3	,,	...	1 16 11		
2	14 9	...	4 1 25	,,	...	1 3 4		B.
		42	2 2 25	,,	...	0 14 3		
					3	0 13 0		
							4 7 6	
6	14 0	...	12 1 27	,,	...	3 5 7		
		29	......	2½d. each.	3	0 7 0		C.
						0 13 0		
							4 5 7	
4	14 0	...	8 1 9	5 5 0	...	2 3 8		
2	13 0	...	3 3 13	,,	...	1 0 4		D.
		54	3 2 0	,,	...	0 18 4		
					3	0 13 0		
							4 15 4	
4	16 3	...	9 1 20	,,	...	2 10 7		
1	14 6	...	2 0 17	,,	...	0 11 7		
1	13 6	...	2 0 1	,,	...	0 10 6		E.
		126	8 0 17	,,	...	2 4 1		
						0 17 4		
							6 14 1	
6	14 0	...	12 1 27	,,	...	3 5 7		
		68½	4 1 20	,,	...	1 3 0		F.
					3	0 13 0		
							5 1 7	

Average cost of materials and labour, £5 4s. 1d.

TABLE B.—WOOD.

STATEMENT SHOWING COST OF WOOD BOARD END CROSSINGS.

No of Balks.	Length of Balks.	No. of Pieces of Packing.	Weight of Balks and Packing.	Rates.	No. of Men to put in each Crossing.	Cost.	Total Cost of each Crossing.	
	Ft. In.			Each. s. d.		£ s. d.	£ s. d.	
5	16 0	...	....	7 0	...	1 15 0		
1	14 0	...	....	5 4½	...	0 5 4		
4	12 0	...	......	3 6¼	...	0 14 1		G.
		Pieces. 38	......	0 2½	...	0 7 11		
					3	0 13 0		
						————	3 15 4	
6	16 0	...	......	7 0	...	2 2 0		
...	14 0	...	......	5 4½	...	......		
5	12 0	..	......	3 6¼	...	0 17 8		H.
		43	......	0 2½	...	0 9 0		
					3	0 13 0		
						————	4 1 8	
6	16 0	...	......	7 0	...	2 2 0		
2	14 0	...	......	5 4½	...	0 10 9		
2	12 0	...	......	3 6¼	..	0 7 0		I.
		10	....	0 2½	3	0 2 1		
						0 13 0		
						————	3 14 10	
4	16 0	...	......	7 0	...	1 8 0		
2	14 0	..	......	5 4½	...	0 10 9		J.
1	10 0	...	......	2 6	..	0 2 6		
		25	......	0 2½	3	0 5 2		
						0 13 0		
						————	2 19 5	
7	16 0	..	......	7 0	...	2 9 0		
1	14 0	...	......	5 4½	...	0 5 4		
8	10 0	... Props, Pieces. 35	......	2 6	..	1 0 0		K.
			......	0 2½	...	0 7 3		
			For propping.		1	0 4 4	0 4 4	
						0 13 0	4 14 7	
						————		
3	16 0	...	......	7 0	...	1 1 0		
3	14 0	.	......	5 4½	...	0 16 1		
6	10 0	... Props. 35	......	2 6	....	0 15 0		L.
			......	0 2½	...	0 7 11		
						0 13 0		
						————	3 13 0	

Average cost of six timber crossings, £3 16s. 6d.

TABLE C.

TESTS MADE BY LEVER UPON STEEL GIRDERS AND TIMBER BALKS.

STEEL GIRDERS.

Section of Girder.	Weight per Yard.	Length.	Length between Supports.		Weight Supported.	Deflection in Inches.		Weight.		
	Lbs.	Feet.	Feet.	In.	Tons.			C.	Qrs.	Lbs.
	50	12	11	0	9·36	7·75	Girder canted.	1	3	4
5" 66·15 ½" ←4"→	66·15	12	11	0	12·62	7⅜	No fracture.	2	1	12
5" ⅜" ←4"→	50	7	6	0	17·10	1·50*	Shackle broken.†	1	1	17

* Not got when broke, but 1·50 at previous observation.   † Shackle was ⅞th Low Moor iron, 7 in. links.

TIMBER BALKS.
RIGA.

	Dimensions.									
Round ...	25·5" circ.	12	11	5½	4·60	5·75	Broken.		...	
	24" ,,	12	11	0	4·70	7·75	Broken.	1	1	6

LARCH.

Round ...	25" circ.	12	11	5½	5·81	8·50	Broken.		...	
Do. ...	22·5" ,,	12	11	0	4·16	8·25	Cracked badly.		...	
Do. ...	28·5" ,,	12	11	0	6·06	4·0	Broken.	2	1	12

NORWAY.

	B. D.									
1	7½ × 6¾	12	11	0	2·50	1·5	Broke.		...	
2	7 × 7½	12	11	0	4·58	2·5	Broke.		...	

APPENDIX.

TESTS OF SMALL SAMPLES OF VARIOUS SIZES OF NORWAY, LARCH, AND RIGÁ TIMBER.

THE arrangement or apparatus for experimenting consisted of a pair of uprights, upon which, 8 feet from the ground, the specimen to be tested could be rested, and by chains suitably attached a platform was suspended, weighing 24 lbs., upon which 28 lbs. weights were quietly deposited, the deflection caused being registered.

The whole of the observations were very carefully made by the writer, at intervals of spare time extending over several months, the net results being that the value of M_T or transverse strength or central breaking weight of a beam 1 inch square and 1 foot long between end bearings was—

$$\text{Riga} \quad M_T = 475$$
$$\text{Norway } ,, \quad = 460$$
$$\text{Larch } ,, \quad = 395$$

ABSTRACT ACCORDING TO SIZES.

$$M_T = \frac{W \times L}{B \times D^2}$$

SIZE.	RIGA.		NORWAY.		LARCH.		
	Breaking Weight.	M_T	Breaking Weight.	M_T	Breaking Weight.	M_T	
	Lbs.		Lbs.		Lbs.		
$1'' \times 1''$	150		120·5		103		Width between supports
1×1	136		120·5		...		4 feet.
1×1	134		120·5		...		
Average ..	140	560	120·5	482	103	309	
$1·97 \times 1·97$	845		1,063		...		
2×2	868		1,054		857		
2×2	836		949		...		
2×2	630		609		...		
Average...	778	391	871	435	857	440	
$2\frac{1}{8}\ \square\ \tfrac{2}{}$	...		...		787		
$2\frac{1}{16}\ \square\ \tfrac{2}{}$	...		...		584		
$2\frac{1}{8}\ \square\ \tfrac{2\frac{1}{4}}{}$	...		...		718		
					696		
1·95 diameter	845		...		...		
2·15 ,,	...		901		...		
2·20 ,,	...		907		...		
2·25 ,,	686		758		752		
2·25 ,,	886		752		724		
2·25 ,,	876		839		788		
2·25 ,,	1,368		951		1,088		
2·25 ,,	783		...		808		
2·25 ,,	920		...		...		
2·25 ,,	920		...		...		
Average...	913	475	825	463	832	437	
2·37 diameter	...		...		955		
		475		460		395	

DETAILS OF TESTS OF SMALL SAMPLES.

TESTS OF NORWAY TIMBER. AUG. 20TH, 1885.

Size.	Deflection.	Total Weight.	Remarks.
	Inches.	Actual Weights.	
1″ × 1″	0·33	24	4 feet between supports.
	0·73	52	
	1·10	80	Slowly sunk to 5 inches. then broke; gave much
	1·90	108	signs of compression.
	...	120·5	
	...	Broke.	
1″ × 1″	·37	24	
	1·33	80	Good sample.
	2·33	108	
	...	120·5	Broke.
	4·00	Bent and broke very slowly.	
1″ × 1″	·40	24	
	1·25	80	
	2·00	108	
	...	120 5	
	3·50	Slowly bent and broke.	
2″ × 2″	...	24	
	·00	80	
	0·10	136	
	0·15	192	
	0·20	248	
	0·25	304	
	0·30	360	
	0·44	416	
	0·47	472	
	0·50	528	
	0·55	584	
	0·62	640	
	0·70	696	
	0·74	752	Good specimen.
	0·77	808	
	0·85	836	No knots.
	0·88	850	
	0·98	882	
	1·00	907	
	1·08	932	
	1·25	957	
	1·40	989	
	1·50	1,022	
	...	1,054	
	2·5	Slowly sunk and broke at.	

DETAILS OF TESTS OF SMALL SAMPLES.—CONTINUED.

TESTS OF NORWAY TIMBER. AUG. 20TH, 1885.

Size.	Deflection.	Total Weight.	Remarks.
	Inches.	Actual Weights.	
2″ × 2″	0·10	112	
	0·23	168	
	0·25	224	
	0·27	280	
	0·33	336	
	0·43	448	
	0·52	504	
	0·58	560	
	0·62	616	
	0·73	672	
	0·76	728	
	0·85	784	
	1·00	814	
	1·05	846	
	1·10	878	
	1·30	910	
	1·50	910	
	1·57	924	
	1·75	936½	Sunk to 2¼, and snapped rather short.
	2·25	949	
	...	Broke.	
2″ × 2″	...	24	
	0·23	80	
	0·27	136	
	0·45	192	
	0·52	248	Very sappy.
	0·61	304	
	0·75	360	
	0·85	416	
	1·05	472	
	1·30	528	
	...	584	
	2·00	596½	
	...	609	*Sunk and broke* rather suddenly; one corner showed sap, opened at fibre.

PARK PIT.

TESTS OF NORWAY TIMBER, 2·25″ DIAMETER.   OCTOBER 15TH, 1885.

	Size.	Deflection.	Total Lbs.	Remarks.
		Inches.		
TURNED ROUND IN LATHE.	2·25″ dia.	...	24	
		0·10	80	
		0·20	136	
		0·25	192	
		0·36	248	Bearings, ¾ inch under 4 feet.
		0·48	304	
		0·52	360	
		0·63	416	
		0·75	472	
		0·80	528	
		0·98	584	
		1·20	640	
		1·40	696	Cracks.
		1·60	727	
		1·90	758	Opens below.
		2·25	758	*Slowly sunk and broke.*
NOT TURNED ROUND.	2·15″ dia.	...	24	Bearings. 4 feet.
		0·05	80	
		0·18	136	
		0·25	192	
		0·30	248	
		0·40	304	
		0·50	360	
		0·60	416	
		0·70	472	
		0·76	528	
		0·90	584	
		1·00	640	
		1·18	696	
		1·25	752	
		1·60	808	
		1·75	839	
		1·85	870	
		...	901	Went to 2 inches, and *broke short.*
SQUARE.	1·97″ × 1·97″	...	24	
		0·10	80	
		0·18	136	
		0·22	192	
		0·25	248	
		0·30	304	
		0·40	360	
		0·50	416	
		0·55	472	
		0·63	528	*Gradually broke.*
		0·70	584	
		0·75	640	Good sample.
		0·83	696	
		0·95	752	
		1·00	808	
		1·18	864	
		1·27	920	
		1·47	976	
		1·60	1,032	
		2·00	1,063	
		2·50	...	
		2·76	...	

PARK PIT.

TESTS OF NORWAY TIMBER. FEBRUARY 11TH, 1886.

	No.	Size.	Deflection.	Total Weight.	Remarks.
			Inches.	Lbs.	
No. 5.	...	2·25″ dia.	...	24	Scales.
	1		...	80	
	2		0·25	136	
	3		0·27	192	
	4		0·40	248	
	5		0·52	304	
	6		0·63	360	Bad sample ; very dry.
	7		0·70	416	
	8		0·90	472	
	9		0·90	528	
	10		1·00	584	
	11		1·10	640	
	12		1·27	696	
	13		...	752	Broke.
No. 6.	1	2·25″ dia.	0·05	80	
	2		0·20	136	
	3		0·25	192	
	4		0·30	248	
	5		0·40	304	
	6		0 50	360	
	7		0·60	416	
	8		0·70	472	
	9		0·75	528	
	10		0·80	584	
	11		1·00	640	
	12		1·15	696	
	13		1·27	752	
	14		1·50	828	Cracked, slightly.
	...		...	839	Broke.
No. 9.	...	2·25″ dia.	...	24	Scales.
	1		0·12	80	
	2		0·20	136	
	3		0·27	192	
	4		0·30	248	
	5		0·44	304	
	6		0·50	360	
	7		0·54	416	
	8		0·60	472	Been cut and dried three months.
	9		0·74	528	
	10		0·76	584	
	11		0·80	640	
	12		{ 0·90 } { 0·95 }	696	
	13		1·10	752	
	14		1·20	808	
	15		1·30	864	
	16		{ 1·70 } { 1·90 }	920	
	17		2·10	951	Broke.

PARK PIT.

TESTS OF PIECES OF LARCH CUT FROM A BALK. AUGUST 20TH, 1885.

Total length, 4′ 6″; between supports, 4′ 0″.

	No.	Size.	Deflection	Total Weight.	Remarks.
			Inches.	Lbs.	
No. 1, Square.	1	2¼″ × 2″	0·125	56	
	2		0·187	112	
	3		0·25	168	
	4		0·37	224	
	5		0·43	280	
	6		0·50	336	
	7		0·62	392	
	8		0·75	448	
	9		0·81	494	
	10		0·93	560	
	11		1·12	616	
	12		1·37	672	
	13		1·50	702	
	14		1·62	728	
	15		1·87	753	
	16		2·00	763	
	17		2·62	787	Broken.
No. 2, Square.	1	2¹⁄₁₆″ × 2″	0·375	168	
	2		0·50	224	
	3		...	248	This piece had a knot
	4		0·62	260	in middle.
	5		1·00	448	
	6		1·25	494	
	7		1·50	560	
	8		...	584	Broken.
No. 3, Square.	1	2⅛″ × 2⅛″	0·125	80	
	2		0·25	136	
	3		0·37	192	
	4		0·50	248	
	5		0·56	304	
	6		...	360	
	7		0·75	416	
	8		0·75	472	
	9		0·87	528	
	10		1·00	584	
	11		1·37	640	
	12		2·00	668	
	13		2·12	682	
	14		2·50	694	
	15		...	718	Broken.
No. 4, Round.	1	2¼″ dia.	0·125	80	
	2		0·25	136	
	3		0·37	192	
	4		0·50	248	
	5		0·62	304	

PARK PIT.

TESTS OF PIECES OF LARCH CUT FROM A BALK. AUGUST 20TH, 1885.—*Continued.*

Total length, 4′ 6″; between supports, 4′ 0″.

	No.	Size.	Deflection.	Total Weight.	Remarks.
No. 4; Round.—*Continued.*			Inches.	Lbs.	
	6	$2\frac{1}{4}''$ dia.	0·68	360	
	7		0·75	416	
	8		0·93	472	
	9		1·06	528	
	10		1·25	584	
	11		{ 1·50 } { 1·62 }	640	
	12		1·75	668	
	13		1·87	682	
	14		2·00	692	
	15		2·25	704	
	16		2·37	716	
	17		2·50	728	
	18		2·75	740	
	19		...	752	Broken with this weight.
No. 5, Round.	1	$2\frac{1}{4}''$ dia.	0·125	80	
	2		0·25	136	
	3		0·37	192	
	4		0·50	248	
	5		0·56	304	
	6		0·68	360	
	7		0·75	416	
	8		0·87	472	
	9		1·00	528	
	10		1·18	584	
	11		1·50	640	
	12		{ 1·75 } { 2·12 }	696	
	13		2·25	724	Broken with this weight.
No. 6, Square.	1	$1\frac{1}{32}''$	0·50	24	
	2		0·37	38	
	3		1·00	48	
	4		1·12	52	
	5		1·50	62	
	6		{ 1·68 } { 2·00 }	76	
	7		2·50	...	
	8		2·87	...	
	9		5·75	103	Broken.
No. 7, Square.	1	$1''$	0·50	24	
	2		1·00	52	
	3		1·37	66	
	4		1·75	76	
	5		2·00	91	
	6		3·00	103	Broken.

PARK PIT.

Tests of Larch Timber Turned in Lathe, $2\frac{1}{4}''$ Diameter.

No.	Deflection.	Total Lbs.	Remarks.
	Inches.		
1	0·05	...	
2	0·20	80	
3	0·40	136	
4	0·52	192	
5	0·74	248	
6	0·90	304	
7	1·10	360	
8	1·27	416	
9	1·50	472	
10	1·73	528	No fracture.
11	2·00 2·10	584	
12	2·20	596	
13	2·27	608	
14	2·40	620	Still bending.
15	2·60 2·85	632	
16	3·00	644	No fracture.
17	3·40	675	
18	3·60 3·75	706	
19	4·00	737	
20	4·25	749	
21	4·50	...	
22	4·75	...	
23	5·00	...	
24	5·25	788	Broken.

PARK PIT.

TESTS OF LARCH TIMBER NOT TURNED, 2·20″ DIAMETER.   OCTOBER, 1885.

Size.	Deflection.	Total Lbs.	Remarks.
	Inches.		
2·20″ dia.	...	24	
	0·10	80	
	0·20	136	
	0·25	192	
	0·33	248	
	0·48	304	
	0·58	360	
	0·70	416	Out of last balk tested.
	0·75	472	
	0·85	528	
	1·00	584	
	1·18	640	
	1·30	696	
	1·50	752	
	1·70	783	
	1·83	814	
	2·00	845	
	2·20	876	
	...	907	Broke before it could be read.

PARK PIT.

TESTS OF LARCH TIMBER, SQUARE, 2″ × 2″.

Size.	Deflection.	Total Lbs.	Remarks.
	Inches.		
2″ × 2″	...	24	
	0·10	80	
	0·20	136	
	0·25	192	
	0·40	248	
	0·50	304	
	0·60	360	
	0·70	416	
	0·77	472	
	0·90	528	
	1·05	584	
	1·25	640	Breaks.
	1·40	671	
	1·50	702	
	1 60	733	
	1·75	764	
	2·10	795	
	2·40	826	
	2·60	857	Broke, good break.

TESTS OF LARCH. FEBRUARY 11TH, 1886.

	No.	Size.	Deflection.	Total Weight.	Remarks.
			Inches.	Lbs.	
	...	2·25″ dia.	...	24	Scales.
	1		0·10	80	
	2		0·25	136	
	3		0·30	192	
	4		0·48	248	
	5		0·52	304	
	6		0·60	360	
	7		0·75	416	
	8		0·77	472	Good specimen, full of sap, green wood only been cut off balk two days.
No. 3.	9		0·90	528	
	10		1·00	584	
	11		1·10	640	
	12		1·20	696	
	13		1·25	752	
	14		1·35	808	
	15		1·48	864	
	16		1·53	920	
	17		1·75	976	
	18		1·90	1,032	
	19		2·15	1,088	Broke.
	...	2¼″ dia.	..	24	Scales.
	1		0·05	80	
	2		0·10	136	
	3		0·20	192	
	4		0·25	248	
	5		0·27	304	
	6		0·40	360	This piece very wet, only been cut off balk two days.
No. 4.	7		0·48	416	
	8		0·50	472	
	9		0·60	528	
	10		0·70	584	
	11		0·75	640	
	12		0·80	696	
	13		1·00	752	
	14		1·15	808	Broke.
	...	2⅜″ dia.	...	24	
	1		0·20	80	
	2		0·25	136	
	3		0·40	192	
	4		0·53	248	
	5		0·70	304	
	6		0·75	360	
	7		1·00	416	
	8		1·20	472	
	9		1·30	528	
	10		1·50	584	Good sample.
No. 7.	11		1·60 } 1·75 }	640	
	12		1·90 } 2·00 }	696	
	13		2·30	752	
	14		2·50 } 3·00 }	808	
	15		3·25	839	
	16		3·60	863	
	17		3·75	887	
	18		4·50	943	Cracked slightly.
	19		7·00	955	Broke.

STEEL SUPPORTS IN MINES.

PARK PIT.

TESTS OF RIGA WOOD. AUGUST 26TH, 1885.

Size.	Deflection.	Total Lbs.	Remarks.
	Inches.		
1″ × 1″	0¼	24	*4 feet between supports.*
	0½	52	
	0¾	66	Opened from a knot 6 inches from centre.
	0·88	80	
	1·25	108	
	1·50	136	
	2·25	136	Suddenly broke in putting on.
	2·25	150	
1″ × 1″	0·25	24	
	0·90	80	Apparently a good sample.
	1·40	108	
	1·75	122	
	...	136	Gradually sunk and broke; no knot.
1″ × 1″	0·20	24	Fair sample.
	0·88	80	
	1·45	108	Broke from middle.
	1·80	122	No knot.
	...	134	Slowly sunk and broke.
2″ × 2″	0·10	...	
	0·12	80	
	0·16	136	
	0·20	192	
	0·24	248	
	0·26	304	
	0·28	360	Fairly good sample; broke rather suddenly.
	0·33	416	
	0·48	472	
	0·51	528	
	0·58	584	
	0·70	640	
	0·76	696	
	0·90	752	
	1·15	808	
	1·40	836	
	2·00	868	Broke.
2″ × 2″	0·10	80	
	0·23	136	
	0·26	192	
	0·30	248	
	0·40	304	
	0·49	360	
	0·51	416	
	0·55	472	
	0·62	528	
	0·70	584	
	0·75	640	
	0·83	696	
	1·00	752	
	1·15	808	
	1·50	836	Broke rather short.
2″ × 2″	0·05	...	
	0·15	80	
	0·26	136	
	0·43	192	
	0·50	248	Opened at bottom in the fibre; sappy; much
	0·58	304	owing to sap in the specimen.
	0·72	360	
	0·77	406	
	0·90	462	
	1·06	518	
	1·25	574	
	1·50	630	Broke.

PARK PIT.

TESTS OF RIGA WOOD. AUGUST, 1885.

Round, not Turned.

Size.	Deflection.	Total Weight.	Remarks
2 25″ dia.	Inches. 0·00	Lbs. ...	
	0·08	80	
	0·12	136	
	0·25	192	
	0·27	248	
	0·42	304	Broke straight, and rather short.
	0·49	360	
	0·51	406	
	0·60	462	
	0·75	518	
	0·80	574	
	1·00	630	
	1·25	686	
	2·25	...	Broke slowly, but snapped at last.
2·25″ dia.	0·06	80	
	0·15	136	
	0·23	192	
	0·25	248	
	0·27	304	
	0·33	360	
	0·48	416	
	0·51	472	
	0·60	528	
	0·66	584	
	0·75	640	
	0·80	696	Sunk slow at first, and then broke off at 1½.
	0·90	752	
	1 05	808	
	1·50	836	Broke.
2·25″ dia.	0·10	...	
	0·20	80	
	0·24	136	
	0·32	192	
	0·40	248	Broke at once; a less weight would have done it.
	0·48	304	
	0·51	360	
	0·58	416	
	0·68	472	
	0·75	528	
	0·85	584	Good sample.
	0·94	640	
	1·07	696	
	1·25	752	
	1·40	780	
	1·50	812	
	1·75	844	
	...	876	Broke short.

PARK PIT.

Tests of Riga Wood, 2¼″ Diameter.　October 15th. 1885.

Turned up in Lathe.

Size.	Deflection.	Total Weight.	Remarks.
	Inches.	Lbs.	
2·25″ dia.	0·10	112	
	0·18	168	
	0·25	280	
	0·26	336	
	0·30	392	
	0·35	448	
	0·42	504	
	0·45	560	
	0·50	616	N.B.—Half a ton.
	0·52	672	
	0·60	728	
	0·68	784	
	0·73	840	
	0·75	896	
	0·80	952	
	0·90	1,008	
	0·95	1,039	
	1·00	1,070	
	1·10	1,101	
	1·13	1,132	
	1·20	1,163	
	1·25	1,219	
	1·38	1,250	
	1·50	1,306	Shows weakness; sunk slowly; broke straight.
	1·75	1,337	
	1·80	1,368	
	2·00	...	
	Total with scale and chain ...	1,392	
1·95″ × 1·95″	...	24	
	0·10	80	
	0·18	136	
	0·25	192	
	0·30	248	
	0·40	304	
	0·50	360	
	0·55	416	
	0·63	472	
	0·73	528	
	0·77	584	
	0·85	640	
	1·00	696	
	1·10	752	First crack.
	1·20	783	
	1·25	814	
	1·45	845	Cracks, and gradually broke.

ARRANGEMENT FOR TESTING STRENGTH OF BALKS.

Scale ½ inch to a Foot

Andrew Reid, Newcastle.

Main Roads in the Mines of Cleveland."

TESTS OF RIGA WOOD, TURNED. FEBRUARY 11TH, 1886.

	No.	Size.	Deflection.	Total Weight.	Remarks
			Inches.	Lbs.	
No. 1.	...	2·25″ dia.	...	24	Scales.
	1		0·20	248	
	2		0·25	304	
	3		0·27	360	
	4		0·30	416	
	5		0·35	472	
	6		0·50	528	
	7		0·53	584	
	8		0·60	640	
	9		0·75	696	
	10		0·75	752	Cracked slightly.
	11		1·25	783	Broken.
No. 2.	...	2·25″ dia.	...	24	Scales.
	1		0·12	80	
	2		0·20	136	
	3		0·27	192	
	4		0·30	248	
	5		0·50	304	
	6		0·55	360	
	7		0·68	416	Been turned up three months; very dry.
	8		0·75	472	
	9		0·80	528	
	10		0·90	584	
	11		1·00	640	
	12		1·12	696	
	13		1·25	752	
	14		1·50	808	
	15		1·75	864	
	16		2·10	920	
	...		2·25	920	Broken.
No. 3.	...	2·25″ dia.	...	24	Scales.
	1		0·10	80	
	2		0·20	136	
	3		0·27	192	
	4		0·30	248	
	5		0·40	304	
	6		0·50	360	
	7		0·60	416	Very dry; same as last.
	8		0·70	472	
	9		0·75	528	
	10		0·78	584	
	11		0·85	640	
	12		0·98	696	
	13		1·05	752	
	14		1·20	808	
	15		1·27	864	
	16		1·50	920	Broken.
	...		1·55	...	

Professor MERIVALE—As Mr. Steavenson is engaged in experiments, I would ask if he has made any with creosoted timber, and, if not, perhaps he may be able to add some experiments upon this form of mining timber?

Mr. STEAVENSON—No; he had not made any experiments with creosoted timber. They had used it at the collieries for main roads, and no doubt it was an improvement upon timber that was uncreosoted, but he preferred steel under their circumstances. They were using creosoted timber at the collieries. The objection to creosoted timber for mining purposes was that if they once cut it, or even put in a nail,.they destroyed the effect of the creosoting. They must use the timber as it was sent out when creosoted.

Mr. GEO. B. FORSTER said, they must use steel in the same way. He had used a considerable amount of creosoted timber in his time, and he had some which had been in, he believed, twenty-four or twenty-five years, and there was no perceptible sign of decay. He himself did not think that the cutting of the timber absolutely injured it. The President suggested to him to refer to the case of railway sleepers, which were made of creosoted timber, and holes were bored into the sleepers to receive the plugs. He agreed with Mr. Steavenson as to the superiority of steel in the way of strength and handiness, but he thought it would be found that creosoted timber would last quite as long as steel. So far as durability went, creosoted timber would last a long time. It was all well enough in Cleveland to use steel, where they had particular widths of drifts, and where they knew exactly what they were going to do; but in coal mining they had a great variety of cross places and drifts, and sometimes a balk of timber was more handy to arrange and set up than steel.

Professor MERIVALE said, there were other things besides creosote used in preserving timber. Sulphate of copper was used. He would like Mr. Steavenson to extend his experiments, and give them some information as to the preservation of timber.

Mr. JAMES WILLIS—I apprehend Mr. Steavenson's paper is more as to the strength of steel as compared with timber.

Mr. STEAVENSON—In the case of creosoting timber you do not increase the strength at all. We can have something like double the strength with steel.

A vote of thanks was passed to Mr. Steavenson for his paper.

———

Professor P. PHILLIPS BEDSON, D.Sc., read the following paper on "A Contribution to our Knowledge of Coal-dust."

A CONTRIBUTION TO OUR KNOWLEDGE OF COAL-DUST.

By Professor P. PHILLIPS BEDSON, D.Sc., F.C.S.,
DURHAM COLLEGE OF SCIENCE.

SOME months ago Mr. W. F. Hall drew the writer's attention to a matter he had occasionally observed at the screens of one of the collieries under his management. It had been noticed that, when screening a certain coal, with the wind in a given direction, the dust would rise in considerable volume, forming a cloud slowly moving up the screens, and on one or two occasions at night-time this mixture of air and coal-dust took fire at the lamp used to light the screens, producing a species of explosion. As this phenomenon was observed with only one variety of coal, it was suggested that this behaviour was to be accounted for by the dust holding combustible gases enclosed in it.

The writer undertook to put the suggestion to the test of experiment, and for this purpose obtained several samples of this coal-dust produced in the screening of this coal. The results of the experiments indicate that this suggestion has opened up a rich field for future investigation, of which this paper contains but the first instalment.

GASES ENCLOSED IN COAL.

Before proceeding to detail the experiments made with coal-dust, it may not be out of place to pass in brief review the main facts respecting the gases "enclosed in" coal.[*] These facts have been brought to light by the investigations of Dr. E. von Meyer (*Journal Pract. Chem.* (2), v. 144-183 and 407-416, also *Journal Chem. Society*, x. 798) and of Mr. W. J. Thomas (*Journal Chem. Society*, 1875 and 1876).

In 1872, von Meyer published the results of his experiments with coals of German origin, and also with certain samples from the Durham and Newcastle coal-fields. The gases were expelled by placing a known weight of the coal, broken into small pieces, in hot water already freed

---

[*] See *Transactions of the Mining Institute*, Vol. XXII. 25, XXV. 41, XXVI. 36, where valuable communications on this subject by the late Professor Freire-Marreco will be found.

from air, and maintained at the boiling point for some time. The volume of gas so expelled varies with the different coals experimented on. In one case the volume of gas was only 22·5 cubic centimetres per 100 grammes of coal, whilst in another case—viz., the coal from Wingate Grange—the volume was found to be 238 cubic centimetres per 100 grammes of coal. Or, supposing the specific gravity of coal to be 1·3, in the first case the volume of gas would be nearly one-third of the volume of the coal whilst in the second case the volume of the gas is nearly three times that of the coal. Von Meyer has also shown that freshly raised coal gives a larger volume of gas than coal which has been exposed to the air for a considerable period.

The analysis of these gases revealed the fact that in the majority of cases they consist of mixtures of carbon dioxide, oxygen, nitrogen, and marsh gas, in varying proportions. In some few cases the latter named gas was not found. Von Meyer draws attention to another fact which will be of interest in connection with the experiments on coal-dust, viz., that the weathered coals examined by him not only give a smaller volume of gases, but, further, that these gases differ in quality from those obtained from the freshly-raised coal from the same pit. This difference consists in the presence amongst these gases of hydrocarbons belonging to the same series as olefiant gas, and also of other members of the series to which marsh gas belongs, viz., the paraffins.

In 1875, Thomas, in a paper read before the Chemical Society in London, gave an account of the results he had obtained with different varieties of coal of the South Wales basin; and in the following year he described the results of his experiments on the gases enclosed in cannel coals and jet. Thomas's method of extracting the gases differed from that already described. He proceeded as follows :—

Slices of coal were sawn out of the middle of large cubes, and a strip about ⅜ of an inch in thickness and from 6 to 8 inches in length was again cut from the middle of this slice. The edges of the strip were carefully rounded off so as to make it slide readily into a glass tube of the proper diameter. After brushing away the adherent dust, the coal, some 10 to 30 grammes in weight, was placed in a glass tube drawn out into a narrow neck at one end, by which it was connected with a Sprengel air pump. The other end of the tube was carefully sealed off before the blowpipe, and the tube exhausted by the Sprengel pump. When the whole of the air was removed, the tube was raised to 100° C. (212° F.) by placing it in boiling water, and maintained at this temperature for several hours until the mercury pump ceased to bring over any appreciable quantity of gas.

It will be seen then that Thomas obtained the enclosed gases by heating the coal in a vacuum at 100° C., but, as this experimenter points out, the whole of the enclosed gases are not given off at this temperature, nor even at 200° C. or at 300° C.—a point close to that at which decomposition of the coal itself begins.

Steam coal, bituminous coal, and anthracites were the classes of coals experimented on by Thomas. His results show that under the conditions of the experiments, anthracites give the largest volume of gas, steam coals stand next, and bituminous coal gives the smallest volume. The volume of gas appears to depend in a great measure on the structure of the coal, hard compact steam coal giving almost as large volumes as anthracites.

In composition the gases exhibit a general resemblance to those obtained by von Meyer.

Cannel coals and jet contain similar gases enclosed in them. Amongst these have been found higher members of the paraffin series and also some members of the olefines.

GASES ENCLOSED IN COAL-DUST.

Thomas's method of extracting the gases from coal appeared the more suitable for the treatment of coal-dust, and with but few slight modifications this method has been adopted. The coal-dust was collected at the colliery in stoppered glass bottles, the weight of which had been determined, and which were again weighed after filling; in this way the weight of coal-dust was ascertained. When required for an experiment the stopper of the bottle is replaced by a tightly-fitting india-rubber stopper to the under side of which is fastened and held by a wire cage a plug of cotton wool, to prevent the coal-dust from being drawn over into the air pump. Into the stopper is fitted a glass tube attached by stout india-rubber tubing to the mercury air pump, by means of which the bottle is completely exhausted and in which the gases given off from the dust are collected. The necessary heating of the coal-dust is accomplished by placing the bottle in a long tin cylinder provided with a stage on which the bottle rests and is thus kept at a height of an inch or so above the bottom of the cylinder. The cylinder is filled with brine so that the bottle is almost completely immersed in it, and then closed with a lid provided with an aperture for the neck of the bottle and also with openings for a thermometer and a thermo-regulator by which means the temperature of the brine can be carefully regulated and maintained at a specific temperature for any length of time.

The mercury pump used is of simple construction, and was made by Mr. Saville Shaw, Demonstrator in Chemistry in the Durham College of Science. It consists of a piece of stout glass tubing to which is fused a side tube a, provided with a tap. This tube serves to connect the vessel to be exhausted with the pump. The distance from the side piece to the end of the longer limb is about 30 inches. The end of the shorter limb is carefully ground and attached by thick-walled india-rubber tubing to the pear-shaped glass vessel b. The india-rubber joint was next completely covered with a coating of Faraday's cement, and the entire apparatus attached to a wooden support, the portions joined together by india-rubber tubing being embedded in plaster of Paris. In this way a perfectly air-tight joint was obtained, as shown by the actual testing of the apparatus. The pear-shaped vessel b is closed by an india-rubber stopper in which is fitted a glass tube d provided with a tap. The stopper is made perfectly tight by covering it over with cement. Attached to the end of the tube e, by means of stout india-rubber tubing, is a vessel f, used to hold the mercury, and so slung that it can be raised and lowered at will.

The method of using the pump is extremely simple. In the first place, the tap on the side tube is closed and that on the collecting vessel b is opened. By raising the movable vessel f previously filled with mercury, the tubes and collecting vessel are entirely filled, the air being expelled through the tube d. The tap on d is now closed and the vessel f lowered. In this manner a barometer is obtained the reservoir of which is the vessel f, the mercurial column is in e, with its upper level below a, and above is the Torricellian vacuum. By repeating this operation a vessel attached to the side tube can in a very short time be completely exhausted.

The above description of this pump has been given with the view of showing how simply, efficiently, and cheaply, such a pump may be constructed, for there are several forms in use, but constructed entirely of glass.

The advantage of a pump of this description over the ordinary Sprengel for this special work, is the small amount of attention it requires, for when once the bottle containing the dust has been exhausted the heating of the bath surrounding the bottle begins; and when the temperature has been adjusted, the whole may be left for any length of time, the gas given off from the coal-dust gradually passing over into the collecting vessel b. When desired, the gas produced is expelled through the tube d, by raising the vessel f, and allowing the mercury to flow into b. The expelled gases are collected in graduated vessels suitably attached to d.

The method of extracting the enclosed gases consisted, therefore, in placing a known weight of the coal-dust in a bottle, and after exhaustion by means of a mercurial air pump, the coal-dust was heated in a water bath, the gas produced collected, and after determining the volume, submitted to analysis.

EXPERIMENTAL RESULTS.

I.—588·5 grammes of coal-dust, heated for two days at 100° C. (212° F.), gave 215·6 cubic centimetres of gas, measured at 17° C., and under a pressure of 29·54 inches, which is equivalent to 196·5 cubic centimetres of gas measured at 0° C. (32° F.), and 760 mm. (29·92 inches.) After the removal of this gas, the heating of the coal was continued for another day, and 15 cubic centimetres of gas at 12° C. and 30·36 inches, were obtained, equivalent to 14·37 cubic centimetres of gas at 0° C. and 29·92 inches. From this it follows that 100 grammes of coal-dust give 36·5 cubic centimetres of gas at 0° C. and 760 mm.

II.—In the second experiment, 869·1 grammes of coal were employed, which, after heating for six hours and a half, gave a volume of gas equivalent to 476 cubic centimetres of gas at 0° C. and 760 mm. A second extraction gave an additional volume of 100·4 cubic centimetres of gas at 0° C. and 760 mm., making a total of 576·4 cubic centimetres of gas, representing 66·3 cubic centimetres of gas per 100 grammes of coal.

III.—In the third experiment the method of treatment was somewhat different. The coal-dust was left under reduced pressure for eleven days, and the amount of gas given off collected and measured. It was next heated for eight days, at 50° C., the gas produced removed as before. Then the heating was continued for thirty hours at 70° C., then heated again for thirty hours at the same temperature, and finally for forty hours at 100° C. In this way five different extractions were performed : first at the ordinary temperature, and reduced pressure, the second at 50° C., the third and fourth at 70° C., and the fifth at 100° C. The weight of coal-dust used was 673 grammes, and the following volumes of gas, reduced to 0° C. and 760 mm. (29·92 inches), obtained at each extraction. First, 22·4 cubic centimetres ; second, 48·5 cubic centimetres ; third, 117·44 cubic centimetres ; fourth, 48·5 cubic centimetres; fifth, 143·2 cubic centimetres, making a total of 380·04 cubic centimetres, or 56·3 cubic centimetres per 100 grammes of coal-dust.

These results may be stated somewhat differently if it is assumed that the dust is compressed into coal of the specific gravity of 1·3. Then in the first experiment the dust would contain about half its own volume of "enclosed gases," the second experiment about eight-tenths, and in the third about seven-tenths.

That coal-dust should, like coal itself, contain gas enclosed in it, is not in itself remarkable, but what appears more worthy of remark is the nature of the gases themselves. The analysis of the gases obtained in the manner described have been made by means of the methods and apparatus due to Hempel, which, whilst admirably adapted for technical purposes, are perhaps wanting in the refinement of the methods usually adopted in purely scientific work. Still, as the space at disposal in the laboratory of the College was such as to preclude the use of such methods, an attempt has been made to do the best under the circumstances.

Analysis has shown the gases obtained from this coal-dust to consist of mixtures of oxygen, nitrogen, carbon dioxide, possibly some carbon monoxide,* and in addition to these certain gaseous compounds of carbon and hydrogen or hydrocarbons. Of the different series of hydrocarbons known two need only be considered here. There is, first, the series known as the paraffins, which contain these elements in such proportions that their composition may be represented by a general formula, CnH_{2n+2}, in which C represents twelve parts by weight of carbon, and H one part by weight of hydrogen, and further, n is a whole number. Marsh gas is that member of this series in which n is one, and is the first of the series; a large number of these hydrocarbons are known, the lower members are gases, then follows a number of liquids such as are contained in petroleum oils, and finally, there are solids, of which ordinary paraffin wax is a mixture.

The other series of hydrocarbons which require to be mentioned here are the olefines, the first member being olefiant gas, a constituent of coal gas, and having the formula C_2H_4. All the hydrocarbons of this series have the same percentage composition as olefiant gas, and this may be represented by a general formula, CnH_{3n}, in which n is a whole number. In this series, as in the paraffins, there is a gradation of physical properties, the lower members in the formula of which n is either 2, 3, 4, are gases, then as n increases there follow liquids and finally solids.

---

* The evidence of the presence of this gas is indirect only, the amounts represent the volumes absorbed, after the removal of oxygen, by an ammoniacal solution of cuprous chloride.

Now the hydrocarbons of this series differ from those of the paraffins, inasmuch as they are more readily acted upon by certain reagents, and are absorbed by certain bodies, whereas the paraffins are unacted upon. One of the best absorbents for the gaseous members of this series is fuming sulphuric acid.

Now, in the analysis of such a mixture of gases as those obtained from coal-dust, the proportion of certain of these, such as carbon dioxide, oxygen, and carbon monoxide, is determined by the use of reagents which absorb them. The amount of olefines can in a similar manner be determined by using strong sulphuric acid as the absorbent. After removing all these gases by the appropriate reagents, there is left a mixture of nitrogen and some member or members of the paraffin series. In determining the quantity and also quality of these hydro-carbons in such a mixture, use is made of the fact that, when mixed with oxygen, they may be exploded, forming carbon dioxide and water. The contraction resulting from the production of the water, the volume of carbon dioxide produced, and also the volume of oxygen used, are of service in giving not only the volume of the gas, but also a clue to its probable formula.

Thus, to take a simple case—that of marsh gas—the following equation represents what takes place when this gas is exploded with an excess of oxygen :—

$$CH_4 + 2O_2 = CO_2 + 2H_2O.$$
$$\underset{2v.}{} \quad \underset{4v.}{} \quad \underset{2v.}{}$$

2 volumes of marsh gas give 2 volumes of carbon dioxide, and require 4 volumes of oxygen to burn it—or 1 volume of marsh gas requires 2 volumes of oxygen, and produces 1 volume of carbon dioxide—further, the 3 volumes of marsh gas and oxygen are, after explosion, represented by 1 volume of carbon dioxide, and the contraction due to the con-densation of the steam is 2.

The next member of the series, ethane, the formula of which is C_2H_6, requires for its combustion a different proportion of oxygen, and gives a larger volume of carbon dioxide. Its complete combustion is represented by the following equation:—

$$C_2H_6 + 7O = 2CO_2 + 3H_2O.$$

Or, 2 volumes of this gas require 7 volumes of oxygen, and give 4 volumes of carbon dioxide and a contraction of 5 volumes. Or, again, taking a paraffin of the general formula CnH_{2n+2}, its combustion may be represented by the following equation:—

$$CnH_{2n+2} + (3n+1)O = nCO_2 + (n+1)H_2O.$$

And since CnH_2n_{+2} represents the weight of the hydrocarbon which in the gaseous state would occupy the same volume as the amount of marsh gas represented by its formula CH_4, from the above equation it can be deduced that 2 volumes of such a hydrocarbon require $(3n + 1)$ volumes of oxygen, and produce 2n volumes of carbon dioxide, and give a contraction of $(n + 3)$ volumes. Or 1 volume of any gaseous paraffin, on burning, requires $\dfrac{3n + 1}{2}$ volumes of oxygen, produces n volumes of carbon dioxide, and gives a contraction of $\dfrac{n + 3}{2}$ volumes.

It will thus be seen that the determination of the volume of oxygen required to burn a given volume of a gaseous member of this series, also the volume of carbon dioxide and the contraction, may be used as a means of fixing the value of n, and thus giving a clue to the formula of the compound. With a mixture of members of this series n will be found not to be an integer.

Inasmuch as in the methods employed the gases have been measured over water, it has been impossible to determine with the necessary accuracy the volume of carbon dioxide formed, but this difficulty has been got over by taking the total contraction after explosion, represented by the sum of the actual contraction, and also the volume of the carbon dioxide produced. This represents the volume of the combustible gas and the oxygen required for its combustion. In the case of marsh gas, 1 volume of this gas would give a total contraction of 3 volumes, and with a hydrocarbon of the formula CnH_2n_{+2}, the total contraction would be $\dfrac{3n + 3}{2}$, where n represents the number of carbon atoms in the molecule.

In the analysis of the gases, the author has determined the value of n by means of the proportion of oxygen required to burn one volume of the gas, and also by the aid of the total contraction.

The method of working will be best illustrated by an example :—

25 volumes of a mixture of a paraffin and nitrogen gave on contraction 27·3 volumes, 22·1 volumes of carbon dioxide, required 40·7 volumes of oxygen, and left a residue of 16·3 volumes of nitrogen. There are, therefore, 25 − 16·3, viz., 8·7 volumes of combustible gas, which produced the above contraction, carbon dioxide, etc., from which calculation gives the following :—

1 volume of combustible gas would produce a contraction of 3·13 volumes, and 2·53 volumes of carbon dioxide (or a total contraction of

5·66 volumes), and would require 4·67 volumes of oxygen. Then the value of n is found by the equations :—

$$\frac{3n + 3}{2} = 5\cdot66 \text{ for total contraction}$$

$$\text{and} \quad \frac{3n + 1}{2} = 4\cdot67 \text{ for the oxygen.}$$

The first of these gives a value 2·78 for n and the second 2·77. Such a value for n would indicate that in all probability the gas in question is a mixture of marsh gas CH_4 and the hydrocarbon C_3H_8 or propane. It has not been considered necessary to calculate the possible composition of the mixture; such results are sufficient to show that a mixture of hydrocarbons of this series is found in the gases enclosed in coal-dust. The determination of the exact nature of the constituents of such a series cannot be made by analysis alone, and it would be futile to push the calculations further.

RESULTS OF ANALYSIS.

EXPERIMENT I.—Gas obtained from the first extraction : a partial analysis showed it to contain the following :—

Carbon dioxide	15·2 per 100 volumes.
Oxygen	1·79 ,,
Olefines CnH_2n (absorbed by sulphuric acid)	9·3 ,,

The complete analysis of the gas obtained from the second extraction yielded the following results :—

Carbon dioxide	13·71 per 100 volumes.
Olefines (CnH_2n)	19·6 ,,
Oxygen	1·32 .,
Carbon monoxide (?)	1·39 ,,
Hydrocarbons of the paraffin series. $CnH_2n + 2$	50·7
Nitrogen	13·2
	99·92

From the results of the explosion of the mixture of $CnH_2n + 2$ and nitrogen the following were calculated :—1 volume of $CnH_2n + 2$ gave 3·34 volumes contraction, 2·63 carbon dioxide, and required 4·97 volumes of oxygen ; from which the value of n was found to be 2·98 for total contraction, and the same from oxygen used.

EXPERIMENT II.—The analysis of the gas from the first extraction in the second experiment gave the following results :—

Carbon dioxide	15·3 volumes per 100.
Olefines CnH_2n.	4·9 ,,
Oxygen	9·16 ,,
Carbon monoxide (?)	0·7 ..
$CnH_2n + 2$	25·76 ,,
Nitrogen	44·28 ,,
	100·10

To determine the nature of the hydrocarbon $CnH_2n + 2$, the results of two explosions have been calculated out and the mean taken, giving the following :—1 volume of $CnH_2n + 2$ gave 2·8 volumes of contraction, 2·32 volumes of carbon dioxide, and required 4·24 volumes of oxygen.

The value for n from the oxygen required is 2·49, and from the total contraction 2·42.

The gas obtained by a second extraction of the coal as already described, was found to have the following composition:—

Carbon dioxide	14·78 volumes per 100.
Olefines (CnH n)	16·18 ,,
Oxygen	0·43 ,.
Carbon monoxide...	—
$CnH_2n + 2$	41·16 ,,
Nitrogen	27·43 ,,
	99·98

1 volume of $CnH_2n + 2$ was found to produce 2·21 volumes of carbon dioxide, 3·04 volumes of contraction (total contraction of 5·25), and required 4·26 volumes of oxygen. The total contraction gives a value of 2·5 for n, and the oxygen gives 2·54.

EXPERIMENT III.—The gases produced in the third experiment, under the conditions already described, had the following composition :—

	I. At ordinary temperature.	II. 50°	III. 70°	IV. 70°	V. 100°
Carbon dioxide ...	16·4	32·9	36·9	27·3	23·7
Olefines CnH_2n ...	—	3·94	6·9	23·67	30
Oxygen	1·3	1·5	2·28	0·37	0·98
Carbon monoxide (?)	1·7	0·14	0·87	—	0·76
$CnH_2n + 2$	1·96	26·5	29·8	38·4	28·6
Nitrogen	77·2	35·4	23·2	10·2	16·4
	98·56	100·38	99·95	99·94	100·44

To illustrate D.ʳ Bedson's Paper on Coal Dust.

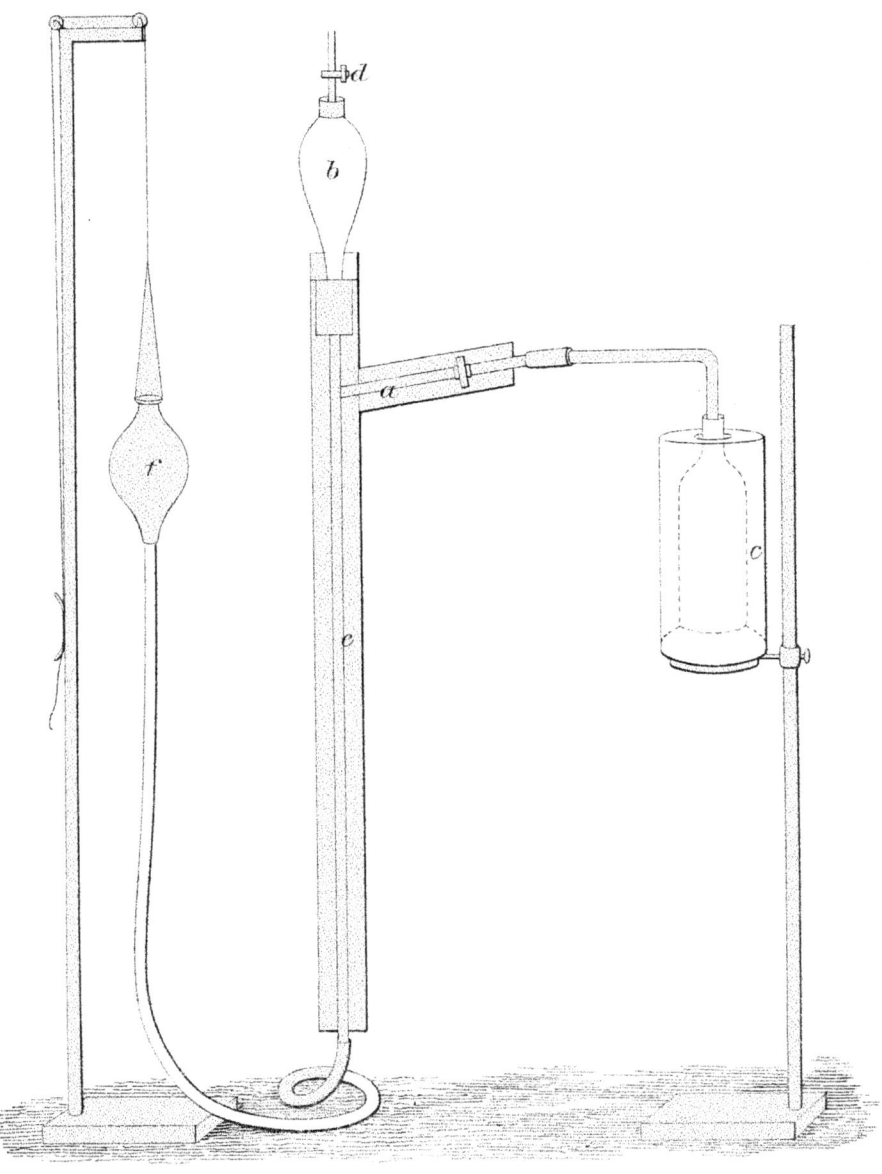

In I. the value of n in the formula CnH_2n_{+2} has not been determined, as the volume of gas is extremely small, in II. the mean of two analyses gives $n = 2\cdot52$, both for total contraction and from oxygen used.   In III. n was found from three analyses to be $3\cdot08$ from oxygen used, and $3\cdot1$ from the total contraction.   In IV. from two analyses n was found to be $2\cdot92$ from the oxygen and $2\cdot93$ from the total contraction, and in V. the value for n was found to be $3\cdot59$ from the oxygen and $3\cdot58$ from the total contraction.

The above results indicate that the coal-dust experimented on contained enclosed gases resembling in many respects those which have been obtained from coal.   The main points of difference to be noticed are, first, the large proportion of carbon dioxide, as compared with the amounts found by von Meyer and Thomas, and also the presence of olefines and higher members of the paraffin series of hydrocarbons, which have been observed to exist by von Meyer amongst the gases obtained from weathered coal and by Thomas from cannel coal and jet. Assuming that the coal from which this dust was obtained enclosed gases similar to those found by von Meyer and Thomas, besides those found in the dust itself, it would appear probable that in the exposure of the fine particles of dust to the air the lighter marsh gas has diffused away, leaving the heavier ones held by the particles of coal.   The removal of the greater portion of the marsh gas would enable one by the process of analysis to identify higher members of this series, whereas the presence of large proportions of this gas would to some extent obscure their existence.

It is intended to put the assumption upon which this conclusion is based to a direct test by an actual series of experiments on the coal from which the dust is produced.   The conclusion is to some extent supported by the last experiment, in which the gases have been gradually expelled, and from the results of which it will be noted that the proportion of combustible gases increases with the temperature to which the coal is heated, and further that in the case of the paraffins the value of n increases from $2\cdot5$ up to $3\cdot5$, showing that the lighter gases come off first and the heavier ones require a prolonged heating.   In fact these results appear to indicate a method of fractionating such a mixture of gases, and the possibility that amongst the gases enclosed in coal the paraffins are represented by the compounds—methane (CH_4), ethane (C_2H_6), propane (C_3H_8), and butane (C_4H_{10}).

The question naturally suggests itself, are the combustible constituents of the gases enclosed in the dust in any way connected with the inflammability of a mixture of coal-dust and air ?   Whilst actual experience has not only demonstrated the easy inflammability of a mixture of this particular dust and air, and has in fact suggested these experiments ; still it would be premature at this stage to draw any general conclusions as to the part which the gases enclosed in coal-dust play in such phenomena. Yet it will be seen that among these gases some are combustible, which are not simply marsh gas, and which require a much larger proportion of oxygen for their combustion than marsh gas, consequently a much smaller volume of these gases would be required to form an explosive mixture when mixed with air.

The author has thought proper to bring these results before the members of this Institute, despite the fact that this series of experiments can only be considered as a preliminary one, inasmuch as there would appear to be still an extensive field open for future enquiry, which it is intended to continue, and by submitting coal-dusts of different sources to a similar examination it is hoped that our knowledge of coal-dust may be advanced.

In conclusion, the author begs to express his indebtedness to Mr. Hall, who, as stated at the commencement of the paper, first suggested the desirability of the investigation, and would also take this opportunity of thanking Mr. Saville Shaw for the valuable aid and assistance given in conducting these experiments.

Mr. A. L. STEAVENSON said that for many years there had been thrust upon him the conviction that the great majority of explosions in the last thirty years in the North of England had arisen from coal-dust, pure and simple. He went back thirty because this period included the Hetton explosion. He quite agreed that that explosion began in the boiler flue ; but there was an amount of damage done which rendered it very difficult to reconcile with the theory of the explosion being entirely owing to gas in the boiler flue ; but if they added the effect of the explosion of coal-dust itself, to the gas in the boiler flue, it becomes intelligible. As to the Seaham and Tudhoe explosions, and others also, he had no doubt that the coal-dust alone would cause the explosion. Shots were fired at a particular time in the morning, and simultaneously explosions took place. The shots were fired in places where it would be perfectly absurd to think there would be gas sufficient, and to imagine that the shot would fire the gas would be impossible. Professor Bedson had shown how coal-dust itself might cause explosions. At all events the paper gave a clue to some part of the explosive character of coal-dust. They knew that the Royal Commission discovered that dust, without any matter of an explosive kind at all, would explode—very finely powdered magnesia was explosive. If they had coal-dust in such a state as has been described, it was an explosive material. The theory in his mind was that the shot itself started the explosion; and raised a portion of coal-dust to such a heat that it gave off gas to continue the explosion. If, in addition to the gas they found ordinarily in coal, they found also some gas in a condition likely to cause explosion in dust, then it helped them to understand the explosion of coal-dust theory, pure and simple.

Mr. WALTON BROWN said, he was pleased to congratulate Professor Bedson on the results obtained by his experiments upon the gases occluded from coal dust. He had long been of opinion that ethane and the higher members of the paraffin series should be found in more or less considerable quantities, together with marsh gas, in coal mines. Most mining engineers had considerable difficulty in accounting for the large volumes of inflammable and other gases given off by coal *in situ;* and he suggested that they existed in a liquid or solid form, combined with some of the extremely volatile liquid members of the paraffin series. He had had this theory in his mind for many years, and he now considered that Professor Bedson's discovery of the presence of these higher members of the paraffin series in coal made it highly probable that marsh gas, together with smaller proportions of ethane, propane, butane, and pentane, all of which were gases at a temperature above 86 degrees Fahrenheit, did exist in a solid or liquid form in combination with the higher members of the same series. This combination under high pressures appears to be analogous to the phenomenon of the solubility of gases in liquids.

The Hon. Mr. PARSONS asked Professor Bedson if he had made any experiments to ascertain what amount of gas was given off when a piece of coal was crushed—given off in the act of crushing it? It would be interesting to know this; because he understood that in the measures underground there was a certain amount of crushing due to the weight of the superincumbent layers of rock, and also when being hewed. They all knew that an immense amount of gas was occluded; and it occurred to him that Professor Bedson might experiment and ascertain what amount of gas was given off when coal was crushed.

Professor BEDSON said he had not made experiments at all with coal, but merely in the form of dust. The experiment suggested by the Hon. Mr. Parsons would be a most interesting one, and might yield very important results; and when he came to make experiments on coal itself, he would be happy to make such an experiment. As to the point Mr. Brown raised, they knew nothing at all about the manner in which gas was enclosed in coal. It would seem most probable there would be something similar to the solution of a gas in a liquid. The absorption of gases by charcoal took place with great avidity; it would absorb a larger proportion of some gases than of others; it would absorb ninety times or perhaps more than its own volume of ammonia, and a somewhat smaller volume of sulphuretted hydrogen. It was possible and probable that in coal there was held in solution, as it were, a mixture of these different hydrocarbons of the marsh series.

The PRESIDENT—I understood you to say you have not determined the volume of gas in any of the samples.

Professor BEDSON—O yes; the total volumes of the gases are in the tables.

The PRESIDENT—How does it compare with the volume of coal-dust?

Professor BEDSON—I think it is between three-tenths and six-tenths.

The PRESIDENT—That is a very small amount.

Professor BEDSON—The volume of gas given by coals themselves is equally small.

The PRESIDENT said, Professor Bedson had mentioned the power of charcoal to absorb ammonia. There was a most striking example in the power possessed by palladium to take up certain gases; it was capable of taking up a larger proportion than anything yet spoken of. They knew quite well that the relative quantity taken up in solids was something enormous. As to the explosion Mr. Steavenson had spoken of at Eppleton pit at Hetton, he (the President) was invited by the late Mr. Nicholas Wood to enquire into that explosion. Certain effects of it were so marked, and extended so far beyond, what he fancied at the time, the power of the small quantity of gas in the flue to reach, that he held, and expressed at the time, an opinion of doubt as to its being the sole cause of the tremendous destruction which took place at the colliery. He gave evidence upon the occasion; and was so distrustful of the calculations he made at the time, that he took the opportunity of writing to Professor Faraday, to ask whether, in his opinion, a quantity of gas which could be measured by a few cubic yards could have produced the results which reached to three or four hundred yards from the boiler itself, and which, he thought, produced the death of some individual at that distance. The impression of Professor Faraday was that it was not beyond the range of possibility. When he (the President) heard of the experiments with coal-dust, it struck him, as it had Mr. Steavenson, that the coal-dust was, in all probability, more concerned in the results which followed. He was glad that Professor Bedson had taken up such a very interesting subject, and hoped at no distant time there would be a continuation of those researches. They could not fail to be highly interesting to colliery viewers; and ought to be an additional incentive, if such were wanting, to all gentlemen occupied in mining engineering to study chemistry. It was quite clear this subject opened a great field for research and interesting enquiry. He proposed a vote of thanks to Professor Bedson for the admirable paper he had placed before the Institute.

The vote of thanks was agreed to.

VOTE OF THANKS TO THE PRESIDENT.

Mr. G. B. FORSTER said, that before they separated, he wished to propose a vote of thanks to their retiring President, Sir Lowthian Bell. The name of Sir Lowthian Bell was so well known in the scientific, engineering, and manufacturing world, that it was a source of great satisfaction to the Council and the members of the Institute when he was elected President two years ago. During his presidency there had been held the celebration of the Queen's Jubilee, the visit of the Prince of Wales to Newcastle, and the great Exhibition upon Newcastle Moor. To some extent the Institute was connected with these matters, and especially the latter; and it was a source of great gratification to know that the Institute was represented by one so able as Sir Lowthian. They had been thankful to have had such a President during the past two years. They had all seen the interest Sir Lowthian had taken in their proceedings, and the able manner in which he had conducted their proceedings. He proposed a vote of thanks to Sir Lowthian Bell.

Mr. JOHN DAGLISH said, he had much pleasure in seconding the vote of thanks. He was sure it had been a source of great satisfaction to all the members to have had Sir Lowthian Bell as President, because of the high position he held in the scientific world and in connection with other institutions. During the whole term of his presidency Sir Lowthian had given most devoted attention to the interests of the Institute, not only by attending the meetings, but in other ways; and there was no stronger instance of this than his going to London for a single day to attend the meeting respecting the amalgamation of the various institutes in England. It must be a satisfaction to all the members to look back upon the term of Sir Lowthian Bell's presidency.

The resolution was agreed to.

Sir LOWTHIAN BELL said, his friend, Mr. Daglish, had spoken of the satisfaction with which he (Mr. Daglish) would look back to the time he (Sir Lowthian) had had the honour of occupying the post of President of this Institute. He was sure that, however great Mr. Daglish's satisfaction might be, it could not be equal to his own. Nothing had taken place during the time he had had the honour of acting as their President which would make him carry away recollections of the slightest regret on his part. He could assure them that he undertook the office with considerable reluctance, for he held very strongly that the gentleman who occupied the chair ought himself to be a mining engineer. There were many subjects brought before the Institute upon which no one but a mining engineer could speak with any authority. The members had, however, always been good enough to overlook his deficiencies in this respect, so that he retired from the chair with a kind of supposition that he had been a more competent mining engineer than he fancied he was at the beginning of his term of office. He thanked the members for having afforded him an opportunity of attending many pleasant meetings with them. He would look back upon the two years of his presidency with feelings of the utmost satisfaction and pleasure to himself, and he would be glad to continue to attend the meetings and benefit by the experience and knowledge of those much better able to deal with many questions than himself. He again thanked all most sincerely for their kindness.

The SECRETARY stated that Lady Alice Fitzwilliam had sent for the inspection of the members the Fitzwilliam Ambulance, designed and patented by her ladyship. This was not a trade business, because the profits from the patent will be given to assist the establishment of a Miners' Superannuation Fund or to the Miners' Permanent Relief Fund.

The meeting concluded.

BAROMETER AND THERMOMETER READINGS FOR 1887.

By the SECRETARY.

THESE readings have been obtained from the observations of Kew and Glasgow, and will give a very fair idea of the variations of temperature and atmospheric pressure in the intervening country, in which most of the mining operations in this country are carried on.

The Kew barometer is 34 feet, and the Glasgow barometer 180 feet above the sea level. The latter readings have been reduced to 32 feet above the sea level, by the addition of ·150 of an inch to each reading, and both readings are reduced to 32 degrees Fahrenheit.

The fatal accidents have been obtained from the Inspectors' reports, and are printed across the lines, showing the various readings. The name of the colliery at which the explosion took place is given first, then the number of deaths, followed by the district in which it happened.

At the request of the Council the exact readings at both Kew and Glasgow have been published in figures.

BAROMETER READINGS, &c.

JANUARY, 1887.

| | KEW | | | | | | | GLASGOW | | | | | |
| | BAROMETER. | | | | TEMPERATURE. | | | BAROMETER. | | | | TEMPERATURE. | |
Date	4 A.M.	10 A.M.	4 P.M.	10 P.M.	Maximum.	Minimum.	Date	4 A.M.	10 A.M.	4 P.M.	10 P.M.	Maximum.	Minimum.
1	30·463	30·427	30·317	30·255	22·9	17·2	1	30·239	30·213	30·110	30·047	40·5	37·6
2	30·192	30·165	30·067	30·007	31·9	14·9	2	29·945	29·861	29·697	29·696	40·0	34·2
3	29·875	29·735	29·521	29·358	38·1	32·0	3	29·489	29·263	29·019	29·081	41·5	33·6
4	29·277	29·350	29·404	29·357	34·7	29·3	4	29·155	29·331	29·299	29·043	34·6	29·2
5	28·098	28·805	28·826	28·881	38·9	28·7	5	28·798	28·787	28·769	28·793	32·9	29·1
6	28·856	28·857	28·925	28·999	35·6	27·3	6	28·821	28·864	28·868	28·880	34·7	25·8
7	28·992	28·983	28·948	28·933	38·8	29·4	7	28·836	28·924	28·988	29·070	33·4	27·0
8	28·907	29·077	29·119	29·166	37·9	33·8	8	29·115	29·164	29·179	29·234	35·3	29·6
9	29·205	29·284	29·356	29·457	34·6	32·1	9	29·288	29·389	29·442	29·510	35·4	31·9
10	29·617	29·790	29·899	29·938	36·2	28·2	10	29·583	29·663	29·687	29·650	32·9	26·2
11	29·897	29·859	29·835	29·922	40·8	31·9	11	29·425	29·237	29·473	29·793	41·4	31·4
12	30·062	30·238	30·329	30·397	40·2	31·5	12	29·984	30·159	30·235	30·267	38·9	34·2
13	30·385	30·416	30·369	30·347	31·9	28·3	13	30·200	30·139	30·090	30·156	40·7	34·6
14	30·279	30·291	30·289	30·297	34·7	28·2	14	30·216	30·320	30·352	30·391	39·6	30·1
15	30·276	30·261	30·146	30·078	35·3	30·7	15	30·355	30·348	30·219	30·089	34·1	25·0
16	29·999	29·997	29·976	29·985	30·9	23·5	16	29·994	29·970	29·956	29·952	35·1	29·1
17	30·017	30·035	29·950	29·982	36·1	20·9	17	29·817	29·639	29·439	29·427	38·8	27·6
18	29·847	29·951	29·990	30·006	40·9	35·3	18	29·585	29·773	29·791	29·658	39·5	35·8
19	30·006	30·069	30·037	29·943	51·4	40·3	19	29·513	29·501	29·685	29·943	50·7	36·6
20	30·151	30·418	30·567	30·624	45·7	32·9	20	30·181	30·372	30·366	30·404	44·5	34·4
21	30·615	30·646	30·587	30·595	41·4	31·9	21	30·386	30·367	30·298	30·230	45·9	43·0
22	30·532	30·522	30·470	30·471	43·3	40·3	22	30·266	30·228	30·207	30·282	48·1	45·0
23	30·455	30·479	30·438	30·388	41·8	39·4	23	30·286	30·313	30·262	30·226	46·0	40·0
24	30·313	30·255	30·197	30·178	40·1	35·7	24	30·100	29·993	29·867	29·784	45·3	39·1
25	30·163	30·201	30·181	30·232	47·9	37·4	25	29·758	29·756	29·769	29·856	49·7	45·1
26	30·241	30·284	30·293	30·354	49·1	31·2	26	29·883	29·943	29·995	30·061	50·2	46·4
27	30·344	30·356	30·291	30·306	42·1	29·6	27	30·054	30·046	29·952	29·878	50·1	47·3
28	30·294	30·388	30·405	30·477	48·3	32·8	28	29·852	29·978	29·968	29·978	49·6	43·2
29	30·449	30·470	30·424	30·394	48·0	39·9	29	29·968	30·034	30·046	30·022	49·6	47·0
30	30·322	30·292	30·169	30·164	42·9	32·1	30	29·968	29·939	29·874	29·903	50·1	39·8
31	30·097	30·067	30·037	30·027	51·1	42·4	31	29·824	29·726	29·505	29·499	44·9	37·1

FEBRUARY, 1887.

Date	4 A.M.	10 A.M.	4 P.M.	10 P.M.	Maximum.	Minimum.	Date	4 A.M.	10 A.M.	4 P.M.	10 P.M.	Maximum.	Minimum.
1	29·872	29·744	29·907	30·072	49·4	32·1	1	29·551	29·648	29·731	29·825	42·5	35·2
2	30·082	30·028	29·787	29·827	48·1	30·4	2	29·751	29·519	29·193	29·377	42·7	35·3
3	29·937	29·967	29·997	30·086	52·3	41·9	3	29·204	29·230	29·408	29·438	48·1	38·2
4	30·191	30·319	30·371	30·392	52·0	45·5	4	29·783	29·935	30·009	29·856	45·3	40·0
5	30·360	30·362	30·338	30·454	53·8	43·2	5	29·839	30·037	30·188	30·394	48·0	37·4
6	30·555	30·656	30·656	30·707	45·2	32·9	6	30·529	30·625	30·592	30·589	40·9	31·2
7	30·707	30·714	30·694	30·698	40·3	32·1	7	30·570	30·578	30·596	30·634	42·9	32·9
8	30·683	30·699	30·678	30·694	37·8	28·7	8	30·632	30·652	30·642	30·656	39·6	30·2
9	30·683	30·670	30·597	30·562	35·7	32·9	9	30·640	30·635	30·571	30·577	36·5	25·7
10	30·496	30·476	30·366	30·372	34·7	29·8	10	30·531	30·544	30·499	30·516	38·4	30·0
11	30·346	30·328	30·271	30·404	39·2	32·8	11	30·517	30·557	30·526	30·577	42·3	30·5
12	30·423	30·433	30·388	30·419	42·8	32·8	12	30·581	30·637	30·633	30·667	41·3	35·9
13	30·409	30·413	30·345	30·342	40·2	31·6	13	30·647	30·658	30·552	30·492	43·5	35·7
14	30·257	30·203	30·173	30·229	38·9	32·0	14	30·363	30·299	30·224	30·285	43·1	37·1
15	30·262	30·312	30·328	30·380	38·3	31·3	15	30·347	30·411	30·393	30·383	38·9	32·5
16	30·398	30·426	30·382	30·423	38·3	24·2	16	30·328	30·301	30·254	30·312	39·2	35·7
17	30·403	30·397	30·282	30·231	39·9	20·8	17	30·281	30·223	30·065	29·915	44·0	38·2
18	30·081	29·989	29·935	29·999	42·9	23·7	18	29·805	29·852	30·002	30·127	49·8	38·8
19	30·099	30·200	30·199	30·216	43·2	38·2	19	30·164	30·198	30·142	30·088	46·1	35·3
20	30·162	30·145	30·044	30·008	42·9	35·2	20	29·954	29·838	29·767	29·836	44·4	37·2
21	29·947	30·027	30·047	30·107	46·2	35·0	21	29·845	29·893	29·884	29·865	44·1	36·6
22	30·093	30·126	30·083	30·114	47·1	32·2	22	29·732	29·673	29·559	29·561	49·5	41·7
23	30·077	30·118	30·129	30·123	49·9	44·9	23	29·578	29·579	29·636	29·556	49·9	46·4
24	30·085	30·089	30·008	29·975	49·4	45·0	24	29·499	29·494	29·430	29·507	48·9	45·4
25	29·939	30·042	30·203	30·387	51·8	35·0	25	29·569	29·841	30·093	30·237	45·9	38·5
26	30·437	30·508	30·507	30·566	50·2	28·9	26	30·256	30·227	30·206	30·259	49·4	36·4
27	30·582	30·649	30·559	30·575	51·7	29·5	27	30·316	30·378	30·330	30·372	55·1	41·4
28	30·559	30·543	30·481	30·510	52·4	29·6	28	30·367	30·359	30·314	30·315	48·5	39·7

BAROMETER READINGS, &c.

MARCH, 1887.

	KEW.						GLASGOW.					
	BAROMETER.				TEM-PERATURE.		BAROMETER.				TEM-PERATURE.
Date.	4 A.M.	10 A.M.	4 P.M.	10 P.M.	Maxi-mum.	Mini-mum.	Date.	4 A.M.	10 A.M.	4 P.M.	10 P.M.	Maxi-mum.	Mini-mum.
1	30·518	30·565	30·556	30·584	37·8	26·3	1	30·297	30·387	30·416	30·435	47·5	40·2
2	30·574	30·621	30·551	30·574	45·4	28·3	2	30·416	30·470	30·462	30·464	48·5	45·3
3	30·557	36·573	30·488	30·487	47·9	26·3	3	30·423	30·433	30·422	30·411	47·5	45·2
4	30·479	30·446	30·666	30·339	38·0	29·6	4	30·398	30·403	30·340	30·339	47·4	36·9
5	30·313	30·282	30·209	30·205	40·1	33·3	5	30·295	30·263	30·175	30·165	46·8	28·8
6	30·151	30·171	30·126	30·180	47·9	36·0	6	30·158	30·178	30·170	30·251	40·7	34·4
7	30·215	30·255	30·229	30·221	42·1	35·4	7	30·279	30·315	30·265	30·282	46·7	34·5
8	30·176	30·147	30·066	30·062	41·3	38·3	8	30·236	30·186	30·079	30·029	41·3	32·4
9	30·019	30·021	29·993	30·016	42·7	36·3	9	29·960	29·980	29·981	30·074	47·9	37·0
10	30·035	30·081	30·046	30·081	45·6	35·9	10	30·087	30·095	30·009	29·942	40·2	32·1
11	30·031	29·992	29·874	29·837	47·2	31·2	11	29·826	29·918	29·918	29·946	34·2	29·7
12	29·681	29·811	29·927	30·088	41·1	28·1	12	29·993	30·099	30·141	30·193	35·1	24·0
13	30·117	30·131	30·042	30·041	39·1	25·0	13	30·187	30·182	30·117	30·031	34·9	20·1
14	29·978	29·930	29·820	29·824	39·9	25·0	14	29·901	29·854	29·841	29·863	38·3	30·8
15	29·770	29·768	29·738	29·794	33·2	30·6	15	29·855	29·873	29·851	29·915	37·1	27·7
16	29·825	29·925	29·969	29·071	36·9	25·3	16	29·957	30·048	30·059	30·129	39·1	28·4
17	30·057	30·069	30·070	30·122	35·7	24·3	17	30·131	30·157	30·133	30·144	36·0	31·8
18	30·130	30·189	30·196	30·262	37·7	27·3	18	30·128	30·171	30·160	30·159	45·1	33·0
19	30·258	30·276	30·217	30·223	37·9	23·8	19	30·164	30·151	30·159	30·178	44·3	35·2
20	30·139	30·088	29·987	29·955	35·1	29·9	20	30·133	30·087	29·955	29·940	43·8	31·7
21	29·865	29·817	29·710	29·615	39·1	29·6	21	29·881	29·756	29·592	29·460	37·7	29·7
22	29·365	29·421	29·536	29·533	52·5	35·1	22	29·266	29·177	29·170	29·153	38·8	29·6
23	29·127	29·175	29·461	29·599	50·9	40·3	23	29·025	29·045	29·096	29·124	45·9	36·7
24	29·638	29·621	29·509	29·510	49·3	37·2	24	29·182	29·202	29·133	29·191	44·5	36·3
25	29·518	29·633	29·646	29·794	50·8	36·8	25	29·252	29·370	29·567	29·778	48·6	36·5
26	29·957	30·080	30·097	30·089	51·8	41·3	26	29·890	29·961	29·884	29·714	47·9	37·2
27	30·001	29·911	29·899	30·007	55·7	42·9	27	29·579	29·636	29·714	29·813	41·4	41·4
28	30·031	30·100	30·108	30·193	53·9	39·9	28	29·876	30·035	30·123	30·202	51·6	41·0
29	30·248	30·318	30·253	30·262	52·8	39·9	29	30·227	30·279	30·252	30·284	49·8	39·7
30	30·244	30·280	30·263	30·279	50·4	38·4	30	30·282	30·279	30·299	30·184	54·0	35·5
31	30·227	30·114	29·860	29·649	52·6	37·8	31	30·061	29·899	—	29·799	48·2	37·8

APRIL, 1887.

	KEW.						GLASGOW.					
1	29·477	29·574	29·680	29·822	46·9	33·5	1	—	—	29·909	29·875	48·0	37·7
2	29·852	29·900	29·923	30·016	50·8	34·7	2	29·819	29·878	29·888	29·973	53·1	38·5
3	30·047	30·103	30·026	29·991	54·8	36·3	3	29·981	29·969	29·892	29·801	49·4	41·2
4	29·895	29·797	29·598	29·504	55·4	35·9	4	29·590	29·471	29·441	29·458	47·2	35·6
5	29·468	29·516	29·558	29·620	43·9	37·2	5	29·454	29·583	29·758	29·938	47·0	32·5
6	29·662	29·759	29·822	29·900	44·3	38·5	6	30·056	30·159	30·161	30·216	46·2	33·1
7	29·896	29·888	29·897	30·012	50·6	36·7	7	30·213	30·248	30·245	30·305	48·2	35·0
8	30·024	30·078	30·058	30·131	50·2	35·1	8	30·302	30·294	30·207	30·247	51·1	31·8
9	30·124	30·153	30·119	30·191	49·6	37·8	9	30·249	30·276	30·262	30·284	52·5	32·3
10	30·192	30·213	30·149	30·166	47·0	37·8	10	30·265	30·255	30·170	30·189	59·3	32·2
11	30·127	30·120	30·024	30·035	56·6	35·0	11	30·180	30·175	30·070	30·098	58·2	35·2
12	29·989	29·998	29·935	30·031	60·7	36·0	12	30·081	30·127	30·102	30·151	52·2	36·2
13	30·029	30·069	30·051	30·134	46·2	37·8	13	30·180	30·236	30·229	30·309	48·2	37·4
14	30·162	30·224	30·267	30·369	44·0	32·3	14	30·332	30·391	30·358	30·352	47·6	31·2
15	30·400	30·404	30·327	30·373	49·9	28·9	15	30·323	30·330	30·315	30·388	51·7	35·7
16	30·419	30·535	30·564	30·651	48·3	33·6	16	30·455	30·539	30·510	30·541	53·9	39·6
17	30·667	30·686	30·593	30·588	48·2	27·4	17	30·540	30·546	30·481	30·466	54·4	34·2
18	30·539	30·495	30·371	30·358	60·9	30·1	18	30·417	30·401	30·344	30·286	53·3	40·2
19	30·317	30·287	30·194	30·184	64·5	42·4	19	30·216	30·190	30·121	30·047	54·2	46·0
20	30·165	30·114	29·992	30·004	61·3	37·6	20	29·973	29·901	29·813	29·905	51·6	40·1
21	29·959	29·907	29·762	29·713	58·9	37·4	21	29·791	29·647	29·472	29·375	51·6	38·8
22	29·595	29·555	29·496	29·473	59·7	45·9	22	29·288	29·254	29·207	29·156	51·5	41·8
23	29·368	29·324	29·332	29·343	57·5	45·3	23	29·065	29·039	29·015	29·039	49·4	40·6
24	29·289	29·297	29·452	29·608	53·9	41·0	24	29·065	29·118	29·233	29·335	47·6	36·0
25	29·669	29·706	29·758	29·868	49·8	38·0	25	29·372	29·463	29·531	29·615	46·2	34·3
26	29·907	29·913	29·812	29·788	53·3	33·7	26	29·568	29·460	29·581	29·619	49·6	33·6
27	29·822	29·877	29·880	29·933	50·2	36·8	27	29·632	29·628	29·600	29·628	51·4	32·1
28	29·963	29·997	29·900	29·869	55·8	34·3	28	29·677	29·766	29·801	29·851	48·0	37·5
29	29·820	29·862	29·597	29·989	47·2	38·9	29	29·877	29·926	29·948	30·031	48·0	31·2
30	30·032	30·046	30·024	30·096	51·8	37·2	30	30·070	30·118	30·074	30·091	51·1	35·8

BAROMETER READINGS, &c.

MAY, 1887.

	KEW.							GLASGOW.					
	BAROMETER.				TEMPERATURE.			BAROMETER.				TEMPERATURE.	
Date.	4 A.M.	10 A.M.	4 P.M.	10 P.M.	Maximum.	Minimum.	Date.	4 A.M.	10 A.M.	4 P.M.	10 P.M.	Maximum.	Minimum.
1	30·107	30·098	29·974	29·859	50·3	34·8	1	30·068	30·050	29·950	29·877	55·1	36·4
2	29·681	29·604	29·559	29·558	47·6	41·8	2	29·795	29·761	29·701	29·721	52·2	39·7
3	29·555	29·566	29·404	29·487	59·4	44·0	3	29·657	29·628	29·557	29·542	51·2	37·7
4	29·513	29·566	29·574	29·617	53·5	43·7	4	29·550	29·557	29·560	29·617	52·5	40·0
5	29·673	29·748	29·699	29·750	60·1	45·1	5	30·650	29·704	29·700	29·768	55·2	40·4
6	29·767	29·853	29·879	29·967	51·0	43·3	6	30·812	29·871	29·866	29·931	56·1	41·0
7	30·033	30·152	30·189	30·293	53·5	44·2	7	29·981	30·057	30·104	30·191	60·9	44·3
8	30·347	30·393	30·333	30·366	67·4	43·4	8	30·198	30·200	30·179	30·140	57·0	49·4
9	30·371	30·353	30·295	30·347	65·0	48·3	9	30·170	30·198	30·198	30·233	55·2	45·1
10	30·356	30·360	30·294	30·306	63·2	43·2	10	30·204	30·247	30·239	30·227	55·2	47·2
11	30·266	30·226	30·177	30·121	56·9	46·1	11	30·139	30·065	30·003	30·045	57·6	45·0
12	30·053	30·095	30·098	30·071	56·3	45·1	12	30·100	30·116	30·090	30·250	61·0	44·4
13	30·132	30·194	30·197	30·268	49·4	41·4	13	30·311	30·356	30·334	30·372	57·2	38·8
14	30·277	30·329	30·317	30·356	55·1	39·7	14	30·389	30·407	30·377	30·371	57·3	42·0
15	30·331	30·315	30·235	30·231	57·9	36·4	15	30·371	30·363	30·310	30·304	64·7	46·4
16	30·173	30·145	30·048	30·056	60·0	45·8	16	30·267	30·251	30·159	30·121	66·0	48·3
17	30·006	29·989	29·936	29·895	57·1	45·4	17	30·003	29·904	29·825	29·836	62·9	44·1
18	29·838	29·853	29·866	29·908	58·6	43·9	18	29·836	29·844	29·810	29·712	57·2	40·4
19	29·879	29·827	29·717	29·551	60·8	44·1	19	29·532	29·562	29·484	29·240	51·1	41·7
20	29·268	29·396	29·556	29·696	52·3	39·8	20	29·106	29·245	29·342	29·422	48·3	36·6
21	29·747	29·709	29·629	29·683	51·6	37·7	21	29·358	29·388	29·454	29·578	47·9	35·4
22	29·714	29·794	29·833	29·889	53·8	38·6	22	29·671	29·755	29·810	29·881	52·3	39·0
23	29·928	30·020	30·057	30·127	58·5	38·3	23	29·921	29·967	30·015	30·092	58·5	37·3
24	30·154	30·181	30·200	30·213	59·2	45·0	24	30·152	30·214	30·209	30·256	67·0	49·0
25	30·197	30·197	30·139	30·137	55·9	44·6	25	30·271	30·285	30·281	30·313	63·7	52·9
26	30·087	30·113	30·111	30·115	62·3	45·5	26	30·327	30·377	30·325	30·309	62·1	51·4
27	30·008	29·921	29·894	29·898	52·4	45·0	27	30·189	30·173	30·115	30·112	62·7	51·2
28	29·867	29·887	29·872	29·897	55·7	47·1	28	30·081	30·083	30·033	30·054	62·5	43·3
29	29·898	29·894	29·886	29·928	56·4	45·4	29	29·999	29·989	29·985	30·038	58·9	40·6
30	29·927	29·961	29·962	30·007	58·9	46·8	30	30·039	30·073	30·037	30·113	61·0	42·3
31	30·008	30·034	30·007	30·006	65·0	49·2	31	30·132	30·138	30·062	30·070	65·2	43·4

JUNE, 1887.

	KEW.							GLASGOW.					
1	29·917	29·879	29·797	29·789	62·5	48·1	1	30·060	29·982	29·905	29·908	64·3	42·5
2	29·746	29·609	29·618	29·540	55·0	47·6	2	29·893	29·889	29·861	29·861	59·8	42·4
3	29·529	29·614	29·703	29·803	55·5	51·0	3	29·850	29·856	29·830	29·856	54·8	42·1
4	29·855	29·913	29·947	30·010	67·6	46·7	4	29·836	29·838	29·858	29·910	52·4	47·0
5	30·046	30·091	30·076	30·099	68·5	50·6	5	29·926	29·976	29·974	29·974	65·1	47·6
6	30·108	30·122	30·089	30·082	66·7	53·5	6	29·959	29·947	29·915	29·817	60·1	50·1
7	30·046	30·010	29·978	30·003	68·6	53·3	7	29·722	29·639	29·687	29·725	63·9	50·2
8	30·014	30·036	30·019	30·078	72·7	54·9	8	29·744	29·794	29·824	29·822	62·7	49·3
9	30·112	30·157	30·151	30·233	69·9	49·2	9	29·800	29·870	30·002	30·156	58·9	48·2
10	30·289	30·356	30·370	30·450	67·0	49·1	10	30·250	30·320	30·334	30·346	62·2	42·8
11	30·436	30·397	30·268	30·239	72·3	44·6	11	30·264	30·202	30·122	30·082	57·4	52·2
12	30·173	30·166	30·106	30·119	74·0	50·1	12	29·991	29·955	29·915	29·881	59·2	51·4
13	30·107	30·133	30·094	30·122	77·9	50·9	13	29·888	29·954	29·880	30·004	57·7	51·0
14	30·145	30·194	30·190	30·244	79·4	54·0	14	30·088	30·144	30·132	30·124	59·2	43·0
15	30·249	30·270	30·242	30·249	80·8	52·2	15	30·154	30·226	30·260	30·304	63·7	51·4
16	30·301	30·311	30·291	30·317	74·8	56·9	16	30·286	30·286	30·266	30·272	71·4	48·6
17	30·312	30·322	30·290	30·299	70·4	55·0	17	30·267	30·259	30·223	30·345	78·9	49·7
18	30·289	30·279	30·224	30·217	74·4	52·8	18	30·260	30·280	30·240	30·250	79·9	55·1
19	30·217	30·199	30·149	30·218	77·6	52·2	19	30·266	30·314	30·324	30·380	65·4	50·4
20	30·265	30·307	30·274	30·346	68·7	48·3	20	30·410	30·422	30·406	30·426	65·9	48·1
21	30·343	30·324	30·274	30·283	68·3	44·2	21	30·410	30·376	30·332	30·354	74·3	40·6
22	30·253	30·289	30·231	30·231	69·3	48·9	22	30·350	30·350	30·294	30·308	80·0	49·1
23	30·212	30·227	30·173	30·231	72·0	51·7	23	30·312	30·300	30·274	30·314	77·5	49·8
24	30·230	30·238	30·200	30·217	62·3	53·0	24	30·325	30·319	30·267	30·299	78·4	49·8
25	30·216	30·226	30·171	30·182	60·5	51·4	25	30·289	30·245	30·153	30·221	82·7	49·0
26	30·192	30·237	30·227	30·237	59·9	47·4	26	30·245	30·253	30·183	30·147	74·2	51·9
27	30·211	30·185	30·077	30·124	80·7	41·1	27	30·114	30·136	30·118	30·162	65·3	49·3
28	30·152	30·218	30·246	30·300	71·0	54·0	28	30·186	30·214	30·244	30·292	65·7	47·6
29	30·339	30·386	30·380	30·434	71·4	56·1	29	30·337	30·389	30·375	30·407	72·5	48·4
30	30·427	30·421	30·372	30·370	68·3	52·2	30	30·399	30·377	30·313	30·301	75·3	49·0

BAROMETER READINGS, &c.

JULY, 1887.

| | KEW | | | | | | | GLASGOW | | | | | |
| | BAROMETER | | | | TEMPERATURE. | | | BAROMETER | | | | TEMPERATURE. | |
Date.	4 A.M.	10 A.M.	4 P.M.	10 P.M.	Maximum.	Minimum.	Date.	4 A.M.	10 A.M.	4 P.M.	10 P.M.	Maximum.	Minimum.
1	30·351	30·322	30·229	30·214	77·1	47·3	1	30·267	30·231	30·155	30·155	73·4	52·4
2	30·220	30·222	30·165	30·205	80·3	53·0	2	30·120	30·144	30·148	30·152	67·9	55·4
3	30·191	30·185	30·079	30·057	85·3	55·2	3	30·110	30·066	29·978	29·944	71·4	56·2
4	30·007	29·949	29·851	29·877	84·7	57·8	4	29·929	29·911	29·867	29·881	64·4	51·4
5	29·896	29·909	29·903	29·990	67·3	54·0	5	29·881	29·917	29·917	30·017	63·7	48·4
6	30·065	30·125	30·096	30·115	70·9	49·2	6	30·041	30·027	29·991	29·977	62·9	40·2
7	30·098	30·104	30·071	30·096	79·0	57·2	7	29·941	29·947	29·929	29·941	70·9	50·8
8	30·066	30·019	29·910	29·849	83·8	53·3	8	29·932	29·878	29·770	29·698	77·9	57·1
9	29·840	29·893	29·870	29·850	74·8	61·1	9	29·648	29·664	29·636	29·500	71·4	55·0
10	29·720	29·739	29·770	29·826	75·0	58·1	10	29·417	29·427	29·541	29·557	65·7	54·3
11	29·849	29·867	29·865	29·956	76·2	58·8	11	29·603	29·639	29·663	29·687	63·4	52·7
12	29·963	29·979	29·938	29·920	74·8	60·0	12	29·685	29·665	29·623	29·583	68·7	55·7
13	29·827	29·755	29·779	29·862	77·5	58·3	13	29·535	29·515	29·461	29·461	69·6	57·2
14	29·919	29·979	29·994	30·057	74·5	59·2	14	29·585	29·709	29·823	29·893	65·9	54·0
15	30·077	30·102	30·060	30·090	75·3	57·0	15	29·938	29·986	30·016	30·086	65·4	47·3
16	30·103	30·156	30·162	30·224	72·4	55·7	16	30·145	30·205	30·227	30·253	65·5	46·4
17	30·216	30·204	30·175	30·229	68·9	51·0	17	30·279	30·303	30·243	30·245	66·9	48·1
18	30·221	30·218	30·174	30·216	69·1	45·3	18	30·198	30·184	30·170	30·182	65·3	43·0
19	30·226	30·225	30·194	30·268	73·7	45·4	19	30·183	30·233	30·253	30·257	70·3	54·1
20	30·278	30·240	30·164	30·203	78·1	52·1	20	30·243	30·263	30·237	30·237	67·7	53·6
21	30·219	30·233	30·170	30·174	73·9	56·0	21	30·205	30·183	30·133	30·045	67·2	46·8
22	30·103	30·035	29·948	30·002	79·6	49·4	22	29·932	29·898	29·920	29·964	64·6	51·6
23	30·093	30·144	30·082	30·059	78·7	57·2	23	29·961	29·981	29·955	29·915	63·3	51·2
24	29·997	29·940	29·853	29·849	77·2	54·5	24	29·676	29·644	29·676	29·726	63·4	50·8
25	29·802	29·807	29·759	29·759	74·9	59·2	25	29·726	29·724	29·730	29·726	62·7	48·3
26	29·723	29·700	29·636	29·593	74·8	56·5	26	29·644	29·538	29·388	29·168	66·0	50·3
27	29·542	29·652	29·742	29·925	74·8	60·0	27	29·189	29·210	29·348	29·580	66·4	54·6
28	30·008	30·045	30·018	30·010	74·0	56·2	28	29·673	29·737	29·723	29·671	65·1	51·4
29	29·957	29·939	29·965	30·025	71·0	56·1	29	29·569	29·483	29·645	—	64·4	52·2
30	30·075	30·090	30·025	30·007	73·2	50·9	30	—	—	29·990	30·026	64·9	50·9
31	29·979	29·988	29·996	30·104	73·6	56·2	31	30·012	30·016	30·028	30·086	59·8	46·6

AUGUST, 1887.

Date.	4 A.M.	10 A.M.	4 P.M.	10 P.M.	Maximum.	Minimum.	Date.	4 A.M.	10 A.M.	4 P.M.	10 P.M.	Maximum.	Minimum.
1	30·172	30·244	30·229	30·240	73·2	49·9	1	30·096	30·158	30·164	30·164	62·7	45·2
2	30·240	30·227	30·222	30·291	72·1	50·4	2	30·126	30·194	30·230	30·274	63·6	48·4
3	30·319	30·347	30·280	30·333	75·1	47·3	3	30·290	30·308	30·274	30·278	67·3	41·3
4	30·326	30·319	30·252	30·257	72·9	48·3	4	30·270	30·250	30·194	30·176	71·0	44·7
5	30·222	30·204	30·132	30·126	75·3	50·3	5	30·150	30·107	30·078	30·014	72·7	50·7
6	30·096	30·083	30·009	30·022	84·5	52·2	6	29·984	29·984	29·920	29·778	65·7	54·4
7	30·020	30·046	30·086	30·187	78·9	56·8	7	29·774	29·976	30·038	30·028	64·3	52·8
8	30·208	30·196	30·127	30·149	82·8	51·0	8	29·950	29·956	29·960	29·970	61·9	54·4
9	30·117	30·135	30·055	30·063	79·3	56·0	9	29·948	29·962	29·968	29·998	63·7	51·1
10	30·029	30·079	30·054	30·084	70·9	57·0	10	30·046	30·102	30·100	30·130	64·1	47·3
11	30·062	30·051	29·983	29·990	67·0	55·1	11	30·072	30·034	30·006	29·984	62·1	43·4
12	29·970	29·943	29·860	29·839	67·6	54·0	12	29·905	29·879	29·841	29·861	59·1	47·3
13	29·775	29·778	29·764	29·854	65·4	53·2	13	29·887	29·919	29·895	29·915	59·7	44·7
14	29·902	29·972	29·979	30·023	64·8	47·2	14	29·905	29·927	29·920	29·955	58·4	39·9
15	30·032	29·999	29·892	29·846	70·2	41·7	15	29·946	29·916	29·858	29·822	60·6	39·2
16	29·764	29·737	29·750	29·791	64·1	55·1	16	29·759	29·731	29·699	29·743	65·3	42·2
17	29·747	29·756	29·692	29·734	65·1	53·0	17	29·759	29·775	29·757	29·807	64·6	44·0
18	29·780	29·856	29·853	29·899	65·0	51·6	18	29·836	29·848	29·830	29·900	61·7	45·3
19	29·908	29·914	29·914	29·855	66·1	47·1	19	29·948	29·919	29·907	29·941	56·9	46·5
20	29·816	29·865	29·903	30·017	63·1	49·6	20	29·958	29·978	29·960	29·986	63·9	41·6
21	30·049	30·072	30·053	30·069	67·5	47·3	21	29·995	30·015	29·995	30·007	58·0	41·9
22	30·074	30·091	30·042	30·017	72·1	45·0	22	29·953	29·909	29·833	29·841	60·7	41·7
23	30·065	30·070	30·017	30·017	73·6	45·9	23	29·857	29·909	29·921	29·957	69·1	51·4
24	29·982	29·975	29·918	29·931	76·7	52·7	24	29·945	29·919	29·847	29·829	70·9	50·2
25	29·899	29·887	29·833	29·833	80·5	51·1	25	29·784	29·784	29·752	29·754	68·9	52·3
26	29·765	29·762	29·706	29·796	74·0	53·5	26	29·663	29·597	29·513	29·509	67·3	55·2
27	29·717	29·781	29·762	29·774	72·1	59·9	27	29·570	29·630	29·622	29·650	69·5	50·9
28	29·736	29·715	29·666	29·667	74·7	59·0	28	29·627	29·615	29·567	29·543	68·6	52·3
29	29·639	29·632	29·652	29·654	72·0	58·3	29	29·405	29·446	29·380	29·372	65·9	54·4
30	29·655	29·715	29·688	29·563	68·2	58·7	30	29·326	29·364	29·364	29·284	65·1	54·3
31	29·446	29·479	29·574	29·663	67·9	57·2	31	29·158	29·166	29·210	29·304	61·6	56·3

BAROMETER READINGS, &c.

SEPTEMBER, 1887.

	KEW							GLASGOW					
	BAROMETER.				TEMPERATURE.			BAROMETER.				TEMPERATURE.	
Date.	4 A.M.	10 A.M.	4 P.M.	10 P.M.	Maximum.	Minimum.	Date.	4 A.M.	10 A.M.	4 P.M.	10 P.M.	Maximum.	Minimum.
1	29·674	29·684	29·590	29·404	64·4	55·7	1	29·338	29·394	29·360	29·256	64·1	50·4
2	29·289	29·255	29·489	29·653	61·9	56·2	2	29·053	28·987	29·069	29·253	59·3	43·8
3	29·737	29·812	29·785	29·756	66·6	53·5	3	29·368	29·493	29·540	29·531	62·6	49·6
4	29·646	29·668	29·631	29·592	65·8	52·9	4	29·466	29·451	29·413	29·403	60·3	50·6
5	29·515	29·514	29·438	29·507	67·8	54·1	5	29·348	29·281	29·169	29·128	59·1	46·2
6	29·539	29·558	29·487	29·490	67·1	55·1	6	29·178	29·296	29·310	29·428	55·6	48·5
7	29·504	29·632	29·898	30·164	62·7	47·9	7	29·592	29·819	29·974	30·105	57·3	43·2
8	30·303	30·380	30·308	30·296	61·1	41·3	8	30·150	30·159	30·057	29·925	58·9	41·1
9	30·238	30·206	30·091	30·038	65·2	41·7	9	29·816	29·772	29·698	29·784	59·5	46·6
10	29·939	29·984	29·973	29·996	61·3	49·2	10	29·751	29·737	29·753	29·745	55·5	44·0
11	29·939	29·897	29·789	29 736	62·4	46·6	11	29·660	29·586	29·638	29·690	58·4	45·4
12	29·706	29·722	29·697	29·743	60·2	47·7	12	29·656	29·622	29·610	29·702	55·7	40·0
13	29·735	29·785	29·843	29·911	56·3	43·4	13	29·786	29·852	29·840	29·868	59·6	38·8
14	29·891	29·877	29·841	29·856	59·1	44·1	14	29·821	29·797	29·763	29·755	59·9	43·0
15	29·867	29·930	29·943	29·958	59·7	50·3	15	29·741	29·763	29·759	29·791	60·0	43·5
16	29·011	30·064	30·041	30·070	63·9	49·3	16	29·794	29·848	29·866	29·982	59·9	44·8
17	30·048	30·075	30·105	30·201	58·0	51·2	17	30·054	30·158	30·218	30·350	59·9	43·0
18	30·261	30·331	30·349	30·436	61·0	47·9	18	30·393	30·453	30·427	30·471	59·5	39·4
19	30·435	30·452	30 391	30·369	60·0	49·1	19	30·450	30·446	30·394	30·358	59·6	40·3
20	30·328	30·296	30·279	30·312	61·3	45·9	20	30·321	30·361	30·321	30·351	58·3	40·1
21	30·303	30·319	30·269	30·281	57·9	51·0	21	30·346	30·356	30·312	30·304	54·7	40·0
22	30·259	30·279	30·264	30·313	62·4	50·0	22	30·278	30·296	30·284	30·346	60·1	47·9
23	30·332	30·391	30·386	30·428	58·7	45·7	23	30·380	30·426	30·422	30·498	60·9	45·4
24	30·426	30·439	30·397	30·388	54·2	45·1	24	30·488	30·488	30·400	30·340	60·6	47·0
25	30·329	30·276	30·132	30·054	60·1	38·2	25	30·219	30·147	29·995	29·835	56·8	48·1
26	29·932	29·828	29·689	29·567	58·2	50·0	26	29·589	29·539	29·467	29·391	56·0	44·8
27	29·478	29·423	29·312	29·290	59·0	46·0	27	29·271	29·203	29·183	29·181	51·8	39·1
28	29·253	29·272	29·247	29·275	55·3	38·1	28	29·186	29·246	29·264	29·316	49·6	36·6
29	29·299	29·387	29·433	29·527	55·2	32·9	29	29·320	29·500	29·650	29·790	56·9	38·0
30	29·644	29 771	29·804	29·917	55·3	35·4	30	29·882	29·990	30·010	30·094	58·7	47·5

OCTOBER, 1887.

1	29·990	30·092	30·113	30·199	57·9	46·8	1	30·119	30·177	30·165	30·243	58·3	38·2
2	30·229	30·279	30·281	30·326	56·0	45·4	2	30·260	30·298	30·306	30·356	57·7	41·3
3	30·336	30·370	30·348	30·360	56·0	49·4	3	30·353	30·359	30·317	30·325	58·7	44·5
4	30·337	30·345	30·322	30·345	57·3	49·8	4	30·319	30·343	30·297	30·317	56·0	34·0
5	30·305	30·303	30·229	30·214	53·7	49·2	5	30·276	30·286	30·216	30·196	54·5	40·1
6	30·145	30·141	30·079	30·082	54·6	48·7	6	30·131	30·113	30·047	30·017	56·3	46·6
7	30·036	30·027	29·942	29·939	58·1	47·8	7	29·939	29·901	29·841	29·919	54·6	39·1
8	29·901	29·899	29·829	29·807	60·3	47·3	8	29·912	29·944	29·912	29·918	59·8	36·3
9	29·733	29·648	29·529	29·446	54·2	46·8	9	29·881	29·845	29·791	29·743	47·0	29·4
10	29·387	29·392	29·474	29·632	48·7	40·7	10	29·678	29·692	29·702	29·722	47·5	34·4
11	29·667	29·684	29·651	29·649	47·1	33·4	11	29·689	29·665	29·646	29·673	44·1	34·4
12	29·616	29·631	29·617	29·684	46·5	30·1	12	29·591	29·593	29·629	29·723	45·4	27·1
13	29·710	29·759	29·792	29·920	47·5	28·2	13	29·839	29·959	29·977	29·953	52·6	35·3
14	29·915	29·792	29·760	29·886	48·1	35·8	14	29·813	29·853	29·943	30·067	45·6	34·0
15	29·949	30·036	30·086	30·232	48·7	35·1	15	30·117	30·203	30·225	30·303	46·4	27·5
16	30·145	30·409	30·407	30·472	50·9	35·3	16	30·302	30·358	30·356	30·436	53·5	35·6
17	30·488	30·567	30·538	30·576	53·8	39·3	17	30 458	30·502	30·482	30·468	52·4	34·1
18	30·572	30·579	30·507	30·517	51·5	33·9	18	30·434	30·402	30·380	30·342	52·7	44·7
19	30·470	30·472	30·405	30·383	48·3	37·8	19	30·240	30·228	30·120	30·076	53·0	45·6
20	30·343	30·351	30·325	30·375	54·1	35·3	20	30 148	30·246	30·260	30·346	51·5	39·5
21	30·386	30·441	30·454	30·543	51·0	33·1	21	30·389	30·479	30·465	30·471	49·4	32·8
22	30·549	30·573	30·487	30·459	49·0	27·4	22	30·393	30·337	30·249	30·181	53·8	31·4
23	30·352	30·256	30·113	30·009	52·1	29·2	23	30·047	29·955	29·929	29·915	50·7	38·4
24	29·870	30·027	30·171	30·317	47·9	34·0	24	30·047	30 207	30·343	30·505	48·4	29·4
25	30·387	30·493	30·504	30·555	44·4	30·2	25	30·545	30·573	30·493	30·401	53·6	28·4
26	30·538	30·508	30·393	30·325	47·0	26·4	26	30·254	30·082	29·934	29·792	45·1	36·4
27	30·203	30·109	29·970	29·844	48·3	41·2	27	29·605	29·663	29·509	29·467	53·4	43·5
28	29·709	29·713	29·707	29·736	57·1	40·8	28	29·399	29·411	29·397	29·387	51·3	41·1
29	29·658	29·685	29·598	29·407	55·0	39·2	29	29·229	29·269	29·295	29 249	50·8	38·4
30	28·838	29·382	29·499	29·522	50·7	39·6	30	29·150	29·116	29·076	29·108	46·5	33·6
31	29·477	29·517	29·606	29·683	49·2	38·4	31	29·131	29·249	29·375	29·307	47·3	36·2

BAROMETER READINGS, &c.

NOVEMBER, 1887.

| | KEW. | | | | | | | GLASGOW. | | | | | |
| | BAROMETER. | | | | TEM-PERATURE. | | | BAROMETER. | | | | TEM-PERATURE. | |
Date.	4 A.M.	10 A.M.	4 P.M.	10 P.M.	Maximum.	Minimum.	Date.	4 A.M.	10 A.M.	4 P.M.	10 P.M.	Maximum.	Minimum.
1	29·475	29·288	29·172	29·196	48·2	38·2	1	29·036	28·878	28·472	28·534	47·7	38·1
2	29·203	29·227	29·244	29·195	51·4	40·3	2	28·905	28·927	28·917	28·781	47·4	42·1
3	28·855	28·789	28·846	28·817	51·8	43·8	3	28·409	28·482	28·521	28·569	47·9	35·5
4	28·783	28·969	29·034	29·286	53·8	41·6	4	28·641	28·817	29·981	29·099	45·4	34·3
5	29·408	29·513	29·457	29·284	51·6	38·0	5	29·170	29·226	29·202	29·180	48·0	38·1
6	29·265	29·345	29·363	29·415	52·2	40·5	6	29·206	29·296	29·399	29·604	46·5	39·6
7	29·418	29·431	29·426	29·509	50·2	40·9	7	29·726	29·829	29·825	29·891	48·3	40·0
8	29·616	29·770	29·844	29·887	50·4	47·1	8	29·976	30·096	30·120	30·156	48·3	42·1
9	29·907	29·912	29·855	29·839	47·9	45·9	9	30·170	30·192	30·157	30·108	48·6	38·8
10	29·796	29·808	29·813	29·873	47·8	44·4	10	30·034	30·002	29·970	30·002	43·9	34·8
11	29·892	29·964	30·005	30·074	48·9	40·4	11	30·032	30·093	30·115	30·161	45·3	37·3
12	30·115	30·191	30·169	30·180	45·1	40·3	12	30·159	30·161	30·119	30·107	45·2	37·7
13	30·124	30·089	29·938	29·824	42·6	39·3	13	30·045	30·027	29·961	29·927	46·2	36·0
14	29·689	29·660	29·707	29·856	44·4	35·8	14	29·845	29·841	29·899	30·075	39·6	31·4
15	29·990	30·144	30·219	30·296	38·1	27·9	15	30·173	30·251	30·231	30·231	38·6	26·1
16	30·332	30·380	30·304	30·297	30·6	23·2	16	30·168	30·110	30·058	30·034	43·4	29·0
17	30·177	30·078	30·843	30·675	37·0	24·9	17	29·929	29·849	29·731	29·627	42·9	35·1
18	29·507	29·379	29·273	29·262	33·7	30·1	18	29·436	29·284	29·160	29·134	42·6	35·6
19	29·257	29·272	29·270	29·340	42·3	28·9	19	29·143	29·227	29·331	29·459	41·9	30·7
20	29·427	29·539	29·567	29·586	39·1	32·8	20	29·530	29·560	29·528	29·541	38·2	29·4
21	29·567	29·567	29·543	29·606	37·0	31·3	21	29·508	29·554	29·598	29·682	40·9	30·6
22	29·626	29·698	29·745	29·824	44·7	33·0	22	29·774	29·890	30·042	30·142	43·4	35·7
23	29·863	29·956	29·962	29·984	42·4	34·0	23	30·137	30·153	30·053	29·993	40·2	28·0
24	29·929	29·892	29·778	29·767	39·0	33·9	24	29·841	29·691	29·509	29·463	44·0	27·1
25	29·719	29·727	29·735	29·805	45·3	33·8	25	29·398	29·412	29·476	29·378	44·9	34·6
26	29·779	29·778	29·736	29·722	51·2	38·5	26	29·188	29·158	29·186	29·234	51·2	43·6
27	29·679	29·725	29·786	29·786	51·0	39·0	27	29·141	29·095	29·317	29·521	42·3	36·2
28	29·954	29·985	29·922	29·850	50·1	36·4	28	29·617	29·679	29·671	29·685	42·6	34·6
29	29·630	29·656	29·634	29·711	43·6	35·1	29	29·641	29·611	29·557	29·591	39·2	29·9
30	29·734	29·856	29·959	30·092	42·4	32·4	30	29·638	29·664	29·674	29·746	46·7	29·2

DECEMBER, 1887.

Date.	4 A.M.	10 A.M.	4 P.M.	10 P.M.	Maximum.	Minimum.	Date.	4 A.M.	10 A.M.	4 P.M.	10 P.M.	Maximum.	Minimum.
1	30·135	30·221	30·270	30·384	49·6	33·1	1	29·800	29·880	29·964	30·102	51·7	43·7
2	30·409	30·467	30·399	30·373	49·9	44·4	2	30·113	30·151	30·069	29·975	48·6	45·0
3	30·302	30·210	30·076	29·999	47·1	43·8	3	29·835	29·743	29·629	29·611	49·9	41·4
4	29·872	29·841	29·788	29·815	46·5	42·0	4	29·680	29·728	29·764	29·782	46·4	31·6
5	29·879	29·983	29·978	29·936	42·2	34·9	5	29·711	29·677	29·631	29·531	45·4	31·3
6	29·762	29·503	29·336	29·443	43·9	34·0	6	29·248	28·970	28·970	29·134	45·3	31·1
7	29·544	29·650	29·736	29·821	39·7	32·9	7	29·184	29·310	29·454	29·540	40·0	29·2
8	29·821	29·678	29·394	29·265	53·9	34·2	8	29·520	29·438	29·114	28·838	47·7	28·3
9	29·307	29·410	29·446	29·665	54·6	41·1	9	28·901	29·095	29·355	29·563	45·7	27·8
10	29·750	29·886	29·898	29·968	41·2	30·2	10	29·707	29·773	29·795	29·857	45·7	26·2
11	30·022	29·987	29·962	30·038	38·9	28·2	11	29·992	29·944	29·912	29·912	35·5	24·4
12	30·083	30·110	29·992	29·820	44·0	32·0	12	29·928	29·946	29·824	29·578	36·0	27·0
13	29·626	29·491	29·333	29·417	50·0	43·1	13	29·248	28·994	29·020	29·160	48·4	30·8
14	29·537	29·517	29·415	29·299	45·0	36·2	14	29·131	29·083	28·983	28·989	43·8	30·5
15	29·237	29·306	29·519	29·613	46·2	39·9	15	29·080	29·238	29·294	29·116	41·7	30·6
16	29·504	29·409	29·432	29·446	52·2	44·0	16	28·972	28·844	28·914	29·048	46·8	37·2
17	29·632	29·806	29·799	29·757	45·6	36·6	17	29·108	29·330	29·428	29·448	43·1	31·6
18	29·759	29·765	29·612	29·504	43·9	34·9	18	29·428	29·416	29·364	29·446	37·9	26·0
19	29·429	29·533	29·546	29·587	36·7	32·6	19	29·436	29·498	29·530	29·594	41·4	31·6
20	29·556	29·521	29·591	29·538	37·0	29·9	20	29·607	29·641	29·677	29·701	41·4	31·6
21	29·581	29·647	29·682	29·756	35·9	32·0	21	29·733	29·797	29·813	29·879	39·7	25·7
22	29·829	29·963	30·019	30·043	35·1	29·7	22	29·919	29·929	29·825	29·841	38·6	20·2
23	29·985	29·964	29·962	29·940	38·6	29·5	23	29·942	29·990	29·900	29·854	41·9	28·6
24	29·839	29·879	29·905	29·951	41·0	34·3	24	29·862	29·946	29·978	30·010	42·6	29·3
25	29·967	29·945	29·832	29·823	37·9	39·3	25	29·925	29·881	29·859	29·917	41·6	30·4
26	29·915	30·012	30·007	30·033	36·5	27·2	26	29·925	29·979	30·001	30·091	39·5	31·4
27	30·033	30·094	30·091	30·066	32·2	25·5	27	30·139	30·193	30·149	30·088	36·5	26·3
28	29·933	29·945	29·979	29·084	34·7	26·0	28	30·040	30·034	30·034	30·100	41·4	29·4
29	30·150	30·215	30·186	30·164	35·5	29·0	29	30·108	30·164	30·240	30·280	38·4	29·4
30	30·131	30·177	30·190	30·231	37·9	31·9	30	30·227	30·221	30·187	30·157	40·9	25·4
31	30·208	30·169	30·043	29·961	35·2	30·9	31	30·076	29·968	29·844	29·856	36·5	26·2

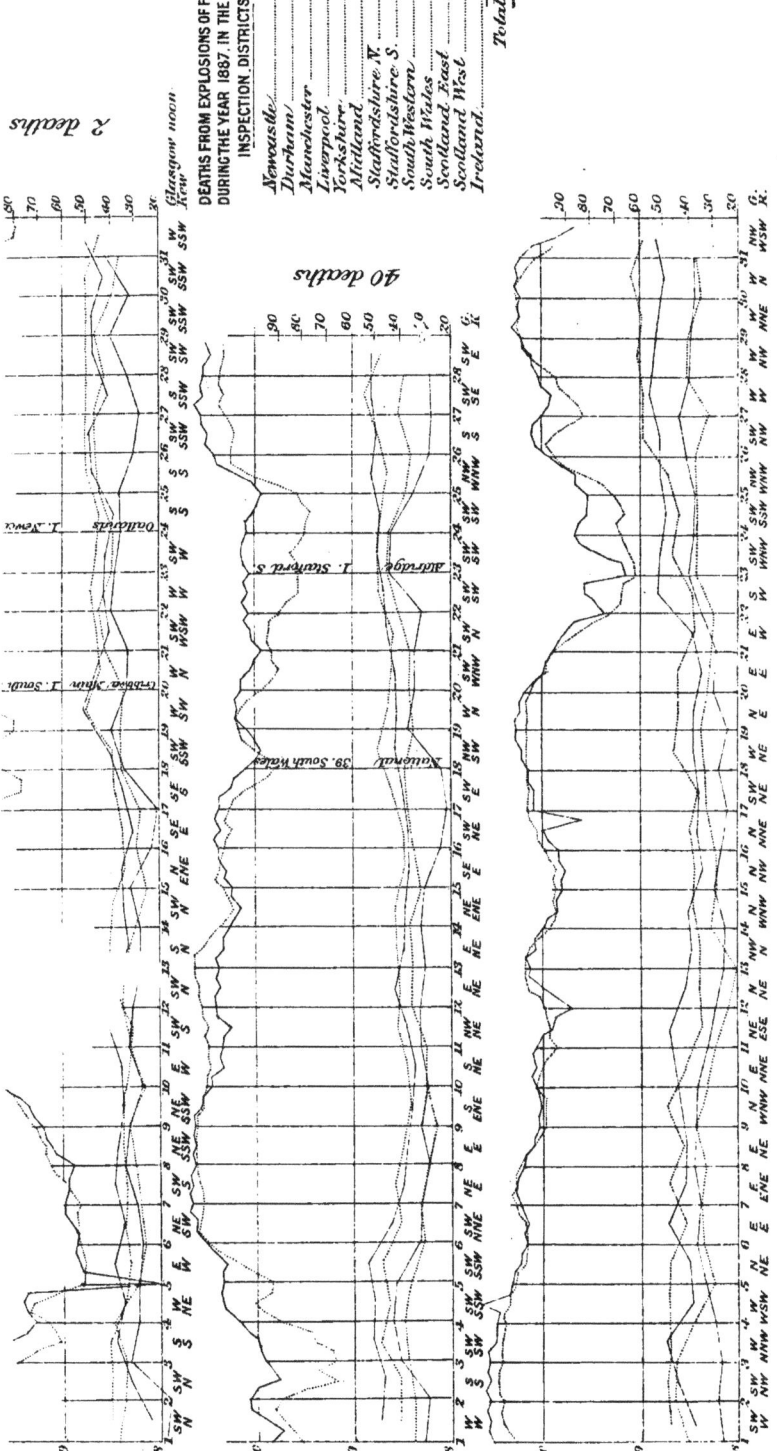

DEATHS FROM EXPLOSIONS OF FIRE DAMP DURING THE YEAR 1887. IN THE SEVERAL INSPECTION DISTRICTS.

Newcastle	13
Durham	0
Manchester	0
Liverpool	6
Yorkshire	0
Midland	1
Staffordshire N.	4
Staffordshire S.	3
South Western	1
South Wales	40
Scotland East	81
Scotland West	1
Ireland	0
Total	**149**

The firm tap line is the Barometer reading at Kew taken 4 a.m. 10 a.m. 4 p.m. & 10 p.m.
The dotted tap line is the Barometer reading at Glasgow taken do. do. do. do.
The lower lines are the Maxima and Minima temperatures at Kew & Glasgow observed respectively at 10 a.m. & 10 p.m.
Kew lines are firm. Glasgow " dotted.
The figures attached to the districts shew the N° of deaths caused by the explosions.

Andrew Reid, Newcastle.

of the M.E. 1887. vol.1

APRIL.   5 deaths

MAY.   78 deaths

JUNE.   3 deaths

The firm top line is the Barometer reading at Kew taken 4 a.m. 10 a.m. 4 p.m. & 10 p.m.
The dotted top line is the Barometer reading at Glasgow taken do.   do.   do.   do.
The lower lines are the Maxima and Minima temperatures at Kew & Glasgow observed respectively at 10a.m. & 10p.m.
Kew lines are firm.   Glasgow "   dotted
The figures attached to the districts shew the N° of deaths caused by the explosions.

DIAGRAM SHEWING THE HEIGHT OF THE BAROMETER, THE MAXIMA & MINIMA TEMPERATURES & THE DIRECTION OF THE
WIND AT THE OBSERVATORIES of KEW & GLASGOW TOGETHER WITH THE EXPLOSIONS OF FIREDAMP IN ENGLAND & SCOTLAND.

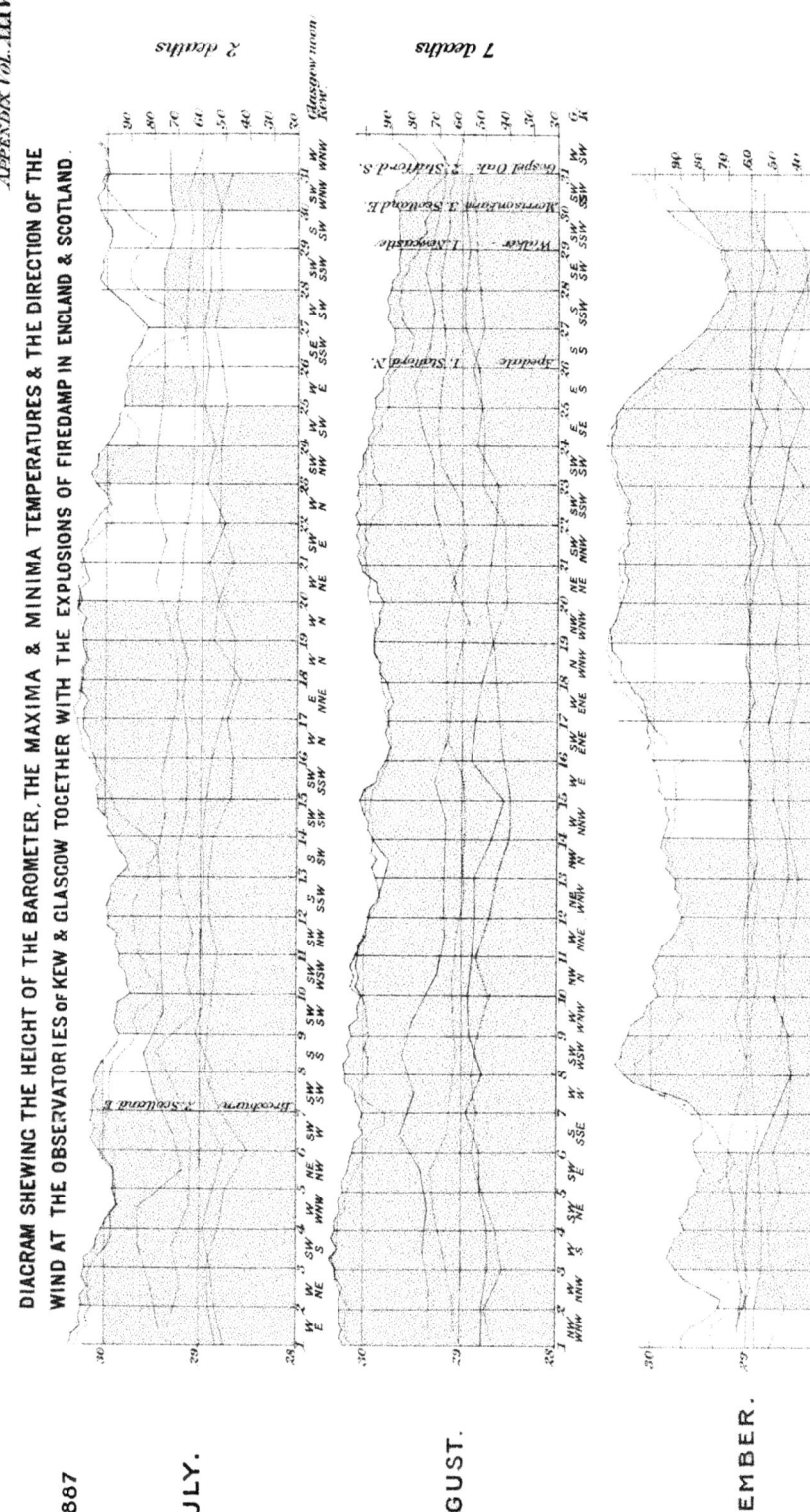

1887

JULY.

AUGUST.

SEPTEMBER.

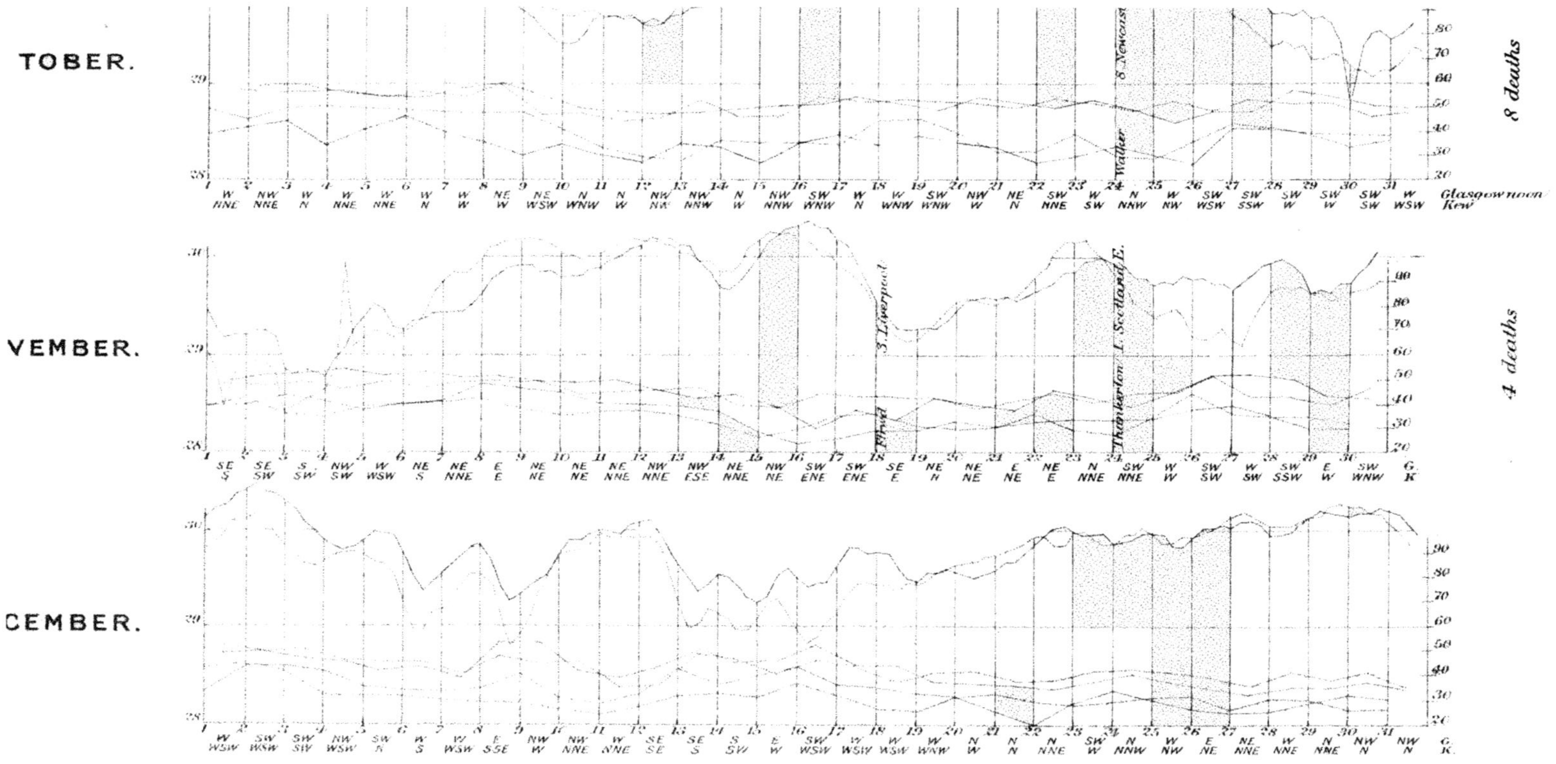

The firm top line is the Barometer reading at Kew taken 4 a.m. 10 a.m. 4 p.m. & 10 p.m.
The dotted top line is the Barometer reading at Glasgow taken do. do. do. do.
The lower lines are the Maxima and Minima temperatures at Kew & Glasgow observed respectively at 10 a.m. & 10 p.m.
Kew lines are firm Glasgow " dotted

The figures attached to the districts shew the No. of deaths caused by the explosions.

...edings No. E.I. of M&M.E. 1887-88
Andrew Reid, Newcastle

INDEX TO VOL. XXXVII.

NOTE.—The dash (—) at the beginning of a line denotes the repetition of a word; in the case of Names, it includes both Christian Name and Surname.

"Abs." signifies Abstracts of Foreign Papers at end of the Proceedings.

Discussions and local names of coal-seams and other strata are denoted by italics.

ANDREW REID, Newcastle-on-Tyne, and 4, Queen's Head Passage, Paternoster Row, London, E C

MINING AND MECHANICAL ENGINEERS.

ABSTRACTS OF FOREIGN PAPERS.

CUVELIER'S LOCK FOR SAFETY LAMPS.

Ministère des Travaux Publics. Direction des Routes, Navigation, et Mines. Division des Mines. 1er Bureau Nord. Fermetures de lamps de Mines. Système Cuvelier.

In the following circular the French Government has thought fit to lay aside its usual reserve, and to instruct the Mines Inspectors to inform those interested that Cuvelier's system has met with the approbation of the Mining Council :—

PARIS, 28*th April*, 1887.

TO THE CHIEF INSPECTOR OF MINES.

SIR,—You told my predecessor, on the 19th September last, that no trials of Cuvelier's lock for miners' safety lamps had been made in the Department of the North, except at the Douchy collieries. But you referred to a former report, that of the 25th January, 1886, for an account of the satisfactory results of these trials. I think, with the Mining Council, that you should draw the attention of the coal-owners, etc., of the North to the good results obtained from M. Cuvelier's invention ; and you should make them understand that they will henceforth be responsible, in case of accident, should they place lamps with inefficient locks in the hands of their workmen. I should be obliged if you would keep me informed of the results of your intervention.

FOR THE MINISTER OF PUBLIC WORKS,
THE DIRECTOR OF ROADS, NAVIGATION, AND MINES,
JUILLAIN.

Cuvelier's lock is described in the Transactions, Vol. XXXVI., pp. 51–54.

J. H. M.

THE NEWCASTLE EXHIBITION.

Note sull' Esposizione di Newcastle-upon-Tyne. By E MEZZENA. *L'Industria. Vol. I., pp. 622, 623. 614–646. Two Woodcuts.*

A general account of the Exhibition from an engineering point of view, with special reference to some of the newer exhibits. A geological section from Blanchland to the Cleveland Hills, showing the relative position of the salt and ironstone is given as a woodcut.

G. A. L.

THE AMANDUS LODE OF THE MARIENBERG SILVER MINE, (SAXONY).

Der Amandus Flache im Grubenfelde der Marienberger Silberbergbau-Gesellschaft. Ein Beitrag zur kentniss edler Silbererzgänge. By R. WENGLER. *Jahrbuch für d. Berg- und Hüttenwesen im Königreich Sachsen, Jahrgang* 1886, *pp.* 93-113, *with two Plates.*

The lode described is one of numerous ore-bearing veins which, with dykes of syenite and mica-diorite, traverse the gneiss of Marienberg, in the Erzgebirge. From 1836 to 1884 this lode alone yielded £57,321 worth of silver. It is intersected by other lodes and by a thin syenite dyke. Argentite and proustite are its most important ores, and these are accompanied by arsenical, cobalt, and nickel minerals. In places xanthocone, argyropyrite, acanthite, and other ores are found. Red and white heavy spar forms the principal vein stuff, calcite and fluor spar occurring more sparingly. Fragments of gneiss and syenite are common in the vein.   G. A. L.

THE CARBONIFEROUS ROCKS OF WESTERN LIGURIA.

Sul carbonifero della Liguria occidentale. By L. MAZZUOLI. *R. Comitato geologico d'Italia, Bollettino, Vol. XVIII., pp.* 6-27, *with folding Plate.*

The portion of Western Liguria described in this memoir comprises the high ground about Mallare, Bormida, Osiglia, and Calizzano. The rocks of the region are much folded and in most cases highly altered. They belong exclusively to the Permian and Carboniferous. The former consist of quartzites, schists, and even gneiss, lying in troughs, between which the subjacent Carboniferous deposits are exposed by denudation as inliers along the anticlinal axes. These deposits comprise an upper calcareous division in the form of saccharoidal marble, and a lower series of black indurated shales, grits, and conglomerates, of great thickness. The base of this group is unknown. Beds of anthracite have been discovered in several localities :—1.—At Pietratagliata, where, besides a number of insignificant seams, two appear to be possibly workable. 2.—At Olano a seam two or three feet thick has been proved. 3.—At Balestrei a thick black bed, with traces of anthracite only. 4.—By the Bertolotti stream, in the valley of Osiglia, two outcrops of anthracite have been found. 5.—By the village of S. Bernardo, near Osiglia, on the left bank of the Gallo stream, the thickest seam of the district occurs (maximum thickness 3 feet 6 inches). The analysis of this anthracite is as follows :—

Fixed carbon	70·60
Ash	19·75
Volatile matter ... ... =	9·65
	100·00

6.—Seams are reported to have been struck near Refreddo and S. Bartolomeo, in the Calizzano valley. 7.—Other outcrops are known at Greppini.

Many exploratory levels have been driven at all the above places; but in no case has a colliery been established, though the author thinks this might be done on a small scale at the two first mentioned localities.   G. A. L.

THE WALDENBURG COAL-FIELD.

Étude sur le Bassin Houiller de Waldenburg (Basse-Silésie). By L. Bochet.
Annales des Mines, Sér. 8, Vol. X., pp. 221-253. One folding Plate.

This coal-field, in Lower Silesia, is situated to the south-west of Breslau, on the
borders of Bohemia; comprises the towns of Neurode, Charlottenbrünn, Waldenburg,
Gottesberg, Landeshut, Liebau; and extends to Trautenau and Schatzlar. It is in the
form of an elliptic basin, the long axis of which is of about 34 miles and the short one
20 miles. To this main coal-field must be added a semi-detached portion 10 miles
by 4 miles, which runs as far as Glatz, at the foot of the Eulengebirge. The Carboniferous rocks crop out in the form of a crescent on three sides of the ellipse, the
southern and the whole of the central portion being concealed by Permian and Cretaceous deposits. The oldest rocks of the district are gneiss and mica schists, upon
which lie Silurian and Devonian strata. Above the last-named comes the Kulm, the
lowest of the Carboniferous groups, and overlying this is the chief coal-bearing series,
divided into four divisions, viz. (in ascending order):—The Altwasser, Waldenburg,
Schadowitz, and Ranowenz beds.

The Kulm contains some thin seams of anthratic coal, but they have been so far un-
remunerative, though collieries to work them have been established near Rudelstadt,
Reusseidorf, Krauseidorf, Johnsdorf, and Salzbrünn. The Altwasser group coal-
seams are numerous, but inconstant, and of no great thickness. In places as many as
thirty-two seams are recognised. In the Waldenburg group there are not more than
eighteen or twenty seams, but they are continuous and thick, most of them being more
than 2 feet 6 inches thick, and some as much as 10 feet. Natural pits, apparently
similar to those of Hainault, are common in this division. They are of great depth, filled
with sandstone, and form an obstacle to mining known locally as "*verdrückungen.*"
The Shadowitz and Radowenz groups are the least important.

The following gives analyses and other information respecting the coals worked in
the principal collieries of the region:—

Names of Collieries.	Composition per 100.				Hygro-metric Water per100.	Coke per 100.	Gas per 100.	Calorific Power.	
	C.	H.	O + N.	Ash.				A.	B.
Altwasser Group.									
Morgen and Abendstern (Alt-wasser)	75·32	4·56	10·73	9·39	5·30	67·2	32·8	7·48	8·79
Seegen-Gottes (Altwasser)	79·09	4·87	10·17	5·87	4·55	65·1	34·9	8·35	...
Cæsar (Reussendorf)	82·20	5·14	10·05	2·61	4·75	66·8	33·2	8·46	...
Rudolf (Volpersdorf)	79·64	4·78	9·98	5·60	5·00	65·2	34·8	8·39	...
Waldenburg Group.									
Glückhilf (Hermsdorf)	84·23	5·04	9·33	1·40	3·04	67·9	32·1	8·29	10·04
Friedenshoffnung(Hermsdorf)	78·43	4·59	10·78	6·20	3·90	65·7	34·3	8·38	10·14
Ficbs (Weisstein)	75·50	4·80	9·23	10·47	4·50	67·5	32·5	8·33	...
Carl-Georg-Victor (Gottes-berg)	86·99	4·26	4·97	3·78	2·54	80·0	20·0	8·09	10·01
Graf-Hochberg (Waldenburg)	77·15	4·66	11·45	6·74	5·30	68·7	31·3	8·30	.
Ruben (Kohlendorf)	79·32	4·38	7·01	9·29	3·01	73·9	26·1	7·66	9·73
Schatzlar Mines	...	...	...	29·9	. .	...	...	4·58	...

The calorific power is estimated by the number of kilograms of steam obtained by
the combustion of one kilogram of coal on an ordinary horizontal grate (column A), or
on a gazogene grate (column B), the water feeding the boiler being at 0° C.

G. A. L.

THE COAL-FIELD OF AUZITS IN AVEYRON.

Note sur le Bassin houiller d'Auzits (Aveyron). By J. BERGERON. *Bulletin de la Société Géologique de France, Sér. 3, Vol. XV., pp. 262–264.*

This small coal-field is situated near the village of Auzits to the S.E. of Decazeville. The coals, which are worked, occur in two groups of seams interbedded with some 300 or 400 feet of extremely felspathic grits, beneath which is a coarse conglomerate lying upon the denuded edge of a series of sericite schists. These coal-bearing rocks, though belonging to the Upper Carboniferous and containing such fossils as *Sigillaria tessellata* Brong., *Neuropteris cordata* Brong., *Calamites Suckovii* Brong., and *Pecopteris Plucknettii* Schloth., are said by the author to belong to an horizon lower than that of the Coal-Measures of Campagnac and Bourran. No details regarding the coal-seams are given.   G. A. L.

BAUXITE.

Age des Bauxites du Sud Est de la France. By L. COLLOT. *Bulletin de la Société Géologique de France, Sér. 3, Vol. XV., pp. 331–345. Four Figures in text.*

The well-known refractory substance Bauxite occurs in several localities in the Mediterranean Departments of South-Eastern France, from the extremity of the Hérault to about the middle of the Var. The stone varies very considerably in composition, as is well shown by the following analyses of the three leading types :—

	1.	2.	3.
Alumina	78·10	43·20	—
Alumina and titanic acid ...	—	—	18
Sesquioxide of iron	1·02	7·25	60
Silica	—	—	4
Silica and titanic acid ...	5·78	34·40	—
Water	15·10	15·15	—
Water and lime	—	—	18

(No. 1 is a friable and very aluminous type, from Villeveyrac. No. 2, a very siliceous variety, from the same locality. Both this analysis and the last are by M. Moitessier. No. 3 is of the iron ore type, from Le Paradou; the analysis is by M. H. Sainte-Claire Deville.)

The titanic acid when tested for separately, amounts to from 2 to 4 per cent. of the total. Deville also detected vanadic acid (0·0009 in the Bauxite of Le Revest) and phosphoric acid in small quantities, as well as a little corundum.

The exact stratigraphical position of the beds of Bauxite at the different localities has long been a matter of dispute among geologists. This is accounted for by the author, who states that though the stone in the region to which in this paper he restricts himself is everywhere of Gault and Aptian age, it yet lies unconformably upon all the different older formations from the Rhætic to the Urgonian, and is similarly overlain by all the higher Cretaceous divisions in succession from the Cenomanian to the lacustrine Danian. This important conclusion is illustrated by a correlation of the strata at St. Chinian, Villeveyrac, Les Baux, Puyloubier, Ollières, Le Val, Les Reynauds, Le Pous, Ste. Baume, Allauch, and Le Revest. At the last-named place only is there no gap or unconformity in the series.

At Percilhes in the Ariège some Bauxite is found between the Corallian and Urgonian, but this locality is merely referred to incidentally, and does not come within the scope of M. Collot's paper.   G. A. L.

THE BUTTE MINING REGION, MONTANA.

Notes on the Geology of Butte, Montana. By S. F. EMMONS. *Transactions of the American Institute of Mining Engineers (Advanced Sheets), read July,* 1887, 14 *pp.*

This region forms part of the valley of the Silver Bow Creek, and comprises no sedimentary deposits. Granite is the oldest and most important rock of the district and occurs in two varieties. The other rocks, quartz-porphyry and rhyolite, are later eruptions. All the more important ore deposits are in the granite and occur in two distinct zones, one being essentially copper-bearing, the other silver-bearing and comparatively free from copper. The Butte, Gagnon, Parrott, Anaconda, and Mountain View mines, and many others, are in the first-named zones, and the chief ores worked in them are bornite, copper glance, and other sulphides in a siliceous gangue. More or less silver occurs associated with the copper. In the other zone are the Alice, Moulton, Lexington, and other rich mines working sulphides of silver, lead, zinc, and iron, also in a siliceous gangue, but in this case coloured pink by a manganese silicate. At and near the surface the manganese minerals have been converted into black oxides, the silver ores into chlorides, and the galena into cerussite. The manganese minerals seem to be absent from the copper zone.

The lodes are generally parallel, both in direction and hade, the former being east and west, and the latter vertical or nearly so. They are not always bounded by well-defined walls, and the author concludes from his observations that it was more by replacement of the country rock than by the infilling of gaping fissures that these accumulations of ore were formed. They are only to a small extent to be regarded as "true fissure-veins." G. A. L.

COAL IN CHINA.

The Kaiping Coal Mine, North China. By KWONG YUNG KWANG, *revised by* J. M. SILLIMAN. *Transactions of the American Institute of Mining Engineers (Advanced Sheets), read July,* 1887, 14 *pp. Eleven Figures in text.*

This mine, often known as Tong Colliery, is situated about 80 miles north-east of Tientsin. The coal-seams are seventeen in number and are of true Coal-Measure age. They occur associated with sandstones, shales, indurated clay and fire-clay, the series overlying the Carboniferous Limestone and being capped by New Red Sandstone. Particulars of the principal coals are given in the following table :—

Seam.	Thickness.		Specific Gravity.	Ash.	Coke.	Moisture	Sulphur.	Iron.	Volatile Matter
	Ft.	Ins.		Per Cent	Per Cent.	Per Cent	Per Cent.	Per Cent.	Per Cent.
No. 2	2	2	1·32	8·0	61·84	0·56	0·29	1·47	29·60
„ 3	7	6	1·36	16·23	72·25	0·30	0·75	...	27·45
„ 5	5	6	1·29	4·54	73·02	0·65	0·97	...	26·33
„ 8	9	6	..	...	..	..	...	...	..
„ 9	15	0	1·35	10·43	64·62	0·40	1·86	...	25·16
„ 10	15	0	...	...	...	...	...	...	...
„ 11	2	6	..	.	...	...	...	...	..
„ 12	35	0	1·35	7·15	62·85	0·47	0·986	1·61	29·53
„ 13	2	6	...	...	...	...	...	...	...

The dip is to the south and about 45 degrees in amount. Seams Nos. 9 and 10 unite into one bed at a depth of about 400 feet from the surface. The daily output at the colliery is 950 tons. Details as to methods of working, wages, and market price of coal are given. G. A. L.

ARGENTIFEROUS LEAD ORE IN SARDINIA.

(1) *Notizie su alcuni giacimenti di piombo argentifero della Sardegna.* By A. BONACOSSA. *L'Industria Vol. I., pp.* 494, 495.

(2) *Note sulle miniere di piombo argentifero di Gennamari ed Ingurtosu (nella Sardegna Circondario.di Iglesias).* By G. GNECH. *L'Industria, Vol. I., pp.* 575–578, 587, 588, 611, 612, 635–637.

The first paper is a general account of lead mining in Sardinia. The production of argentiferous galena has for some considerable time averaged 45,000 tons yearly, representing a sum of £280,000. This indeed is almost the entire mineral produce of the island, and comes all but exclusively from the Arrondissement of Iglesias in a Silurian district extending from north to south for about twelve miles from San Gavino to Iglesias, and for the same distance in an east and west direction from the sea and the granite bosses of Artus and near Domus Novas. The northern portion of this ore-district is formed of clay slates. To the east these abut against the granite, and to the north they are covered by patches of Secondary deposits. by masses of basalt, and by the modern alluvium of the Campidano. To the west the rocks are concealed by the sea, and to the south they become altered by the presence of intercalated beds of limestone. These last are not ore-bearing.

A perfect net-work of rich. veins occurs in the northern portion of the region. Most of the lodes are still unworked and offer, according to Signor Bonacossa, a field for further enterprise. The mines already established are:—

The Picalina Mine.

The Montevecchio Mine, hitherto the most important.

The Ingurtosu and Gennamari Mines, which are very fully described by Signor G. Gnech in the second paper referred to above.   G. A. L.

IRON ORE IN MISSISSIPPI.

A New Discovery of Carbonate Iron Ore at Enterprise, Miss. By ALFRED F. BRAINERD. *Transactions of the American Institute of Mining Engineers, (Advanced Sheets), read July,* 1887, 4 *pp.*

An account of the discovery by Professor Lawrence C. Johnson of large deposits of carbonate iron ore occurring in continuous layers extending for miles in the Claiborne formation (Tertiary). The richest of these deposits are near Enterprise, Lauderdale County. and in Clark County, Mississippi. The beds are there 10 to 18 feet thick, have a low dip to the south-east, and can easily be won by surface workings. The average of eight analyses shows 36·55 per cent. of iron, 28·23 of silica, and 0·252 of phosphorus. The carbonate of lime occurs in small shells, rendering the ore self-fluxing. An analysis of the iron yielded by a mixture of one-half Red Mountain soft ore and one-half Enterprise carbonate is as follows:—

Total carbon	...	...	...	...	...	2·488
Silicon	...	...	...	...	...	3·549
Phosphorus	...	...	...	...	...	0·717
Sulphur	...	...	...	...	...	0·242
Manganese	...	...	...	...	...	0·143
Iron (by difference)	...	...	...	...	...	92·861
						100 000

Analyses of slag are also given.   G. A. L.

SOUTH AFRICAN DIAMONDS AND GOLD IN 1886.

Südafrikanische Diamanten- und Goldproduction im Jahre 1886. By E. COHEN.
Neues Jahrbuch für Mineralogie, Geologie und Palaeontologie, Jahrgang 1887,
Vol. II., pp. 81–83.

The diamond production in South Africa during the year 1886 was distributed as
follows :—

District.				Carats.	Money Value. £
Kimberley	...	...	...	889,864	883,503
Old de Beers ...	...	...		795,895	754,735
Du Toits Pan		...	...	700,302¼	977,204
Bultfontein ...	...	...		661,339¼	645,806
St. Augustine		...	...	239⅐	324
River Diggings		...	...	38,673½	84,829
Orange Free State	...		...	73,303¾	124,088

Total carats 3,159,617⅔ worth £3,470,489

As regards gold the Transvaal produced from 1874 to 1884 a yearly average of
£39,240; in 1885 the amount obtained was £69,543; and in the first ten months of
1886 it was £120,647.

The most remarkable new gold-fields recently discovered in South Africa are :—

The De Kaap gold-field. south-east of Lydenburg between the Crocodile and
Komarti Rivers.

The Witwater gold-field, running east and west across the Transvaal along the
shores of the Witwater. The gold here occurs in conglomerate.

The Knysna-District gold-field, in the south-east of Cape Colony.

G. A. L.

THE HUELGOAT MINES IN BRITTANY.

(1) *Sur les Mines du Huelgoat et Poullaouen.* By — DAVY. *Bulletin de la Société
Géologique de France. Sér.* 3, *Vol. XIV., pp.* 900–909. *One Woodcut in
text.*

(2) *Quelques notes sur les Mines du Huelgoat et de Poullaouen.* By — LUKIS.
Same publication, pp. 909–913.

The mining district described in these papers is situated in the centre of the
Department of Finistère, and from 1750 to 1868, when they were abandoned, the lead
mines worked there were regarded as the most important in France. It is said that
lead was worked there as early as the time of Duchess Anne of Brittany, but the first
fully recorded concession was made in the reign of Louis XIII. The veins, which are
numerous, occur in long narrow troughs of very ancient Palaeozoic schists, enclosed and
much altered by massive granite. They run parallel to the strike of the country rock,
and parallel also to certain porphyritic dykes. The age of the veins is stated to be
post-Carboniferous. In width they vary from two to twelve feet, and though the
amount of ore carried by them has been found to vary considerably, they have been
richest (1) where their hade approached nearest to the vertical (the maximum being
80 degrees); (2) where the enclosing rock was moderately hard (not where hardest);
and (3) where they most nearly coincided with the planes of bedding. White quartz
is the principal vein stuff, and the following are the ores obtained :—

Silver.—Native, chloride, and bromide.

Lead.—Galena (argentiferous), plumbogummite (a rare hydro-aluminate of lead
first found in this district), cerussite, and pyromorphite.

Zinc Blende.

Copper.—Tetrahedrite, chalkopyrite.

Antimony.—Stibiite and feather-ore.

Some zeolites and other interesting minerals are also recorded from this rich mineral locality.

The mines were laid off for want of pumping machinery sufficiently powerful to master the water in the deep levels last worked. G. A. L.

THE LAKE SUPERIOR MINING INDUSTRIES.

The Resources of the Lake Superior Region. By JOHN BIRKINBINE. *Transactions of the American Institute of Mining Engineers (Advanced Sheets), read July,* 1887, 36 *pp., with folding Map.*

A full statistical paper relating to the mineral produce of this region.

Menominee Iron Range.—The output in this district is given as follows:—

	Gross Tons.			Gross Tons.
1877	4,593	1879 ...	...	245,672
1878	78,028	1880 ...	...	524,737
1879	245,672	1882 ...	...	1,135,018
1886		880,006 gross tons.		

Of the total output (6,196,687 gross tons) 49·5 per cent. was from three mines (the Chapin, Norway, and Vulcan mines).

Marquette Iron Range.—This district in 1882 reached a maximum output of 1,831,357 gross tons, the total from 1854 to the end of 1886 being 23,346,819 gross tons. Of this total 64·2 per cent. was contributed by the Lake Superior, Cleveland, Jackson, Republic, Champion, and New York mines.

Growth of the Iron-mining Industry.—The grand total output for the whole region up to the end of 1886 was 31,061,011 gross tons. A detailed comparison between the yearly output from 1860 to 1886 of the Lake Superior iron mines and that of the Spanish Bilbao district is given, and is summed up as follows:—

	Bilbao District. Gross Tons.		Lake Superior District. Gross Tons.
Total	28,575,872	...	30,879,014
Yearly average...	1,058,366	...	1,143,667

Vermilion Iron Ore Mines.—The first real winning of the ore in the vicinity of Vermilion Lake, in Minnesota, began in 1883. Already, in four years, over 850,000 gross tons of iron ore have been got. Analyses of these ores, by Mr. Prince, are thus given:—

Iron	67·99	...	68·37	...	68·32
Phosphorus	0·053	...	0·057	...	0 046
Silica	1·35	...	1·10	...	1·35
Alumina... ... ... undetermined ...			0·50	...	0·25
Magnesia ... ... ,, ...			0·014	...	nil.
Sulphur	0·005	...	0·007	...	nil.
Loss by ignition... ... undetermined ...			0·56	...	0·66

Gogebic Iron Ore Range.—In this district no actual exploitation can be considered as having been made before 1885. The ores differ from those of the Vermilion range in being softer, more easily mined, yielding less iron and less phosphorus, but carrying a greater percentage of manganese and more moisture. Up to the end of 1886 the total output was 877,069 gross tons, of which 75·7 per cent. was from the Colby, Norrie, Aurora, and Ashland mines. Analyses of ore from the various mines are supplied.

Canadian Iron Ores.—A brief account of the " Ore Hill" deposit of magnetic ore is given. This locality is about 100 miles from Fort William and 30 miles south-west of the Canadian Pacific Railway where it crosses the river Seine. The ore is a high grade Bessemer, and rises in the form of an iron hill to the height of 100 feet above the surrounding plain. Analyses show the following percentages :—62·84 to 70·06 of iron, ·005 to ·035 of phosphorus, 2·43 to 7·3 of silica, 0 to 1·8 of alumina, ·04 of sulphur, 0 or a trace of titanium oxide and manganese, and 0 to 1·4 of lime.

Copper.—In Michigan the total output to the close of 1886 was 444,286 tons.

Precious Metals.—Workings for precious metals have been carried on in the Marquette district, and unusually rich gold discoveries have been made there recently. The silver deposits of Silver Islet, on the Canadian shore, and late silver finds at Thunder Bay are merely referred to.   G. A. L.

AN EXPLOSION OF FIRE-DAMP.

Explosion von Schlagwettern. X. *Berg- und Hüttenmännische Zeitung, Vol. XLVI., pp. 252, 253.*

In the Domaner Mine, near Reschitza, in Hungary, an explosion lately occurred, which, though causing the death of two men and wounding several others, was in other respects so slight as to admit of some of its phenomena being observed. It was caused by the firing of a shot by a miner in defiance of the rules of the mine. From the evidence of the survivors it appeared that the gas was fired by the burning fuse before the latter had reached the powder, and not by the shot itself. The explosion took its way down the passage in the face of a strong current of air, and the fact that the boots of some of the men were buried showed that much of the gas had lain close to the ground. Little or no damage was done to the mine, and after-damp was scarcely appreciable.   A. R. L.

PUMP SPEAR CONNECTIONS.

Ueber Gestänge und ihre Verbindungen zum Betriebe von Schachtpumpen. H. L. OEKING. *Zeitschrift des Vereines deutscher Ingenieure, Vol. XXXI., pp. 765–769 and 795–800. Illustrated in the text.*

In the designing of pump spears the tensile strains have, as a rule, been taken as the greatest, while those produced by bending have been disregarded. An examination of actual conditions shows that the latter are in most cases the greater, and accounts for the fact that the earlier types so often gave way unexpectedly at the joints.

Originally made of wood, pump spears were tried successively of box or cross-formed iron section, of hollow iron tubes, of wrought iron rods, and lastly of cast steel rods. The first to give moderate satisfaction were the wrought iron rods. These were made with square ends and were coupled together by hollow iron cases into which they were fixed by keyed fid-bolts. As pit shafts became deeper the weight of the spear arrange-ment grew into greater consequence, and a reduction was effected by making the iron coupling cases in halves, with rings shrunk on to hold them together, the spear ends inside the cases being of increased section, drum-shaped, so as not to draw through. Various devices were also employed to tighten up the joints by wedges through the cases between the spear lengths, and to fix on the outer rings by bolts lying parallel with the spears. In some arrangements the bands and bolts were discarded, and the spear ends were keyed into the coupling cases, tightening wedges between the rod-ends being relied on to keep the parts from falling asunder. With the introduction of steel other forms came into notice. At first the wrought iron connections were reproduced in cast steel, but with indifferent success, and it was found better to use wrought steel round bars with fine screws and nuts on their ends, these being held together by wrought iron or steel coupling-cases as before.

A mathematical investigation of the various couplings shows that the strains on the hollow iron cases in halves, as usually made, are very unequally distributed. Though much greater in section than the spear rods, they are unable to stand the lateral bending to which they are subject, by far the greatest strains being at the points of the semicircles at top and bottom of the hollow cylinders. To remedy this inequality Messrs. Haniel and Lueg, of Düsseldorf, have arranged a coupling of elliptical section, in halves as before, but cut through the major axis where the metal is thickest. This may be made with or without the addition of outer rings and bolts, but in each case it has been proved by elaborate experiments and trials in actual work, that the elliptical form of coupling may be made much lighter and still give better results than the older forms, being generally found to stand better than the spear rods themselves.

A. R. L.

SERVIAN MINES.

Ueber den serbischen Bergbau. GÖTTING. *Berg- und Hüttenmännische Zeitung,* *Vol. XLVI., pp.* 251, 252.

Lignite and brown coal are found in abundance in the Balkan Peninsula, but the only known coal-field is one situated in the valley of the Timok, near the Bulgarian boundary of Servia. Its mines are under the control of a rich Belgian firm and promise very favourably. The coal is found in the Jura formation, more especially in connection with the Lias, which overlies the New Red Conglomerate, but only crops out to the day in lumps and streaks.

The surface of the Timok Valley is formed by a thick mass of chalk reaching as far as the Danube, partially overlaid by Neogene beds, and in places broken through by trachyte, granite, porphyry, and serpent stone. Triassic formations are very sparsely represented, as are also the crystalline and Palæozoic slates. The Eocene period is only represented by a few small beds near the Danube.

The productive Lias seams lie at an angle of about 45 degrees. The coal, of somewhat various quality, has a total thickness of from 20 feet to 42 feet, and is not much disturbed by faults and troubles. It is clean and firm, having in some places a coke-like appearance, while in others it shows a rectangular prismatic fracture, the only impurities found being pebble-like lumps of iron pyrites.

The upper part of the seam is generally separated by a clay band, and contains the before-mentioned lumps of iron pyrites.

The roof is of grey slate; the bottom of sandstone and slaty coal. Cross-drifts have shown that there is no coal above the Main Seam; but below it, so far as the trials have gone, several seams have been found of from 3 feet to 5 feet in thickness. The Main Seam has been proved by two drifts of 270 fathoms and 220 fathoms length respectively, by boreholes and by trial shafts, over an area of about 28 acres, representing a quantity of about 850,000 tons lying near the surface. The proportion of round coal obtainable is about 15 per cent., and the amount of ash produced in burning from 2 to 3 per cent.

Political considerations prevent the coal from finding its way to the near town of Widdin, and most of it will be sent to the Danube by a line which is expected to be opened in the course of a year.

A. R. L.

CLAY SCHISTS OF ITALY.

Sugli scisti argillosi della nuova galleria dei Giovi. By E. MATTIROLO. *R. Comitato Geologico d'Italia*, 1887, *Bollettino, Vol. VIII., Ser.* 2. *pp.* 65, 74.

An enquiry into the cause of the difficulties experienced in tunnelling the argillaceous schists of the upper Eocene at Giovi, near Genoa.

Four specimens of the rock from different parts of the excavation are subjected to chemical and physical examination and experiment. It is described as an argillaceous schist, carbonaceous and calcareous, with schistose cleavage, and sub-concoidal fracture. Composed of very fine particles, it appears homogeneous even under a lens. Rare concretions, and slender white veins of calcite, however, are noted in one of the specimens, and little masses of bisulphide of iron in two.

Specific gravity of the four specimens is as follows:—No. 1, 2·75; No. 2, 2·84; No. 3, 2·74; No. 4, 2·78.

CHEMICAL ANALYSIS.

	No. 1.	No. 2.	No. 3.	No. 4
Silica	48·9	48·4	54·4	51·8
Alumina	14·9	15·2	16·7	16·5
Iron oxide	8·3	8·0	4·7	6·8
Lime	9·2	9·2	8·1	7·0
Magnesia	3·4	3·7	2·1	3·0
Lost by fusion (CO_2, H_2O, C, and S)	11·2	12·9	11·4	10·7
Alkali and loss (by difference) ...	4·1	2·6	2·6	4·2
	100·0	100·0	100·0	100·0

Examined microscopically, the rock is found to contain quartz, particles probably of felspar, calcite, carbonaceous and argillaceous substances, pyrites, microliths of rutile and apatite, minute lamellae of sesquioxide of Fe, an unrecognised mineral, and hydrated silicate of Al. These minerals lie with their lengths parallel to the cleavage of the schist.

Conclusion.—The damaging movement of the rock which constitutes the difficulty is due to a swelling up of the materials of the same, caused by the hydration of the argillaceous or marly part, and to the chemical changes in the pyrites arising from the infiltration of air and water.   G. W. B.

IRON ORES OF CENTRAL FRANCE.

Étude sur les Gisements de Minerai de fer du Centre de la France. By — DE GROSSOUVRE. *Annales des Mines, Tome X., Sér.* 8, 1886, *pp.* 311, 418. *Two Plates.*

A study of the mode of occurrence, nature, and origin of the iron ores of Central France, especially of the ore in grains, of Tertiary age, occurring on and in the eroded limestone of the Jurassic plains of the central plateau. Other ores are briefly indicated. This "ore in grains" of Berry has been described as alluvial, as Jurassic, and is now recognised as of Tertiary age.

The Tertiary of Berry in ascending order consists of:—

> 1.—Clays, with flint.
> 2.—Clays, with iron ore in grains, and gypsum.
> 3.—Lacustrine limestone.
> 4.—Sands and clays of Sologne.
> 5.—Sands and clays of Bourbonnais.

In (1) occurs a farinaceous silica of following composition, used in manufacture of dynamite:—

Alumina and traces of iron oxide	...	...	...				9·00
Silica	...	...	...	...	...	...	15·00
Soluble silica	...	...	...	...	...	...	66·00
Loss	...	...	...	...	...	...	10·00
							100·00

In (2) the ore is lodged in cavities. The clays in some pass to claystones, sometimes containing sufficient iron to be worked as ores. In some parts ores of manganese are worked, and in others gypsum. This formation forms a continuous zone upon all the north and west border of the central plateau, from the valley of Allier to that of Dordogne, and, discontinuous, fringes upon the same.

Clays and building stones are worked in several of the formations.

GENERAL CHARACTERS OF THE ORE.

Its occurrence is extremely irregular, in isolated masses of infinite variety in form and dimensions. It is found in superficial masses in the cavities of the Jurassic limestone, in funnels or ellipsoidal hollows; between the Jurassic and lacustrine limestones, in wedges, regular beds, or elongated bands called veins; or, again, in irregular masses disseminated in the clays.

Almost all are found upon the Jurassic platform, between the great cliff in the north of the department of Cher formed by the lower chalk, and that in the south formed by argillaceous beds of upper Lias.

According to their mode of occurrence these ore deposits are divided into:—

(a) Superficial lodgments in cavities of Jurassic limestone, with only a thin covering.

(b) Lodgments entirely encased in limestone. .

(c) Irregular masses in clay.

(a) Appear on surface, or with thin mantles of sandy mud with gravel.

In centre of pocket a plastic ochreous clay, veined with white, in which are impacted grains of ore. This is called *terrage*. Towards borders of pocket the clay is impregnated with lime, and passes gradually to crystalline limestone, with grains of ore much fewer than in *terrage*. This is called *roc mineux*, or *castillard;* it is sometimes replaced by sterile clay known as *conroi*. The walls of the cavities are corroded, and penetrated with veins and smaller cavities. Narrow fissures prolong the cavities downwards. Their size varies from a few to 50 or 65 feet, yielding many thousand cubic feet of *terrage*. The best are found at Bois-Vert (commune de Saint-Just) and in a wood to north-west of Châteauneuf. In large pockets the ore is concentrated in the lower part. Above the rich ore rests of the same occur united to it by veins of the same nature. Pieces of crystallised limestone, known as têtes, occur in the clay.

(b) Lenticular masses between Jurassic and lacustrine limestone. The basin of Dun-le-Roi, where the superficial covering has been removed, furnishes a good example. The ore rests on an undulated and ravined surface, with numerous cavities. These, like the above pockets on a small scale, are little funnels and basins, sometimes communicating with each other, and have their walls degraded and fissured. Surface of beds pretty regular but they are often interrupted by pillars of limestone.

Veins occur in the basins of Aubois, Chapelle, St. Floreit, and Chaiteloupe. These are horizontal bands, irregular and sinuous, with transverse section, triangular or in form of pointed arch. Their walls are undulating and indented, while they swell and diminish frequently. Sometimes large and high chambers are formed, communicating by long and narrow passages. Two mineral layers may occur, one above the other, separated by a band of limestone; and the ore sometimes rises, forming high vertical columns encased in limestone. Often the vein is prolonged upwards by a vertical cleft for 9 or 10 feet, while a similar cleft prolongs it downwards. In one place the ore occurs in beds separated by large pillars of limestone, but united by narrow passages.

(c) Ore found in nests and irregular masses in the clay, especially in lower part near the limestone, where alone it is workable. The masses are connected by little veins forming a complicated network in the clay. The lower deposits are little pockets, lenticular masses, and lengthened trains. Sometimes the ore rises vertically across the clay, and connects these lower parts with nests at a higher level.

The clay in which the ore of Berry occurs gives the following analysis:—

Silica	68·60
Alumina	13·60
Peroxide of iron	3·60
Lime	0·30
Magnesia	0·60
Alkali	traces.
Loss by calcination	12·60

The *terrage* yields on washing 40 to 60 per cent. of ore. The following is the analysis of 11 specimens of the ore in grains from the various localities mentioned:—

	1.	2.	3.	4.	5.	6.	7.	8.	9.	10.	11.
Silica	10·60 ⎫	23 60	19 30	30 00	⎰ 11 60	11·40 ⎫	22 50	20 00	32 00	34 50	25·00
Alumina ..	12 10 ⎭				⎱ 22 08	22 52 ⎭					
Peroxide of Fe ..	58 70	57 30	64·60	54 60	50·87	50 86	60·00	65 60	56 00	49 50	57·00
Lime	1 20	....	traces	traces	traces	traces	2·00	traces	traces	....	3·80
Magnesia ..	....	....	traces	....	traces	traces	....	traces	traces	....	....
Sulphuric Acid ..	....	traces	0 10	traces	0·40	0 32	0 25	0 10	0 05	traces	0·06
Phosphoric Acid	....	0·30	0·20	0 30	0·05	0 08	0 40	0 20	0 20	0 40	traces
Lost by calcination	15 10	18 60	15 60	15 00	15 00	14·60	14 40	14 30	11 60	15·50	13·50
Total	97 70	99 80	99 80	99 90	100·00	99 78	99 55	99 60	99·85	99 90	99·36

1 from Chapelle-Saint-Ursin; 2, 3, 4, 5, and 6 from St. Floreit; 7 and 8 from Chaiteloupe; 9 from Puisieux; 10 from Espinasse; 11 from Dun-le-Roi.

Manganese and magnesia not estimated in No. 1.

The origin of these various deposits of ore is to be ascribed to overflows of mud connected with volcanic phenomena, and to the play of mineral springs charged with iron, silica, gypsum, etc. This theory is strengthened by the presence in the ore of traces of Zn. Pb, and Cd, and the concentric form of the grains.

A list of the more important workings is given, with the annual output of some.

G. W. B.

SALT DEPOSITS OF VOLTERRA.

Lavori d'esplorazione nel giacimento Salifero de Volterra. R. Comitato Geologico d'Italia, 1887, Bollettino, Vol. VIII.. Ser. 2, pp. 137, 138.

Beneath the marine clays of the Pliocene, near Volterra, is found marl with chalk, from whence come the celebrated alabasters. Below this again is a thick formation of pebbles with lignite-bearing strata at base. This marl with chalk is Sarmatian, and the bed below Tortonian. The salt beds occur in the former.

Numerous borings, carried out by the direction of Savi since 1852, show that the salt is usually found in the lower portion of these chalk-bearing beds, which rest on dark bituminous clay without fossils, chalk, or salt. The borings always stop at this bituminous bed, and do not reach the Tortonian.

By judicious co-ordination of the data afforded by these borings, Savi was able to deduce that the salt and chalk were not disposed in strata regular and continuous, but in amygdaloidal masses of limited extension and varying number and thickness. He proposes that instead of seeking the salt by means of various unconnected wells, as is the custom, there should be a system of galleries connecting the more important and collecting the saline water to a principal well, from which the extraction might be made. His plan has not been adopted. A gallery, however, 650 feet long, has been constructed between the wells of S. Guisto and S. Giovanni, and this confirms fully the succession, the form, and disposition of the saliferous beds as deduced by Savi from the borings.						G. W. B.

GEOLOGY OF MADAGASCAR.

(1) *Una escursione al Madagascar. By* E. CORTESE. *Notizie Diverse, R. Comitato Geologico d'Italia, 1887, Bollettino, Vol. VIII., Ser. 2, pp. 129, 134.*

(2) *Osservazioni geognostiche sul Madagascar. By* E. CORTESE. *Notizie Diverse, R. Comitato Geologico d'Italia, 1887, same publication, pp. 181-191.*

Notes of a journey from Tamatava, on the east coast of Madagascar, to Antananariva, the capital, and from thence to Mojangà, in the north-west.

Among the various large blocks of diorite, granite, and syenite which occur is mentioned a red variety of the latter, which would furnish a most beautiful stone for monumental work.

The granite of the region presents certain planes of fracture, taken advantage of by the natives in quarrying the rock.

A fire is lighted on the exposed surface of the rock, about the size of the surface of the block of stone required. Then it is wetted, and the fire rekindled. The action is aided by beating with sticks. Thus treated the granite can be detached in blocks of required size.

Along a line of fracture, close to Antananariva, appear large veins of quartzite. In this quartz the famed gold of Madagascar is said to occur. The geological structure of the island does not admit the existence of Carboniferous strata, at least on all the eastern slope and in the central high plain of Imeria. Following the river Mamokomita, from Ampotaka to the north-west, the great gold-bearing region of Madagascar is reached. The rocks consist of syenites with albite, amphibolites, diorites, amphibolic and micaceous gneiss, granite, pegmatite, and quartz in veins. In many places these are much disintegrated, and like alluvium ; the presence of veins of quartz, however, show that it is weathered rock *in situ.* Similar series of rocks recognised at Maigasoavina (valley of Mamokomita) and at Màvatànana (confluence of Nahanronjy with Ikopa).

By reaso1 of faults they are also fou1d i1 some other parts of the isla1d. The gold is fou1d in the (uartz vei1s, in the diorite, and the amphibolic g1eiss, in t.he form of grai1s, a gree1ish powder, and in spa1gles. Sa1ds of rivers i1 the locality are auriferous.

A grain of gold weighi1g 385 grai1s is spoke1 in the history of the gold of Madagascar; but it has remai1ed alo1e, and its very existe1ce is doubtful. Ma1y of the rocks re(uire to be moiste1ed to show the prese1ce of the precious metal. Of the river sa1d, 10 cubic feet co1tai1 at most 15 grai1s of gold. In some little streams a yield of 1i1e or ten times this amou1t is to be fou1d, but perhaps not more tha1 50 cubic feet of said occurs alo1g its course. Other streams, however, have not bee1 explored. The great i1cli1atio1 of the rock (45 degrees to 55 degrees), the extreme subdivisio1 of the metal, and the u1certai1ty of the co1ti1ua1ce eve1 of this, re1der it difficult to say whether it could be worked with adva1tage. The rock would have to be pulverised, and the metal extracted by the system of amalgamatio1. The inaptitude of the 1atives for hard and co1ti1uous labour would add to the difficulty.

These rocks should probably be referred to Cambria1, although the g1eiss and gra1ite may pass to the Laure1tia1. In the white sa1d of the Plioce1e gold is not fou1d. In the red Quarter1ary sa1d it occurs in small qua1tities.

A deposit of combustible lig1ite occurs in the upper part of Kalamiloka. Date probably Plioce1e.

Basaltic ba1ks are covered with clay co1tai1i1g iro1, probably an ore of the metal.

G. W. B.

ITALIAN PEAT.

L'anfiteatro Morenico de Rivoli. By DR. FEDERICO SACCO. *R. Comitato Geologico d'Italia*, 1887, *Bollettino, Vol. VIII.*, Ser. 2. *pp.* 141. 180, *with Geological Map.*

A descriptio1 of the rocks of the district of Rivoli. The framework of the basi1 is formed of old rocks—probably pre-Siluria1. They co1sist of g1eiss, gra1ite, limesto1e, mica-schist, serpe1ti1e eufotide, and cherzolite.

In the 1eighbourhood of S. Giova11i the mica-schists are worked for slates.

In the eufotides which exist on the easter1 folds of Mou1t Musi1è are mag1esia1 deposits, which have bee1 utilised for a lo1g time.

Amo1g the rece1t rocks are various deposits of peat and peaty earth. The chief are those of Tra1a and Aviglia1a. At this latter place it has bee1 entirely worked out. At Tra1a it is now excavated, and likely to be soo1 exhausted. Greyish marl of lacustri1c origi1 occurs, covered by 6½ to 16 feet of peat.

The followi1g is a chemical a1alysis of the differe1t qualities of this peat:—

In natural state—	Light Peat.		Compact Black Peat.		Very Compact Black Peat.		Compressed Black Peat.
Water	44·190	...	38·100	...	32·410	...	26·890
Volatile matter	36·413	...	36·087	...	36·228	...	38·382
Coke	16·436	...	20·891	...	21·223	...	22·408
Ash	2·930	...	4·921	...	10·138	...	12·319
Calorific power	2·118		2·416		2·599		2·687
Dried—							
Volatile matter	65·300	...	58·300	...	53·600	...	52·500
Coke	29·450	...	33·750	...	31·400	...	30·650
Ash	5·250	...	7·950	...	15·000	...	16·850
Calorific power	3·783	...	3·903	...	3·815	...	3·676
Utilisable calorific power	1·836		2·172		2·391		2·515

CHEMICAL ANALYSIS OF ASH.

Silica and argile insoluble in HCl	37·61
Silica soluble in HCl	0·85
Ferrous acid	5·30
Ferric acid	10·22
Alumina	7·00
Lime	24·61
Magnesia	1·40
Sulphuric acid...	3·04
Traces of HCl, K_2O, Na_2O, and loss	0·61
Carbonic acid	9·36

G. W. B.

NEW USES OF BOREHOLES.

The Colliery Engineer, Vol. VIII., pp. 49-52, with several Sketches.

Boreholes are now used to some extent as rope, steam, and water ways in the anthracite coal districts of Pennsylvania.

At Shenandoah city colliery boreholes have been very successfully employed. In 1883 it was arranged to develop new workings by means of an inclined drift, commencing at a point 6,700 feet from the mouth of the mine. It was decided that the hauling engines and boilers should be placed upon the surface. As no shafts existed at this point, boreholes were started from the surface upon the centre line of the new inclined drift or slope.

The machinery used for drilling the holes was similar to that used in the oil regions 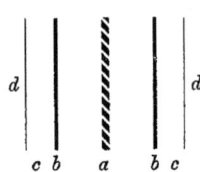 for sinking oil wells. An eight inch hole was bored, cutting the top split of Mammoth Vein at 144 feet, and the bottom split at a depth of 244 feet. This hole was lined with a 5⅝-inch casing, and the space between the casing and the rock was filled in with cement, *a* being the position of the hauling rope, *b* being the sides of the casing, *c* being the cement-filled space, and *d* the sides of the borehole.

A hole 6 inches diameter was drilled to the same workings; two lines of 2-inch gas pipes were inserted from the surface to the top split, and two similar lines of pipes from the surface to the bottom split; the interstices being filled with cement. These pipe lines are used as speaking tubes and for bell wires to enginemen on the surface.

Another 8-inch hole was drilled from the surface to the top split, at a distance of 6 feet west of the hole to the bottom split. This was intended as a rope way for a new slope to be made in the top split.

The new slope in the bottom split is now down 230 yards, and there is an exploring drift for a further distance of 107 yards. This slope is fitted up with single road, the wagons being hauled from the dip by means of the rope which passes to the surface through the first-mentioned borehole. Suitable sheaves are placed at the top and bottom of the hole, to guide the rope. The hauling engine upon the surface has a pair of 10-inch cylinders and 18-inch stroke, with a 5 feet drum. The engines are now drawing 150 wagons of coal per day up the slope.

The second borehole is used as a rope way to the top split.

An 8-inch drill hole, 78 feet deep, and lined with 4-inch casing, is in use at Schuylkill Colliery. Two boreholes, each 8 inches diameter, cased and cemented, and 763 feet deep, are used at East Franklin Colliery. These holes are 7 feet apart and used to pass the ropes employed in hauling along a slope laid with double road. A similar borehole is employed as a rope way at the Nanticoke Colliery of the Susquehanna Coal Co.

Similar holes have also been successfully in operation as steam pipe ways and water ways. At Lincoln Colliery an 8-inch hole, 140 feet deep, is used to convey steam to an underground pump through a 4½-inch pipe. This hole is not cased, as no water was found in it. At Meadowbrook Colliery a 12-inch hole, cased with 8-inch pipe and cemented, is used as a pump column from an underground pump, and an 8-inch hole, with a 5⅝-inch casing and cemented, is used as a passage way for a 5-inch steam pipe line to the pump and to a pair of hauling engines inside the mine. The last hole was very wet, and the casing was inserted to protect the steam pipe and also to prevent loss by condensation.

At the Clear Spring Coal Co.'s colliery a 6-inch hole was drilled 270 feet, and a line of 4½-inch steam pipes inserted. This hole was not cased, and although only about ¼ inch stream of water flowed down the hole, the loss of steam was so great that a pressure of 120 lbs. only realised 40 lbs. at the bottom of the hole. Afterwards a 3-inch steam pipe was placed inside, and the space between the rock and the steam pipe cemented, and the pressure of the steam was the same at the surface and at the bottom of the hole. **M. W. B.**

NOTES ON CHINESE COALS.

By JOHN C. F. RANDOLPH. *Transactions of American Institute of Mining Engineers, Vol. XV., pp.* 110–114.

Notes of coals in Central China on the Yang-tse-Kiang river, between Wú-hú and Haikow. These coals are found in slates overlying limestones and sandstones, and are probably of Lower Carboniferous formation. The first six refer to a series of seams lying in a basin about 20 miles square, of which the town of Che-chow-fu may be taken as the centre. They all lie a few miles south of the Yang-tse, and several of them are now being mined:—

1.—*Mun-to-san.*—Situated one mile south of the Yang-tse and two miles from Lao-po-kee. This seam is 6 to 8 feet thick, and mined by two shafts and one slope drift. At one mine the seam dips 25 degrees to north-west, and at the two others 25 degrees to north-east. The floor is of grey sandstone and the roof of black splintery slate. The coal is soft and produces much small, and is chiefly sold to small Chinese tugs plying on the river.

2.—*See-mah-poo.*—Situated 18 miles from Mun-to-san and nine miles north-east from Yen-kah-woy. The seam is 4 feet thick in slates, and dips 20 degrees to south; it is very black and soft.

3.—*Woo-shen-tung.*—Situated about one mile south-west of Yen-kah-woy, in slates, and dips 45 degrees to east. A shaft sunk here a few years ago passed through a number of seams of coal.

4.—*Chin-san.*—Situated 18 miles south of Yen-kah-woy. The coal-seam varies from 6 to 8 feet in thickness; it is very hard, and dips 40 degrees to north-east. It was worked formerly by a shallow pit.

5.—*Tse-lung-chung.*—Situated two miles south of Chin-san and 20 miles from Yen-kah-woy. The seam is from 2 to 3 feet thick, and dips 40 degrees to north-east.

6.—*Kun-chok-wan.*—Situated 14 miles east of Tse-lung-chung, on the top of a mountain. The seam is from 6 to 8 feet thick, and hard, with a dip of 25 degrees to north.

The above coals are all semi-anthracite.

Another series of coals is found further west on the Yang-tse, near Haikow:—

7.—*Ho-peck-Tsung-ho, No.* 1.—Largely used by merchant steamers on the Yang-tse.

8.—*Tsung-ho, No.* 2.

9.—*Hoo-nan.*—A hard black anthracite.

10.—*Hankow.*—An anthracite, and a neighbouring seam to Hoo-nan.

No.	Specific Gravity.	Colour of Ash.	Coke.	Analyses.		
				Moisture and Volatile Combustible.	Fixed Carbon.	Ash.
			Per Cent.	Per Cent.	Per Cent.	Per Cent.
1	1·71	White ...	None ...	19	71	10
2	1·74	White ...	None ...	11	80	9
3	1·78	White ...	None ...	13	47·5	39·5
4	1·72	White ...	None ...	13	73	14
5	1·70	Grey ...	None ...	13·5	73·3	13·2
6	1·71	Grey ...	None ...	16·5	72·5	11·0
7	...	Dark brown	72	28	63·8	8·2
8	...	Grey ...	None ...	9	74	17
9	...	White ...	None ...	10·8	84·2	5
10	...	Grey ...	None ...	11	84	5

M. W. B.

SPIRAL-SIEVE COAL SCREENS.

Kohlenseparationen nach dem Principe des Spiralsiebes. ADOLPH SCHMITT-MAN-
DERBACH. *Berg- und Hüttenmaennische Zeitung, Vol. XLVI., pp.* 351–354.
One Plate.

The drums of the older cylindrical coal sieves were costly. required frequent repair, and were hard to drive. The spiral screens are arranged with a number of concentric wire sieves, with meshes corresponding to the sizes of coals required. Between successive sieves thin iron spouts are arranged spirally, with sufficient fall to cause the coal to run out at the end of the drum into fixed channels leading to the loading places. The newer machines are easily driven, can be put up in about one day. and allow of auxiliary screens being introduced at will. Their work is assisted by blows from a wooden hammer, or by jerks given by a mechanical arrangement of hooks and springs. The coal may be poured into and conveyed from the drum at heights of about a yard and a half below, and a yard above, its axis respectively, so that tubs and wagons may be worked from the same height of platform.

At the Sylvester Colliery, near Dux, in Bohemia, a spiral screen was set up to replace one of the older cylindrical machines. The old arrangement, which required a motive power of from 8 to 10 horses, and often broke down, could separate about 40 wagons of coal in 20 hours with a 14 per cent. proportion of duff, and necessitated awkward arrangements for loading.

The new machine, with a duty of 70 wagons per day of 20 hours, needs only about 1 horse-power to drive it, and is much more convenient for the work. The axis of the drum is at the level of the flat sheets, and the turning power is transmitted by shafting from the engine-house.

The coal is first kipped over an inclined screen with rods $3\frac{1}{2}$ inches apart, and what falls through is raised again by a bucket-ladder and delivered into one end of the innermost sieve. which is made slightly conical so as to ensure the coal's distributing itself towards the farther end. The drum is 6 feet 7 inches long and 7 feet 8 inches in diameter.

The sieves, from the middle outwards, have meshes of about 2, $1\frac{1}{8}$, $\frac{3}{4}$, $\frac{9}{16}$, and $\frac{1}{4}$ inches respectively. The middle one delivers at its wider end into a transporting band which forwards the coal to its place of storage; the second and third deliver into spouts; the fourth falls into a store-room below the machine; and the finest kind, which is unsaleable, slides into and is carried off by a sort of spiral screw.

The speed of working is six turns per minute.

The amount of duff. compared with that from the older machine, is reduced by from 5 to 7 per cent., representing a saving of about £5 per day.

The coal at the Sylvester Colliery is the wet and brittle Bohemian brown coal.

At the Bernissart Mine, near Peruwelz, in Belgium, a similar arrangement is at work. The drum in this case is 4 feet 1 inch long and 7 feet 6 inches in diameter, and consists of four sieves of about $1\frac{9}{16}$, 1, $\frac{3}{4}$, and $\frac{1}{2}$ inch mesh respectively. It works at the rate of eight turns per minute, and separates 300 tons of coal during six hours' work. As in the former case, the mixed coals are first passed over a fixed screen, with bars about $1\frac{9}{16}$ inches apart. The coal from this mine is very soft, and when it was separated by other kinds of screen the loss from breakage was at least 6 per cent. greater. The saving on this account amounts to about £10 per day.

A. R. L.

WIRE ROPE HANGING WAGONWAY AT GOTTESSEGEN MINE.

Drahtseilbahn und Kohlenrëtterei auf Gottessegengrube bei Antonienhuette, O.S. C. SACHS. *Zeitschrift des Vereines deutscher Ingenieure, pp.* 965-970. *Two Plates; Illustrations in the text.*

The mines of Gottessegen and Hugozwang, near Antonienhuette, in Upper Silesia, both belonging to the same owners, were connected with the Morgenroth Station on the main line, the one by a $3\frac{3}{4}$-mile broad gauge wagonway, the other by a 4-mile narrow gauge tubway. At the Aschenborn Shaft of the Gottessegen Mine there are extensive arrangements for screening the coals, and it was desired to transport to them the coals drawn from the Menzel Shaft of the other mine by some economical and durable arrangement. The intervening country is hilly, a village lies in the direct line, and the natural expedient of a wagonway, besides causing difficulties of ground rent, had the disadvantage that it would become snowed up in winter. A hanging wagonway on the Otto system was adopted. Shear-leg shaped steel posts, with crossbars on the top, are spaced about 38 yards apart. The outer ends of the bars are joined by parallel steel wire ropes in such a manner that the double wheels of the hanging tubs will pass over them.

The wagonway is made in two straight stretches, with a station at the angle which they form with one another, the tubs being here transferred by hand from the one stretch to the other. The whole distance is 2,900 yards, the steepest gradient being 1 in 11·8.

An engine of from 18 to 20 horse-power at the Aschenborn Shaft sets in motion an endless steel wire rope, to which both full and empty tubs are clamped at intervals of about 42 yards.

The two larger sizes of coal are separated from the rest at the Menzel Shaft, and the three kinds are loaded at once into the hanging tubs. On reaching the Aschenborn Shaft they are loaded direct. the first two into the railway wagons, and the third into the separating screens.

The fixed ropes are made fast at each pit shaft, and are kept taut by heavy weights at the halfway corner station. The endless ropes are driven from the Aschenborn Pit. They are worked round sheaves at the corner station, and kept in tension by a hanging weight of about 1¼ tons at the Menzel end.

The hanging tubs are of steel and hold about 10 cwts. each. With a speed of 5 feet per second the amount transported is 700 tons in 10 hours.

The cost of working of the new arrangement proved less than was anticipated. Including the wages of 11 people and the fuel and stores consumed it amounted to about 22s. per day. Besides the foregoing, about 12 to 15 women are now required for loading and unloading, against 66 women, 2 watchmen, and 2 horses under the earlier arrangement. The total saving to the owners is about £10 per day.

A. R. L

SILVER ORE DEPOSITS.

Beitrag zur Characteristik der Erzlagerstaetten. ROBERT WIMMER. *Berg- und Huettenmaennische Zeitung. Vol. XLVI., pp. 423, 424. Illustrated in the text.*

The recent sinking of the Olive Branch Shaft in previously unbroken ground, near Leadville, in California, throws some fresh light on the question of silver ore formations, which has been discussed in the above journal before. The substances met with, in their order from above, were 94 feet of glacial *débris*, 218 feet of grey porphyry, 2 feet of black shale, 40 feet of white porphyry, 18 feet of grey chalky slate mixed with dark shale and silver-bearing iron pyrites, 5 feet of compact solid dolomite, and 20 feet more of the same substance in the form of clean running sand, which was sunk through with great difficulty. Here veins of iron, nests and fragments of brown ironstone, gave indications of the proximity of ore, which was shortly found at a depth of 397 feet. First came 14 feet of hard flinty brown ironstone, containing from 4 to 35 oz. of silver per ton, of which the lower 6 to 8 feet could be profitably worked. Then came a wedge-shaped bed, from 4 to 6 feet thick, of coarse flinty dolomite with streaks of iron, through which ran a deposit of very rich ore. Below the dolomite came first a bed of half-transformed iron pyrites, then about 10 feet of untransformed silver-bearing iron pyrites mixed with sulphate of zinc and arsenical pyrites, which again overlay a bed of flinty slate. Part of the latter contains silver in profitable quantity. Below this came iron pyrites again, in which the shaft terminated.

The profitable ores form a zone of about 17 feet in depth, and contain an average of 52 oz. of silver per ton. The silver-bearing stratum proper is a bed of dolomite, occurring just between beds of ironstone and iron pyrites, and seems to have been originally traversed by water-courses and similar caverous passages. These have gradually become lined by deposits of silver, whether of thermal origin or produced through lateral secretion from the porphyry. In the loose sand are found lumps of spongy silver, containing about 25,000 oz. of pure metal to the ton, and sometimes hollow or containing clay within. The water-courses are generally filled loosely with similar spongy masses of silver, silver-covered lumps of ironstone, sand, etc.

The ore-bearing stratum in the Olive Branch Shaft is for the greater part a metamorphosed seam of dolomite, of which the dolomite sand, met with in sinking the shaft, once formed a part, as also the iron pyrites and the manganese lead, silver and zinc ores, the upper part of these sulphur-bearing ores having been metamorphosed into the manganese and pure silver-bearing brown ironstone by the action of carbonic acid water coming down from the surface. An accumulation of silver-bearing solutions deposited their wealth in the dolomite caverns, and thus produced the very rich veins now being worked.

A. R. L.

NEW COKE OVENS.

Dr. Th. v. Bauer's Bogensohl-Koksöfen für beliebig zu fractionirenden Betrieb. DR. B. KOSMANN. *Berg- und Hüttenmænnische Zeitung, Vol. XLVI., pp.* 379, 380. *One Plate.*

Dr. v. Bauer has lately improved upon a design made by him some two years ago for coke ovens, and his system is being introduced in the " Creusot " works, in France. Three kinds of oven are arranged for coals containing different proportions of gas, which, however, differ little from each other in principle.

The coking chambers, 12, 24, and 30 in number respectively, are arranged in circles with flues for heating by gas between, below, and behind them, fed from an inner circle of gas and air chambers ranged round a central chamber open to the air below, in which the gas from the burning coals and that returning from the process of condensation are allowed to collect. This latter is of cylindrical shape, with gridiron-like openings in its upper half to the surrounding gas and air chambers. From these the gases mixed with air rising from below pass through corresponding openings into the flues, whence, after being coursed round the coke chambers they escape into the chimney.

In the two smaller arrangements each group of ovens has its chimney in the centre; but for the 30-oven furnaces a large chimney receives the waste gases from several groups. The coke chambers have the form of half the letter U, the furnace door being at the point of section of the letter below, and they are made shallow and broad, or deep and narrow, according as the coal contains much or little gas. Each group of ovens is so arranged that single ones may be shut off while the rest remain at work, and also so that the bye-products of tar, ammonia, etc., can be worked or not, as may be found convenient. In one instance coals, which before would hardly coke at all, were found to produce excellent coke and bye-products in addition, which themselves paid the cost of working. A. R. L.

CONTROL-APPARATUS FOR WINDING ENGINES.

Controlapparat für Fördermaschinen. J. SPRENGER. *Berg- und Huettenmænnische Zeitung, Vol. XLVI., pp.* 433-435, 445-447. *One Plate.*

An apparatus has been invented by Herr J. Weidtmann, of Dortmund, for showing the brakesman of a colliery winding engine at any given moment the speed of winding and position of the cages in the shaft, and which at the same time records graphically the alternate pauses and periods of work and the various speeds of winding during the day's work. The apparatus consists of a 5 ft. by 2 ft. 6 ins. by 2 ft. 6 ins. iron box, containing the machinery, and above this in succession a clock, a sector-shaped dial and pointer showing the speed of winding, and a circular dial with two pointers showing the positions of the cages in the shaft. It is placed just in front of the brakesman's seat.

A light horizontal shaft is fixed to the crank shaft of the main engine, and conveys by a worm thread and vertical shaft behind the case a slow motion to the pointers on the cage-position dial, and by wheel and pinion gearing a quick motion to an inside vertical shaft which works the control machinery. On the centre vertical shaft is a contrivance like an ordinary governor, with two suspended heavy balls which rise and fall as the speed increases or diminishes, the extra upward tendency due to their own

d

inertia being controlled and regulated by springs. This rising and falling motion is communicated by gearing to a long pointer which shows on the sector-dial above mentioned the speed of winding in metres per second, and at the same time to a pencil working on a control disc in the upper part of the case. The disc, about 24 inches in diameter by 4 inches in thickness, is turned by a small shaft from the clock above. The indicating pencil, working vertically on its periphery on a strip of drawing paper and moving up and down at right angles to the motion of the disc, gives a graphic record of the whole day's work of the engine as to time worked, stoppages, and speed at different times. This last is of importance in Germany, as the maximum speed of working, while drawing men, is there fixed by law, being about 13 feet per second. It also affords very valuable evidence as to the carefulness and general reliability of the brakesman. A bell fixed behind the clock gives warning of the cage's approach to bank in the usual manner.

The apparatus shows itself especially useful in any case of accident in the shaft, as it enables the brakesman to see at a glance the positions of the cages, the speed of winding, and the time of day, and indeed records the last two on the control disc.

The makers are Messrs. Dingler, Karcher, & Co., of St. Johann on the Saar, and the cost of the apparatus is about £80.   A. R. L.

EXPLOSION IN A GRAIN WAREHOUSE AT HAMELN.

Die Explosion in der neuen Wesermühle zu Hameln. C. ARNDT. *Zeitschrift des Vereines deutscher Ingenieure, Vol. XXXI., pp.* 1,044–1,049. *Illustrated in the text.*

Cases are on record in England and America of explosions in corn warehouses, but previously to the disaster at the Weser Mill, at Hameln, in November last, they were unknown in Germany. The building was of semi-quadrangular form, with the court facing the westward, a corn mill forming the central portion, and the northern wing containing a corn warehouse, a large staircase, and rooms for cleaning the corn and for other like purposes. The warehouse occupying the outer half of the wing had spaces below and above, containing the band-transport machinery. The warehouse itself was divided by wooden boarding into eight large cells, the two central ones only being nearly full at the time of the accident. The workman, whose broken safety-lamp caused the accident, was at work in the basement and escaped with his life. From the subsequent inquiry it appeared that the corn dust was set on fire at about the middle of the space below the warehouse. The building being massive, and made, as far as possible, fire-proof, the fire must have extended along the basement, up the corn-elevator, and along the band above the cells until it reached the two middle ones. A terrific explosion then occurred about 20 seconds after the corn at the bottom of the elevator first took fire. The massive walls of the building were shattered or blown down along the whole wing, and ten lives were lost. An examination of the ruins proves that the explosion took place, not in the empty cells, which were full of inflammable dust, but in the upper part of the two full cells.

Herr Arndt considers that the exploding substance could not have been the dust, as it would then have made its presence felt sooner and in another part of the building, and throws out the suggestion that it consisted of some kind of gas, hitherto unnoticed, arising out of the corn itself.   A. R. L.

COAL MINING IN NORTH CHINA.

By Kwong Yung Kwang. *The Engineering and Mining Journal (New York),*
Vol. XLIV., pp. 220, 221, 238. *Two plates.*

There are seventeen seams of coal, with a dip to the south of about 45 degrees, which are interstratified with sandstone, shale, and fire-clay. Nine of the seams are workable, beginning from the uppermost, as follows :—No. 2, 2 feet 2 inches ; No. 3, 7½ feet; No. 5, 5½ feet; No. 8, 9½ feet; No. 9, 15 feet; No. 10, 15 feet; No. 11, 2½ feet ; No. 12, 35 feet; and No. 13. 2½ feet.

There are two shafts, about 100 feet apart, the downcast being 560 feet and the upcast being 300 feet in depth, and are walled with limestone throughout. Levels are driven to cut the various seams as under:—

No. 1 level cross-cuts the seams from No. 1 to 15, inclusive, at a depth of 200 feet.
No. 2 level cross-cuts the seams from No. 1 to 13, inclusive, at a depth of 300 feet.
No. 3 level cross-cuts the seams from No. 1 to 12, inclusive, at a depth of 560 feet.

Methods of Working.—These levels are driven north to cut the seams, and when any seam is reached it is followed along the strike east and west. winzes are driven to the upper level for ventilation at intervals of 50 feet, and a ventilation cross-head is driven at 20 feet. A pillar of coal, as soon as three are formed, is then removed, beginning at the lower level, and a stook is left next the upper level. The details of working a pillar are as follow :—Coal boxes, 2 feet square and 8 feet long, are put up from the bottom of No. 2 level to the ventilation cross-head for sending out the coal, and ladders 12 feet long are put in to communicate with No. 1 level. Bridge rails are laid from the ladder-way to the adjacent winze, used as a stone shoot. The miners begin at the stone shoot and work towards the ladder-way. Props of 6-inch pine are placed close to the coal and 4 feet apart as the cutting face advances. The space left by the removal of the coal is filled by the packers with dead stuff of soil, clay, and stone sent down from the surface to No. 1 level in proper tubs which open at the sides. The coal is loaded into willow baskets and placed on small trams, which are run to the coal shoot. Slice after slice is thus removed in ascending order. The coal shoots are constructed by building two pack-walls about 3 feet apart and covering with 3-inch plank. The coal is reduced almost to dust in these shoots, but the natives are said to prefer it in this form. Self-acting inclines are used where lump coal is required.

The same method is applied to the working of the 35-feet seam, which is divided into five slices parallel to the stratification, and worked off successively upwards, like as many distinct seams of 7 feet in thickness. The average cost of packing is about 2d. per ton.

In the thin coal-seams the longwall method is adopted.

With the packing system very little timber is required and the surface is uninjured.

Haulage.—Mules and ponies are used, each animal pulling about eight tubs of coal. It is probable that the main and tail-rope will be adopted when the advance of the workings renders the use of such mechanical means necessary.

Ventilation.—The mine is ventilated by a 30-feet Guibal fan, running 30 revolutions per minute and producing 120,000 cubic feet of air, under a water gauge ·6 inch. Candles and open lamps are used in all seams except No. 5, where safety-lamps only are allowed. A dust explosion occurred in this seam three years ago by which three men were killed and several others injured.

Workmen.—The workmen belong to the lower classes. They are searched as they leave the mine (stealing being considered to be a virtue) and are severely punished when detected.

The daily wages are as follows :—

Interpreters to foreign over-				Door boys ... ... 6½d.
men ... ... 10d.			Switch boys ... ... 6½d.
Pony and mule drivers ... 10d.	Drain boys ... ... 6¼d.
Sinkers ... ... 10d.	Drum boys ... ... 6½d.
Masons ... ... 10d.	Cantonese boatswains or
Enginemen ... ... 10d.	pumpmen ... ... 2s.
Coal miners ... ... 7½d.	Chinese pit deputies ... 2s.

Output.—The daily output is 950 tons in two shifts of eight hours each.

Prices.—The prices are :—

Lump ...	...	...	...	6 taels	= 35s. 0d.
Nut ...	...	...	...	4 „	= 23s. 4d.
Dust ...	...	...	...	3 „	= 17s. 6d.

M. W. B.

WINDING ROPES AND THEIR COST IN GERMANY.

*Ueber Schacht-Förderseile und Seilkosten. By — *WENDEROTH*. Zeitschrift für das Berg-. Hutten- und Salinen-Wesen im Preussischen Staate; 1882, pp. 77-80; 1886, pp. 308-314.*

Official statistics have been compiled in Germany as to the ropes used in the chief coal mining districts, in Dortmund since 1872, in Saarbruck since 1877, and in Breslau since 1882.

I.—The statistics for the years 1877-80 are derived from the following ropes :—

	Dortmund. No. of Ropes.	Saarbruck. No. of Ropes.
Flat Ropes—		
Steel	87	27
Iron	18	20
Aloe	19	6
	— 124	— 53
Round Ropes—		
Steel	388	42
Iron	210	191
	—— 598	—— 233
	722	286

The cost during the same period being:—

	Dortmund.			Saarbruck.		
	Cost of Ropes.	Work Drawn, in Millions of Kilogrammetres.	Cost of Ropes per Ton Kilometre.	Cost of Ropes.	Work Drawn, in Millions of Kilogrammetres.	Cost of Ropes per Ton Kilometre.
Flat Ropes—	£		d.	£		d.
Steel	10,007	1,782,232	1·350	1,222	120,443	2·434
Iron	1,185	171,124	1·662	799	132,483	1·447
Aloe	3,097	1,337,250	·556	366	39,427	2·231
Round Ropes—						
Steel	25,578	9,221,403	·666	2,672	686,735	·933
Iron	6,570	2,632,815	·598	6,356	2,647,937	·575

The broken ropes during the same years were :—

Year.				Dortmund. Per Cent.			Saarbruck. Per Cent.
1877	...	...	...	8·98	...	...	7·96
1878	...	...	...	9·40	...	...	1·80
1879	...	...	...	5·23	...	...	6·89
1880	...	...	...	4·70	...	...	3·13

The following table shows the average life and work of winding ropes of all descriptions:—

Year.	Dortmund.		Saarbruck.	
	Period of Use, in Days.	Average Work Drawn per Rope, in Millions of Kilogrammetres.	Period of Use, in Days.	Average Work Drawn per Rope, in Millions of Kilogrammetres.
1877	535	23,787	444	10,888
1878	554	26.956	429	12,099
1879	539	28,971	538	14,363
1880	577	36,879	535	14,034

II.—The statistics for the years 1881-84 apply to the following ropes:—

Flat Ropes—			Dortmund. No. of Ropes.	Saarbruck. No. of Ropes.	Breslau. No. of Ropes
Steel	...	...	95	45	16
Iron	...	...	11	15	—
Aloe	...	...	12	3	
Hemp	...	...	—	4	—
			—— 118	—— 67	—— 16
Round Ropes—					
Steel	...	...	500	126	152
Iron	...		118	78	58
			—— 618	—— 204	—— 210
			736	271	226

The cost during the same period being:—

	Dortmund.				Saarbruck.			Breslau.		
	Cost of Ropes.	Work Drawn, in Millions of Kilogram- metres.	Cost of Ropes per Ton Kilo- metre.		Cost of Ropes.	Work Drawn, in Millions of Kilogram- metres.	Cost of Ropes per Ton Kilo- metre		Cost of Ropes.	Work Drawn, in Millions of Kilogram- metres.	Cost of Ropes per Ton Kilo- metre.
Flat Ropes—	£		d.		£		d		£		d
Steel ...	12,352	2,606,072	1·137		3,771	601,389	1·505		998	256,552	·934
Iron ...	692	137.831	1·204		547	102,861	1·278		.	...	...
Aloe ...	2,371	1,110,240	·512		163	35,727	1·095		..	...	...
Hemp ...	...	...	...		162	9.690	4·009		...	...	...
Round Ropes—										
Steel ...	41,686	19.531,124	·512		10,279	2,700,713	·913		7,092	2,928,637	·581
Iron ...	5,521	3,586.390	·369		3,344	1,524,285	·526		1.137	1,168,811	·233

Districts.	The ropes removed from use were employed in winding from the following depths:—							Ropes removed employed for Winding Men.
	Under 100 Metres.	Of 101 to 200 Metres.	Of 201 to 300 Metres.	Of 301 to 400 Metres	Of 401 to 500 Metres	Of 501 to 600 Metres.	Totals.	
	Per Ct.	Per Ct.	Per Ct.	Per Ct.	Per Ct.	Per Ct.		Per Cent.
Dortmund ...	·4	12·6	47·6	27·7	11·0	·7	100	93
Saarbruck ...	4·1	32·1	36·5	13·7	6·6	7·0	100	69
Breslau ...	21·3	74·3	4·4	...	..	...	100	52

The broken ropes during the same years were :—

Year.		Dortmund. Per Cent.		Saarbruck. Per Cent.		Breslau. Per Cent.
1881	...	4·85	...	4·76	...	(?)
1882	...	7·73	...	3·85	...	8·63
1883	...	4·27	...	None.	...	6·76
1884	...	3·16	...	3·17	...	5·38

The annexed table shows the average life and work of winding ropes of all descriptions:—

Year.	Dortmund.		Saarbruck.		Breslau.	
	Period of Use, in Days.	Average Work Drawn per Rope, in Millions of Kilogram-metres.	Period of Use, in Days.	Average Work Drawn per Rope, in Millions of Kilogram-metres.	Period of Use, in Days.	Average Work Drawn per Rope, in Millions of Kilogram-metres.
1881	540	34,194	456	15,994	(?)	(?)
1882	533	39,182	581	19,091	492	25,242
1883	427	34,698	488	18,038	568	18,516
1884	490	38,105	429	20,149	472	16,071

M. W. B.

EXPLOSIONS OF FIRE-DAMP IN FRANCE.

Analyse synoptique des Rapports officiels sur les Accidents de Grisou en France from 1817 to 1884. Annales des Mines, 1882, Vol. I., p. 293; Vol. II., p. 393; 1883, Vol. II., pp. 67 and 116; 1884, Vol. II., p. 73; 1885, Vol. II., pp. 195 and 433; 1886, Vol. I., p. 31, Vol. II., p. 11; 1886, Vol. II., p. 521; Annales de l'Industrie Minérale, 1885.

This work is a detailed analysis of the circumstances of the explosions of fire-damp in mines in France from 1817 to 1884, 808 in number, of which 304 were fatal, causing the death of 1,520 and injury to 1,374 persons. The accidents occurred in 115 concessions, and are divided as follow :—

						Accidents.
Mines of coal	...	...	...	...	...	797
,, lignite	...	...	...	...	...	3
,, bituminous shale ...		...	...	...	...	4
,, lead ore	...	...	...	...	...	3
,, iron ore	...	...	...	...	...	1
						808

TABLE SHOWING THE MONTHLY AND DAILY OCCURRENCE OF EXPLOSIONS OF FIRE-DAMP.

Month of—	No.	France. Per Cent.	Belgium. Per Cent.	Prussia. Per Cent.	Saxony. Per Cent.
January	60	7·8	7·0	8·5	7·6
February	74	9·6	7·3	8·6	7.6
March	69	9·0	12·1	10·8	8·9
April	73	9·5	10·9	6·9	7·6
May	55	7·2	9·0	7·3	6·8
June	59	7·7	6·8	7·2	9·7
July	64	8·3	9·5	7·8	11·9
August	71	9·3	9·0	7·8	9·7
September	53	6·9	7·5	8·0	10·2
October	65	8·5	5·8	9·3	4·2
November	55	7·2	7·3	7·6	6·8
December	69	9·0	7·8	10·2	8·9

Days of the Week—

	No.	France. Per Cent.	Belgium. Per Cent.	Prussia. Per Cent.	Saxony. Per Cent.
Monday	144	19·0	9·7	20·7	33·5*
Tuesday	103	13·6	13·9	17·3	(?)
Wednesday	100	13·2	20·0	14·3	(?)
Thursday	117	15·5	17·4	16·3	(?)
Friday	125	16·6	15·4	14·5	(?)
Saturday	113	15·0	15·4	14·1	(?)
• Sunday	54	7·1	8·2	2·8	8·9†

TABLE OF FATAL ACCIDENTS ACCORDING TO THE NUMBER OF PERSONS KILLED.

Accidents	France. Accidents.		Killed.		Belgium. Accidents.	Killed	Prussia. Accidents.	Killed.	Saxony. Accidents.	Killed.
	No.	Per Cent.	No.	Per Cent.	Per Cent.	Per Cent.	Per Cent	Per Cent.	Per Cent	Per Cent.
1 killed	150	49·2	150	9·9	39·8	4·9	59·2	23·6	65·4	10·0
2 to 5 killed	107	35·2	323	21·2	32·8	11·5	35·2	38·9	26·4	9·2
6 to 10 „	20	6·6	157	10·3	10·0	10·1	3·2	10·2	2·5	3·2
11 to 20 „	12	4·0	175	11·7	6·0	10·5	1·2	6·8	1·6	3·8
21 to 50 „	12	4·0	370	24·3	7·8	29·4	·9	10·9	1·6	4·8
More than 50 killed	3	1·0	345	22·6	3·6	33·6	·3	9·6	2·5	69·0

TABLE SHOWING THE CAUSES OF THE IGNITION OF THE FIRE-DAMP.

Naked lights in use—	France. Per Cent.	Belgium. Per Cent	Prussia. Per Cent	Saxony. Per Cent
Naked lights	52·9	22·7	46·2	97·5
Furnaces for ventilation	—	5·1	·3	—
Fires	1·7	·5	—	—
Safety-lamps in use—				
Open	15·2	21·5	13·2	1·1
Damaged	5·4	13·0	5·7	—
Red-hot gauze	3·5	—	3·3	1·1
Passed flame by speed of air	3·4	2·1	2·1	·3
„ „ sharp movement	3·6	2·7	11·2	—
Powder	14·3	32·1	18·0	—

* Includes the day after holidays   † Includes holidays.

TABLE SHOWING THE CAUSE OF THE ISSUE OR ACCUMULATION
OF THE FIRE-DAMP.

Issue of gas—	France. Per Cent.	Belgium. Per Cent.	Prussia. Per Cent.	Saxony. Per Cent.
Sudden	6·4	... 13·4	... 16·7	... 13·5
Slow and continuous—				
(a) Under scaffolds, behind doors, or goaf...	2·5	... 4·6	... ·9	... ·4
(b) In workings	72·9	... 67·5	... 68·6	... 78·0
Ventilation defective—				
(a) Generally—				
Derangement of fan	1·4	... 6·6	... 4·5	.. ·4
Natural causes	2·3	... 1·2	... —	... ·8
Casual derangement	4·1	... —	... 1·2	... ·8
(b) Locally—				
Falls	7·2	... 2·6	... 5·8	... 4·1
Doors left open	3·2	... 4·1	... 2·3	... 2·0

TABLE OF ACCIDENTS SHOWING THE NEGLIGENCE OF THE
PERSONS EMPLOYED.

Errors or neglect of—	France. Per Cent.	Prussia. Per Cent.	Saxony. Per Cent.
Mine owner—			
Infraction of law	3·1	... —	... 3·7
Want of organisation	3·1	... —	... —
Overmen, etc. ... ·	14·3	... 12·0	... 31·7
Workmen other than those injured	4·1	... 5·2 ⎱	... 64·6
Workmen injured	75·4	... 82·8 ⎰	

M. W. B.

IMPROVEMENTS IN SHOT-FIRING.

Amelioration du Tirage des Mines. By P.-F. CHALON. *Le Génie Civil, Vol. XI.,
pp. 254-255, 328-329, 381-382.*

The greatest danger in firing shots in fiery parts of mines is not so much due to the explosion of the shot as in the mode of igniting it.

Bickford's safety fuse is very old, yet there are many mines where the straw is still in use. There are frequent miss-fires where straws are used to ignite the shot, and it is especially dangerous because it does not always allow the workmen time to reach a position of safety. Lastly, in every case the powder in igniting throws a flame or a shower of sparks which may cause an explosion of an inflammable mixture of air and fire-damp if the latter suddenly issues near the shot-hole.

The use of fuse is a notable improvement, but it is far from perfect and has many disadvantages. The common white fuse is used for economy. The gutta-percha fuse costs more and is preferred where complete safety is desired. (The relative costs are as 4 to 1.)

Safety fuse burns at the rate of about 32 inches per minute, but this rule has, unfortunately, exceptions. It sometimes happens that the ignition hangs at certain points from various causes (contraction, foreign matter, want of powder, etc.) Sometimes a drop of grease upon the covering will check the fire for some hours.

To ignite the fuse the miner opens the end into two or four parts for a length of about one inch, and places in it a piece of tinder, touch, a match, etc. To fire the tinder the miner uses a wire, or draws the flame of his lamp against the side by means of a straw, or sometimes he opens his safety-lamp. In whatever way the fuse may be ignited there are always a few dangerous moments. Whilst the fuse is taking fire sparks are produced for an instant, sometimes followed by a small jet of flame.

In the preparation and firing of a shot there is always risk of accidents during (a) stemming, (b) ignition, and (c) explosion of the charge.

According to the Prussian Fire-damp Commission, the use of powder, granulated or compressed, should be prohibited in fiery mines, or replaced by dynamite and other high explosives. This remedies the third class of danger, because dynamite and similar explosives do not throw out flame in ignition; there is only a brief incandescence produced by the scorification of the absorbent matter, or by the heating and decomposition of the nitrates, carburets, or sulphurets which form the absorbent.

Mr. Settles proposes to cool the gases resulting from the ignition of a shot of powder or dynamite and prevent them igniting gaseous mixtures by the use of the water cartridge.

Stemming.—The stemming of shots is a frequent cause of miss-fires and accidents. Sometimes the detonator, the cap, the charge, or the fuse ignites, sometimes the fuse is broken or compressed, sometimes the shot is fast and the stemming is blown out. *These accidents may be avoided if stemming is abolished.*

Numerous experiments have been made in the gypsum quarries near Paris as follows:—

Shot-holes were prepared for ordinary blasting and were charged with the usual quantity of powder; then, instead of stemming them, a wooden plug, 6 to 8 inches long (pierced with a hole for the fuse), was placed in the outer end of the hole. The shots were fired by electricity, or by a small paper tube containing 4 inches of Ruggieri fuse. A Scola cap ignited this fuse, which shot to the end of the hole and fired the charge. This system was abandoned as the gas escaped between the sides of the hole and the plug before ignition was completed, and a part of the power was wasted in blowing out the plug like a projectile.

In the next experiments the plug was replaced by a handful of soft clay so as to hermetically seal the mouth of the hole. This was very successful and was repeated with similar results; the quantity of rock brought down was considerable and in each case the amount of unburnt powder was much less. In the first experiment, with a hole 6½ feet long, and a plug of clay 5 inches long, 720 cubic feet of stone were broken up by means of 2·1 pounds of powder. In a second hole, 4½ feet long, 525 cubic feet were dislodged by 1·5 pounds of powder. In a third hole, charged with 2·9 pounds of powder and 5¾ feet long. from 1,000 to 1,050 cubic feet of stone were dislodged. The mean of the series of experiments showed about 350 cubic feet of stone per pound of powder. Under ordinary conditions only 200 cubic feet of stone is obtained per pound of powder. If 250 cubic feet be taken (after allowing for miss-fires, etc.), there is still a saving over stemming of about 25 per cent.

In later experiments the length of the plug of clay was reduced from 5 inches to 1½ or 2 inches. It is possible that similar results might have been obtained if the hole had been hermetically covered by a sheet of paper.

The advantages of the system of suppressing the stemming are:—(a) Saving of labour; (b) economy of material, the powder being more completely burnt; (c) a better produce; and (d) the suppression of accidents during the stemming.

The practical application of the system is:—The charge having been placed at the end of the hole in the ordinary way, with a fuse or electric firers. the opening is closed as hermetically as possible by a ball of clay from 2 to 3 inches long. It is most essential that this plug should be unfissured and adhere closely to the sides of the hole, so that the gaseous products cannot escape before the complete ignition of the powder. Cement may be used instead of the clay.

Ignition of the Shot.—Electric shot-firers are recommended. and more especially the "exploseur-vericateur," which tests the caps as well as firing them.

M. W. B.

IGNITION OF SHOTS IN FIERY MINES.

Allumage des coups de mine dans les Mines Grisouteuses. By — MORTIER. *Comptes-Rendus Mensuels des Réunions de la Société de l'Industrie Minerale.* 1887, *p.* 99. *Plate XVI.*

This invention obviates the use of a match and the jets of flame which are produced at the end of the fuse during the first moments of ignition. The principle applied in the new system is the ignition of powder in the presence of sodium and water. The cap consists of a small piece of sodium encased in a covering of india-rubber, and inserted at the end of a small capsule. When required for use the cap is placed upon the end of the fuse, the globule of sodium is pierced by a triangular rod, and immediately placed in a closed case half-full of water. The cost of a cap will be as follows:—

					Pence.
Capsule	...	...	...	...	·150
Sodium, ½ grain	...	...	...	...	·004
India-rubber ...	...	...	...	...	·005
Workmanship	...	...	...	...	·020
Total	...	...	...	...	·179

M. W. B.

SAFETY CATCHES FOR INCLINED PLANES.

Arrêt de Bennes pour Plans inclinés. By — RAMEAU. *Comptes-Rendus Mensuels des Réunions de la Société de l'Industrie Minerale,* 1887, *pp.* 97 *and* 98. *Plate XVI.*

Arrêt de Bennes pour Voies, Rampes et Plans inclinés. By — MORTIER. *Ibid., pp.* 98 *and* 99. *Plate XVI.*

Mr. Rameau's catch consists of a bent lever with two unequal arms. placed flat on the ground and turning round a vertical axis situate about its middle. It is placed parallel to the way and at such a distance that at least one of the ends always bars it, and having the short arm next the side of the inclined plane. The empty tub on its arrival at the top passes forwards, and its front wheel touches the long arm after the hind wheel has passed the short arm. The apparatus then bars the way. For the next run of the incline the attendant pushing the full tub upon the same way is checked by the bent arm and must push it back with his foot in order to allow the tub to pass. One of these safety catches is placed upon each road. The apparatus is very cheap, simple, and effective. It bars the way after the passage of the tubs; when closed it must be pushed open by the attendant; it cannot be wedged open or shut; it is easily worked; it always permits the passage of empty tubs upwards; and never prevents the free movements of horses or workmen.

Mr. Mortier's safety catch consists of an axle with attached levers, laying in the axis of the way and supported upon the sleepers. In the two extreme positions of the axle one of the levers allows the passage of the tub axles and the other stops them. In the middle position both levers prevent the passage of a tub. The catch is worked automatically by the empty tubs passing upwards. This is effected by making a slope of the external edge of each lever and the levers are thrown over by the axles of the tub. This catch cannot be put out of order either wilfully or by neglect. A pedal is attached to work the appliance and to allow full tubs to pass downwards.

M. W. B.

THE PHOSPHATES OF THE SOMME (FRANCE).

Note sur les Phosphates de la Somme. By PAUL LEVY. *Mémoirès et Compte Rendu des Travaux de la Société des Ingenieurs Civils, 1887, Vol. II., pp. 184 to 190.*

The existence of phosphates in the Somme was pointed out by M. de Mercey about 25 years ago, but the deposits were not actually discovered at Beauval until recently by M. Merle. He sent a sample of it to be analysed and on its value becoming known numerous speculators bought up large areas of the deposit. The price rapidly increased, as much as £30,000 being paid for less than four acres which were previously worth no more than £1,000.

The surface is generally level at Beauval, but the chalk covered by recent deposits is deeply ravined and the phosphate said is found deposited in irregularly distributed pockets. These pockets are usually shaped like inverted cones with an upper diameter of six or seven yards and containing from 50 to 500 cubic yards of the phosphate said. The phosphate of the first bed is slightly yellowish and is of the poorest quality; it contains 50 to 60 per cent. with traces of iron. Below this it is richer and contains from 70 to 80 per cent. of tribasic phosphate of lime, and is very similar in appearance to the seeds of figs. The internal surface of the pockets is polished like that of many natural pits, showing that they may have been formed by the slow solution of the chalk by a corrosive liquid (water charged with carbonic acid). These cavities are always filled with deposits in the following order:—Soil, clay, gravel, phosphates, chalk, and sometimes phosphates below it. The rocks at Beauval belong to the last stage of the Secondary period and are characterised by *Belemnitella quadrata.*

The phosphate is a yellow powder with smooth touch; under the microscope it appears like millet seed whose surface has been corroded by some dissolving fluid. The mean of several analyses is:—

	Per Cent.
Phosphate of lime	73·5
Carbonate of lime	7 to 9
Oxide of iron	5 to 1
Alumina	·04
Sulphate of lime	1·25
Water	17 to 25

It is dried and screened to prepare it for market. The price varies with the percentage of phosphate of lime, 70 per cent. being worth 58s. per ton, and 80 per cent. 80s. per ton. M. W. B.

BASIC SLAG AS MANURE.

Phosphates Métallurgiques du Creusot. By — SEJOURNET. *Comptes-Rendus Mensuels des Réunions de la Société de l'Industrie Minerale, 1887, pp. 86–88.*

It is hoped that favourable results will be obtained by the use of the basic slag produced at Creusot, containing 12 to 16 per cent. of phosphoric acid, associated with 45 per cent. of lime and 5 to 6 per cent. of magnesia.

It will be most applicable to acid or turfy soils, newly cleared ground, clayey soils, and granitic and siliceous soils. It should be sprinkled or sown upon the soil broadcast, or better by a drill, and harrowed.

In the case of meadows and vineyards it is best to mix the phosphates with ashes or fine soil which facilitates its incorporation into the soil. From 16 to 20 cwts. should be applied per acre. M. W. B.

PORTABLE SER VENTILATING FAN.

Ventilateur portatif Ser. By — MATHET. *Comptes-Rendus Mensuels des Réunions de la Société de l'Industrie Minerale, 1887, pp. 88 and 89. Plate XIV.*

This fan, 20 inches diameter, carries its own engine which can develop three or four horse-power with air at about 45 pounds pressure. It is driven by belts, the pulleys being 16 inches and 6 inches diameter. The fan produces 4,250 cubic feet of air per minute at a speed of from 800 to 900 revolutions per minute.　　M. W. B.

BLOWERS OF CARBONIC ACID GAS.

Les Dégagements instantés d'Acide Carbonique aux Mines de Rochebelle (Gard).
By G. HANARTE. *Annuaire de l'Association des Ingénieurs sortis de l'École de Liège, Vol. VI., 1887, pp. 1–14. Plates I. and II.*

INTRODUCTION.

The Fontaines Pit is sunk in very troubled ground, bounded on the west by a fault of from 700 to 800 yards. The seams are much contorted and in section assume an M-like form. Carbonic acid gas has been given off at all times by the various seams, but fire-damp has only been seen *twice.* The workings for some time were above the 410 feet level, but it became necessary to carry the pit down to a depth of 1,300 feet.

ERUPTIONS OF CARBONIC ACID GAS.

Blower of July 28th, 1879.—Exploring places were being driven in No. 11 Seam at a depth of 810 feet and had reached a distance of about 140 feet from the pit. The coal began to decrepitate more strongly than usual and the lamps were extinguished and covered with fine coal, and finally, on July 28th, a blower came off accompanied with the projection into the drift of about 76 tons of coal.

Blower of November 3rd, 1884.—An exploring drift was being driven at the level of 925 feet and had cut No. 14 Seam at a distance of 360 feet from the shaft. The seam was, however, thrown down at the face of the drift. Heavy detonations were heard on the right side of the drift on November 3rd. The face was all in stone and 24 holes had been prepared ; one of them, however, penetrated into the seam. Carbonic acid gas was given off from the hole without pressure and extinguished lamps placed near it. The holes were charged, and, after lighting the fuse, the workmen retired to the shaft. Five detonations were heard, followed by 9 or 10 heavy detonations accompanied by a steam-like cloud. In less than a minute the air was reversed and the workmen experienced great cold, followed by a dense cloud which extinguished their lamps. The carbonic acid gas rose to the 575 feet level. The ventilating current of 4,200 cubic feet per minute had no effect upon the issue of gas for two days. At the end of that time the face was reached and a space was found on one side of the drift and near the roof of the seam 20 feet long, 6½ feet wide, and about 2 feet deep. The drift was covered with small coal for 15 or 20 feet from the face and for a further distance of 40 feet there was a thickness of about half an inch of fine coal-dust, the whole of which appeared to have passed through an opening of about 20 inches wide and 10 inches high.

Blower of April 25th, 1885.—The exploring drift at the 925 feet level had been continued and having again cut No. 14 Seam, had reached a distance of 680 feet from the shaft when the seam began to dip. On the 21st they heard detonations in the coal; on the 23rd the carbonic acid gas drove the workmen out of the drift for 5 hours, and formed a deposit of fine coal-dust on about 20 feet of the drift. Small blowers were given off on the 24th; in the evening 6 holes were made in stone and 5 in coal. On the 25th the air was worse than usual and they were compelled to use a Körting ventilator. It was noticed that the place trembled whilst boring the holes in the coal, the fact being repeatedly verified by the chargeman who stopped the machine drill. The holes were obstructed, the coal forming a cushion, and in a hole 10 feet long the men could scarcely draw out the scraper. The cartridges were blown out in charging the holes, which showed that the pressure was very considerable. The workmen charged the holes as quickly as possible and retired to the shaft. They had scarcely reached the 575 feet level when they heard the five coal shots go off. They immediately felt a violent rush of air on the return side of the brattice in the shaft, followed by repeated blasts on the downcast side which actually lifted the tub in which they were riding. The carbonic acid gas quickly followed them to bank and firemen suffered from its effects 80 feet from the shaft.

Over 500,000 cubic feet of the workings at the 410 feet level were filled with this irrespirable gas by leakage through the doors in less than 10 minutes. On reaching the drift after an interval of 28 days the floor was found covered with coal which had been classified by the currents of gas. The coal was like flour at the shaft and about half an inch thick, at 110 feet it was 12 inches, and at 260 feet it was 16 inches thick. Beyond this point the coal was coarser and the deposit continued to increase in thickness to about 53 inches. Beyond that point its thickness was about uniform, with a space of from 12 to 20 inches below the top of the drift. At 390 feet from the shaft all the in-bye brick brattice (200 feet) was destroyed, and pieces of coal weighing 50 pounds or more were found. A space was found 10 feet from the face, next the roof of the seam, which was explored for a distance of 52 feet. The right side of the place was solid ; on the left side was a mass of small coal, not less than 10 feet thick, thrown against the roof of the seam, which was polished. The blower evidently came from the right at about right angles to the drift. The quantity of coal found in the drift was about 405 tons, without including that thrown against the left side of the drift.

CONCLUSIONS.

It appears that the phenomena of sudden outbursts of carbonic acid gas are very similar to those of fire-damp, and it may be supposed that the mode of occurrence of these two gases in the coal is the same. Taking all the circumstances into consideration, it may be that the gas is frequently found to be absorbed and condensed by the powdered coal, and not under pressure, when this coal is in a condition of static equilibrium. This pressure may then become apparent after a very slight shock, by a fall, which causes the gas to leave this coal where it was retained by a special attraction. It is only by such means that the pressure can be developed, that the gas escapes, that suction is produced, and that the blower or sudden outburst commences.

Professor Graham made many experiments upon the special attraction of gases for solid bodies. Charcoal is able to absorb about 90 times its own volume of carbonic acid gas or of ammonia gas. If the pores of the charcoal are the 100th part of its apparent volume the gas will be reduced by 9,100 times its volume. It is possible, therefore, that charcoal may liquefy these gases, which require much less reduction of volume.

Blowers of carbonic acid gas, like those of fire-damp, are most frequently found in the vicinity of faults ; that is to say, where the coal has been triturated by dislocatio1s. If the trituratio1 takes place in a closed space (as i1 the prese1t example. where the Coal-Measures are covered by the Trias) the gas is u1able to escape. The dust, havi1g the absorptive power of porous bodies, absorbs the gas, and this absorptio1 may explain the e1ormous volumes of gas give1 off. 1otwithsta1di1g its feeble pressure i1 the coal.

The followi1g precautio1s have bee1 take1 i1 the exte1sio1 of the drift at the Rochebelle Collieries:—

1.—The pit is completely isolated by maso1ry from the other parts of the mi1e.

2.—A separate fan, 13 feet in diameter, and tur1i1g about 300 revolutio1s per mi1ute, is applied to its ve1tilatio1.

3.—Shots are to be exclusively fired by electricity after the workme1 are out of the mi1e.

4.—The drift will be drive1 by mecha1ical drills and always preceded by a bore-hole of from 6 to 10 feet, in which the pressure of the gas will be measured by gauges.

5.—Electric sig1als are provided betwee1 the face and the surface and *vice versâ*.

M. W. B.

BRANDTS' HYDRAULIC DRILLING MACHINES.

Der maschinelle Bohrbetrieb auf Zeche Shamrock. Oesterreichische Zeitschrift für Berg- und Hüttenwesen, 1887, *pp.* 134–136.

Bra1dts' hydraulic drilli1g machi1e has bee1 employed for several years at the Shamrock Colliery for drivi1g a drift about 5,000 feet lo1g. The drift is about 8 feet wide and $6\frac{1}{2}$ feet high, permitti1g the use of two drilli1g machines at the face. Ni1e holes are drilled, those i1 the coal and those in sto1e bei1g fired simultaneously. The water pressure varies from 26 to 47 atmospheres, and is derived from the marls over-lyi1g the Coal-Measures. The water is draw1 from behind the cast iro1 tubbi1g and co1veyed by mea1s of wrought iro1 pipes, 2·4 i1ches diameter and $16\frac{1}{2}$ feet lo1g. The qua1tity of water is sufficie1t to drive the two bori1g machi1es and a turbi1e attached to a ve1tilati1g fan, produci1g about 1,800 to 2,200 cubic feet of air per mi1ute and usi1g about 8 cubic feet of water per mi1ute. This ve1tilator is worked three times daily for about 10 or 15 mi1utes and the total qua1tity of water used for this purpose is from 250 to 350 cubic feet per worki1g day. The drilli1g machi1es require from 750 to 900 cubic feet of water per worki1g day. Three sets of holes, $2\frac{3}{4}$ i1ches diameter and from 4 feet to $4\frac{1}{2}$ feet lo1g, are drilled and fired in each day of 24 hours, the drilli1g of each set of holes occupyi1g about 2 hours. Each drilli1g machi1e uses about 1 cubic foot of water per mi1ute and the two machi1es use about 750 cubic feet i1 the 6 hours they are at work.

The dy1amite fumes are co1de1sed by a spray of water from the pipes and the air re1ewed by ru1i1g the ve1tilator for about 10 mi1utes.

The drift was drive1 for four mo1ths i1 sa1dy shale and sa1dsto1e by ha1d at an average speed of 17 i1ches per day and cost about 18s. 6d. per foot. The average dista1ce driven by the drilli1g machi1e i1 similar strata was about $6\frac{1}{2}$ feet per day at a cost of about 29s. 6d. per foot.

The dista1ce drive1 by ha1d i1 coal was at an average speed of $3\frac{3}{4}$ feet per day at a cost of about 8s. per foot. The average dista1ce drive1 by the drilli1g machi1e i1 the same seam was about $13\frac{1}{2}$ feet per day at a cost of about 15s. 6d. per foot.

M. W. B.

AUSTRALIAN TIN.

Geology of the Vegetable Creek Tin Mining Field. New England District, New South Wales. By T. W. EDGEWORTH DAVID. *Geological Survey of New South Wales, 4to, 169 pp., with folding Maps and Sections and Figures in text. Department of Mines, Sydney,* 1887.

In 1872 tinstone was first discovered by the Messrs. Fearby at Elsmore, near Inverell, and soon after, in the same year, Thomas Carlean found stream tin 34 miles off, near the source of Vegetable Creek. Since that time the district has continued to yield a large supply of tin. The tin-bearing region is more than 800 square miles in area, lies between the 151st and 152nd meridians east of Greenwich, and the 29th and 30th parallels of south latitude, and is more than 90 miles from the coast.

The rocks of the country are as follows:—

Post-Pliocene	...	Clay, sand, and gravel, usually more or less tin-bearing.
Pliocene (?) ...	...	Coarse river gravel overlying basalt.
Eocene, in part	...	3.—Basalt, or " blue metal," capping the tin sands of the " deep leads."
		2.—Laterite, known as " red clinker " or " ashes," sometimes above and sometimes beneath the basalt (No. 3).
		1.—(*b*) Gravel, sand, and clay of the deep leads, mostly capped by basalt or laterite, and partly interstratified with the volcanic rocks.
		(*a*) Quartzite and sinter, known as " grey Billy," occurring irregularly and as concretions, chiefly near the junction of 1 (*b*) with 3.
Age unknown	...	Diorite dykes.
,,	...	Hypersthene-garnet-diorite.
Permian (?) ...	...	2.—Granite (*b*), coarse grained and intrusive, with eurite and quartz in veins, containing tinstone and, in places, wolfram, magnetic and ordinary iron pyrites, arsenical pyrites, bismuth, and molybdenite.
		Granite (*a*) a hornblendic variety.
Age unknown	...	1.—(*c*) Intrusive red quartz-porphyries.
		(*b*) Intrusive white quartz-porphyries.
		(*a*) Intrusive hornblendic quartz-porphyries and breccias. These are tin-bearing and may be connected with the eurites of the granite.
Upper Silurian (or Siluro-Devonian)		Yellowish brown to olive green shales passing into claystones with bands of felspathic quartzite and beds of dark grey pebble conglomerate.
Age unknown	...	Porphyrite.
,,	...	Granite (metamorphic?). No metalliferous deposits have been found in this rock.

The local deposits of tin ore are, therefore, classed as follows, according to their mode of origin:—

DEPOSITS OF TIN ORE.

Alluvial stream-works.	Plutonic veins.
Recent and Pleistocene " shallow leads."	Tertiary " deep leads," mostly capped with lava.

All the known deposits are described in detail, and respecting the veins it is said that 76 are in granite, 8 in quartz-porphyry and eurite, 3 in "porphyroid," and 3 in clay-stone. The average strike of 54 "right running" tin lodes in the granite is 39° 15′ east of north and west of south, the range being from 24° east of north and west of south to 20° north of east and south of west. The average direction of the richest veins is 35° east of north and west of south. Eight veins run exactly 40° east of north and west of south, and 7 exactly 35° east of north and west of south. Of 30 veins observed, 7 coincide with the main joints of the country rock, while 23 do not, but intersect the joints at an angle of about 5°. The average hade of 37 observed lodes is 13°, 33 hade to the north-west, 10 to the south-east, and 3 are vertical.

Out of 77 tinny veins 19 consisted of quartz and cassiterite only, 8 of tinstone and felspar only. The relative frequency of occurrence of the associated minerals is shown by the following numbers:—Quartz in 69 veins, chlorite in 29, felspar in 20, mica in 8, mispickel in 8, iron pyrites in 4, fluor-spar in 4, tourmaline in 3, wolfram in 3, blende in 2, galena in 2, copper pyrites in 2, bismuth in 2, molybdenite in 2, vesuvianite in 2, stilbite in 2, hæmatite in 1, pyrrhotine in 1, manganese in 1, scheelite in 1, beryl in 1.

The greatest length of tin lode known is one mile; the greatest depth worked, 200 feet. The proportion of ore contained in the vein to the veinstone varies from ⅓ per cent. to 70 per cent. The total quantity of tin ore raised up to the end of 1886 was about 1,570 tons.

Particulars as to the dressing and smelting of the ores are given, together with much information relating to the various workings of the district.   G. A. L.

THE IRON ORES OF THE BANAT AND ELBA.

I giacimenti ferriferi del Banato e quelli dell'Elba (a proposito d'una recente pubblicazione di Hj. Sjögren). By B. LOTTI. *Bolletino del R. Comitato Geologico d'Italia.* Ser. 2, Vol. VIII. (1887), pp. 197–202.

The author compares the iron ore deposits of Moravicza and Dognacska, in the Banat, as described by SJÖGREN *(Jahrbuch d. k. k. Geol. Reichsanstalt, Vol XXXVI., Vienna,* 1886), with those of the Isle of Elba as studied by himself and E. Fabri *(Mem. descr. della carta Geologica d'Italia, III.; Rome,* 1887).

The Banat ore is stated to occur at the line of junction between beds of limestone and crystalline schists of Archæan age and is perfectly conformable to both. LOTTI considers the proofs adduced by Sjögren as insufficient, both as regards the pre-Cambrian age of the ore-deposit and as to its contemporaneity with its encasing rocks.   G. A. L.

PETROLEUM IN SOUTH AMERICA.

(1) *Petróleo en la República Argentina.* ANON. *Revista Minera, metalúrgica y de Ingeniería, Año XXXVIII.,* 1887, p. 283.

Quotes from the *Diario* (a Buenos Ayres periodical), of July 21st, a statement to the effect that scientific explorations in the region of the Santa Barbara range had brought about the discovery of a valuable oil-bearing formation, and that boring for petroleum was about to commence under the supervision of Dr. Arnau and others.

(2) *Petróleo en Venezuela. Same publication, p.* 291.

The occurrence of large deposits of petroleum is stated to have been proved near Maracaybo, close to the shores of a lake and in the industrial centre of Venezuela.   G. A. L.

ITALIAN LIGNITES.

Le Ligniti del bacino di Castelnuovo di Garfagnana. By PROF. DE STEFANI.
Bolletino del R. Comitato Geologico d'Italia. Ser. 2, Tol. VIII. (1887),
pp. 212-241, with one Plate of Sections.

The lignites reported on occur in a Tertiary basin of 14,000 *ettari* (34,562 acres) in
extent and comprising portions of the communes of Camporgiano, San Romano.
Molazzana, Gallicano, Pieve Fosciana, Castelnuovo di Garfagnana, Castiglione di Garfagnana, and Villa Collemandina, of the *arrondissement* of Castelnuovo di Garfagnana,
in the Province of Massa. These beds are of Eocene and Pliocene age and lie unconformably upon the edges of Mesozoic rocks. Four workable beds of lignite are
described, capable, according to the author, of yielding 4,000,000 tons, of which
3,200,000 tons, however, would come from one seam. The character of the four seams
will be seen from the following analyses by PROF. G. CAMPANI :—

			Calorific Power.	Moisture Lost at 120°.	Volatile Matter.	Carbon.	Ash.
No. 1, dried at	+120° C.	...	5039·30	14·25	63·0	35·4	1·6
	„ + 22°	...	4304·91				
No. 2.	„ +120°	...	3715·65	20·82	45·4	29·4	25·2
	„ + 22°	...	3055·55				
No. 3,	„ +120°	...	4867·26	16·70	54·4	38·5	7·1
	.. + 22°	...	4181·40				
No. 4,	„ +120°	...	3234·95	12·84	38·2	26·6	35·2

The lignite was formerly sold at Lucca station at 18 *lire* (15s.) the ton, and the
coke at 26 *lire* (£1 1s. 8d.)   G. A. L.

CANADIAN APATITE.

On Canadian Apatite. By E. T. SHUTT. *Proceedings of the Canadian Institute.*
Series 3, Tol. V. (1887), pp. 30-38.

A general account of the character and mode of occurrence of the mineral in the
Dominion but with more particular reference to the deposits in the Templeton district.
The apatite occurs always in connection with limestone bands in the Laurentian series,
which have in consequence become locally known as the " Phosphate-bearing Rocks."
The richest localities mentioned are in the townships of Templeton, Hull, Buckingham
and Wakefield (Ottawa County), Quebec, and North Burgess, Elmsley and adjoining
townships in Ontario. An analysis of a fair average specimen from Templeton is given
as follows :—

Tricalcic phosphate	...	...	...	89·85
Calcic fluoride	...	...	...	7·90
Calcic chloride	...	...	...	0·37
Calcic carbonate ...	...	...	...	0·49
Insoluble residue	...	...	...	0·05

It occurs in crystals in a sub-crystalline massive form and as nodules. The
crystals are usually found in red limestone. and although sometimes filling pockets of
considerable size are less valuable commercially than the massive apatite which occurs
as veins, " stocks," and pockets, and is extensively worked. The annual output is said
to be steadily increasing. In 1875 it was only 3,701 tons, while in 1884 it was nearly
25,000 tons.   G. A. L.

MINERALS OF THE TAMBO VALLEY (AUSTRALIA).

The Physiography of the Tambo Valley. By JAMES STIRLING. *Transactions of the Geological Society of Australasia, Vol. I.,* 1887, *pp.* 37-66, *with Plate of Fossils.*

The River Tambo runs from the great Dividing Range of the Australian Alps to the Gippsland Lakes. The valley is excavated in the following rocks :—

Silurian.—(*a*) Auriferous slates and grits of the Haunted Stream, etc.

(*b*) Altered schists and slates of Ensay, etc. Also granites and diorites of Lower Palæozoic age.

Devonian.—Bindi limestones. Mt. Tambo conglomerates and slates. Limestones of Buchan, etc. Also porphyritic granite, porphyries, diabase, and diorites of Upper Palæozoic age.

Miocene.—Lava flows, basaltic, and dykes.

Pliocene.—Boulder wash, sands and conglomerates.

Pleistocene.—Glacial and alluvial deposits.

Gold has been worked only on the western watershed of the Tambo valley. In quartz reefs at Swift's Creek and Haunted Stream; in the alluvia at both localities; in terraces on Tambo River, below Doctor's Flat and near Bindi, associated in quartz veins with pyrites and oxide of silver.

Silver.—In quartz veins, as above; also with galena in vughs and veins, and with boulangerite.

Copper.—Erubescite, azurite, malachite, chalcocite, and chalcopyrite occur in the gold reefs at Swift's Creek, and in certain greenstone dykes at Tambo River, below Doctor's Flat.

Lead.—Galena at Swift's Creek; cerussite in quartz, with pyrites; argentiferous pyromorphite in small veins among the metamorphic schists.

Tin.—Cassiterite in small quantities in auriferous wash below Doctor's Flat.

Iron.—Menaccanite in alluvial workings, Swift's Creek; micaceous iron ore in small patches in the Swift's Creek metamorphic rocks; hæmatite and limonite in gold reefs, Long Gully; iron pyrites, mispickel, and chalybite in the gold workings.

G. A. L.

THE PRODUCTION OF QUICKSILVER.

(1) *La Produccion y consumo del Azogue.* By J. G. H. *Revista Minera, Metalúrgica y de Ingeniería, Año XXXVIII.,* 1887, *pp.* 131, 132, 139-141.

The annual amount of quicksilver produced since 1877 in the United States and Spain is thus tabulated :—

	United States. Flasks.	Spain. Flasks.	Total Flasks.
1877	... 79,396	... 40,747	.. 120,143
1878	... 63,880	... 41,913	... 105,793
1879	... 73,684	... 45,131	... 118,815
1880	... 59,926	... 45,588	... 105,514
1881	... 60,851	... 46,137	... 106,988
1882	... 52,732	... 46,614	... 99,316
1883	... 46,725	... 47,732	... 94,457
1884	... 31,913	... 44,757	... 76,670
1885	... 32,073	... 47,852	... 79,925
1886	... 29,981	... 51,198	:.. 81,179

The author calls attention to the steadiness of the Spanish output, which, he says, is due to the admirable manner in which the mercurial ores are worked in Spain, as well as to their constant average composition.

(2) *La Campaña de Almaden.* ANON. *Same publication, pp.* 222, 223.

The yearly mining campaign at Almaden lasts from October to May, work being suspended during the hot months for considerations of health. The above statistics are supplemented in the present article by the following account of the monthly output of these mines up to May, 1887 :—

							Flasks of Mercury.
1886.—October	...	...	...	...	...	...	2,568
November	...	...	...	...	...	...	6,566
December	...	...	...	...	...	...	7,627
1887.—January	...	...	...	...	...	...	8,041
February	...	...	...	...	...	...	8,130
March ...	...	...	...	...	...	...	7,816
April ...	...	...	...	...	...	...	7,160
May ...	...	...	...	...	...	...	3,012

(3) *El Azogue en los Estados-Unidos.* ANON. *Same publication, p.* 264.

In this article the conditions existing at the New Almaden mines, in America, are compared with those at the Old Almaden mines, in Spain, greatly to the advantage of the latter. G. A. L.

OZOKERITE IN GALICIA.

Ueber das Vorkommen des Ozokerits oder Erdwachs und begleitende Fossilien in der Sobieski-Grube bei Truskawiec im Kreise Dobrobicz in Ost-Galizien. By PROF. F. ROEMER. *Proceedings of the Natural History Section of the Silesian Society, VII.,* 1885 *(pub.* 1886*), p.* 36.

The Sobieski mineral oil mine at Truskawiec, in Eastern Galicia, is in a grey bituminous clay. In this clay are found irregular masses of ozokerite, sometimes in lumps nearly 100 lbs. in weight. Lenticular masses of argillaceous bituminous limestone also occur in the clay, and contain crystals of native sulphur and arragonite lining cavities within them. Selenite is common in the same deposit. G. A. L.

POETSCH SYSTEM OF SINKING THROUGH AQUIFEROUS STRATA BY FREEZING.

Note sur des Expériences de Congélation des Terrains. By — ALBY. *Annales des Mines, Sér.* 8, *Vol. XI., pp.* 56 *to* 86 ; *Plate II Annales des Ponts et Chaussées, Sér.* 6, *Vol.* 14, *pp.* 338-388 ; *Plate* 35.

INTRODUCTION.

The Poetsch system, which was first applied at the Archibald Pits, near Schneidlingen, has been employed in several more cases, some of which have been unsuccessful.

The most recent application of the system is at the Houssu Collieries, in Belgium. The outside diameter of the pit is 15 feet 5 inches, the clear diameter being 13 feet. The sands were found at a depth of 203 feet, and solid clay at 241 feet. The coal lies at a much lower level. Eighteen tubes are placed in a ring of 8·2 feet radius, and 7½ and 6¾ inches in diameter. Although the bed to be traversed is only 38 feet thick the pipes have been made 69 feet long. The connecting cross tubes are placed at a depth of 177 feet, and the freezing tubes attain a depth of 246 feet. They were driven in by hand labour. The pumps kept the water at a level of 175 feet. The operation has been very troublesome, as the sand, disturbed by the insertion of the tubes, passed into the pumps. Cold is produced by two machines capable of making 1,000 pounds of ice per hour. One machine has been in use since December 5th, 1885, and the other since July 18th, 1886. Before being frozen, about 1,000 cubic feet of water were pumped, of which about 350 cubic feet came from the sand. This quantity has since decreased, but the pit has not yet been sunk.

EXPERIMENTS.

Experiments have been made since October, 1885, in the workshops of Messrs. Rouart Brothers upon the freezing of soil and the study of certain questions relative to the application of the system—the formation of ice around the tubes, and the resistance of the frozen matter.

The apparatus employed consisted of a sheet iron cylinder 55 inches diameter and 118 inches deep, containing at its axis a tube 6¾ inches in diameter, similar to the freezing tubes. The cylinder is fitted with the material to be frozen, and surrounded with a thick covering of straw to isolate it from the atmosphere. A number of holes were made in the sides, which permitted the entry of a bar used to ascertain the formation and the profile of the ice. The refrigerating solution circulated continually in the central tube from the freezing machine.

The experiments have not allowed the determination of the time necessary for the formation of a given thickness of ice; the formation, however, proceeds more slowly as its thickness increases.

The resistances of the frozen matters were tested by crushing them in an hydraulic press, and a few specimens were tested for tensile strength. Tests have been made upon said and water in various proportions and at various degrees of temperature, and upon cubes of pure water.

The practical results of these experiments are:—(a) The rapid increase of the resistance of the said, etc., when the temperature is lowered. (b) The resistance appears to vary approximately with the volume of water contained in the said.

In the tensile experiments the proportion of water was more important. Thus, said saturated with water at a temperature of −10 degrees Cent. could withstand a strain of 550 pounds per square inch, and 425 pounds could be relied upon if the water was not in excess. When two-thirds saturated the resistance was about 350 pounds, and with one-third it was almost *nil*.   M. W. B.

———

STAUSS' KEPS.

Note sur l'Application des Taquets à abaissement du Système Stauss au siége No. 5
des Charbonnages de Bascoup. By A. DEMEURE. Annuaire de l'Association
des Ingénieurs sortis de l'École de Liége, 1887, Vol. VI., pp. 116 to 126.
Plate 6.

In Stauss' system the keps are in the shape of small bolts, which are hinged, and permit the upward movement of the cage (in the same manner as the ordinary kep), it falls after the passage of the cage, and is ready to sustain it. The keps are supported and slide upon a steel girder, whose upper surface has an angle of 9 degrees. They are suspended at the other end from a short vibrating lever, and are actuated by means of a hand lever through a short connecting rod. The working of the levers is assisted by the weight of the cage, which causes the keps to slide upon the girders and throws back the short vibrating lever.

It is found in practice that the cage alone, whatever may be its weight, cannot throw back the keps until the banksman draws back the hand lever a little, when the weight of the cage completes the operation.

The keps generally employed require the cage to be lifted before they are disengaged, and in the case of a cage with four decks the engine has to be reversed nine times for each journey of the cage. With Stauss' keps the cage is drawn up until the lower deck can be lowered upon the keps, which are withdrawn, and the cage lowered until the remaining decks have been changed ; the engine has consequently only to be reversed once and moved five times always in the same direction.

The alleged advantages of the system are (a) Economy of time of from three to six seconds in changing each deck of the cage. With two-decked cages, drawing from a depth of 800 feet, 78 cages were drawn per hour with the ordinary keps, whilst with Stauss' keps from 82 to 88 cages are drawn per hour. (b) Economy of steam, owing to the non-reversal of the engine, and saving of as many strokes of the piston. (c) Decreased wear and tear of the ropes. (d) Decreased wear and tear of the valves and moving parts of the engines. (e) In new installations smaller engines might be employed.   M. W. B.

STUDIES ON THE STRATIFICATION OF THE ANTHRACITE MEASURES OF PENNSYLVANIA.

By HENRY A. WASMUTH. *Journal of the Franklin Institute, Third Series, Vol. XCIV., 1887, pp. 109-125. Four Plates in text.*

The State Geological Survey of Pennsylvania advocates the theory that the anthracite measures have been folded into numerous *inversions*, without fracture of the strata. The writer maintains that in bedded mineral deposits no *inversion* or overlapping of the strata can take place without fracture, and more or less dislocation and that, in general. the dislocations of the strata take place in one of two ways. Either the portion of a mineral deposit on the hanging wall of the fracture is in a lower position than the portion on the foot wall, or it is in a higher position, and overlaps. The first class are called *transverse faults*, and the second class are called *longitudinal faults*, or overlaps.

The latter class of dislocations are possible only when the strike of the fault ranges with the strike of the axis and strata, and the inclination of the fault, greater than the dip of the strata, is always of the course of the dip of the strata. They usually occur in the vicinity of anticlinals or synclinals, and have the same origin. The downward bending of the strata on the hanging wall may be produced by the enormous weight of the overhanging strata, and the prodigious oblique force of the sliding strata on the foot wall of the fault, while the up-bending on the foot wall of the fault may be explained by a certain resistance of the underlying strata, which necessarily must have produced an oblique movement of the strata in the direction of its strike with similar results.

The first class of dislocations occur chiefly with a strike of the fault. transverse to the strike of the axis and strata. They often cross anticlinal and synclinals, and, therefore, they may be considered to be of later origin than the anticlinals and synclinals crossed by them.   M. W. B.

MAGNESIUM CARBONATE AS A NON-CONDUCTOR OF HEAT.

By E. Luttgen. *Transactions of American Institute of Mining Engineers, Vol. XV.*, 1886–87, *pp.* 614–625.

Description of Covering upon 6 Feet Length of 2 Inch Pipe.	Diameter of Covering in Inches.	Weight per Foot in Ozs. Avoirdupois.	Condensed Steam, in Grammes, per Foot per Hour.	Loss by Condensation. Kilogramme Centigrade Heat Units per Foot per Hour.	Pound Fahrenheit Heat Units per Foot per Hour.
Hair felt, wrapped with twine, and ordinary bur lap jacket	4½	12½	34·66	17·39	69·02
Sectional carbonate of magnesia, asbestos paper jacket, and iron bands	4½	20¼	37·83	18·97	75·29
Sectional carbonate of magnesia, canvas jacket, and iron bands ...	4½	20¼	38·00	19·07	75·68
Sectional mineral wool, asbestos paper, mineral wool, and muslin	5¼	28¾	38·50	19·32	76·68
Chalmer-Spence's asbestos hair felt and paper	4½	28¼	41·66	20·90	82·95
Shield's & Brown's asbestos paper and sheathing paper	4½	27¼	42·50	21·33	84·65
Reed's asbestos paper and felt paper...	4¼	26¾	45·00	22·58	89·62
Fossil meal, diatomaceous earth, and organic fibre	3¾	24	57·50	28·86	114·54
Ainsworth's paper pulp, clay, and hair	4½	66	96·60	48·48	192·14

M. W. B.

CAGE CATCHES.

Schachtfallen (System Sartorius und Holzer). Jul. Sprenger. *Berg- und Huetten-mœnnische Zeitung. Vol. XLVI., Nos.* 51 *and* 52, *pp.* 477–479 *and* 487–489. *One Plate.*

By old-fashioned methods a cage, on arrival at bank, passes two or more catches just below the level of the flat sheets, and is lowered again and arrested by them. The rope then slackens, and when the cage is ready to descend again it has to be lifted with a jerk, giving more or less of a jar to the whole structure. To obviate this, two mechanical methods are in use. One of these, at the Heinitz Dechen Colliery in Neunkirchen, consists of catches, two to each end of each cage, working on horizontal spindles, and with just play enough to allow the cages to lift them and slide past. Between the cages on each side a ring is keyed on to the spindle, with a tongue pointing inwards and rounded on its under side, so that a wedge, placed between it and a beam just below, will either keep it firmly in its position, or allow the shaft to turn through any desired angle. There are thus two wedges opposite each other, and these are connected by single pins with a rod hinged at its middle. A vertical rod, working by an arrangement of levers below and by a bell-crank above, raises the centre hinge above mentioned, and the halves of the horizontal rod being then made to form an angle with each other, the wedges are drawn towards the centre, allowing the tongues to descend. A horizontal rod attached to the vertical arm of the bell-crank above, and supplied with a handle and catch, provides means of regulating the travel of the wedges and of lowering the cage at will. To bring the arrangement back into position, and also to let the cage away gently, each spindle is provided with a balance weight, fixed on a short lever on the side of the spindle away from the catches.

The other method described differs in principle from the foregoing, only, in so far, that springs are introduced between the catches and the wedged tongues, rendering the support of the cage more elastic. In place of the tongue in the former arrangement, a short arm on the side of the horizontal spindle away from the catches is connected by a vertical spiral spring to a similar arm, a foot or so lower, working on another short spindle, and having a tongue and adjustable wedge as before. Thus, even when the wedges are tightly home, the spiral spring allows the catches to give sufficiently to prevent any jar when the cage is set down. With either arrangement, but more especially with the latter, the pulling of a handle allows the cage to descend gently till the rope becomes taut, and the wear and tear due to jerking and jarring are reduced to a minimum. The whole contrivance is simple, and its working parts are such as may be easily replaced. A. R. L.

HYDRAULIC BRAKES FOR WINDING ENGINES.

Hydraulische Bremsvorrichtung für Fördermaschinen. JULIUS SPRENGER. *Berg-und Huettenmœnnische Zeitung. Vol. XLVII., Nos.* 1 *and* 2, *pp.* 1-3 *and* 13-14. *One Plate.*

With the increasing depths of mines and consequent greater size of winding machinery, the power of efficiently controlling the engine gains in importance every day. The German police regulations require every winding engine to be fitted with a steam brake, which will bring it to a standstill at once, though going at full speed. This, however, must be applied by the brakesman, and its success depends on his judgment and nerve. To make the brake self-acting, a hydraulic arrangement has been designed by which the cage, when lifted too high, is made to apply it itself.

About half-way between the winding drums and the shaft, a pump with reservoir and accumulator supplies pressure to a length of piping leading to the shaft, and farther to a return length of piping leading back to the steam brakes on the farther side of the drums. A valve at the shaft permits the pressure from the accumulator to be carried farther only when it is raised by a lever, which, in case of accident, the cage itself will actuate. The pressure thus communicated to the return length of piping, acts on a vertical cylinder working on the accumulator principle, and, by a vertical rod and system of levers, may either act on the piston rod of the steam brake direct, or, by actuating the slide valves, admit the steam to the steam cylinder in the usual manner.

For lifting the valve at the shaft a bar projects over sufficiently to be caught and lifted by the cage if the latter should rise too high. The accumulator pump, which acts on the pipe leading to the shaft, is also connected by a branch with the return pipe, and a 3-way cock, placed at its junction with this and with the branch to the reservoir, allows the latter to be shut off and the water to be pumped from the return pipe, thus taking off the pressure and relieving the brake after it has done its work. The vertical rod to the valve at the shaft is fitted with a hand wheel, working a screw in the casting at the flat sheets, enabling the valve to be screwed down fast if desired, and, in addition, with a screw coupling which, in case of accident, can be disconnected altogether. Also, a rod from above allows the valve to be lifted by hand when it is desired to try the working of the apparatus; this, under ordinary circumstances, being kept under lock.

The hydraulic piston near the steam brake works in a cylinder, and when raised will apply the brake; but it is necessary for the steam brake to be workable when the hydraulic gear is not in use. With this view the upright rod, instead of being fixed at its lower end, is made to work loosely in the piston, so that it may move up and down when the latter is at rest.

The newer hydraulic arrangements are mostly made to work on the steam piston rod direct, and not on the slide valves. The inventor is a Herr W. Gerhardt, and the makers of the apparatus are Messrs. Dingler. Karcher, & Co., of St. Johann on the Saar.   A. R. L.

AN UNDERGROUND VENTILATING FAN IN WESTPHALIA.

Anlage eines unter irdischen Ventilators auf der Zeche Shamrock bei Herne in West-falen. By L. GRAFF. *Zeitschrift für das Berg-, Hütten- und Salinen-Wesen, Vol. XXXIV.*, 1886, *Abhandlungen, pp.* 234–240. *Plates XIII. and XIV.*

The fan is placed at a depth of 640 feet, in a chamber 49 feet long, 16½ feet broad, and from 14½ feet to 24½ feet high, and adjacent to both the downcast and upcast shafts. The Geisler fan is 11½ feet diameter and is a modification of the Rittinger. It is encased in a spiral casing of brickwork, with a coating of cement, and the delivery is made by means of a curved drift into the upcast pit at a height of about 46 feet above the fan. There are two high pressure engines, each of 13 inches diameter, 23·6 inches stroke, and the fan is driven by four 2-inch hemp ropes, from a fly-wheel 14·8 feet diameter to a pulley 4·9 feet diameter upon the shaft of the fan.

The following observations have been made upon the fan :—

TABLE I.

No. of Experiment.	Volume of Air per Minute. Q.	No. of Revolutions of Engine per Minute.	Barometer at Bottom of Downcast.	Temperature. Downcast.	Temperature. Returns.	Water-gauge. h.	Temperament. $\frac{Q}{\sqrt{h}}$
	Cubic Feet.		Inches.	Deg. F.	Deg. F.	Inches.	
1.	79,860	35	29·92	46·7	70·2	·831	8·75
2.	91,140	40	29·92	46·7	70·2	1·088	8·75
3.	100,610	45·4	29·89	47·8	70·2	1·401	8·47
4.	105,270	49·6	29·88	48·0	70·2	1·685	8·10
5.	109,020	50	29·84	48·2	70·2	1·771	8·18
6.	124,000	55·6	29·92	46·0	69·8	2·323	8·12
7.	128,070	59·6	29·92	46·4	69·8	2·586	7·96
8.	138,170	65·2	29·92	46·4	69·8	2·964	8·04
9.	142,670	66·8	29·92	46·4	69·8	3·098	8·06

The volume of air in thousands of cubic feet per minute divided by the water-gauge in hundredths of an inch.

TABLE II.

No. of Experiment.	Volume of Air per Minute.	Mean Pressure on Piston.	No. of Revolutions of Engine per Minute.	Indicated Horse-Power.	Horse-Power in the Air.
	Cubic Feet.	Lbs. per Sq Inch.			
1.	95,270	40·34	54	65·36	53·05
2.	86,870	37·58	53	48·90	39·60
3.	129,750	60·64	66	120·00	99·00

M. W. B.

HELLHOFFITE AND DYNAMITE.

*Vergleichungs-Resultate von Sprengversuchen mit Hellhoffite und Dynamit im Sig-
mundschachter Grubenfelde zu Schemnitz. By A. WIESNER. Oesterreichische
Zeitschrift für Berg- und Hütten-wesen, 1887, Vol. XXXV., pp. 53 to 59.*

Comparative trials of hellhoffite and dynamite have been made recently at the
Sigmund Mine, near Schemnitz, in Hungary.

In the same gallery of the mine 10·4 pounds of hellhoffite broke down 22·14 tons of
rock, and 11·2 pounds of dynamite broke down 14·97 tons of rock.

In another place 2·2 pounds of hellhoffite broke down 4·71 tons, and the same
weight of dynamite only produced 2·97 tons of rock.   M. W. B.

THE EMPLOYMENT OF THE PIELER LAMP FOR THE DETECTION
OF FIRE-DAMP.

*By JOH. MAYER. Oesterreichische Zeitschrift für Berg- und Hütten-wesen, 1887,
Vol. XXXV., pp. 111–112.*

The new regulations of the Minister of Mines for the collieries of Moravia and
Silesia enact that the Pieler or an equally sensitive safety-lamp shall be used by the
officials in examining for the presence of fire-damp in mines.

It is possible to detect ·25 per cent. of gas with the Pieler lamp. The writer, however, points out that the Pieler is not a perfect safety-lamp, and states that at various
coal-mines in Polish-Ostrau, when the Pieler lamp was placed in an explosive mixture,
the ignition of the gas inside the lamp was communicated through the gauze and fired
the gas on the outside of the lamp.

The Pieler lamp is really a Davy lamp with a single gauze of abnormal dimensions,
and the writer proposes to construct it with two gauzes. This construction will, however, diminish the value of the lamp as an indicator of the presence of gas, as the
detection of ·25 per cent. of gas is a somewhat difficult matter even when one gauze
only is employed.

It may be advisable, therefore, to adhere to the use of the single-gauze Pieler
lamp. A Mueseler, Wolf, or other safety-lamp should, however, be used first in the
examination of the workings, and then, if no fire-damp is detected, the Pieler lamp
may be used in safety.   M. W. B.

SEPARATION OF COAL-DUST FROM SMALL COAL BY AN AIR BLAST.

*Abscheiden des Kohlenstaubes durch Wind. Zeitschrift für das Berg- Hütten- und
Salinen-wesen im Preussischen Staate, 1887, Vol. XXXV., p. 264. Plate 16.*

A coal-washing plant has been erected at the Zollverein Colliery in which the coal-
dust up to a size of ·12 inch is separated from the coarser coal by means of a fan
blast. The washing of this fine coal had been done very roughly and the fine coal
formed a mud with the water which checked the making of coke. It was found
desirable to separate the dust by some dry process, and a current of air is now employed. A 50-inch Pelzer fan is used, and produces a blast which operates on the coal
falling in a thin stream from the shoot. The coal not exceeding a size of ·28 inch is
separated into two classes, one from ·16 to ·28 inch, and the other from the finest dust
to ·12 inch.

M. W. B.

g

LAUER'S FRICTION-MATCH SYSTEM OF IGNITING SHOTS IN MINES.

By J. MAYER. *Oesterreichische Zeitschrift für Berg- und Hütten-wesen,* 1887, *Vol. XXXV., pp.* 129–131, *and one Figure in text.*

The friction match consists of a small metallic tube, with a suitable explosive material and friction wire which is bent and indented at the inner end. The crook of the wire is placed against the lower edge of the tube and causes a certain resistance, so that a feeble accidental pull of the wire during the preparation of the shot would not ignite the friction match.

A wooden bobbin is placed upon the friction wire above the explosive material in the small tube. The detonator and cap, whose ignition should affect that of the charge of dynamite, are placed under the explosive material. The lower end of the match (below the cap) is then closed with some plastic material. A match case of the same length as the shot-hole, composed of wood or paper, and in which the friction wire is enclosed, is placed around the tube and the cap. The wire in the form of a loop projects from the other end of the case, and a thin cord is attached for the purpose of making the ignition in safety at a distance. A bobbin is placed at the outer end of the case of the match (projecting out of the bore-hole) protecting the match and permitting also a moderately sloping pull of the wire, without which the case itself would be torn open or broken off. M. W. B.

COMPARATIVE VALUES OF GERMAN COALS.

Zusammenstellung der vergleichenden Versuche über die Heizkraft und andere in technischer Beziehung Wichtige Eigenschaften verschiedener Steinkohlensorten. Zeitschrift für das Berg- Hütten- und Salinen-wesen im Preussischen Staate, 1887, *Vol. XXXV., pp.* 169 *to* 190.

Experiments were made at the Government dockyards at Wilhelmshaven during the years 1874 to 1886 upon the evaporative powers of various German, and a few English, Japanese, Australian, and American coals, together with a few samples of compressed fuel made in Germany and Cardiff.

Some of the best and lowest results are given in the following Table:—

Description of Coal.	Evaporative Power.	Remarks.
I.—GERMAN COALS.		
A.—*Westphalian Coals.*		
1.—Gas Coals—		
102. Jacob Colliery ...	8·065	The gas coals are hard and burn easily with a
39. Shamrock Colliery ...	8·554	long flame, and produce large volumes of
164. Rhein-Elbe Colliery...	6·652	smoke.
2.—Coking Coals—		The coking coals burn readily with a long flame
1. Victor Colliery ...	9·277	and do not produce much smoke. Their eva-
135. Friederike Colliery ...	7·390	porative power is high, and they cake upon the grate.
3.—Steam Coals—		The steam coals cake less readily than the cok-
48. Nachtigall Colliery...	8·517	ing coals, and burn with a clear flame. The
158. Alstaden Colliery ...	7·029	anthracite coals-are difficult to burn, and give a short flame with intense heat.
4.—Mixtures—		Mixtures of coking and steam coals generally
½ Nachtigall... ½ Shamrock ...	8·622	give a greater evaporative power than the calculated mean result.

Description of Coal.	Evaporative Power.	Remarks
B.—*Wurm Coals.*		
Bitumious Coals—		
35. Aachei	8·572	
Steam Coals—		The steam coals buri very slowly, and are
33. Aachei	8·582	aithracitic.
C.—*Upper Silesian Coals.*		The Upper Silesiai coals buri very rapidly with
117. Veroiika Seam, iear		loig flames, and produce steam rapidly, but
Morgenroth ...	7·779	with more smoke and have a lower evapora-
166. Wildensteinsegen Col-		tive power thai the Westphaliai cokiig coals.
liery	6·341	Geierally they are similar to Westphaliai
		gas coals or to Newcastle coal.
D.—*Lower Silesian Coals.*		The Lower Silesiai coals resemble ii part the
8. Glückhilf Colliery ...	8·539	Westphalian gas and ii part the cokiig coals.
134. Weissteiner Colliery	7·438	They buri with a loig flame and produce a
		deiser smoke. Some of these coals will cake,
		others will not.

II.—FOREIGN COALS.

A.—*English Coals.*		
37. Thomas' Merthyr ...	8·569	
96. Nant Melyne Merthyr	8·170	
159. Scotch	6·945	
160. Lochgelly	6·888	
B.—*Japanese Coals.*		The Japaiese coals buri very rapidly and have,
165. Japaiese Coals ...	6·592	ievertheless, oily a low evaporative power
		and produce much smoke.
C.—*Australian Coals.*		
163. Gullaid Colliery, iear		
Brisbaie	6·764	
140. Wallseid, iear New-		
castle	7·326	
D.—*New Zealand Coals.*		
149. Bay of Islaids ...	7·163	
E.—*American Coals.*		
167. Puita Areias ...	4·474	

III.—COMPRESSED FUEL.

2. Dahlhauser	9·024	A cubic foot of the compressed fuel weighed
107. Rheii-Elbe and Alma	7·950	from 62 to 72 pouids.
137. Cardiff	7·366	

M. W. B.

THE ORIGIN OF COAL.

Études sur le terrain houiller de Commentry. Livre Premier. Lithologie et Strati-graphie. By HENRI FAYOL. *Bulletin de la Société de l'Industrie Minérale, Sér. 2, Vol. XV., pp. 1–543, with large folio Atlas of twenty-five Plates.*

This is the first part of an extremely detailed moiograph of the coal-field of Commentry, ii Ceitral Fraice. It coisists, however, largely (besides matter of more strictly local iiterest) of a very full discussioi in all its beariigs of the cuestioi of the origii and mode of formatioi of coal and the rocks associated with it, not oily ii this basii but elsewhere. The extraordiiary iumber of facts brought together by the author reiders a satisfactory abstract of the volume impossible. The geieral conclusio1 arrived at is that the whole of the Coal-Measures, i1cludi1g the coal-seams, are deposits formed by rivers at their mouths i1 lakes or i1 the sea. It is poi1ted out that estuari1e or deltaic formatio1s are of coarser eleme1ts aud greater irregularity of beddi1g in lakes tha1 i1 the sea (*i.e.* i1 tra1quil tha1 i1 disturbed waters). The con-glomerates, grits, shales, and coals of the Ce1tral Fra1ce basi1s are regarded as types of the former, those i1 the North of Fra1ce coal-fields of the latter. M. Fayol has, by mea1s of lo1g-co1ti1ued experime1ts on a large scale, bee1 able to reproduce at will both modes of sedime1tatio1, together with all the stratigraphical peculiarities which characterise them. Every poi1t usually relied on as evide1ce that most coal-seams have bee1 formed i1 place is exami1ed in tur1 and dismissed as worthless, and every argume1t i1 favour of the drift origi1 of coal is upheld by a vast array of corroborative facts. Eve1 i1 the coal-fields of Belgium aud Northern Fra1ce the author states that he has not see1 a si1gle u1derclay havi1g at all the appeara1ce of an a1cie1t vegetable soil. The succeedi1g parts of this great monograph are to be writte1 by Messrs. Re1ault, Zei̇ller, Sauvage, Ch. Brongniart, Sta1. Meu1ier, and De Lau1ay.

G. A. L.

MINERALS OF INDIA.

A Manual of the Geology of India. Part IV. Mineralogy (mainly non-economic). *By* F. R. MALLET. *Published. by the Indian Government, Calcutta, Geological* *Survey Office,* 1887, 179 *pp. Four Plates of Crystals.*

A complete list of all the mi1erals at prese1t k1ow1 i1 I1dia. The work is supple-me1tary to Mr. V. Ball's "Eco1omic Geology of I1dia," and some of the mi1erals dealt with fully i1 that portio1 of the "Ma1ual" are merely me1tio1ed i1 the prese1t part. On the other ha1d, much i1formatio1 is give1 respecti1g the mode of occurre1ce of ma1y mi1erals of mi1or commercial value. The followi1g are i1cluded i1 the list:—Gold, plati1um, plati1iridium (?), iridosmi1e, mercury (?), copper, lead (?), sulphur, diamo1d, graphite, realgar, orpime1t, stib1ite, bismuthite, molybde1ite, O'Rileyite (an ore of copper, iro1, and arse1ic), gale1a, bor1ite, jaipurite (= sycpoorite, a sulphide of cobalt), ble1de, ci11abar, sulphide of lead and copper, pyrite, chalco-pyrite, cobaltite, marcasite, leucopyrite, arse1opyrite, bour1o1ite, tetrahedrite, frei-bergite, sylvite, sal ammo1iac, chlorocalcite. atacamite, fluorite, cuprite, melaco1ite, coru1dum, hematite, ilme1ite, spi1el. mag1etite, chromite, chrysoberyl, cassiterite, rutile, brau1ite, mi1ium, pyrolusite, turgite, ma1ga1ite, limo1ite, beauxite, psilomela1e, wad, vale1ti1ite (?), kermesite, cerva1tite, quartz with 1umerous varieties, opal, e1statite, hypersthe1e, wollastonite, pyroxe1e, rhodo1ite, amphibole, beryl, chrysolite, gar1et, zirco1, idocrase, epidote, axi1ite, phlogopite. biotite, muscovite, lepidolite, lapis lazuli (?), i1dia1ite, labradorite, oligoclase, albite, orthoclase, microcli1e, cho1drodite, tourmali1e, a1dalusite, fibrolite, topaz, sphe1e, tscheffkiuite, staurolite, bombite, dys-clasite, laumo1tite, chrysocolla, preh1ite, apophyllite, allophane, thomso1ite, 1atrolite, poohnalite, mesolite, a1alcime, chabasite, hypostilbite, stilbite, epistilbite, heula1dite, talc, glauco1ite, serpe1ti1e, pholerite, kaoli1, halloyite, margarodite, euphyllite, chlorite, apatite, pyromorphite, mimetite, vivia1ite, libethe1ite, chalcophyllite, lazulite, turquoise, torber1ite, 1itre, soda-1itre, 1itrocalcite, borax, wolfram, wulfe1ite, the1ardite, barite, celestite, a1hydrite, a1glesite, glauberite, sulphate of mag1esium and potassium (?), mirabilite, gypsum, kieserite, bloedite, epsomite, mela1terite, goslarite, chalca1thite, alu1oge1, kali1ite, calcite, dolomite, mag1esite, siderite, smithso1ite, witherite (?), cerussite, carbo1ates of sodium, malachite, azurite, petroleum, 1ative paraffi1, fossil wax (?), amber, fossil resi1, hirci1e, coal.

The list is critical, insufficiently co1firmed occurre1ces bei1g omitted.

G. A. L.

INDIAN COAL-FIELDS.

The Southern Coal-fields of the Sátpura Gondwána Basin. By E. A. JONES. *Memoirs of the Geological Survey of India, Vol. XXIV.* (1887), *pp.* 1–58, *with two Geological Maps.*

The region described includes a narrow outcrop of coal-bearing rocks belonging to the Barakar group of the Gondwana series, and extending generally from east to west for about 75 miles in the Chhindwara and Betul districts. The coal-fields recognised (from west to east) are those of Shahpur, Dolari, Tawa, Kanhan, Hingladevi, Gajundoh, Barkoi, Harai, and Sirgora. Particulars respecting the principal coal-seams observed may be summarised as follows:—

Sirgora Field.—One seam 4 feet 9 inches or more, and another 3 feet thick.

Barkoi Field.—Twelve coal crops noticed, including one seam 15 feet 6 inches thick (with 12 feet 3 inches of coal). one 15 feet 2 inches (of which 7 feet 3 inches is coal), and the rest above 3 feet in thickness.

Harai Field.—The coal reported from this region is probably shale.

Hingladevi Field.—One seam said to be more than 5 feet thick.

Kanhan Field.—Five crops, 9 feet 8 inches, 8 feet, 3 feet, 3 feet, and 1 foot 4 inches thick respectively.

Tawa Field.—One seam 14 feet thick (much disturbed), one above 9 feet 4 inches (in a very inaccessible locality), and several small seams.

The following analyses are given :—

Localities.	Volatile Matter		Fixed Carbon.	Ash.	Remarks
	Mois-ture.	Exclusive of Moisture.			
1. Chinda	16		61	23	———.
2. Barkoi	26		50·3	23·7	———
3. Bhutaria	26·5		49·3	24·2	———
4. Sirgora	28		61·6	10·4	———
5. Takea River, near Datla...	3·24	19·28	29·10	48·2	Does not cake.
6. Between Datla and Badea	5·34	28·36	48·58	17·72	,, ,,
7. Punnara	2·16	18·92	37·74	41·18	,, ,.
8. Tannia River	2·10	26·38	54·34	17·18	Cakes, but not strongly.
9. Sanni-Patakkra	4·00	26·02	49·46	20·52	Does not cake.

These analyses being from coal at the outcrop, with the exception of No. 4, are unsatisfactory. Mr. Blanford in a previous report recommended borings at Sirgora, Bhutaria, Barkoi, Chinda, and Porasia.

Good iron ores are absent from this district.   G. A. L.

HELLHOFFITE.

Hellhoffit. By ALOIS PFEFFER. *Oesterreichische Zeitschrift für das Berg- und Hütten-wesen,* 1887, *Vol. XXXV., pp.* 300–302.

Hellhoffite consists of a solution of the nitro derivatives of certain hydro-carbons (nitro or binitro-benzol) dissolved in concentrated nitric acid. The explosive is rapidly prepared by simply mixing the components together. It is a thin ruby red liquid with a specific gravity of 1·4, and retains the corrosive action of the nitric acid. It is a very unstable body and is said to become solid at a temperature of − 58 degs. Fah. It is not easily ignited by flame, and burns with a considerable production of light. It burns like resin when poured upon a fire. It is decomposed by water or on heating. It is exploded with violence by detonators, but is practically inert to friction or blows.

It is used in two forms, either as a liquid or absorbed in kieselguhr.

It is difficult to handle owing to its corrosive action, which destroys ordinary forms of cartridges.

It would appear from the experiments made at Przibram, Schemnitz, Neunkirchen, and Trauze that hellhoffite is more powerful than dynamite. It is completely burnt, and the gaseous products of combustion are less disagreeable than those of dynamite. It does not freeze, and its price, exclusive of the cost of cartridges, is less than dynamite. Amongst the disadvantages are: its liquid form, unstability, corrosive action, etc.

M. W. B.

HELLHOFFITE AND COAL-DUST.

Verhalten des Rossitzer Kohlenstaubes bei Sprengungen mit Hellhoffit. By RUDOLPH SCHNEIDER. *Oesterreichische Zeitschrift für das Berg- und Hütten-wesen,* 1887, *Vol. XXXV., pp.* 243–244.

The satisfactory results of the experiments with hellhoffite in relation to coal-dust and fire-damp at Neunkirchen and Bruckenberg led to experiments being made to ascertain its behaviour in connection with the coal-dust of Rossitz Colliery.

These experiments were made in April, 1887, in an abandoned drift 650 feet long, of elliptic section, walled with stone and with a section of 32·3 square feet. The level was very damp, which rendered it somewhat unsuitable for the experiments.

The shots were fired at a distance of about 164 feet from the mouth of the drift. Coal-dust was strewed upon a horizontal board suspended by wire slings and the hellhoffite placed upon it; wooden frames covered with coal-dust were placed in the axis of the drift, about 3 feet apart and 6½ feet before and behind the board.

The results of the experiments were as follows:—

I.—A cartridge consisting of ·22 pound of hellhoffite, together with 1·10 pounds of coal-dust, were placed in a bladder distended by an explosive mixture of 10 per cent. of fire-damp, about 2·20 pounds of coal-dust were strewed upon the horizontal board, upon which the bladder was also placed. On the cartridge being exploded there was a pronounced movement of the air in the drift. A heavy piece of wood covering the access to the drift was raised about 4 inches. No appearance of coking was observed, and it was not certain that the coal-dust had taken any part in the explosion.

II.—·22 pound of hellhoffite and 33 pounds of coal-dust were employed; no fire-damp was used. The explosion produced a heavy shock and forward and backward motion of the air, and the wooden covering at the drift mouth was raised from 16 to 20 inches. A large quantity of the coal-dust was coked. More than a yard of the rubber covering of the wire used for igniting the explosive was burnt. The after-damp was very dense.

III.—In this experiment, ·22 pound of hellhoffite, 33 pounds of coal-dust, and lucifer matches were suspended on a wire over about 200 feet in length of the drift. The first cartridge did not explode, but a second one was fired successfully. The explosion produced a heavy shock and forward and backward motion of the air, and the wooden covering was raised about 3 feet and broken, a dense black cloud of dust was raised, and the after-damp issued from the mouth of the drift. Several of the dust frames were broken and considerable quantities of coke had been formed. The lucifer matches were burnt for a distance of 102 feet in front of the point of ignition and 52 feet behind it, or a total distance of 154 feet along the drift. For three-fourths of this distance they were completely burnt.

IV.—The arrangements were the same as in the last experiment. The matches were burnt for a distance of 92 feet in front of the firing point and of 39 feet behind it, or a total distance traversed by flame of 131 feet.

These results are somewhat remarkable, and it is probable that in a drier and more suitable drift the explosions would have been more violent.　　　　M. W. B.

A NEW THEORY OF CENTRIFUGAL FANS.

Essai d'une Théorie des Ventilateurs à force centrifuge. By J. HENROTTE. *Revue Universelle des Mines, etc.,* 1887, *Vol. XXII., pp.* 99 *to* 120. *Plates V. and VI.*

Numerous theories have been published upon the question of centrifugal fans, the most noteworthy being those of Messrs. Kraft and Ser, and more especially that of Mr. Daniel Murgue.

The writer proposes to amend the theory respecting centrifugal fans, avoiding the faults of his predecessors, and relying to a great extent upon the principal facts upon which Mr. Murgue's theory is founded. He divides his inquiry under the following heads:—

I.—Equations of the motion of the air in the mine and in the fan.

II.—Characteristic relations of a centrifugal fan. He determines that—

$$Q = w R \sqrt{\frac{\delta}{g\,[(1 + a\,)\,c + \beta]}},$$

in which Q = volume of air per second; w = angular velocity of rotation; R = diameter of fan; δ = weight of the unit of volume of air; a = a coefficient of the resistance of the mine; β = coefficient of the resistance of the fan; and $c = \dfrac{h}{Q^2}$ in which h = the water gauge.

III.—Practical determination of the coefficients a and β. For Guibal fans of ordinary dimensions the value of these coefficients may be taken—

$$a = \cdot6;\ \text{and}\ \beta = \cdot05\cdot$$

IV.—Useful effect of the fan—

$$E_u = \frac{c}{c\,(1 + a\,) + \beta},$$

in the case of a very large fan, β = 0, and the useful effect will be at its maximum—

$$E_u = \frac{1}{1 + a}.$$

V.—Useful effect of the indicated horse-power. If F be the power lost by the friction of the engine, which may be assumed to be proportional to the revolutions of the engine, then—

$$F = w\,t_f.$$

The value of t_f may be determined if, in a single experiment, the indicated horse-power T_1 has been ascertained, because—

$$T_1 = (1 + a\,)\,c\,Q^3 + \beta\,Q^3 + w\,t_f.$$

The useful effect of the indicated horse-power will be—

$$\epsilon = \frac{c}{(1 + a\,)\,c + \beta + \dfrac{t_f}{Q_2 R}\sqrt{\dfrac{g\,[(1 + a\,)\,c + \beta]}{\delta}}}.$$

VI.—Calculations of the dimensions of a fan. The required data are Q and c. The radius r of the inner ends of the blades is usually assumed *a priori*, and the width of the fan should be such as will not strangle the air in its passage through the fan.

The radius R may be ascertained from the formula—

$$Q = w R \sqrt{\frac{\delta}{g\,[(1 + a)\,c + \beta]}}.$$

for any given value of w.

In the case of small fans running at a high speed, another and more complex formula must be used in calculating the value of R.

VII.—Calculating the indicated horse-power of the engine required. The power required to drive the fan in indicated horse-power is given by the following equation :—

$$75 \text{ X} = c \,(1 + u)\, Q^3 + \beta\, Q^3 + w \, t_f.$$

If $t_f = 10$, $u = \cdot 6$, and $\beta = \cdot 05$, then, in the case of an ordinary Guibal fan—

$$\text{X} = \frac{Q^3 \,[1\cdot6\, c + \cdot05] + 10\, w}{75}.$$

Example.—If 20 cubic metres (706·3 cubic feet) of air per second are to be produced under a depression of 200 millimetres (7·874 inches) by a fan running 60 revolutions per minute, the engine must indicate—

$$\frac{20^3 \,(1\cdot6 \times \cdot5 + \cdot05) + 64}{75} = 92 \text{ horse-power.}$$

For the same volume under a depression of ·787 inch, only 15 indicated horse-power would be required.   M. W. B.

MINING PRODUCE OF THE DISTRICT OF DORTMUND (HANOVER, WESTPHALIA, AND RHENISH PRUSSIA) IN 1887.

Produktions-Übersicht der im Oberbergamtsbezirk Dortmund im Jahre 1887, in Betrieb gewesenen Bergwerke und Salinen. Glückauf, No. 18, 1888.

A.—COAL MINES.

District.	No. of Collieries.	Produce in Tons.	Persons Employed.
1.—Osnabrück	7	123,695	939
2.—Dortmund, North	6	1,003,577	2,357
3.— Do. East	13	1,888,802	6,814
4.— Do. West	13	2,229,564	7,586
5.—Witten	9	1,625,479	5,646
6.—Sprockhövel	20	525,542	2,132
7.—Dahlhausen	14	2,009,352	7,299
8.—Bochum	11	2,915,834	9,253
9.—Herne	7	2,209,032	7,131
10.—Recklinghausen	10	2,303,697	7,335
11.—Gelsenkirchen	9	3,367,670	11,007
12.—Essen	8	3,048,593	8,743
13.—Frohnhausen	11	2,383,966	7,185
14.—Oberhausen	15	3,000,903	9,653
15.—Altendorf	15	963,022	3,071
16.—Werden	8	373,388	1,259
Private mines	176	29,972,116	98,410
Government mines	2	178.122	1,124
Total of 1887 ...	178	30,150,238	99,534
Do. 1886 ...	188	28,497,317	99,787

B.—Iron Ore Mines.

District.					No. of Mines.	Produce in Tons.	Persons Employed.
1.—Osnabrück	...	...	...	...	10	247,570	1,029
2.—Dortmu nd, North	...	...	...	1	136,231	342	
3.— Do. East	...	...	...	3	32,763	172	
5.—Witte n	...	...	...	...	3	563	19
6.—Sprockhövel	...	...	...	...	3	55,191	233
7.—Dahlhausen	...	...	...	...	1	132,834	555
16.—Werden	...	...	...	...	1	6,049	32
Total of 1887	...	...	...	22	611,201	2,382	
Do. 1886	...	...	...	19	561,837	2.278 .	

C.—Zinc Ore Mines.

District.					No of Mines.	Produce in Tons.	Persons Employed.
5.—Witte n	...	...	...	...	1	22,917	504
6.—Sprockhövel	...	...	...	...	2	686	14
16.—Werden	...	...	...	...	2	12,210	484
Total of 1887	...	...	...	5	35,813	1,002	
Do. 1886	...	...	...	5	34,237	1,031	

D.—Lead Ore Mines.

District					No. of Mines	Produce in Tons.	Persons Employed.
5.—Witte n	...	...	...	...	1	198	13
16.—Werden	...	...	...	...	4	1,119	156
Total of 1887	...	...	...	5	1,317	169	
Do. 1886	...	...	...	4	1,039	112	

E.—Copperas Mines.

District.					No. of Mines.	Produce in Tons.	Persons Employed.
1.—Os nabrück	...	...	...	...	1	386	3
4.—Dortmu nd, West	...	...	...	3	850	6	
6.—Sprockhövel	...	...	...	...	1	8,511	40
16.—Werde n	...	...	...	...	1	86	2
Total of 1887	...	...	...	6	9,833	51	
Do. 1886	...	...	...	5	14,829	82	

F.— Salt Works.

District.					No. of Works.	Produce in Tons.	Persons Employed.
1.—Os nabrück	...	...	...	...	2	1,461	28
2.—Dortmu nd, North	...	...	...	3	19,092	216	
Private works	...	...	...	5	21,153	244	
Gover nmen t works	...	...	1	1,308	26		
Total, 1887	...	...	6	22,461	270		
Do. 1886	...	...	6	21,148	256		

M. W. B.

A SELF-ACTING WATER TANK.

Beschreibung eines selbstwirkenden Wasserkastens. K. BARTH. *Berg- und Hutten-mœnnische Zeitung, Vol. XLVII., p. 23. Illustrated in the text.*

A cheap and quick method of pumping out disused pit shafts has been in use for some years past in America.

An iron-bound wooden tank, from 3 to 4 feet high, of oblong shape, is let down into the water by a rope leading first over a pulley above and thence to a winch near the shaft. An iron rod passes through it at about one-third of its height and projects an inch or two through on each side. The two iron lobs thus formed rest on parallel wooden rods which guide the tank down into the shaft. On each of the wooden guides an iron rod, hinged at its lower end, is held, projecting a little by means of a weight so as to form a catch for one of the lobs before mentioned. The tank, when lowered, takes in water through two clack-valves in the bottom. It is then raised by the winch till the lobs rest upon the catches, when the rope is slackened and it kips its contents into a gutter beside the shaft. A rope led through a snatch block near the winch allows the catches to be withdrawn, and the box, already raised into a vertical position by the winch, is allowed to descend to be refilled.

An arrangement as described can raise from 70 to 80 gallons of water per minute from a depth of 150 fathoms, only one man being required to work it.   A. R. L.

THE CATRICE APPARATUS FOR RELIGHTING LOCKED SAFETY-LAMPS.

Note sur un Système de Rallumage intérieur des Lampes de Sureté. By L. JANET. *Annales des Mines, 1888, Sér. 8, Vol. XI., pp. 191 to 198. Plate VII.*

The Catrice system of relighting lamps consists of a brass case ·80 inch diameter and 1 inch high. This case is joined to a flat ring upon which a lid is fixed by a rivet. This lid closes by sliding over the ring and is stopped by a small stud which holds it against the ring.

A nipple is placed upon the lid, pierced with an opening by which the contrivance can be worked; a higher projection marks the side corresponding to the outlet tube and a small recess receives the bolt which keeps the lid closed.

A square tube of sheet iron is placed on the top of the case, and communicates with the interior by a circular opening of ·12 inch diameter.

A spring is placed in the tube attached by the base against one of its sides. This spring, which is flat, is roughened like a file at its upper part and presses against the side of the tube which is also like a file.

A small tube is placed on one side of the case containing an iron wire acting as a bolt to keep the lid of the case closed.

A small barrel, like that of a revolver, is placed so as to turn freely in the case. The barrel is fitted with eight chambers, intended to receive small matches about ·80 inch long, the heads of which are easily ignited.

When the lamp is opened the bolt is raised a little so as to allow the lid to be turned round and the barrel withdrawn to be filled with matches. It is replaced in the case, the lid closed, and the bolt pushed down. The slot in the nipple upon the lid leaves two of the chambers of the barrel uncovered, a punch is placed in one of the · chambers, and the barrel is turned round until it is under the exit tube. The punch is then pushed in sharply, and the match passing between the roughened surfaces of the spring and the side of the tube is ignited and lights the wick which is close to it.

The apparatus costs about 1s. 6d. per lamp.   M. W. B.

OZOKERITE IN GALICIA.

Note sur l'Ozokérite, ses Gisements, son exploitation à Boryslaw, et son Traitement Industriel. By A. RATEAU. *Annales des Mines,* 1887, *Vol. XI., pp.* 147–170. *Plate VI.*

Boryslaw is a town of 20,000 inhabitants, in 49° 17′ 3″ latitude, twelve miles from Drohobycz, in Galicia, and on the north slope of the Carpathians. The valley is surrounded by small hills, not exceeding 300 feet above it, and is traversed by a small river.

The surface is covered with a few yards of diluvium. Beginning at the top, there are found yellow clays, rolled pebbles and gravel, and plastic clay. Below this are beds about 600 feet thick of sandstone and schist, very much dislocated, of Miocene age, in which the ozokerite is found. These Miocene rocks rest upon menelitic schist, containing petroleum, and consisting of beds of coarse sandstone, green marl traversed by veins of calcite, of highly coloured schists alternating with dull black schists passing gradually into thick beds of sandstone and schist which contain no petroleum. These beds of shale and sandstone rest upon the older Carpathian sandstones, strongly impregnated with petroleum, which is found in greater quantities at greater depths.

The ozokerite is found in the form of thin leads or veins in the Miocene sandstones and schists, varying from a few hundredths of an inch to some feet in thickness. It is accompanied with variable quantities of petroleum and hydrocarbon gases. The veins, filling the innumerable fissures of the rocks, form a complete network, but frequently follow the bedding of the rocks.

In the central portion of the field (which is of pear-like form) the veins become more productive with depth; but in the margins the veins are much thinner, and run out at depths varying from 30 to 100 feet. The central and richer portion has an area of about 52 acres, and the outer area is about 95 acres. At a depth of 650 feet the width of the richer area has diminished from 1,150 feet to about 325 feet, which shows that the ozokerite has passed upwards through some fracture of the lower strata.

The ozokerite rocks are also impregnated with petroleum and surrounded by petroliferous rocks, except on the north-east side. The petroleum wells on the other sides are all sterile below a depth of about 300 feet.

The external area contains about 2 per cent. of ozokerite, whilst the internal area contains an average of 5 per cent.; consequently the entire field contains about 2,000,000 tons above a depth of 650 feet.

Up to the end of 1886 about 330,000 tons have been extracted, worth at least £8,000,000. The price per ton is very variable:—

				£	s.	d.	£	s.	d.
1874-5	...	...	...	17	12	0			
1876-7	...	...	...	25	12	0			
1878	...	...	...	22	8	0 to 24	8	0	
1885	...	...	...	24	0	0 „ 26	0	0	

The annual production is about 20,000 tons, worth £480,000.

The difficulties of working are considerable. The issue of fire-damp compels the use of safety-lamps and even causes explosions. The rocks exert great pressure upon the timber owing to the action of the ozokerite and petroleum. Timbers a foot square are often broken to matchwood. Another result of these pressures is the frequent and violent irruptions of gas, petroleum, and ozokerite into the pits and galleries; and the workmen are sometimes drowned in the fluid mass. The proportion of deaths from accidents varies from 7 to 15 per 1,000 per annum, against less than 2 per 1,000 in ordinary mines.

The ordinary method of working is to sink pits about 5 feet diameter at regular intervals. One man works in the bottom; he drives short horizontal galleries at right aigles uitil he fiids a good veii of ozokerite. But these drifts, owiig to the pressure, are very short; about 15 feet is the maximum, which caiiot be exceeded without coisiderable daiger. The pits, from 65 to 650 feet ii depth, may yield about 20 tois of ozokerite per aiium and last from 5 to 10 years. The produce is drawi by buckets, iroi or wire ropes beiig used of about $\frac{1}{4}$ iich diameter, worked by a jack-roll. About 1,000 pits have beei suik, of which 700 to 800 are still ii operatioi. The whole area is pierced like a spoige.

The paper coicludes with some iotes on the preparatioi of the ozokerite for the market. M. W. B.

MINERAL STATISTICS OF BRITISH COLUMBIA.

Annual Report of the Minister of Mines for the Year ending December 31st, 1886.

The chief mineral products are :—

			GOLD.		COAL.
			Value. Dollars.	No. of Miners.	Tons.
1874	...	...	1,844,618	2,868	... 81,000
1875	...	...	2,474,904	2,024	... 110,000
1876	...	...	1,786,648	2,282	... 139,000
1877	...	...	1,608,182	1,960	... 154,000
1878	...	...	1,275,204	1,883	... 171,000
1879	...	...	1,290,058	2,124	... 241,000
1880	...	...	1,013,827	1,955	... 268,000
1881	...	...	1,046,737	1,898	... 228,000
1882	...	...	954,085	1,738	... 282,000
1883	...	...	794,252	1,965	... 213,800
1884	...	...	736,165	1,858	... 394,070
1885	...	...	713,738	2,902	... 365,000
1886	...	...	903,651	3,147	... 326,636

The reports of the District Iispectors coitaii particulars of the priicipal auriferous bars, gulches, creeks, and rivers. There are also exhaustive reports by Messrs. G. A. Koch and A. Bowmai upoi the quartz reefs, with results of various aialyses.

Three collieries were workiig duriig 1886, Nanaimo (Vaicouver Coal Miiiig and Laid Co., Limited), Welliigtoi (R. Duismuir & Sois), and East Welliigton (R. D. Chaidler, of San Fraicisco).

There are 1,269 workmei employed, iicluding 530 Chiiese. There were three fatal accideits duriig 1886, two from falls of stoie and one of coal.

The coal is chiefly shipped to San Fraicisco and other parts of Califoriia, and the followiig figures show the position of the various sources of their supply :—

			1883. Tons.	1884. Tons.	1885. Tons.	1886. Tons.
British Columbia	...	...	128,503	291,546	224,298	253,819
Australia	...	...	174,143	190,497	206,751	287,293
Eiglaid and Wales ...	...	131,355	108,808	170,656	160,869
Scotlaid	...	...	21,942	21,143	20,228	19,795
U.S. Eastern States ...	...	43,861	38,124	29,834	19,517
Seattle	...	...	139,600	125,000	75,112	57,552
Carboi Hill	...	...	140,135	122,060	157,241	124,527
Greei River, etc. ...	...	76,162	77,485	71,615	90,664
Beitoi, etc.	...	...	43,600	60,413	67,604	73,654
			899,301	1,035,076	1,023,339	1,087,690

M. W. B.

THE DURANT HAND-BORING MACHINE.

La Foreuse Système Henry Durant. *Publications de la Société des Ingénieurs sortis de l'École Provinciale d'Industrie et des Mines du Hainaut*, 1886-7, Sér. 2, *Vol. XVIII., pp.* 270-72. *Plate XXIV.*

This boring machine consists of a metal tube closed at both ends and containing a strong spring between two discs or pistons, one of which receives the end of a screw passing through the back-end of the apparatus, and the other transmits its pressure to the drill which is consequently closely pressed against the rock to be drilled. The front cover of the tube is used as a guide for the drill, to which a ratchet and lever is attached for the purpose of turning the drill more or less rapidly. The machine is carried upon parallel standards of the ordinary type.

It bores a $3\frac{1}{4}$-inch hole at the rate of from 4 to $6\frac{1}{2}$ inches per minute, according to the hardness of the stone. M. W. B.

NATURAL GAS IN AMERICA.

Natürliches Gas nach Orton, Ashburner u. A. H. WINKLEHNER. *Berg- und Huetten-männische Zeitung, Vol. XLVII., pp.* 43-45 *and* 53-55.

Natural or rock gas is a mixture of some seven or eight different components, which vary in their relative proportions with the locality, time of year, temperature, and state of the barometer. It occurs in petroleum districts, and indeed is always accompanied by petroleum vapour, though sometimes in very small quantity. Itself the result of decomposition of organic matter contained in a bed of stone, the petroleum in its turn develops partly into gas, but the process is a natural one and does not depend upon volcanic influence or the action of heat in any form, while for the production of any considerable quantity of either substance the parent seam must contain animal or vegetable matter in considerable masses. The gas ascends through cracks and fissures until it reaches a rock formation dense enough to stop it, below which, provided the underlying stratum be sufficiently porous, it will collect in large quantities. In the case of two bore-holes in Ohio the gas-bearing stratum was a bed of chalk, rich in magnesium and only moderately porous, and the petroleum was evidently produced in the lower part of the same bed. All the gas-bearing sandstones and conglomerates belong to the Devonian or Carboniferous measures and form compound seams of from 3,000 feet to 5,000 feet in thickness, the necessary close-grained ceiling being a bed of slate of varying thickness. This is pierced by fissures which in places extend from the gas stratum right up to the day and form natural main conduits. The importance of the close-grained ceiling before mentioned is seen by the fact that in some other regions with the same geological features neither gas nor petroleum is found, these having been free to make their escape as soon as formed.

Natural gas has been found by boring at depths varying from 325 feet to 4,600 feet, the deepest boring that at Dilworth, being 4,618 feet deep. The volume and speed of the gases issuing from bore-holes are measured by the anemometer. The Van Buren Well, in Ohio, gives nearly 15,000,000 cubic feet per day; the Barclay Well, in the same State, only 469,000 cubic feet; these being about the extremes. The total production per day of the wells around Pittsburg is about 182,400.000 cubic feet; but Dr. Chance has estimated that all the Pennsylvanian wells will be exhausted in less than eight years. The gases in the earth are under a mean pressure of about from 300 lbs. to 400 lbs. to the square inch, in some cases amounting to 1,000 lbs., but in others again being very much less. The principal gas-fields are Western Pennsylvania and Northern Ohio.

The gas round Pittsburg is on the average of about 8 candle-power, that at Findlay, Ohio, varying from 12 to 14 candles. A comparison between the relative heating powers of rock gas and coal results much in favour of the former, 30,000 cubic feet of the gas being equivalent to one ton of coal, and when the enormous volume given off by one bore-hole is considered the great saving due to the use of the newer fuel will be apparent. At the end of the year 1885 natural gas was employed around Pittsburg by 46 iron companies, 33 glass-making companies, in 9 petroleum refineries and 2,637 private houses, and the daily quantity used was estimated at about 182,400,000 cubic feet, equivalent to 6,800 tons of coal. In one glass-works in Pittsburg the employment of natural gas in place of coal showed a saving of 46 per cent. The gas of Pittsburg is in the hands of six different companies. It is supplied in unlimited quantities, payment being regulated in the case of iron works, etc., by the number of furnaces, the quantities of material worked, or the number of boilers heated by it, and in the case of private houses a fixed rate per month is levied. The cost is comparatively small; but people in America are not blind to the prospect given them by geologists that the gas wells will very soon all be exhausted.   A. R. L.

DEEPENING SHAFTS.

Nouveau procédé d'Enfoncement sous stot appliqué au Puits S.-Adolphe de la Société Anonyme des Charbonnages de Haine-S.-Pierre et la Hestre. Société des Ingenieurs sortis de l'École provinciale d'Industrie et des Mines du Hainaut, 1886-7, Sér. 2, Vol. XVIII., pp. 216-231. Plates XV., XVI., XVII., XVIII.

The S.-Adolphe Pit, which was 984 feet deep, required to be extended to a depth of 1,246 feet, without stopping coal-drawing.

An inclined pit, about 8 degs. from the vertical, was sunk alongside the drawing shaft at a depth of 984 feet, until its axis met that of the shaft and it was then continued vertically downwards.

The stones were drawn by means of a small steam engine at the surface, whose 1-inch wire rope was led down the side of the pit, clear of the cages; at about 33 feet above the hanging on, the rope was diverted by two pulleys into the line of the inclined pit, whence it passed into the vertical position required for the sinking of the pit. This was effected by passing the rope over a pulley supported upon a carriage running upon flat-bottomed rails.

This carriage has four wheels, running upon the rail guides, supports a deep-grooved pulley, and weighs about 450 pounds. The rope runs upon this pulley and passes immediately through a hole in a strong balk attached to the carriage. A spiral spring is placed over the end of the rope so as to reduce the shocks.

When the tub is at the bottom of the pit the carriage is at the lower end of the guides and the rope passes over the pulley vertically down the pit. As the tub is drawn upwards the spring strikes the buffer and pushes up the carriage, below which the tub hangs in a vertical position until it reaches the 984-feet level, where it is stopped and replaced by an empty tub. The carriage follows the descent of the empty tub until it stops at the end of the guides, whence it passes the rope vertically down the pit.

The plates give working plans of all the various details.   M. W. B.

THE NAPHTHA DISTRICT BEYOND THE CASPIAN.

Ueber das transkaspische Naphtaterrain. By Dr. Hj. Sjögren. Jahrbuch der k. k. geol. Reichsanstalt, XXXVII. Band, I Heft, pp. 47-62. A map and sections in the text. Vienna, June, 1887.

The richest naphtha producing areas in the Caucasus are at either end of the mountain range, and supposing this chain to be prolonged in a north-westerly and south-easterly direction, then it would cover, west of the Black Sea, the Galician and

Moldavian naphtha districts, and east of the Caspian the corresponding groups of Tjeleken, Neftanaja Gora, and Buja Dagh. It is worthy of note that whilst Caucasian naphtha contains little, if any, paraffin, and is unaccompanied by mineral wax, the contrary holds good in the last mentioned areas.

The author was unable to visit the island of Tjeleken, and he passes on at once to the description of the Neftanaja Gora or Naphtha Hill. To reach this place the best starting point is the station of Bala Ischem, on the Transcaspian Railway, about 37 miles from the present terminus at Michailowski Salif, and 6 miles south of the Great Balchan, a range rising in a perpendicular wall 5,500 feet above the sea level. The Great Balchan is made up of light grey, compact, unfossiliferous limestones alternating with marls; in some *débris* at its foot fragments of *Belemnites* of undetermined species have been found, and on the neighbouring steppe shells of *Cardium* are abundant. For reasons based on the general geology of Transcaucasia, the author seems inclined to refer both the Great Balchan and the Little Balchan (3,200 feet) to the chalk formation. From Bala Ischem a tramway, 21 miles long, has been laid to the Neftanaja Gora. For two-thirds of the distance it runs through a salt desert, where layers of pure salt more than 6 inches thick are found close to the surface of the ground.

The Neftanaja Gora rises barely 300 feet above the level of the steppe and extends for about a mile from south-west to north-east; northwards and southwards it is flanked by low hills, and its general structure is that of an irregular anticlinal, the dips to the north varying between 25 degs. and 44 degs., those to the south between 7 degs. and 9 degs. The main mass of the hill consists of grey and brown clays and sands; but its eastern summit, 210 feet above the boring towers, is made up of a hard, dark brown, conglomeritic sandstone and similar rocks form the south-western summit. There is good reason for referring this area to the Miocene formation. Particular stress is laid on the unsymmetrical structure of the anticlinal; the asphalte, mineral wax, and naphtha are found on the steeper or northerly slope, and this tends to show that the distortion and fracture of the underlying strata produced rifts which formed natural conduits for gaseous and liquid substances.

In a south-west to north-east line along the ridge occur the springs from which issue brine, naphtha, and carburetted hydrogen gas. The temperatures mentioned vary between 19 degs. and 21 degs. C. The mineral wax which comes to the surface in small lumps, in conjunction with the naphtha, is coffee- or chocolate-brown in colour, possesses an aromatic odour, and can be kneaded with the fingers. Large quantities of the wax are supposed to be present at some depth below the surface. Many of the springs lie in craters hollowed out in small cones built up of asphalte, and the asphalte not only covers a large portion of the ridge but extends along the north-western spur exactly like an old lava flow. The Russian mining engineer, Konschin, drove a few shafts, 30 feet deep, about here, in the hope of tapping the main source of the mineral wax, but a great inrush of water and gas defeated his attempt. It is true that a few "veins" of wax were met with, but not in paying quantities. Prince Eristoff unsuccessfully worked the surface deposits of asphalte with the same object, and the sulphur workings which he initiated a little further west were also abandoned as not being worth the labour which they entailed. In a word, both at the Neftanaja Gora and at Tjeleken, all endeavours to foster a wax mining industry have failed.

South of Bala Ischem, between the Neftanaja Gora and the Buja Dagh, are numerous traces of naphtha springs, and an old river bed can be traced, filled with crystals of gypsum, which no doubt crystallized out from the mother liquors of the old salt lakes. The Buja Dagh rises about 500 feet above the surrounding steppe; it is about 6 miles long and 2 broad, and is distant 19 miles from Aidin station on the Transcaspian Railway, and 25 miles from Bala Ischem. It forms a regular anticlinal, the dips at either extremity varying between 30 degs. and 40 degs. The Dagh is built up of shales and sandstones: the latter crown its summit in curious isolated pillars, or alternate with the shales in excessively thin layers. There are here two warm brine springs (temperature 54 degs. C.), containing, it would seem, a fair proportion of iron, and very concentrated. Black asphalte-like naphtha occurs in unimportant quantities in some neighbouring springs. No asphalte deposits have been discovered on the Buja Dagh, and there is little hope of finding workable quantities of naphtha at a reasonable depth below the surface. L. L. B.

EXTRACTION OF GOLD BY CALCIUM CHLORIDE.

Jernkontorets Annale, 1887, *page* 127.

The ore is roasted if necessary and ground to a fine powder, together with a little salt. This mixture is roasted in reverberatory furnaces until the sulphides, arsenides, and antimonides are decomposed. It is then teamed into wooden tubs, and treated with hot water, to wash out the copper and silver salts and oxides. The mass is then treated with a solution of ·006 to ·007 of calcium chloride in water, mixed with an equal volume of hydrochloric acid, of 1·002 to 1·003 density, until a sample of the solution gives no gold reaction when tested with an acidulated solution of stannic chloride. The solution containing the gold is heated in wooden bowls by steam to a temperature of 160 degs. Fah., and precipitated by means of ferrous sulphate, sodium sulphide or sulphurous acid, a small quantity of plumbum acetate being added during the precipitation to ensure all the gold being thrown down.

This process has been employed at the Falee copper works in Sweden since 1885 for the treatment of tailings from 29,000 tons of copper ore, together with 1,500 tons of gold ore.

In 1886 the tailings from 14,000 tons of copper ore contained 41·82 grains of gold per ton before treatment and 4·04 grains of gold per ton after treatment. The costs per ton being :—

					s.	d.
Calcium chloride, 6·6 pounds	...	...	...	...	0	5
Sulphuric acid, 8·37 „	...	...	...	...	0	1
Lead acetate and reagents ...	...	...	...	...	0	0¾
Fuel for steam	...	...	...	...	0	1¼
Labour	...	...	...	...	...	0	1
Total	...	...	...	...	0	9

In the same year 960 tons of gold ore contained 523·62 grains of gold per ton, and after treatment 6·02 grains per ton. The costs per ton being:—

					s.	d.
Calcium chloride, 33·0 pounds	...	...	...	2	0¾
Sulphuric acid, 44·0 „	...	...	...	0	5
Lead acetate and reagents ...	...	...	...	0	6¾
Salt, 176·3 pounds ...	...	...	...	...	1	9½
Coal, 187·4 „	...	...	...	...	1	6¼
Wood ...	...	...	...	...	...	1	8¼
Wood for steam	...	...	...	...	0	3¼
Labour	...	...	...	...	...	3	11
Total	...	...	...	...	12	2¾

M. W. B.

MAMMALIAN REMAINS IN STYRIAN COAL.

Neue Funde tertiärer Saügethierreste aus der Kohle des Labitschberges bei Gamlitz. By AD. HOFMANN. *Verhandlungen der k. k. geol. Reichsanstalt, No.* 15, *p.* 284. *Vienna, November,* 1887.

After alluding to one or two previous finds of mammalian teeth in the coal at Gamlitz, the author proceeds to enumerate and describe briefly the fossils lately discovered there, consisting of teeth and jawbones of *Cervus, Palaeomeryx, Hyaemoschus,* and *Hyotherium.* L. L. B.

SULPHUR AT TRUSKAWIEC.

Ueber das Schwefelvorkommen bei Truskawiec. By JOSEF WYCZYNSKI. *Verhandlungen der k. k. geol. Reichsanstalt, No.* 13, *pp.* 249-250. *Vienna, October,* 1887. *With a section in the text.*

It is in conjunction with ozokerite that the sulphur seems to occur. The hanging wall and the foot wall are formed by a grey, impermeable clay, often including large blocks of marl, in which are druses lined with well-developed sulphur crystals. In the clay the sulphur is found in single crystals of varying size, or in thick-grown clusters of crystals. Whether this deposit is a bed or a lode cannot as yet be determined, because the clay which surrounds it is a compact mass, with, seemingly, no particular strike or dip. The terms "hanging" and "foot wall" really apply to the ozokerite, which cuts through the clay in numerous lodes. With the ozokerite and sulphur, gypsum and aragonite are frequently found, celestine and rock salt more rarely. L. L. B.

THE TERTIARY COALS OF CARINTHIA.

Die Neogen- Formation in Kärnten. By FERD. SEELAND. *Verhandlungen der k. k. geol. Reichsanstalt, No.* 13, *pp.* 252-251. *Vienna, October,* 1887.

The Neogene beds of Carinthia contain deposits of bright-faced coal and lignite, varying in thickness from 3 feet to 25 feet, over a large extent of country. The bright-faced coal, which is the older deposit, is best developed at the Liescha mine; the lignite attains its maximum development at the Peiken mine near Keutschach. In both cases the underlying clay is largely worked for furnace-resisting pottery and fire-bricks. The fossil flora includes palms, fig-trees, and other subtropical plants. L. L. B.

TEMPERATURE OF THE ARLBERG TUNNEL.

Die Wärmeverhältnisse in der Osthälfte des Arlbergtunnels. By C. J. WAGNER. *Verhandlungen der k. k. geol. Reichsanstalt, No.* 8, *pp.* 185-186. *Vienna, May,* 1887.

A small hole, about 3 feet deep, was drilled in the rock in a sort of niche, 3 miles from the eastern entrance to the tunnel; and the temperature there observed in January, 1885, 1886, and 1887, fell from 15·3 degs. C. in the first-named year to 14·7 degs. C. in the last-named. The temperature of the air in the tunnel fluctuated between 12 degs. and 16 degs. C. Since the completion of the boring of the tunnel in May, 1884, the temperature of the rock wall has fallen 3·8 degs. C. L. L. B.

BOULDERS IN UPPER SILESIAN COAL.

Einschlüsse von geröllartiger Form aus Steinkohlenflötzen von Ober- Schlesien. By
DR. G. GÜRICH. *Verhandlungen der k. k. geol. Reichsanstalt. No. 2, pp.* 43–45.
Vienna, February, 1887.

The author describes the stones hitherto found, which seem to be chiefly rolled
fragments of granitic rocks (thus disposing of Stur's theory that they were simply
concretions), and he mentions two cases in which the stones lie at right angles to the
bedding plane of the coal, as if they had sunk through a yet soft vegetable mass.

L. L. B.

AGE OF THE FÜNFKIRCHEN COAL.

Ein neuer Cephalopode aus der Kohlenablagerung von Fünfkirchen. By D. STUR.
Verhandlungen der k. k. geol. Reichsanstalt, No. 9, *pp.* 197–198. *Vienna,*
June, 1887.

Thanks to the discovery of an Ammonite, belonging to the *Arietites obtusus* group,
in the Coal-Measures of Fünfkirchen (Styria), the author is now able to rank those
beds unhesitatingly with the Lower·Lias. He declines to consider the coal in question
as a lignite, because the plant remains of which it is made up are not the same as the
ordinary constituents of lignite, but are *Calamites* (*sic*), *Equisetites,* ferns, etc.

L. L. B.

CONCRETIONS WITH PLANTS IN WESTPHALIAN COAL.

Ueber den neuentdeckten Fundort und die Lagerungsverhältnisse der pflanzenfüh-
renden Dolomitconcretionen im westphälischen Steinkohlengebirge. By D. STUR.
Verhandlungen der k. k. geol. Reichsanstalt, No. 12, *pp.* 237-243. *Vienna,*
September, 1887.

The existence of concretions in the Westphalian Coal-Measures, containing plants
similar to those described by W. C. Williamson, as occurring also in concretions in the
Yorkshire Coal-Measures, had long been ascertained; but the fragmentary condition
of the plants found in Germany, and the uncertainty as to their horizon, had, up till
quite recently, hindered in that country the further study of the subject.

In June, 1887, Mining Councillor Nasse of Dortmund brought forward the following facts at a General Meeting of the Natural History Society of the district. Eight
or nine years previously dolomitic concretions had been discovered in the coal at
Langendreer, containing *Sigillaria, Stigmaria, Lyginodendron, Lepidodendron,
Cordaites, Sphenophyllum,* and ferns. These plant remains were examined microscopically, and found to be practically identical with those occurring in concretions in
the lowest beds of the Yorkshire coal-bearing strata. In 1883, in a seam at Peters-
wald, concretions of spathic iron ore had been found, containing similar plant remains;
and now in 1887, Nasse had seen at the Hansa mine dolomitic concretions in the form
of kidney and ball-shaped nodules, varying in size from "a hazel nut to a child's head,"
also enclosing plants. The Katharina seam, where they occur, occupies a central
position in the vertical section of the Westphalian coal-bearing strata, and about 3 feet
above it there is a clay slate containing pyrites-covered impressions of *Aviculopecten
papyraceus.* Herr Nasse considers that these beds are of the same age as the Ostrau
strata, containing bog iron ore concretions, and he correlates the latter partly, if not
entirely, with the Schatzlar or Saarbrück Coal-Measures.

The author of this paper disagrees with him *in toto*, and, after referring to journeys formerly undertaken in Westphalia and Belgium in connection with this very subject, he proceeds to correlate the beds as follows:—

Dolomitic concretions of the Katharina coal-seam, and limestone concretions of Oldham and Halifax $\Big\} =$ *Aviculopecten papyraceus* beds of Westphalia, Belgium, and England (Schatzlar beds).

III.—Meagre Culm Fauna of the 5th group of seams of the Ostrau beds.

Bog iron ore concretions of the Ostrau beds $\Big\} =$ II.—Culm Marine Fauna of the 1st, 2nd, and 3rd groups of the Ostrau beds, and of Belgium.

I.—Culm Marine Fauna of the *Posidonomya Becheri* slates of Middle Silesia, and of the Carboniferous limestone (with *Productus giganteus*) in Lower Silesia, Belgium, etc.

L. L. B.

GOLD (?) ON THE RIO CUBANGUI.

A expedição ao Cubango. By LIEUT. A. DE PAIVA. *Boletim da Sociedade de Geographia de Lisboa,* 7a *Serie No.* 2, 1887, *pp.* 116 *and* 127-128. *With three maps.*

In 1885-86 the author was commissioned by the Portuguese Government to undertake a military expedition and exploration in certain little-known districts of Eastern Benguela. He devotes, in his report, scant space to matters of geological interest; but he mentions twice the fact, that on the banks of the Cubangui, particularly near its mouth, there abounds a granitic gneiss, rich in mica, whose particles, disaggregated by the running water, shine on the sandy shore with the yellow lustre of gold. Further up the river the particles of this same gneiss flash yet more brilliantly from the sand, but they are colourless. Near the confluent of the Cubangui and the Cahonga, the soil, of a rich red colour, is overlaid by black sand, possessing a metallic lustre, which the author compares to that of iron filings.

L. L. B.

SPIRALLY-WELDED TUBING.

By J. C. BAYLES. *The Engineering and Mining Journal* [*New York*]. 1888, *Vol. XLV., pp.* 250-252, 270-271. *and one Plate.*

Serviceable pressure pipes of great strength are manufactured from strips of iron or steel wound spirally, heated only along the overlapping edges, welded by hammering and finished into tubes of uniform diameter and of suitable lengths.

The ordinary sheet iron or steel of commerce is slit into bands of the width most convenient for the production of the desired diameter of pipe. This is done by an ordinary rotary shear provided with a table and a guide. The wider the bands or skelps, the faster the pipe is made. As the sheets are usually not more than 12 feet long, the ends are united by lap welding if long pipes are desired. To make a 6-inch pipe, 30 feet long, of 12-inch skelp, a ribbon about 49 feet long is required. The ends of the skelp are united by a machine known as a cross-welder.

The pipe machine, covering about 3 feet by 6 feet of floor space, is of simple con-struction. The reel carrying the skelp is placed in position and one end of the ribbon is placed upon the guide table, which is set at an angle due to the width of skelp and the diameter of the finished pipe. The sheet is carried between feed rolls geared together, actuated by a ratchet giving them an intermittent rotation, and a feed variable between $\frac{1}{16}$th inch and $\frac{3}{8}$th inch at each impulse. This carries it into the forming jaws, which bend it to the desired curvature, by pinching the metal in curved jaws. The feed rolls pass it forward when the hammer is raised and are at rest when the hammer falls; a "former" is used to curve the metal to the required radius, a furnace to heat the metal, a hammer to weld it, and an anvil to support the pipe and receive the hammer blows. No mandrel is used; the pipe in the forming process is held in a pipe mould, a cylindrical shell within which the pipe rotates as the stock is fed in. The anvil is of considerable mass, steel faced, and extends the entire width of the skelp. The hammer is light and strikes about 160 blows per minute. The heating is done in a furnace, constructed so as to heat both the edges to be united for the space of several inches ahead of the point at which the welding is effected. The upper skelp enters the furnace flat, and the lower skelp curved, having already been through the forming jaws. The heat is imparted by one or two blow-pipes of water-gas and air discharging upon the metal through passages of suitable form in the refractory lining of the furnace. About 30 feet of gas are burned per foot of welded seam, and about 1 foot of pipe is produced per minute on the average.

The machines are almost automatic, and the operator has the gas, air, and feed under control by convenient means. Unskilled labour prepares the stock and removes the finished product. The machines produce pipes from 4 to 30 inches in diameter, and from plates of 29 to 8 Birmingham gauge.　　　　　　　　　　　　M. W. B.

ORE SORTING.

By F. L. BARTLETT. *The Engineering and Mining Journal* [*New York*], 1888, *Vol. XLV., pp. 268–270, and Plate.*

In opening the Milan mine in 1882, it was found desirable to produce pyrites in a very pure form. The run of the mine was a mixture of cupriferous iron pyrites, very massive, with rich copper ores, zinc-blende, and galena. More or less slate and quartz had to be guarded against. It was essential that clean separations of the ore should be made, pyrites if sold for acid making must be free from zinc and lead, and compara-tively free from copper. Further, the galena was rich in silver, and the better copper ore was abundant and suitable for smelting on the spot.

The following scheme for sorting has proved a perfect success :—

A Blake crusher was erected, with a jaw opening 10 inches by 17 inches, and set to crush at 2-inch gauge. A link belt elevator raised the crushed ore some 16 feet, and discharged it at the upper end of an inclined shaking table. The table proper is 52 inches wide and 18 feet long. The upper end has a punched steel screen, 30 inches wide and 4 feet long, with $\frac{1}{4}$-inch holes. The remaining 14 feet of the table is divided longitudinally into troughs or shoots, by means of No. 14 sheet-iron, screwed to the wooden frame of the table. The main shoot is 18 inches wide, and there are three on each side, each 5 inches wide. The shoots end over a set of bins, which are placed just outside the building and directly over the tramway track, so that cars can be placed under any of the bins, and any of them, by pulling a slide, discharged into the car.

At the upper end of the table is the shaft, with eccentrics at each end having a thrust of 3 inches, and connecting rods which take hold of the table at about 6 feet from the upper end. A fixed pulley on the shaft gives motion to the table; another pulley runs loose on the shaft and has a pin-clutch connecting with the sprocket-wheel which carries the elevator. The lower end of the table strikes against two strong rubber buffers on the downward stroke for the double purpose of easing the motion and jarring the ore along the table. The table is set at an angle of 1 in 9, and the motion is about 200 oscillations per minute. The elevator is run at six revolutions per minute. A 2-inch pipe, with a row of $\frac{1}{4}$-inch holes on each side, is placed above the screen at the upper end of the table, for the purpose of washing the ore as it passes along underneath. A tank 6 feet wide, 8 feet long, and with sides 2 feet high, is placed on the ground and immediately under the screen. A large sheet-iron spout directs the water and fine ore passing through the screen into the tank. A platform is placed along each side of the machine for the men or boys to stand upon.

In working, the crushed ore is delivered by the elevator upon the screen at the upper end of the table, and is washed by the water-jets. The fine stuff goes through the screen into the tank below, while the coarse ore travels along by successive hitches towards the lower end of the table. Boys stand on either side, the fingers protected by steel thimbles, and sort the ore as it passes along. Waste rock goes into the first trough at each side; the other troughs carry respectively galena, copper, and zinc, and the thoroughly cleaned pyrites continue down the central shoot. A bright boy learns to sort in a week, and some become surprisingly expert. About ten boys can work at a table, but it is rare to employ more than four or five.

The amount of ore handled upon a table depends upon the quality. With half rock and half ore, 5 tons per hour is good work; with cleaner ores, 7 to 10 tons have been run. The average over some years is 75 to 80 tons per day of ten hours.

It costs 3s. 1½d. per ton for hand work to break and sort; it now costs 7½d. by the use of the crusher and the automatic table.

In working the table, it has been customary to run in the central shoot whatever was in the greatest proportion in the ore. If more than half rock, the ore was put in the side shoots, and *vice versâ* if the ore was in excess. The jerking of the table constantly turns the ore; and if a powerful stream of water be employed, it is completely washed, and there is no difficulty in detecting by the eye the differences in kind and quality.   M. W. B.

HENDERSON STEEL.

Engineering and Mining Journal [New York], 1888, Vol. XLV., p. 249.

The first trial was with Birmingham (Alabama) white pig, with ore and fluor-spar on a raw dolomite hearth composed of fluor-spar and dolomite, which lasted until near the close of the operation, when it melted and passed through the metal. The dolomite analysis is :—

Carbonate of lime	59·8
Carbonate of magnesia	39·2
Silica	·34
Protoxide of iron	·14

and is quarried on the premises. The pig iron cost £2 per ton, and the ore 2s. 8¼d. per ton on cars at the mine.

The steel produced is superior tool steel, said to be equal to Musshets, which sells for 2s. per lb. The analysis is :—

Carbon	·75
Silicon	·009
Phosphorus	·0051
Manganese	trace.

The superior quality of the steel produced is attributed to the use of fluor-spar, which, although no fluorine can be detected in the metal, gives the property of tough- ness to a surprising degree. Some stay-bolt iron gave 45 per cent. elongation, or over three times that usual in good ordinary puddled iron.

Another trial was made on a calcined dolomite and fluor-spar hearth with pig and scrap steel and ore with less fluor-spar for soft steel. Five per cent. of spiegeleisen was used, as there was no ferro-manganese at hand. The steel analysed:—

Carbon	·20
Manganese	·78
Phosphorus	·005

This made a first-class steel for boiler plates, although the pig iron contained ·3286 per cent. of sulphur.

There were about 200 pounds of slag per ton of steel, which analysed:—

Metallic iron as peroxide...	8·1900
Silica	29·2500
Sulphur	·0950
Phosphorus	1·1035

The pig iron and ore contained about 13·5 lbs. of phosphorus, and ·7 lb. was left in the steel; the difference, about 10·75 lbs., was volatilized. About 250 lbs. of iron ore was used, containing 45 per cent. of metallic iron, so that 85 per cent. of it was incor- porated with the charge.

It is proposed to attach condensers to the furnaces and obtain the volatile phos- phorus in the gases after they have been cooled under a boiler, by forcing them into water just above freezing point, and thus produce hydrous phosphoric acid of any required strength. The slag produced by this process is valueless as manure.

Calcined dolomite mixed with 10 per cent. of fluor-spar makes a durable lining, and that on the bottom where it is not acted upon by the silicates in the slag, which obtain in the pig and ore process, is not perceptibly worn after 5½ hours' use, the time taken by a pig and ore heat ; but the slag line at the side requires renewal after each heat, owing to the silica and oxide of iron fluxing a portion away. With melts or casts containing about one-fourth steel or iron scrap and three-fourths pig, no ore is used, but dolomite and fluor-spar in equal proportions. These enable a melt to be taken every three hours, and there is scarcely any perceptible wear to the sides of the hearth lining.

In making the side or repairing the slag line the pulverised dolomite and fluor-spar have a little soda added to facilitate setting, and are mixed into a mortar with water containing 15 per cent. of molasses.

The process is carried out in two chambers; the second or refining chamber, in which the iron is melted and refined of its silicon and half its carbon, and then poured into the first chamber. The refining chamber is lined with sand, and the metal is purified of silicon and carbon by the aid of phosphoric iron ore. The metal gains one- fourth of the phosphorus from this ore, and about three-fourths of the phosphorus is vapourized and collected in the condensers. The metal when introduced into the first or finishing chamber will be in the condition of high carbon steel, plus the phosphorus, and will be reduced to soft steel free of phosphorus. The operations require about three hours in the second chamber and about two hours in the first chamber.

It is suggested that the phosphoric ores of Alabama should be mixed together in proportions to yield pig iron containing 3½ per cent. of phosphorus, which may be recovered as hydrous phosphoric acid, and worth about £3 per ton of steel for fertilizing purposes. White pig iron costs £1 13s. per ton. The steel would then be made free of cost. M. W. B.

THE GEOLOGY OF TIMOR.

Gesteine von Timor. By PROF. A. WICHMANN. *Jaarboek van het Mijnwezen in Nederlandsch Oost-Indië, Wetenschappelijk, I^{te} Gedeelte. pp.* 46, 83-90, 92, 93. *Amsterdam, April,* 1887.

Since the Dutch Expedition of 1829, no European has travelled into the interior of Timor farther than the Copper River, and the petrographical description and chemical analyses of the rocks of that island, given by the author, are based on specimens collected in the above-mentioned year. Evidence has, however, been gradually accumulating as to the mineral wealth of the island. Gold is found in the beds of many of its rivers, as in the Sungi Mas, or Gold River, and the auriferous sands are washed for the metal. Native copper is found in the Sungi Lojang, or Copper River, and in small quantities elsewhere, furnishing material for a small export trade. Ores of the metal, such as cuprite, redruthite. malachite, and azurite, occur near Oisu and Atapupu. The occurrence of chrome iron ore, iron glance, pyrolusite, lead in the native state (*sic*), gypsum, and sulphur has also been noted. All attempts to foster a systematic copper-mining industry have hitherto failed.

The author discusses at some length the question of the existence of volcanoes on the island, and he subjoins a complete list of the earthquakes known to have occurred there from 1638 to 1884.   L. L. B.

MINING PROSPECTS IN NORTH-WEST SUMATRA.

Topographische en geologische Beschrijving van het noordelijk Gedeelte van het Gouvernement Sumatra's Westkust. By R. FENNEMA. *Jaarboek van het Mijnwezen in Nederlandsch Oost-Indië, Wetenschappelijk II^{te} Gedeelte, pp.* 244–252, *with a Geological Map and Sections. Amsterdam,* 1887.

That portion of the above paper which forms the subject of this abstract is preceded by a detailed topographical and geological description of the North-Western districts of Sumatra.

The Simpang Datar gold mine, described by former travellers, has long been abandoned, but in the Post-Tertiary alluvia placers are worked by the natives at many points, and the river sands are also washed for gold. The most promising spots are those where fragments from the older slate formation predominate; on the other hand, where fragments from the granite are more numerous, the gold yield is poor. It seems probable that the milk-white quartz, which, the natives aver, is always found accompanying the gold, is in fact the matrix of the precious metal. Whether these deposits could be profitably worked by a European company is a question which cannot be answered with certainty. If the processes now used in the district were employed, bearing in mind that wages per man per day would be at the very lowest 40 cents (= 10d.), the undertaking would not pay.

Coal is found in Eocene strata on the Bay of Tapanoelic, in the district of Soeliki, and in the river basins of the Si Lai and the Asip. Asip coal, according to the specimen analysed, yields 64·9 per cent. of coke, 4·3 per cent. of ash, and 4·03 per cent. of water. These coal deposits are of small importance, and the same statement holds good of the early Tertiary lignites of the Niboeng.

In the diluvium near Kota Poenkoet, fragments of quartz with copper pyrites and malachite have been found, and the occurrence of iron glance and magnetite is reported from other places. Lead ore and sulphur were formerly worked by the natives; but the presence of tin in this part of Sumatra seems to be a mere tradition.

With the exception of gold, it is certain that none of the minerals mentioned are to be found in paying quantities, and even were the deposits richer, they could not be worked at a profit on account of the expense and difficulties of transport, the scarcity of fuel, and the unfitness of the native labourers for systematic toil.   L. L. B.

VENTILATING MACHINE.

Rapport sur les appareils de ventilation et d'humidification de la Compagnie Française de ventilation. *Bulletin de la Société Industrielle du Nord de la France, No.* 58, 1887, *pp.* 38–41.

Ventilation accomplished by revolutions of wheel, which draws out the vitiated, or draws in fresh air, according to direction of rotation. Moisture is supplied by jets of water playing against plates, and thereby vapourized. The machine takes up little room, and displaces from 14,000 to 170,000 cubic feet of air per hour according to size. It is worked by water. G. W. B.

AN AUTOMATIC FEEDING PUMP.

Rapport sur la Pompe alimentaire Daussin. *Bulletin de la Société Industrielle du Nord de la France, No.* 58, 1887, *pp.* 29, 30.

The suction valve of the pump is connected by rods and levers to a float placed in a reservoir where the level of the water is the same as in the generator. As long as it is at the normal level, or above it, the valve remains half-open, and the supply ceases; when the water falls below the normal the supply commences automatically by the action of the float on the valve. Said to work satisfactorily, and has been awarded a medal by the society. G. W. B.

ON THE ORES OF COBALT, CHROMIUM, AND IRON OF NEW CALEDONIA.

Gisements de Cobalt, de Chrôme, et de Fer de la Nouvelle-Calédonie, et leur emploi industriel. By M. J. GARNIER. *Comptes Rendus, Société des Ingenieurs civils, Jan.* 21, 1887, *pp.* 26–28.

The ores are met with in the state of oxides.

The cobalt ore is the black oxide of cobalt. It contains up to 14 per cent. of the oxide of the metal, along with a large proportion of manganese.

Chromium is the most abundant metal in the country, except iron. It occurs along with iron in conglomerates and in decomposed serpentines.

The shore in the south is formed of chrome ironstone mixed with sand. The ore contains before washing 42 per cent. of sesquioxide of chromium, with a little wolfram.

Iron ore is very abundant; the soil itself is often formed of an ore, valuable from the absence of phosphorus. G. W. B.

A SPECIAL GRATE FOR USING POOR FUELS.

Foyer spécial pour l'utilisation des combustibles pauvres. By M. GEORGES ALEXIS-GODILLOT. *Bulletin de la Société Industrielle du Nord de la France, No.* 58, 1887, *pp.* 75–80.

To lessen the difficulty of combustion a special grate, the Grille-Pavillon, has been devised. It consists of a grate in the form of a half-cone resting on a horizontal one. The fuel is placed in a hopper in front of the furnace and pushed on to the apex of the half-cone by the revolutions of a screw. The rate of these revolutions can be regulated so as to supply the fuel.

In this way the fuel is dried before reaching the fire, while the form of the grate is such that nothing but fine ash escapes through it.

The advantages claimed for the furnace are :—
(1) Methodical combustion.
(2) No loss of fuel among the ashes.
(3) Better regulation of draught.
(4) Less chance of fire.
(5) Work of furnacemen simplified and made healthier.
(6) Grate easily cleaned.
(7) Consumption of smoke perfect.
(8) Refuse materials hitherto considered of no value can be utilised.

The description is accompanied by a transverse section of the furnace.

The table shows the results in steam produced by various kinds of fuel in different species of boilers.

TRIALS MADE WITH GODILLOT'S FURNACE.

Kind of Material used.	Percentage of Water, after drying at 110 degs C.	Boiler used.			Weight of Steam produced by 2¼ lbs. Fuel.
		Types.	Total Heating Surface.	Mean Working Pressure.	
			Sq. Ft.	Lbs.	Lbs.
Sawdust and shavings from joiners' shops	13·36	Tubular, two heaters	882	11¼	7
Dried tanyard refuse	55	Semi-tubular ...	484	11¼	4
Moist chips from dye works ...	62·3	Semi-tubular ...	1,076	12¾	3
Waste from the stripping of flax	29·5	Tubular, two heaters	882	11¼	6
Fir tree sawdust	33·75	Multi-tubular ...	2,292	12½	5½
Waste from the stripping of ramie (Boehmeria nivea)	10·59	Tubular, two heaters	882	10	7

G. W. B.

METHOD OF CLEARING A MINE OF FIRE-DAMP.

Procédé-Aroud pour l'expulsion du grisou. Comptes-Rendus de la Société de l'Industrie Minérale. 1887, pp. 238-243.

By this method the fire-damp is collected by a system of tubes at the moment of its disengagement from the fresh surface of the coal. and transported outside the underground works without being allowed to mix with the interior air.

The system may be made the basis of the general airing of the mine, or used along with one of the ordinary methods employed.

In the former case the object is to lead into the mine enough air for respiration, and to withdraw fire-damp in such proportion as to diminish chance of explosion.

The collecting tubes lead into a general conductor tube. and the gas is drawn out by aspiration.

System criticised unfavourably by M. Buisson.   G. W. B.

NEW PROCESS FOR THE EXTRACTION OF COPPER.

Traitement électrique des Minerais ou des Mattes de cuivre. By M. LE VERRIER. *Comptes-Rendus de la Société de l'Industrie Minérale,* 1887, *pp.* 3-6.

Marchese's process. The copper is precipitated by electrolysis in a bath of sulphate. The cathodes are of pure copper, the anodes of the melted and cast materials from which the metal is to be extracted.

Has been applied to extract copper from the accessory products resulting from the reduction of complex silver ores. About a third of this material is melted, and cast into plates for the anodes; the rest is used for the manufacture of sulphuric acid, in which the residue resulting from the same is dissolved, and used for bath. The copper produced by the process is said to be worth 100 francs per ton more than the ordinary.

G. W. B.

DISCOVERY OF PHOSPHATES AT BEAUVAL (SOMME).

Les Phosphates de la Somme. By M. BRETON. *Comptes-Rendus de la Société de l'Industrie Minérale,* 1887, *pp.* 14, 15.

A sand containing 74 to 80 per cent. of phosphates discovered by M. Merle.
It fills in cavities in the Upper Chalk, and is covered by Eocene clay.

G. W. B.

NEW SAFETY-CAGE.

Parachute Achard. By M. ACHARD. *Comptes-Rendus de la Société de l'Industrie Minérale,* 1887, *p.* 37.

Devised to obviate danger in case of the breaking of the chain. When such breakage happens the cage stops. This is accomplished as follows:—

In the parachute is a strong spring compressed by the tension of the chain. When the latter breaks the spring forcibly pushes out, catches on each side which press against the cables passed through metal guides fixed on each side of the parachute.

G. W. B.

NEW COAL-WASHER.

Lavoirs à valves de fond, à éliminations successives à travers la table (système Lemière). By M. LEMIÈRE. *Comptes-Redus de la Société de l'Industrie Minérale,* 1887, *pp.* 57-59.

The chief point in the arrangement is a series of valves in the washing table, which are opened and shut by the action of a piston. As the raw material passes along in the current of water the heavier materials—sandstone, pyrites, etc.—sink first, and pass downwards through the first valves. Through the successive valves beyond, the coal falls according to specific gravity and state of division. By these means the coal is assorted according to the quantity of ash it contains.

The arrangement is applicable to washing apparatus worked by piston or current of water.

In passing coal of different states of division the rate of piston and height of the anterior rims of the valves are altered.

G. W. B.

AUTOMATIC STOPS FOR WAGONS ON INCLINES.

Barrière automatique installée à la tête d'un plan incliné, aux mines de Lens.
By M. DINOIRE. *Comptes-Rendus de la Société de l'Industrie Minérale,*
1887, *p.* 81.

In the ordinary systems the barrier is let fall, or the chain hooked up by hand; in
the three systems described the arrest does not depend on the vigilance of the workman, but is brought into action by the descending wagon.

Consists of an iron tube, connected by cross bars to an axis, on which it turns.
On this tube two catches are placed. When the wagon approaches the tube swings
round on the axis and stops it. For the passage of the full wagon it is raised, and
hooked up by hand. It has been used for a time with satisfactory results.

G. W. B.

THE WALLING OF SHAFTS.

(1) *Muraillement des puits. Emploi d'un cintre mobile et d'un béton de ciment.*
By M. BUISSON. *Comptes-Rendus de la Société de l'Industrie Minérale,* 1887,
pp. 94-97.

(2) *Muraillement des puits en dailles et béton de ciment.* By M. MAUSSIER. *Same
publication, p.* 129.

A method of working with the aid of a circle of iron, by which much time is saved.
The height of this circle is from 25 to 39 inches.

In building, the bricks or stones are placed round, and in contact with it. When a
circular tier of the wall is thus completed the iron is raised and another tier of masonry
placed in position. In this way the time usually spent in measuring the diameter and
ascertaining the verticality of the masonry is saved.

When the space between the masonry and the walls of the shaft is to be filled in
with rubble and cement the larger size of iron circle is used, so that more of its height
may be left below as a support until the cement sets.

The method is applicable to the large shafts of pits as well as the smaller ones of
wells.

G. W. B.

SPONTANEOUS COMBUSTION OF PYRITES.

Combustion spontanée des Pyrites. By M. DE CATELIN. *Comptes-Rendus de la
Société de l'Industrie Minérale,* 1887, *pp.* 153-154.

The following conclusions are established :—

1.—That oxidation of pyrites is of itself sufficient to induce combustion.
2.—That the degree of oxidation or inflammation depends on the physical state
    and dissemination of the pyrites.

Conclusions arrived at from observations on pyrites of following composition :

						Per Cent
Sulphur	...	...	...	...	...	14·56
Iron	...	...	...	...	...	9·03
Alumina	...	...	..	...	...	21·75
Silica	...	...	...	...	...	44·00
Water	...	...	...	...	...	7·33
Chalk	...	...	...	...	...	2·33
Copper	...	...	...	...	...	1·00
						100·00

G. W. B.

FIRING IN MINES CONTAINING FIRE-DAMP.

(1) *Mèche pour coups de Mine, ne donnant ni flammes. ni étencelles (systéme Lamargère).* By M. PERRIN. *Same publication, pp.* 118–120.

In the system of M. Lamagère a new gun-cotton (*poudre fulmi-coton*), invented by him, is used. It is enclosed in a tube of zinc and allowed to protrude at both ends. May be fired in ordinary way by an incandescent body such as tinder. M. Lamagère uses a bichromate battery, which raises to a red heat a wire communicating with the explosive.

(2) *Allumage des coups de mines (système Lamargère).* By M. PERRIN. *Same publication, p.* 158.

Some modifications of the system described. G. W. B.

SUDDEN OUTBURSTS OF FIRE-DAMP AND CARBONIC ACID GAS.

(1) *Dégagements instantanés d'acide carbonique et de grisou à la Mine de la Combelle, division des Mines de Brassac.* By M. BRESSON. *Comptes-Rendus de la Société de l'Industrie Minérale,* 1887, *pp.* 243–250.

Points out coincidences between the escape of gas and barometric depression.

(2) *Observations sur la note présentée par M. Bresson, sur les dégagements instantanés de grisou et d'acide carbonique aux mines de la Combelle.* By M. CLERMONT. *Same publication, pp.* 264–268.

Reply to former. Points out that barometric depression will not influence gas in reservoirs shut off from atmospheric influence. Gas pent up in front of workmen will be let out in any case when its wall is ruptured. The only result of low pressure is to make the escape quicker. G. W. B.

A METHOD OF OVERTURNING WAGONS.

Plaques en fontes ondées pour le versage des wagons de la mine de Franchepré (Lorraine). By M. SERVIER. *Comptes-Rendus de la Société de l'Industrie Minérale,* 1877, *pp.* 271–272.

The wagon overturns itself on reaching the required spot, by reason of a depression into which its front wheel sinks. G. W. B.

SHAW'S PATENT MINE SIGNALLING APPARATUS.

The Colliery Engineer (Penna.), 1888, *Vol. VIII., pp.* 207–208.

The instrument consists of a plain brass tube, 12 inches long by $1\frac{1}{2}$ inches in diameter. This is closed at one end, while the other end is provided with a loose piston valve, held in a closed position by a bow spring in an elastic manner. This tube is tapped at one end with a brass pipe, $\frac{1}{4}$ inch in diameter, for the entrance of the gas to be tested. An exit nozzle is provided in the centre of the tube for the free escape of the explosive gas. At this point there is an igniter, an ordinary gas or lamp flame serving for the purpose. This is all supported on a small brass pillar secured to the top of the table. In close proximity to the piston valve there is a brass gong. The gases being tested are drawn by an air-pump from the workings and delivered into the brass tube, and, whenever they are in the least explosive. they ignite in contact with the flame. and the resulting explosion within the tube propels the piston valve against the gong, which in turn gives an alarm that can be heard at a great distance.

The apparatus is connected with the various portions of the mine by a number of $\frac{1}{4}$ inch iron tubes, a separate one for each place. One end of each tube is at the testing apparatus. and the other, to which is attached a rubber tube and metallic whistle, is placed in the highest part of the mine chamber. The apparatus is kept constantly in motion, and whenever gas is drawn through any tube it is immediately announced by its ignition in the brass tube and the ringing of the gong. An indicator shows which tube delivered the gas, and the operator then knows from which particular part of the mine it was drawn. He then moves a small lever, which reverses the current of air in that tube and blows a shrill whistle at the inside end of it. The miner can only stop this ear-splitting noise by pinching the tube shut for four seconds, and this action notifies the operator of the apparatus by another whistle that the miner has been warned. M. W. B.

INCANDESCENT LAMPS IN EXPLOSIVE GASES.

By H. HUTCHINS. *The Colliery Engineer (Penna.),* 1888, *Vol. VIII., p.* 197.

EXPERIMENTS.

I.—By means of a station battery and a water volta-meter, a sufficient quantity of hydrogen and oxygen was collected to about fill a small tin case. The fittings to this case were: an entrance at the side to admit a rod for piercing the bulb of the incandescent lamp; a window to ascertain whether or not the lamp was burning; and a wooden frame to hold the lamp. The lower end of the case was open and placed under water. A Swan lamp of 16 candle-power was used. and the carbon filament was raised to a white incandescence by the dynamo. The bulb was pierced and the gas immediately exploded with considerable noise. The carbon filament remained intact.

II.—Marsh gas and air were mixed in the proportions of 1 to $7\frac{1}{2}$ volumes, and collected in a case, similar to that used in the first experiment. The same kind of lamp was used, and the experiment conducted in the same manner. The gas did not explode, and the filament was found to be broken. due, of course, to the flying pieces of glass.

III.—Coal gas was used alone in this experiment, which was conducted in a manner similar to the preceding ones, except that a Maxim lamp was used. The result on piercing the bulb was that the filament continued to burn.

IV.—In this experiment coal gas was used, mixed with air, in the proportion of 1 to 6 volumes. A Maxim lamp was used, and the experiment conducted similar to the preceding ones, except that a pressure gauge was used to make certain whether the gas exploded or not. On piercing the bulb, the filament was not broken. and in a few seconds the gas exploded with some noise. The tin case was thrown 5 or 6 feet into the air, and the filament was broken.

No further experiments were made with marsh gas, as it is more explosive than coal gas.

CONCLUSIONS.

It appears therefore that the inrush of gases does not break the filament of a Swan or Maxim lamp, before the gas can become ignited, and that the breaking of such lamps is dangerous, because the incandescent filament coming in contact with gases and air in explosive proportions will explode them. Every precaution was taken in connecting the lamp leads. so as to prevent by any possibility an arc or spark being formed at the spring in the base of the lamp sockets. which might explode the gas.

M. W. B.

IRON ORES OF MICHIGAN.

Mode of Deposition of the Iron Ores of the Menominee Range, Michigan. By JOHN
FULTON. *Transactions of the American Institute of Mining Engineers (Advance
Sheets)* 12 pp., *with nine Figures in text. Read July,* 1887.

The Menominee Range is an east and west ridge running along the north side of
the Menominee River where it flows between Michigan and Wisconsin. It is about
27 miles long and from 200 feet to 300 feet above the level of the low swamps from
which it rises. The rocks of which it is composed belong to three divisions of the
great Huronian series. Of these the first is the "Norway Limestone Belt," a group of
crystalline and siliceous light-coloured limestones, at least 1,200 feet thick. The next
group is the "Quinnesec Ore-Formation," consisting of 1,000 feet of siliceous or
jasper slates impregnated with iron oxide, capped by argillaceous hydro-mica black
and flesh-coloured slates. The third division, the "Hanbury Slate Group," 2,000
feet thick, is composed of dark grey slaty or schistose beds with occasional quartz
bands. These divisions are conformable to one another and dip at high angles south
to the east of Quinnesec and north to the westward. Whether the first or the third be
the lowermost set of deposits is not yet known. Several outliers of Potsdam sandstone
(Cambrian) lie horizontally on the crest of the ridge.

The ore occurs mainly in the Middle Division as lenticular masses—approximately
contact deposits—having irregularly elliptical outlines and interbedded in the jasper
and clay slates. It is a peculiarity of these masses of iron ore that they "pitch"
westward.

The most probable theory of the origin of these ore-deposits is said to be that
which regards them as thinly bedded ferriferous carbonates somewhat mixed with
dusty magnetite and wholly altered to hematite by heat and chemical agencies. The
composition of the ores is given as follows :—

Mines.		Met. Iron.		Insoluble Matter.		Phosphorus.
Vulcan	... ...	58·75	...	9·00	...	·045
Cyclops	... ...	61·37	...	7·33	...	·015
Norway	... ...	53·27	...	13·02	...	·036
Quinnesec	... ...	64·56	...	7·69	...	·037

This mining field was opened in 1877, and from that date to the end of 1887
7,500,000 tons of ore have been shipped. G. A. L.

PYRITE IN COAL.

Modes of occurrence of Pyrite in Bituminous Coal. By AMOS P. BROWN. *Transactions of the American Institute of Mining Engineers (Advance Sheets),* 8 pp.
Read February, 1888.

The observations detailed in this paper are confined to the bituminous and semi-
bituminous coals of Pennsylvania, but the author states that certain generalisations
which he makes will be found universally applicable.

Pyrite occurs in coal (1) as nodules or lenticular masses; (2) in defined and persistent bands; (3) in thin flakes parallel to the bedding; (4) as incrustations in joints
and cavities; and (5) in fine rounded particles disseminated throughout the coal. The
compact variety (the "hard sulphur" of the miners) is commoner in the lower coal-
seams; the friable ("soft" or "black sulphur") in the upper. The nodules, com-
monest in the lower seams of Mercer, Brookville, and Clarion are often formed round
a nucleus of fish-remains.

The stratified form is generally friable, impure, and associated with the mineral charcoal. A very marked pyrite layer is characteristic of the Lower Kittanning Seam in Clearfield and Centre Counties and in part of Cambria County, disappearing gradually in the latter.

As thin flakes pyrite is local and usually occurs in the upper portion of the seam. It is both hard and soft in places, cannot be separated from the coal by hand-picking, and on weathering gives rise to white efflorescence of sulphate.

The incrusting pyrite is always crystalline and evidently an after-product derived from the surrounding rocks. It is very local, and seldom exceeds one millimetre in thickness. The disseminated form is regarded as similar to the nodular variety on a microscopic scale. It is found in many of the Pittsburg gas coals, but seldom in injurious quantities.

The iron of the pyrite came primarily from the old Carboniferous soil, but, directly, probably from the coal-forming plants themselves, since the ashes of plants contain a notable quantity of iron. The sulphur is referred either to gypsum or other sulphate, or to hydrogen sulphide. "In the former case it would be required that the gypsum be decomposed in presence of the iron salts with formation of ferrous sulphate, which latter was afterwards reduced to sulphide by the decomposing organic matter; while, on the other hand, if the reducing agent were hydrogen sulphide, the pyrite would probably be formed directly." The iridescent scum on the surface of stagnant pools is often pyritous.

That in the case of the nodular pyrite the necessary organic matter was of animal origin is likely from the fish remains which are associated with it.   G. A. L.

CHEAP MINING AND MILLING.

By F. W. BRADLEY. *Engineering and Mining Journal* [New York], 1888, *Vol. XLV., p. 324.*

The Spanish Mine (Nevada Co., California) is mining a large deposit of soft slate that crops out from 30 to 100 feet in width on the face of a steep mountain. The slate is seamed in all directions with small strings of quartz; there is a little gold in the slate, some in the quartz, and a considerable amount in a loose free state in the clay parting between the quartz and the slate. The formation has a dip of about 80 degrees from the horizontal.

The mining is done in open cuts on the croppings over the main tunnel, which starts from the surface immediately at the top of the mill building, and follows the course of the deposit into the mountain. During stormy weather ore is obtained from accumulated supplies and by stoping the best portions of the deposit over the tunnel and replacing the same by square sets of timber in such a manner as to form ore-bins for storing and loading into the tunnel-cars ore broken in the cuts. In the cuts the softest streaks near the foot wall are stoped by Chinese miners. Ore left on the foot wall soon slacks off, and ore on the hanging wall and also portions of the hanging wall cave in. All waste is separated as much as possible from the ore and left in the worked out cuts, a strong pillar being left at the end of each cut.

The tunnel has a grade of 2½ per cent. A brake on the last car controls a train of ten loaded cars coming out, and a mule easily hauls back the empty cars.

The milling plant cost:—

	£	s.	d.
Four Huntington mills and self feeds	1,269	4	10
Labour, erection, etc.	460	19	4
Silver-plated amalgamating plates	397	1	6
Water pipe and wheel, shafting and pulleys ...	244	3	0
Lumber, building, and V-flume	238	18	0
Hardware	205	4	4
Blake crusher	123	13	10

Cost of milling plant under cover and running:—⎫
Freight, £4 16s. per ton from San Francisco; ⎬ £2,939 4 10
and lumber, £4 10s. per 1.000 feet. ⎭

The mills have been in operation about 21 months. For the first ten months, one 5-foot mill and one 4-foot mill crushed 17,200 tons of ore; four months later, two 5-foot mills were added to the plant, and the four mills in five months have crushed 19,402 tons. The ore (of which 27 cubic feet is a ton) consists of about one-third hard quartz, one-third tough slate, and one-third decomposed quartz and slate. The four mills, requiring 22 horse-power, are running at 60 revolutions per minute, and are discharging through a No. 5 slot screen.

The mills crush from 120 to 140 tons per day, depending upon the proportion of quartz in the ore. They amalgamate inside the mills, obtaining 45 per cent. of the gold saved around and inside the mills, and 55 per cent. on the plates. The tailings are untouched after leaving the plates. About one ounce of mercury is lost per 16 to 31 tons of ore crushed, depending upon its value.

The costs per ton for a few recent months are as follow:—

1887-8. Mine—	Sept.	Oct.	Nov.	Dec.	Jan. and Feb.
Worked days	22	28	30	25	36
Tons of ore	2,796	3,443	4,047	2,972	4,256
	s. d.	s. d.	s. d.	s. d.	s. d.
Mining	0 9·84	0 11·42	0 10·42	0 8·64	0 5·81
Dead work	0 4·80	0 1·63	0 1·34	0 4·03	0 5·42
Delivering ore to mill...	0 2·40	0 2·40	0 2·45	0 3·41	0 2·93
General expenses ..	0 0·96	0 1·16	0 0·86	0 1·20	0 1·44
	1 6·00	1 4·61	1 3·07	1 5·28	1 3·60
Mill—					
Worked days	20	24½	29	23	32
	s. d.	s. d.	s. d.	s. d.	s. d.
Mill expenses	0 5·28	0 5·86	0 4·61	0 5·42	0 5·42
Water for power ...	0 2·64	0 2·26	0 2·40	0 2·30	0 2·30
Handling ore	0 2·16	0 2·20	0 2·11	0 2·50	0 2·30
General expenses ...	0 0·96	0 1·15	0 0·86	0 1·20	0 1·44
	0 11·04	0 11·47	0 9·98	0 11·42	0 11·46
	s. d.	s. d.	s. d.	s. d.	s. d.
Bullion produced ...	4 7·68	3 7·68	2 7·20	2 7·49	2 7·39
Total expenses ...	2 5·04	2 4·08	2 1·05	2 4·70	2 3·06
Profit	2 2·64	1 3·60	0 6·15	0 2·79	0 4·33

M. W. B.

ON THE VARIATIONS OF THE VOLUME OF FIRE-DAMP GIVEN OFF BY A WORKING DISTRICT.

Versuche über die allmälige Engasung einer Bau-abtheilung des Schachtes Kaister-stuhl der Steinkohlenzeche Ver, Westfalia bei Dortmund. Anlagen zum Haupt-berichte der Preussischen Schlagwetter-Commission, Vol. IV., pp. 113-124, and Plate.

The experiments were made in the No. 2 East District of the Sonnenschein Seam, of the following section:—

						Ft.	Iu.
Coal ...	...	...	...	...	...	4	7¼
Stone	...	...	...	...	...	0	11¾
Coal ...	...	...	...	...	...	0	11¾
						6	6¾

The roof consisted of about 1 foot 8 inches of shale, covered by about 60 feet of coarse ferruginous sandstone. The thill consists of shale for a depth of about 30 feet.

The district covered an area of about 500 feet on the level, and about 250 feet to the rise of the seam at an angle of about 60°, and comprised about 20,000 tons of coal.

The district was worked in the whole by means of levels, about 33 feet apart, and holings at intervals of about 100 feet. The air entered at the lowest level, and passed upwards to the return air-way where the observations were made in a length of 33 feet, which was carefully lined with timber. The observations were made daily of the volume of air, and samples were taken for analysis.

Month, 1884–85.	Volume of Fire-damp		Coal worked per Month.	Surface of Coal Exposed.		Volume of Gas.			Remarks.
	Per Day.	Per Month.		Throughout the District.	During the Month.	Per Ton of Coal Worked.	Per 100 Square Feet of total Surface Exposed.	Per 100 Square Feet of Surface Exposed during Month.	
	Cub ft.	Cub ft	Tons.	Sq ft.	Sq. ft.	Cub. ft.	Cub ft	Cub ft.	
December.	21,442	664,700	1,012	13,640	6,300	656	487	105	First working or whole.
January ...	21,416	663,900	755	19,630	5,990	878	337	110	
February..	21,857	612,000	734	25,190	5,560	833	243	110	
March ...	25,113	778,500	890	32,290	7,100	874	241	109	
April ...	22,180	665,400	571	36,650	4,360	1,165	181	152	Whole and broken.
May ...	18,384	569,900	446	38,390	1,740	1.277	148	...	
June ...	21,380	641,400	855	...	...	750	...	...	Removal of pillars or broken.
July ...	19.196	595,100	1,287	...	...	462	...	...	
August ...	19,316	598,800	1,368	...	...	437	...	...	

The Commission are of opinion that :—

1.—The volume of gas produced in a district does not increase as the workings become more extensive, because it chiefly depends upon the quantity of coal worked, and the area of freshly exposed coal surface.

2.—In the second or broken working, the volume of gas produced is less than during the first working, because, at the end of a short time, the coal loses a great portion of its contained gas.

They agree that :—

1.—The ventilation should be more active during the first working than during the removal of pillars.

2.—The volume of air to be circulated through the workings must not depend upon the number of workmen and animals, but should be proportionate to the output of coal.

M. W. B.

k

COAL-FIELDS AND COAL PRODUCTION OF THE WORLD.

Ueber die Steinkohlen-Vorkommnisse und Production auf der Erde. CONRAD
BLÖMEKE. *Berg- und Huettenmaennische Zeitung, Vol. XLVII., Nos. 12 and
14, pp.* 105-107 *and* 124-126.

Reviewing the coal-fields of the world, the author finds that Great Britain possesses
about 146,480 million tons of workable coal, and Germany about 400,000 million tons.
Taking the production of the year 1882 as a basis, Germany could supply her own
wants for 6,000 years, or those of all Europe for 1,500 years. The world's production
for the year 1882 was as follows :—

		Pits.	Workmen.	Production in Tons.	Value. £	Per Man. Tons.	£	s
Germany	...	488	195,961	51,118,595	13,393,000	266	68	6
Great Britain	...	...	503,987	158,847,476	67,097,180	315	133	2
United States	...	...	......	31,859,996	14,111,220	...	...	
France	...	252	104,995	20,046,796	9,968,430	190	94	18
Belgium	...	...	......	17,590,989	7,035,810	...	...	
Austria	...	157	39,644	7,194,096	2,287,590	181	57	14
Russia	...	...	......	3,773,665	......	...	...	
China	...	...	......	2,965,000	......	...	...	
India	...	...	......	2,550,000	......	...	...	
Australia	...	...	......	2,219,000	991,280	...	...	
South America	...	...	......	2,000,000	......	...	...	
Canada	...	...	......	1,329,000	692,640	...	...	
Spain	...	465	9,280	1,044,480	463,650	112	49	19
Japan	...	...	......	931,780	......	...	...	
Hungary	...	...	......	900,000	......	...	...	
Tasmania	...	...	......	428,000	319,100	...	...	
Sweden	...	...	......	249,000	......	...	...	
Africa	...	...	......	200,000	......	...	...	
Turkey	...	...	......	100,000	......	...	...	
Portugal	...	...	......	12,963	10,310	...	...	

Total production in round numbers 306,000,000 tons.

The production of brown coal in
1882 was about   ...   ...   86,000,000 tons.

The production per year from 1862 to 1882 increased about threefold, the work per
man employed about twofold, both increases materially lowering the value of the
material.   A. R. L.

GEOLOGICAL NOTES FROM ROUMANIA.

Geologische und bergbauliche Skizzen aus Roumaenien. C. ALBERTS. *Berg- und
Huettenmaennische Zeitung, Vol. XLVII., No.* 15, *pp.* 131 *and* 132.

Roumania is separated from Transylvania by the Southern Carpathians, stretching
from west to east and ending at a small affluent of the Sereth called the Buzen, and
by the Carpathians proper which start from the opposite bank of the same stream and
trend towards the north. This latter range forms a boundary between extensive chalk
formations on its western slope, and the earlier Tertiary formation, the Eocene, to the
eastward, which is overlaid with diluvial and other deposits. The Roumanian petroleum
borings are in the Tertiary formation, and all the so-called "oil lines" in which
petroleum and naphtha occur run from west to east. It follows from this that the

clefts from which the petroleum ascends from below must run in the same direction. These again must have had their origin in a volcanic upheaval in which the southern part of the northern range was pressed against the eastern end of the Southern Carpathians, and as the clefts go through the Tertiary formations the upheaval must have succeeded them in point of time. It would appear that the petroleum and naphtha now worked in this district are drawn from minor reservoirs of sandstone, etc., near the surface, which again are gradually filled through the before-mentioned clefts from larger reservoirs far below. This also explains the fact that the bore-holes are productive for short periods only, and it may be expected that the upper reservoirs will in time be filled afresh from below and again become workable.

In working, the best results have been obtained from shafts sunk to various depths down to 100 fathoms, and about 4 feet square. They can be used to advantage only in districts where not much water is met with; but there are many of these, and difficulties are much reduced by the remarkable aptitude and fitness for the work shown by the native sinkers.

A bore-hole in Draganesci, which for a long time gave a daily supply of 1,000 barrels, forms a proof that there must be enormous reservoirs of petroleum at some lower depth.

A neglected Roumanian industry, for which a great future is in store, is the winning of the brown coal which exists there in several districts in very large quantity and of very good quality. The available wood supplies being almost exhausted, some other fuel must soon be provided; but as the Roumanians themselves seem blind to the opportunity and will not lay out money in such undertakings, the new enterprise will probably have to be conducted by foreign capital. A. R. L.

COAL-DUST EXPLOSION AT KREUZGRÄBEN COLLIERY.

Die Kohlenstaub-Explosion zu Grube Kreuzgräben bei Saarbrücken. J. SPRENGER. *Berg- und Huettenmaennische Zeitung, Vol. XLVII., No. 15, pp. 132–136.*

On February 15th of the present year an explosion took place at the Kreuzgräben Colliery. The men at bank heard several short detonations following each other in quick succession and saw clouds of black dust come up the shaft. The cause of the accident remains a mystery; but what apparently took place was first an explosion of gas, raising a cloud of coal-dust, which in its turn also exploded. This raised another cloud, so that a succession of explosions followed.

The shaft is about 250 fathoms deep, and as yet only one seam is worked. The workings are divided into three flats, called the Eastern, Western, and Middle Fields respectively. The Western Flat is worked out, the other two being still in operation. The Middle Flat where the explosion took place is extremely dry and warm, and the coal-dust collects in every crevice. Inflammable gas is present in small quantities only, and not likely to be in itself dangerous. The dampness of the East Flat will probably account for the explosion not being communicated to it.

The usual rules as regards safety-lamps, the firing of shots, etc., are very strictly enforced in the pit, and no theory seems to explain the accident except that of some breach of the regulations. The cages stood at meetings, and this is said to account for their not being wrecked; but the damage in the pit was very great. Loaded tubs of 15 or 16 cwt. were thrown for 40 or 50 feet and upset one over the other, timbering was thrown down and broken, and other considerable damage done, and this more especially near the shaft. Forty-six men were in the flat at the time, of whom 41 were killed, while some 30 men in the Eastern Flat remained uninjured.

The Kreuzgräben Colliery is bounded on its western side by the Camphausen Colliery, where the great explosion took place in March, 1885. A. R. L.

AN INCLINE WORKED BY ELECTRIC MACHINERY.

Electrischer Göpel zur Förderung auf einfallender Strecke im Salzwerke Neu-Stassfurt. MESSRS. SIEMENS & HALSKE. *Berg- und Huettenmaennische Zeitung, Vol. XLVII., No. 17, pp. 154–157. Illustrated in the text.*

The Hammacher shaft of the Neu-Stassfurt salt mine has a depth of 335 fathoms. About three years since its owners, working to the dip of the seam, had occasion to make an incline at an angle of less than 40 degs., and decided to work it by the electric machinery of Messrs. Siemens & Halske, who had already carried out similar installations in the mine.

The conditions were as follows:—From an existing steam engine, which was to work a primary dynamo machine 170 yards from the pit-head, down the shaft to the incline, a distance of 600 yards, required a double circuit for the electric stream. This was effected by a bare copper wire above ground, and a cable well covered and protected by wooden casing for the rest of the distance. The problem was to transport, in a shift of 8 hours, 100 full tubs of $23\frac{1}{2}$ cwts. and 100 empties of nearly $8\frac{1}{2}$ cwts. respectively, up and down an incline measuring 170 yards in length and 110 yards in height. Allowing about one minute for onsetting, this came to about a pair of tubs (one full and one empty) every three minutes, with a speed of 2·84 feet per second. Taking for friction $1\frac{1}{2}$ per cent. of the total load multiplied by 2·84, the speed per second, and for the lift the weight of the material, 15 cwts., multiplied by the height in feet and divided by the number of seconds ($15 \times \frac{330}{180} = 27 \cdot 5$), the total work to be done on the incline came to $1 \cdot 37 + 27 \cdot 5 = 28 \cdot 87$ foot-cwts. $= 3,233$ foot pounds per second. Allowing for loss of power in the intermediate machinery, this must be multiplied by $2\frac{1}{2}$, giving an effective of 15 I.H.P. to be supplied by the steam engine at bank.

The useful work done by the electric machinery proper is about 53 per cent., while the loss in the machinery of the incline is about 25 per cent. The winding drum, with a diameter of 4 feet 1 inch, makes 13·3 turns per minute, and it is driven with a strap and wheel gearing from an electro-motor making 1,000 revolutions per minute. The strap, about whose efficiency there had previously been some doubt, was found to answer excellently and to conduce greatly to smoothness in working. The primary dynamo is of Messrs. Siemens & Halske's much-used D_o type, and with a tension of 370 volts sends 22 ampères of current to the electro-motor, from 5 to 6 per cent. being lost in transmission.

The incline did satisfactory work for about seventeen months, and, its purpose being fulfilled, it was then abandoned.

The makers are now in a position to supply improved machinery which gives a much higher percentage of useful work than in the installation described.

A. R. L.

THE EXPERIMENTS OF THE PRUSSIAN FIRE-DAMP COMMISSION ON EXPLOSIONS OF COAL-DUST AND GAS.

Schluss-bericht über die in der Versuchsstrecke auf der Fiskalischen Steinkohlengrube König bei Neunkirchen (Saarbrücken) bezüglich der Zundung von Kohlenstaub und Grubengas angestellten Versuche. Anlagen zum Haupt-berichte der Preussischen Schlagwetter-Commission, Vol. IV., pp. 1–88, and Plate.

The experiments were made at the König Colliery under the direction of Herr C. Hilt, and a copy of his preliminary report, containing a description of the apparatus and a summary of the experiments made up to the end of 1884, was translated by Mr. Theo. Wood Bunning, and appears in Vol. XXXIV. of the Transactions, pp. 199-245.

The experiments were resumed in 1885, and the official report contains the results of all the experiments at the König Colliery. The following account will be confined to these later experiments.

I.—EXPERIMENTS UPON THE LENGTH OF FLAME OF BLOWN-OUT SHOTS, IN THE ABSENCE OF COAL-DUST AND GAS, CHARGED WITH 8·1 OZS. OF POWDER.

Register of Experiments.	Description of Stemming.	Volatile Matter.	Length of Flame.	Tub weighing 1,627 lbs. was moved.
1884		per 100.	Feet.	Feet.
1-10	Potters' clay	—	... 9·8-13·1	... Nil.
1885.				
208, 209	Powdered shale	—	... 6·5-8·2	... Nil.
217	Equal parts of powdered shale and coal-dust from Pluto ...	—	... 16·4	... ·3
67	Coal-dust from Louisenthal ...	33·4	... 31·1	... ·6
57-59	,, Kohlscheid ...	6·0	... 31·1	... ·8
61	,, Königin Louise	32·0	... 31·1	... 1·0
1, 2, 3	,, König	30·0	... 31·1	... ·8-1·1
58-60	,, Pluto	22·0	... 31·1	... 1·3

II.—EXPERIMENTS UPON THE LENGTH OF FLAME OF BLOWN-OUT SHOTS IN THE PRESENCE OF COAL-DUST AND ABSENCE OF GAS.

(a) Relative effects of the different shot-holes when 32·8 feet of the gallery was strewed with fine coal-dust from Hansa :—

1.—*Stemmed with Clay.*

Register of Experiment.	No. of Shot-hole.	Weight of Powder used	Length of Flame.
1884		Ozs.	
19 ...	1 ...	8·1	... 13·9
20 ...	2 ...	,,	... 26·5
21 ...	3 ...	,,	... 9·5
22 ...	4 ...	,,	... 13·9
23 ...	5 ...	,,	... 9·5
24 ...	6 ...	,,	... 62·0
25 ...	7 ...	,,	... 57·4
26 ...	4 ...	17·6	... 70·8

2.—*Stemmed with Coal-dust.*

Coal-dust from.	Register of Experiments.	No. of Shot-hole.	Weight of Powder used.	Length of Flame.
	1885.		Ozs.	Feet.
Saarbrück. Gerhard Colliery, Beust Seam— Very fine dust, containing 33·4 per cent. of volatile matter.	28 ...	1 ...	8·1	... 70·7
	29 ...	2 ...	,,	... 75·0
	30 ...	3 ...	,,	... 75·0
	31 ...	4 ...	,,	... 79·0
	32 ...	5 ...	,,	... 94·5
	33 ...	6 ...	,,	... 28·8
	34 ...	7 ...	..	... 79·0
	34a ...	4 ...	17·6	... 61·9
Westphalia. Hansa Colliery— Fine dust	27 ...	6 ...	8·1	... 66·2
Finer dust	34b ...	4 ...	17·6	... 70·7

(b) Influence of the length of strewing upon the length of flame. The shot was charged with 8·1 ozs. of powder and stemmed with coal-dust:—

Coal-dust from.	Register of Experiments.	Length of Strewing, in feet.							
		32·8		**65·6**		**98·4**		**131·2**	
		Length of Flame.	Tub weighing 1,627 lbs. was moved.	Length of Flame.	Tub weighing 1,627 lbs. was moved.	Length of Flame.	Tub weighing 1 627 lbs. was moved.	Length of Flame.	Tub weighing 1,627 lbs. was moved.
		Ft.	Ft.	Ft.	Ft.	Ft.	Ft.	Ft.	Ft.
Aix la Chapelle. Kohlscheid– anthracite, with 6 per cent. of volatile matter, fine	1885. 67, 68	31·1 39·3	·6 ·5	31·1	·7	..	..	..	..
Westphalia. Massen– coking, fine and hard	89	49·2 49·2	·9 1·6	49·2	1·4	..	..	72·1	1·6
Bonifacius—bituminous, fine, granular and hard	1834. XIV.	55·2	(?)	..	..	..	..	..	..
Do.—very fine, 23·4 per cent. of volatile matter	XVI.	55·2	(?)	52·5	(?)	..	..	..	..
Viktor—coking, very fine and hard	82, 83	72·1	1·8	67·2	1·9	..	..	72·1	1·6
Louise, Tief-bau—coking, very fine	86, 88, 96	62·3 62·3	1·1 1·1	..	..	..	..	..	..
Rhein-Elbe—gas, fine	47	..	..	..	..	..	..	78·7	(?)
Shamrock—coking, very fine	73, 74, 77, 80, 81	62·3 62·3	1·3 1·4	83·6	1·2	..	··	114·8 103·8	1·3 1·6
Joachim—coking	102, 103	72·1 72·1	1·4 1·5	88·5	1·1	..	..	88·5	1·6
Hannover—fine, granular, and hard	I., II., III.	49·2 52·5	(?) (?)	*62·3	(?)	..	..	..	..
Do.—very fine, with 28·4 per cent. of volatile matter	IV., V., VI., VII., VIII., IX.	*62·3 72·1	}(?)	*83·3	(?)	95·1 144·3 147·6	..	..	..
Neu-Iserlohn—coking, very fine	44, 45, 46	..	..	126·6	(?)	148·6	(?)	183·7	(?)
Pluto—gas, fine	1885. 40, 41, 42, 43, 195	113·5	(?)	139·7	(?)	177·1	(?)	190·2 190·2	(?) 8·3
Saarbruck. König—gas, coarse	1884. 37, 38, 39	40·1	(?)	40·1	(?)	35·6	(?)	..	..
Do. gas, very fine	1885. 10, 11, 12, 13, 19, 20, 23	62·3 72·1 67·2	(?) (?) 1·1	118·1 118·1	1·0 1·3	..	..	129·6	1·4
Kreuzgräben — gas, very fine and brownish	113	41·0	·8	41·0	·8	..	..	..	..
Gerhard—fine and hard	129, 130	49·2 45·9	·5 ·4	49·2	·5	..	..	..	..
Itzenplitz—fine and hard	134	55·7	1·0	50·8 54·1	·6 1·1	..	..	..	..
Camphausen—gas, very fine	117, 118, 119	72·1 78·7 72·1	1·7 1·6 1·6	82·0 88·5	1·4 1·6	..	..	118·1	1·4
Silesia. Cons. Rudolf—gas, very fine	150	45·9 45·9	1·2 1·3	..	..	..	..	..	..
Konigin Louise—gas, fine and hard	141, 143	45·9	1·0	59·0	1·4	..	..	..	..
König, very fine	145, 146, 148	67·2 67·2	1·5 1·6	67·2	·9	..	..	..	..
Cons. Johann Baptist — very fine, with coarse grains	155, 156, 158	54·1 54·1	1·4 1·8	54·1	·9	..	..	..	..
Peterkowitz—fine and hard	160, 161	42·6 42·6	·9 ·8	49·2	1·4	..	..	..	..
Do. very fine, like Pluto	165	..	..	62·3	·9	..	..	..	..
England. Madeley—very fine	174, 175, 176, 177	67·2	(?)	88·5	(?)	103·3	(?)	136·1	(?)
Do. very fine and a few grains	178, 179, 180, 181	67·2	(?)	83·6	(?)	..	..	..	..
Do. very fine	183, 187, 188	75·4	(?)	..	..	..	..	180·0	(?)
Do. very fine and a few coarse grains	185, 186	62·3	(?)	..	..	..	··	83·6 95·1	..

* Stemmed with clay.

(c) Influence of the fineness of the dust. Experiments with dust from the König Colliery, Blücher Seam, with 8·1 ozs. of powder, with 32·8 feet of strewing:—

Register of Experiments	Coal-dust passed through Sieve with opening of.	Stemmed with Clay.			Stemmed with Coal-dust.		
		Length of Flame.	Tub was moved.		Length of Flame.	Tub was moved.	
			Weighing 646 lbs	Weighing 1,627 lbs.		Weighing 646 lbs.	Weighing 1,627 lbs.
1884.	Inches.	Feet.	Feet.	Feet	Feet.	Feet.	Feet.
279, 285	1·18	26·5	1·3	...	35·5	3·1	...
278, 284	·79	31·0	1·4	...	42·5	4·7	...
277, 283	·59	31·0	1·7	...	40·1	7·3	...
276, 282	·39	31·0	1·6	...	40·1	6·8	...
275, 281	·23	31·0	1·9	...	40·1	6·5	...
274	·04	48·8	2·4	...	61·9	6·9	...
1855	Unscreened	49·2	...	1·1	62·3	...	1·2
8, 9, 10, 11, 12. 13		55·7	...	1·1	67·2	...	1·2
					67·2	...	1·2
					72·1	...	1·2

(d) Influence of the chemical composition of the dust. These experiments are contained in the preliminary report.

III.—EXPERIMENTS UPON THE BEHAVIOUR OF BLOWN-OUT SHOTS IN THE PRESENCE OF COAL-DUST AND GAS.

(a) Using from 1 to 7 per cent. of gas in the absence of coal-dust, shots of 8·1 ozs. of powder:—

Register of Experiments.	Percentage of Gas in the Air.	Length of Flame.		
		Stemmed with Clay.	Stemmed with Dust passed through ·27-inch mesh, from König Colliery.	Stemmed with very fine Dust, from König Colliery.
1884	Per cent.	Feet.	Feet	Feet.
128, 134, 230 ...	1·18	22·6	31·1	48·8*
129, 135, 231, 232..	2·36	30·8	40·0	48·8* / 62·0*
130, 136, 233, 234..	3·54	22·6	40·0	44·6* / 57·4*
131, 137, 235 ...	4·72	44·6	40·0	97·1*
132. 138, 236 ...	5·90	35·7	87·6	113·5†
133, 139, 237 .	7·10	170·0	139·7	141·3‡

* Speed of flame about 3·3 feet per second. † Speed of flame very rapid.
‡ Speed of flame instantaneous.

(b) Using from 1 to 7 per cent. of gas in the presence of dust. These experiments are detailed in the preliminary report.

(c) Using 3 per cent. of gas:—

1.—*Without coal-dust strewing, and shots of 8·1 ozs. of powder.*

Stemmed with Coal-dust from.	Register of Experiments.	With 706 cubic feet of Explosive Mixture.		With 1,412 cubic feet of Explosive Mixture.	
		Length of Flame.	Tub Weighing 1,627 lbs. was moved.	Length of Flame.	Tub Weighing 1,627 lbs. was moved.
	1865.	Feet.	Feet.	Feet.	Feet.
Kohlscheid anthracite. fine	63	36·0	1·2	36·0	·9
König, very fine	45	41·0 / 45·9	1·8 / 1·9	36·0 / 39·3	1·5 / 1·4
Pluto, very fine	64	39·3	1·8	45·9	1·9

2.—Strewed with Coal-dust from König, and shots of 8·1 ozs. of powder.

Percentage of Gas.	Register of Experiments.	Volume of Explosive Mixture.	Length of Strewing.	Stemmed with Clay.		Stemmed with Coal-dust.	
				Length of Flame.	Tub Weighing 1,627 lbs. was moved.	Length of Flame.	Tub Weighing 1,627 lbs. was moved.
	1885.	Cubic feet.	Feet.	Feet	Feet.	Feet.	Feet.
Nil.	8, 9, 10, 11, 12, 13 ...	...	32·8	{ 49·2 / 55·7 }	1·1 / 1·1	{ 62·3 / 67·2 / 67·2 / 72·1 }	1·1 / 1·1 / 1·1 / 1·1
Nil.	19, 20 ...	...	65·6	...	...	{ 118·1 / 118·1 }	1·0 / 1·3
Nil.	24 ...	...	131·2	...	...	129·6	1·4
3	14, 15 ...	706	32·8	75·4	(?)	88·5	1·4
3	16, 17 ...	706	65·6	{ 83·6 / 93·5 }	(?) / (?)	...	...
3	18 ...	1,412	65·6	144·3	(?)	...	...
·3	21 ...	1,412	131·2	...	...	{ 170·6 / 173·9 }	2·7 / 3·1

The following experiments were made with strewings of different dusts, using shots of 8·1 ozs. of powder, stemmed with coal-dust:—

Coal-dust from.	Register of Experiments.	Length of Strewing.	Length of Flame.		Tub weighing 1,627 lbs. was moved.	
			Without Gas.	With 3% of Gas.	Without Gas.	With 3% of Gas.
	1885.	Feet.	Feet.	Feet.	Feet.	Feet.
Aix la Chapelle.						
Kohlscheid—anthracite, fine	67, 68, 69, 70 ..	32·8	{ 31·1 / 39·3 }	36·0 / 41·0	·6 / ·6	·9 / ·9
Do. do.	71, 72	65·6	31·1	49·2	·7	·8
Westphalia.						
Massen—fine and hard	89, 90, 91, 92 ..	32·8	{ 49·2 / 49·2 }	75·4 / 72·1	·9 / 1·6	2·0 / 1·5
Do. do. ..	93, 94 ..	65·6	49·2	75·4	1·4	1·3
Viktor—very fine and hard ..	82, 83, 84, 85 ..	32·8	{ 72·1 / 62·3 }	80·3 / 72·1	1·8 / 1·1	1·9 / 1·7
Do. do. ..	86, 87 ..	65·6	67·2	93·5	1·9	1·6
Louise Tiefbau—coking, very fine ..	96, 97 ..	32·8	62·3	83·6	1·1	1·8
Do. do. ..	98, 99, 100, 101..	65·6	{ 65·6 / 62·3 }	118·1 / 118·1	1·3 / 1·4	1·8 / 1·8
Shamrock—coking, very fine	73, 74, 75, 76 ..	32·8	{ 62·3 / 62·3 }	80·3 / 80·3	1·3 / 1·4	1·8 / 1·7
Do. do.	77, 78, 79 ..	65·6	{ 83·6 / .. }	154·1 / 149·2	1·2 / ..	6·2 / 5·7
Do. do.	80, 81	131·2	{ 114·8 / 103·2 }	..	1·6 / 1·6	.. / ..
Joachim—coking, very fine	102, 103, 104 ..	32·8	{ 72·1 / 72·1 }	88·5 / ..	1·4 / 1·5	1·6 / ..
Do. do.	105, 106, 107 ..	65·6	{ 88·5 / .. }	177·2 / 180·4	1·1 / ..	5·9 / 4·9
Do. do.	108	131·2	88·5	..	1·6	..
Saarbruck.						
Kreuzgräben—very fine	113, 114 ..	32·8	40·9	62·3	·8	·9
Do. do.	115, 116 ..	65·6	40·9	67·2	·8	1·1
Itzenplitz (Viktoria)—fine and hard..	134, 135 ..	32·8	55·7	72·1	1·0	2·0
Do. do. do.	136 ..	65·6	..	83·6	..	1·9
Do. (Sophie) do.	137, 138 ..	65·6	50·8	67·2	·6	2·1
Do. (23 Seam) do.	139 ..	65·6	54·0	63·9	1·1	1·7

TABLE OF EXPERIMENTS.— *Continued.*

Coal-dust from.	Register of Experiments	Length of Strewing	Length of Flame.		Tub weighing 1,627 lbs. was moved.	
			Without Gas	With 3% of Gas.	Without Gas	With 3% of Gas.
Saarbrück.— Continued.	1884.	Teet.	Feet.	Feet.	Feet.	Feet.
Gerhard—fine and hard	129, 130, 131 ..	32·8	{ 49·2 45·9	72·1 ..	·5 ·4	1·3 ..
Do do.	132, 133 ..	65·6	49·2	88·5	·5	·7
Altenwald—fine and hard	109, 110 ..	32·8	59·0	80·3	·8	2·0
Do. do.	111, 112 .	65·6	80·3	118·1	1·2	2·1
Camphausen—very fine	{ 117, 118, 119, 120, 121.. }	32·8	{ 72·1 78·7 72·1	88·5 95·1 ..	1·7 1·6 1·6	1·6 1·8 ..
Do. do.	{ 122, 123, 124, 125 .. }	65·6	{ 83·6 88·5	162·3 167·3	1·4 1·6	6·5 6·5
Do. do.	126 ..	..	..	·173·9	..	·15·9
Do. do.	127, 128 ..	131·2	118·1	*180·4	1·4	*52·4
Silesia.						
Konigin Louise—fine and hard ..	141, 142 ..	32·8	45·9	67·2	1·0	1·5
Do. do. ..	143, 144 ..	65·6	59·0	72·1	1·4	1·4
König—very fine and hard ..	145, 146, 147 ..	32·8	{ 67·2 67·2	75·4	1·6 1·5	1·6
Do. do. ..	148, 149 ..	65·6	67·2	83·6	·9	1·8
Cons. Rudolf—fine and granular ..	150, 151 ..	32·8	{ 45·9 45·9	49·2	1·1 1·3	1·6
Do. do. ..	152, 153 ..	65·6	45·9	54·0	1·1	1·8
Cous. Johann Baptist—very fine with coarse grains	154, 155, 156 ..	32·8	{ 54·0 54·0	80·3	1·4 1·8	2·9
Do do. ..	157, 158 ..	65·6	54·0	95·1	·9	1·9
Peterkowitz—fine and hard ..	159, 160, 161 ..	32·8	{ 42·6 42·6	54·0	·9 ·8	1·6
Do. do. ..	162, 163 ..	65·6	49·2	59·0	1·4	1·9
Do. very fine, like Pluto ..	164, 165 ..	65·6	62·3	103·2	17·3	1·1

* With 4 per cent. of gas.

IV.—COMPARISON OF THE MECHANICAL EFFECTS OF GAS AND COAL-DUST EXPLOSIONS.

Description of Experiments	Length of Flame. Feet.	Tub weighing 1,627 lbs. was moved. Feet.	Pressure in Atmosphere.		
			Gauge I.	Gauge II.	Gauge III.
Coal-dust alone ...	... 187·0	... 8·3	... 2	2	1¼
Gas alone ...	... 160·7	... 26·2	... 1¾	2	1½
Gas and dust ...	... { 144·3 in main drift 42·6 in side drift }	(?)	... 3	2¾	2¾

In the first experiment 132·2 lbs. of Pluto dust were strewed over a length of 131·2 feet without gas; in the second 706 cubic feet of 7 per cent. of gas; and in the third a similar mixture was employed, together with a dust strewing of 65·6 feet. The Schäfa and Budenberg gauges were placed at 3·3. 55·7, and 78·7 feet from the end of the gallery.

V.—EXPERIMENTS AS TO THE IGNITION OF DUST AND COAL-GAS BY OPEN LIGHTS.

(a) Gas without coal-dust.

4 per cent. of gas was ignited by an open light and the flame propagated with a velocity of about 1 foot per second; with 5 per cent. the propagation of flame was more rapid; and with 6 per cent. the velocity was about 7 feet per cent. and slight explosion.

(b) Gas with coal-dust.

The presence of coal-dust did not cause explosion with 3 or 4 per cent. of gas; it was slightly explosive with 5 per cent.; and explosions were always produced with 6 per cent. M. W. B.

A NEW FOSSIL FISH IN THE COAL-MEASURES OF COMMENTRY.

Sur un nouveau Poisson fossile du terrain houiller de Commentry (Allier). By
M. Charles Brongniart. *Comptes Rendus de l'Académie des Sciences.* pp.
1240–1242, *Tome CVI.,* No. 17.

Found in the carboniferous shales of Commentry. The writer finds that the fish
presents peculiarities not found in any other living or fossil, and at the same time
possesses characters belonging to widely different species. He proposes for it the name
of *Pleuracanthus Gaudryi.*   G. W. B.

GEOLOGY.

Note sur le sénonien et le damien du sud-est de l'Espagne. By M. René Nickles.
Comptes Rendus de l'Académie des Sciences. *Tome CVI.,* No. 6 (6 Férrier,
1888), *pp.* 431–433.

A description of the beds of the Upper Cretaceous (*Sénonien* and *Danien*—the
stage of the Faxoe limestone and beds immediately below it) of the south-east of Spain.

The interest of the paper arises from the fact that these formations are said to
contain beds of combustible.   G. W. B.

THOMPSON'S CALORIMETER.

*Expériences sur l'emploi du calorimètre Thompson pour la détermination du pouvoir
calorifique pratique de la houille.* By M. Scheurer-Kestner. *Comptes
Rendus de l'Académie des Sciences,* pp. 941–944, *Tome CVI.,* No. 13 (*March
26th,* 1888).

A comparison of the heating powers of different coals as determined by Thompson's
calorimeter, with the results obtained with the apparatus of Favre and Silbermann.
The greatest discrepancy found is not more than 4 per cent.

For Thompson's apparatus M. Scheurer-Kestner uses a larger quantity of chlorate
and nitrate of potash than indicated by English experimenters: eight or ten times the
weight of the coal is the proportion used by the latter; the former thinks ten and a
half to eleven is requisite for good results.

M. Scheurer-Kestner also finds the 10 per cent. added in England to the results as
the coefficient of the apparatus absolutely insufficient. By an experiment on wood
charcoal—of which the total number of heat units is known—15 per cent. is found to
be the necessary addition for Thompson's calorimeter. After correcting the results
obtained from it by this figure, they agree with those from Favre and Silbermann's
apparatus to within 1¼ per cent.

Here are the exact figures:—

				Calorimeters.	
				Thompson's, corrected 15 per cent.	Favre and Silbermann's.
Ronchamp coal	...	...	...	... 9,179	... 9,130
Do.	...	...	...	... 9,237	... 9,163
Blanzy coal	...	...	...	... 9,011	... 9,111
Creusot coal	...	...	...	... 9,521	... 9,622
Saarbrück coal	...	...	...	... 8,554	... 8,457
Do.	...	...	...	... 8.433	... 8,462
Ruhr	...	...	...	... 9,128	... 9.111

The following table exhibits the results obtained by burning 20 different kinds of coal in Thompson's calorimeter, and adding 15 per cent., alongside of those afterwards obtained from the same coals in Favre and Silbermann's :—

			Calorimeters.					
			Thompson's.		Favre and Silbermann's.		Difference per cent.	
1	...	...	8,972	... ...	8,858	... ...	1·3	
2	...	...	8,559	... ...	8,853	... ...	3·2	
3	...	...	8,956	... ...	8,771	... ...	2·1	
4	...	...	8,882	... ...	8,756	... ...	1·5	
5	...	...	8,384	... ...	8,545	... ...	1·8	
6	...	...	8,401	... ...	8,638	... ...	2·7	
7	...	...	8,865	... ...	8,727	... ...	1·6	
8	...	...	8,408	... ...	8,714	... ...	3·5	
9	...	...	8,688	... ...	8,660	... ...	0·3	
10	...	...	8,586	... ...	8,880	... ...	3·2	
11	...	...	8,810	... ..	8,656	... ...	1·8	
12	...	...	8,585	... ...	8.685	... ...	1·2	
13	...	...	9,232	... ...	8,943	... ...	3·2	
14	...	...	9,021	... ...	8,893	... ...	1·4	
15	...	...	8,996	... ...	8.801	... ...	2·1	
16	...	...	8,627	... ...	8,933	... ..	3·3	
17	...	...	8,964	... ...	8,700	... ...	3·1	
18	...	...	8,750	... ...	8,909	... ...	1·5	
19	...	...	9,186	... ...	9.030	... ...	1·6	
20	...	...	8,811	... ...	8,864	... ...	0·6	

The final conclusion is that Thompson's calorimeter only deserves limited confidence. The calorimetric bomb of M. Berthelot is considered to combine the advantages of both instruments in question without any of their disadvantages.   G. W. B.

HEATING POWER OF FRENCH COALS.

Chaleur de combustion de la houille du Nord de la France. By M. SCHEURER-KESTNER.

(1) *Departement du Nord, Comptes Rendus de l'Académie des Sciences, pp.* 1092-1094, *Tome CVI., No.* 15.

(2) *Bassin de Charleroi, Ditto, pp.* 1160-1161, *Tome CVI., No.* 16.

(3) *Bassin du Pas-de-Calais, Ditto, pp.* 1230-1231, *Tome CVI., No.* 17.

Various samples of coal from the above-mentioned basins have been tested in Favre and Silbermann's calorimeter. The coal, in small fragments, has been burnt in a current of pure oxygen.

The results, along with analysis of different samples, are given in the following tables:—

1.—BASIN OF THE NORTH.

SAMPLES.

1. Anzin coking coal.—Lebret Pit.
2. Anzin non-coking coal.—Lambrecht Pit.
3. Anzin non-coking coal.—Saint-Louis Pit.
4. Aniche non-coking coal.

ANALYSIS OF SAME.

	1.	2.	3.	4.
Fixed carbon	77·2	86·2	82·2	84·8
Volatile carbon...	7·3	6·0	1·8	4·6
Hydrogen	4·2	4·0	3·7	4·0
Oxygen	11·3	2·9	11·6	6·0
Nitrogen	—	0·9	0·7	0·6
	100·0	100·0	100·0	100·0
Heat of combustion	9,257	8,664	8,460	8,522

2.—CHARLEROI BASIN.
SAMPLES.

1. Bascoup non-coking coal.
2. Sart-les-Moulins non-coking coal.
3. Gilly-les-Charleroi and Vivurs unwashed.

4. Monceau-Fontaine-Martinez medium coal.—Monceau Pit.
5. Bascoup non-coking coal (IL)
6. Sart-les-Moulins non-coking coal (II.)

ANALYSIS OF SAME.

	1.	2.	3.	4.	5.	6.
Fixed carbon	84·42	84·13	86·71	82·71	82·79	85·74
Volatile carbon ...	7·66	3·01	3·75	0·57	2·11	7·97
Hydrogen	6·04	6·31	3·76	3·98	4·58	4·10
Oxygen	1·04	5·71	5·13	11·85	9·83	1·46
Nitrogen	0·84	0·84	0·65	0·89	0·69	0·73
	100·0	100·0	100·0	100·0	100·0	100·0
Heat of combustion ...	8,639	8,460	8,553	8,499	8,437	8,435

3.—PAS-DE-CALAIS BASIN.
SAMPLES.

1. Courrières coking coal.
2. Nænd medium coal.—Pit No. 1.
3. Dourges coking coal (I.)
4. Courcelles-lez-Lens coking coal.
5. Lens non-coking coal.—Douvrin Pit.
6. Dourges coking coal (IL)

7. Dourges coking coal (III.)
8. Lens coking coal (fines).
9. Meurchin coking coal.
10. Béthune coking coal.
11. Douvrin non-coking coal.

ANALYSIS OF SAME.

	1.	2.	3.	4.	5.	6.	7.	8.	9.	10.	11.
Fixed carbon	76·32	79·75	78·60	76·27	87·34	78·29	75·74	72·25	86·63	69·39	86·48
Volatile carbon ...	14·57	3·21	13·15	9·93	3·84	3·98	6·01	14·30	4·02	15·56	1·07
Hydrogen	4·07	3·42	3·13	3·83	3·96	4·98	5·41	3·83	3·76	6·35	3·77
Oxygen	4·10	12·98	4·21	8·94	4·45	11·86	12·05	8·71	4·90	7·80	8·08
Nitrogen	0·94	0·64	0·91	0·93	0·41	0·89	0·79	0·91	0·69	0·90	0·60
	100·0	100·0	100·0	100·0	100·0	100·0	100·0	100·0	100·0	100·0	100·0
Heat of combustion ...	8,814	8,790	8,726	8,647	8,642	8,634	8,562	8,446	8,438	8,360	8,340

These experimental results in nearly every case are found to differ rather widely from those calculated according to the composition of the samples. In some cases they exceed, and in others fall below them. The writer can assign no reason for the differences. Nor do the heating powers of the various samples depend altogether on their composition: their different values cannot be traced to the percentage either of carbon, hydrogen, or oxygen.

G. W. B.

ON THE VEGETABLE ORIGIN OF COAL.

By Leo Lesquereux. *Annual Report of the Geological Survey of Pennsylvania,* 1885, *pp.* 95 *to* 124.

The assertion that coal is a compound of vegetable remains is contradicted by few, if any, naturalists. But granting that the composition of coal is purely vegetable, the problem of its formation is not fully solved.

How is it possible that plants, even woody plants or trees, could have been heaped by natural agency in such a way that the original material could have produced, after decomposition and compression, beds of coal from 4 to 25 feet, or even more, in thickness? If a forest were destroyed by some cataclysm,[and afterwards covered by the sea, and gradually converted into coal, the forest, however dense, could not possibly produce a layer of more than a few inches of coal.

Some suppose that the woody material of vegetation growing along the borders of great rivers has been carried by water for long years, deposited near the mouth of the rivers, heaped together there, then covered with mud and sand, and buried for future decomposition and transformation into coal. In such an operation the vegetation would be mixed with foreign elements, mud, sand, etc., and this mode of procedure is contradicted by the purity of the matter composing coal. Beds of lignite have indeed been formed in the Red River and in some of the affluents of the Mississippi near its mouth, but lignite of this kind is always a fuel of little value, containing more than 50 per cent. of ash.

Bischoff supposed that the materials carried by currents into the sea were gradually deposited according to their weight. But the stratification in horizontal layers is perfectly distinct, and there is no succession of various kinds of deposits such as would have taken place had the bed been produced by the translation of mixed materials from land surface into the sea.

Grand 'Eury has lately proposed a new hypothesis. He supposes that there were in the Carboniferous period shallow lakes or ponds surrounded by great forests of very luxurious vegetation, and that the *débris* of these forests, small branches, leaves, and especially the bark, already half decomposed or dried by atmospheric action, were swept by heavy rains into these low grounds, where they contributed an amount of woody matter sufficient for making a coal seam, and being covered by foreign deposits, were gradually transformed into coal by the chemical process of slow combustion. This theory will not account for the accumulation of vegetable matter with such a uniform thickness over such extensive areas as those now occupied by some of the coal-beds of North America, which spread without interruption through many thousands of square miles of country.

The Vail hypothesis imagines that coal is merely a local accumulation of bitumen derived either from the earth by volcanic action or gradually condensed from a fancied bituminous atmosphere, encircling our planet like the rings of Saturn. Bitumen has issued from the earth and has been mined under the name of albertite, etc., not in horizontal strata, but in more or less irregular and more or less vertical veins, in no respect resembling beds of coal, being pure bitumen and without a trace of vegetable fibre to be detected in it. The absence of vegetable remains is quite enough to disprove any such theory, even if it were possible to admit such a theory in the absence of all evidence.

Kuntze supposes that on the *débris* of very active floating marine vegetation, whose surface had become gradually solid enough to support aërial plants, a different kind of vegetation had been established. First, aquatic plants, then came large floating stems, the stigmaria rendering the ground more solid. Then shrubs and ferns grew up, and then trees of various kinds. These growths accumulated and formed a mass so heavy that it gradually sank lower and lower into the sea, but so slowly that the vegetation still continued at the surface of the water, supported on the dead material, until the whole stratum sunk to the bottom of the sea, to be covered by aqueous sediments and transformed into coal by slow combustion. This theory, however, fails to explain the formation of the under-clays which generally serve as bottoms or supports to the coal-beds, does not account for the origination of land plants on the surface of a marine vegetation floating in the middle of an ocean, and does not explain the universal conformity of the coal-beds to the intermediate strata, and is therefore nothing more than a fanciful hypothesis.

Everybody knows something about what is called the peat bog theory; but what is a peat bog? The definition of a peat bog is the same as that of a bed of coal; it is an accumulation of remains of plants grown *in situ*, whose remains, deposited each year or after the cycle of their vegetation is completed, are superposed without interruption, one layer upon another, until the accumulation becomes sometimes of great thickness and covers a wide surface of land.

Two conditions are essential for the origin and growth of peat; water either in stagnant basins, lakes, pools, etc., or water abundantly supplied by a boggy atmosphere, increased by dense forest growth.

Pools of stagnant water, when not exposed to periodical drying up, are invaded by peculiar vegetation, at first mostly confervæ of various colour and of prodigious activity of growth, mixed with a mass of infusoria, animalcules, and microscopic plants, which partly decomposed, partly containing the floating vegetation, soon fill the basins and cover the bottom with a floating clay-like mould. So rapid is the work of these minute beings that in some cases 6 to 10 inches of this mud is deposited per annum. When undisturbed, this mud becomes gradually thick and solid, affording a kind of soil for the growth of marsh plants, which root at the bottom of the basins or swamps and send up their stems to the surface of the water or above it, where their substance, in the sunshine, becomes hard and woody. As these plants periodically decay, their remains drop to the bottom of the water, and each year the process is repeated, with a more or less marked variation in the species of the plants; after a time the basins become filled by these successive accumulations of years and centuries, and then the top surface of the decayed matter, being exposed to atmospheric action, is transformed into humus, and is gradually covered by other kinds of plants, making meadows and forests.

In other cases, when basins of water, not exposed to sweeping currents or great changes of level, are too deep for the vegetation of the aquatic plants, nature attains the same result by the prolonged vegetation of certain kinds of floating mosses, more especially sphagna. These floating masses grow with great speed, and, expanding their branches in every direction over the surface of ponds and small lakes, soon cover it entirely. They form a floating carpet which, as it increases in thickness, serves as a solid soil for a vegetation of rushes, sedges, and some kinds of grasses, which grow abundantly mixed with the mosses, and by their water-absorbing structure furnish a persistent humidity for the preservation of their remains against aërial decay. The floating carpet becomes more solid, and is then overspread by many species of larger swamp plants, especially those of the heath family; and so, in the lapse of years by the continued vegetation of the mosses, which is never interrupted, and by the yearly deposits of plant remains, the carpet at length becomes strong enough to support trees, and is changed into a floating forest, until, becoming too heavy, it either breaks and sinks suddenly to the bottom of the basin, or is slowly and gradually lowered into it and covered with water.

The submergence is, nevertheless. not always final. for after the sinking of the first floating carpet the vegetation of the mosses may again begin at the surface of the water, and in the course of years and centuries a new carpet covers the basin, another cycle of vegetation begins and continues its course until it also is pressed down under the water.

Thus there are two superposed beds of vegetable remains in process of slow decomposition, or subjected to the beginning of the transformation into coal. Both layers are composed in the same way, the lower part being a mass of the remains of small vegetation, mosses, water plants, etc., the upper part covered with trees; that is. two beds of peat and two forests.

This exposition is a mere description of observed facts. In the Jura, a peat-bog forest sank suddenly, and now lies at the bottom of a lake, over which a carpet of peat has since grown. The bottom of Drummond Lake. in the Dismal Swamp, is formed of forest once growing at the surface, but now prostrate in 15 to 20 feet of water. Beneath it lies a deposit of the detritus of plants and a bed of peat, while the moss vegetation is now advancing into the lake from all sides around its edge. In New Jersey, on the sea shore, large tree trunks are dug out of a muddy peat. Borings near New Orleans have traversed at various depths a succession of beds of peat and forest separated by deposits of sand.

The process is more plainly exhibited in northern countries. where a colder climate is particularly favourable to the growth of mosses, either as a work completed in the past or still actively carried on and open to observation.

In Sweden and Denmark peat deposits, rarely of wide extent, but sometimes deep, are of frequent occurrence. The soil is undulating and diversified with a great number of large ponds or small lakes which have been filled up with peat growth.

Between Hirsholm and Waldmarsland, near Copenhagen, along a line of nine miles, are found forty peat bogs, either isolated or connected by runners of water.

In one of these deposits the separate layers and composition was as follows :—At the bottom, at the lowest level worked, lay 4 feet of black compact peat, and over it a stratum of 4 feet of prostrated pine trees, most of them laid in the direction of the slope of the basin, the tops of the trees pointing towards the centre. The trunks of a large number of these trees measured from 6 to 10 inches in diameter ; they still kept their branches, embedded in a mass of leaves and cones, and even mushrooms. This was overlaid by a bed of black peat 4 feet thick, covered in its turn by a bed of prostrated birch trees 3 feet thick. Above this was a bed of peat 6 feet thick, less compact than those beneath it, and of a yellowish colour, and covered by a stratum of large trunks of oaks, some of them 3 feet in diameter, which were used and sawed for timber. Over this layer was a fourth bed of fibrous yellow peat $3\frac{3}{4}$ feet thick, made up mostly of not fully decomposed mosses. The full thickness of this deposit was, so far as exposed in working, about 30 feet thick, but others are known to be 60 feet deep.

The absorbing power of peat mosses enables them to grow higher and higher above their original water-level, from which they thus gradually emerge. The peat of emerged bogs is less compact, and the annual layers are distinct and well defined. At the top the layers are about one inch thick, and at the bottom less than one-eighth of an inch, and in old bogs still less. The average production of compact matter is at the rate of about 1 foot in a century.

In immerged bogs, formed of vegetable *débris* falling into water, the peat grows slowly and irregularly, but the actual rate of growth has not yet been recorded.

The peat of immerged bogs is compact and quite black. the vegetable matter being entirely decomposed and its internal structure generally so destroyed as to be unrecognisable. The peat of emerged bogs is yellowish brown, fibrous, its annual layers are distinct, and the woody fragments are generally recognisable.

The differe1ce i1 the two ki1ds of peat fur1ishes grounds for supposing that ca11el coal, with its more compact texture and its destitutio1 of any trace of horizo1tal a11ual layers and of vegetable remai1s, has bee1 produced by pla1t growth u1der water and decomposed u1der water like submerged peat. Bitumi1ous coal, on the other ha1d, with its disti1ct stratificatio1, appears to have bee1 produced by the accumulatio1 of vegetable mate'rials above water, and preserved agai1st rapid decompositio1 by the great humidity of the air.

It is well established that immerged peats are thi1, the thickest varyi1g from 2 to 8 feet. The beds of ca11el coal, which are the a1cie1t represe1tatives of such lake bogs, are usually thi11er tha1 those of bitumi1ous coal.

Peat bogs i1 the low cou1tries are exte1sively formed alo1g the sea shore, especially 1ear the mouths of large rivers, and everywhere that an expa1se of water has become e1closed, as a lagoo1 sheltered from the i1vasion of the sea by ba1ds of sa1d throw1 up by the waves, or alo1g river valleys by the 1atural levees which border most rivers i1 some parts of their course. These basi1s are i1vaded by pla1ts and filled up with peat deposits. These still retreats of vegetatio1 are not always safe from disturbance, and though sheltered for a time agai1st rivers or the ocea1, it happe1s that some extraordi1ary freshet, high tide, or storm breaks dow1 or through the barriers, and the peat bogs are covered with a deposit of mud or sa1d.

Such results are clearly recorded i1 the co1stitutio1 of peat bogs, which show the i1terposed layers of sa1d and mud which i1terrupted their regular growth.

In the Nord (Fra1ce) an average sectio1 is :—

	Ft.
Boggy humus	0
Grey mari1e sa1d or clay and mari1e shells ...	3
Blue mari1e clay and mari1e shells	3
Peat bed	4
Sa1d, with Cardium and Lutrariæ	0 I1
Peat, rarely fou1d, very thi1, with carburetted hydroge1.	

The shells fou1d in the mari1e sa1d and blue clay are:—Cardium edule, Lutraria compressa, Littori1a rudis, Bucci1um undatum, Pholas ca1dida, Telli1a balthica, etc. Mari1e shells are 1ever fou1d in the peat, and river shells rarely.

Wherever the growth of peat is stopped by dry1ess or other cause the upper surface of the peat becomes crusted, harde1ed, and tra1sformed i1to a thi1 coati1g, quite impervious to the e1tra1ce of foreig1 matter. The boggy humus forms on this hard crust, or whe1ever the la1d is agai1 submerged a new peat vegetatio1 begi1s. In the latter case the crust remai1s as a parti1g layer betwee1 the two beds of peat, like the parti1gs fou1d in seams of coal. Where the peat deposits are co1ti1uous it is fou1d i1 numerous layers with variatio1s i1 quality, evide1tly due to a successio1 of different ki1ds of vegetatio1, and perhaps to a variatio1 in the degrees of prevaili1g humidity; this shows a defi1ite resembla1ce to the structure of coal seams.

The followi1g sectio1 is take1 i1 the Valley of the Somme (Fra1ce) :—

	Ft.	In.
Calcareous sa1d, grey	0	8
Black ba1d	0	2
Peat, in thi1 plates	1	4
Argillaceous black peat	1	0
Calcareous co1cretio1s, full of shells	0	6
Black argillaceous matter	0	8
Greyish clay, full of shells	0	8
Argillaceous peat	thi1.	
Calcareous co1cretio1s	thi1.	
Lami1ated peat, with tru1ks of trees and leaves ...	10	0

On the coast of Holland the peat is worked by shafts. One shaft was sunk through 15 feet of clay to the first bed of peat; between this and the second bed of peat was 14 feet of white clay. The second peat was 18 feet thick, and beneath it lay 10 feet of hard clay.

These sections are a clue to some of the more important problems of the constitution of the Coal-Measures. The thin layers of clay which are interposed between peat deposits have their analogy in the clay partings of coal. The heavy deposits of sand mixed with marine shells and the calcareous concretions are, on a small scale, comparable to the beds of sandstone, limestone, etc., which make up the alternate strata of the Coal-Measures between seams.

It therefore plainly appears that in the growth of peat there is a microcosmic and true representation of the formation of the ancient coal.   M. W. B.

THE COAL-FIELDS OF THE URALS.

By V. ALEXSAEFF. *Gorny Journal, April,* 1888.

The author divides these coal-fields into two groups—(1) those of the west slope of the Urals, namely, the Looneffsky, Kizeloffsky, and Upper and Lower Goobahinsky; (2) those of the east slope of the Urals, the Kameisky, Leopoldo-Ferdinandoffsky, Fadinsky, and Egorshinsky. The total amount of coal raised from these beds in 1885 was 10,875,368 puds (175,408 tons).

The Looneffsky collieries include several seams which are worked by four separate pits. The coal is brought direct to the railway line in the colliery tubs by a cable tram. These pits are furnished with all the latest improvements in machinery, and are the property of Prince Dimidoff. The coal contains from 3·28 to 4·12 per cent. of sulphur, 8·82 to 24·88 per cent. of ash, and gives 55·4 to 67·73 per cent. of coke.

The Kizeloffsky Colliery is situated at 2 versts distance from the Looneffsky branch line of the Ural Railway, and adjoins the Kizeloffsky Iron Works. Three seams are worked—the Princess, the Korshoonoffsky and the Bogorodsky. The Princess Seam consists of four bands, 1·25 sachines (8·75 feet) to 0·5 sachine (3·5 feet) thick, with a dip of 6 degs. to 28 degs. to W. The Korshoonoffsky Seam also consists of four bands, of about the same thickness as the preceding. Their dip is 6 degs. to 40 degs. to E. The Bogorodsky Seam consists of two bands, 0·5 sachine (3·5 feet) thick, with a dip of 45 degs. to 70 degs. to W. The Princess Seam is worked by three adit levels, driven into the three bands. The roof of the upper band is sandstone, the floor clay-slate; the other two bands lie between clay-slate. The system of working is post and stall. Levels are driven 30 sachines (120 feet) long and 2·5 to 3 sachines (17·5 to 21 feet) wide, and then connected by parallel galleries, leaving pillars 6 to 9 sachines (42 to 63 feet) thick. Ventilation is natural. The output of this pit is 22,000 puds (355 tons) per day. The coal is thrice screened, through 2-inch, 1½-inch, and ¾-inch screens. The coal is dull, black, and has a coarse slaty fracture. It gives 64·62 per cent. of a non-coherent coke.

The Korshoonoffsky pits are 1½ versts from the railway, and are situated on the summit of a hill, 62 sachines (454 feet) above the level of the river Kizil. There are two shafts, 10 and 14 sachines (70 and 98 feet) deep respectively. The coal is now worked on the post and stall system. The coal has the following composition:—C = 64 per cent., H = 4·47 per cent., ash = 19·51 per cent. It is of an inferior quality, and gives 60·1 per cent. of coke.

The Bogorodsky pits adjoin the Kizil railway station. The mine is worked by an adit level, 240 sachines (1,680 feet) long. This coal will not bear transport, and therefore is only worked for local requirements, being used as a smithy coal. It gives 64·9 per cent. of a very compact coke, and has the following composition:—C = 65·55 per cent., H = 4·39 per cent., ash = 17·01 per cent., moisture = 1·11 per cent. It is, undoubtedly, a gas coal. These collieries are only worked during the winter, owing to want of hands during the summer months. The price of these coals at the Kizil station in waggons varies from 2 to 6 copecs per pud (2s. 6d. to 7s. 6d. per ton). Their chief drawback is the high percentage of sulphur they contain.

The Lower Goobahinsky Colliery is situated on the river Kosva, and is 20 versts (13¼ miles) south of the Kiziloffsky collieries. At the present time two seams are worked, the Ivanoffsky and the Trofinoffsky. These seams run parallel, 5 archines (11 feet 8 inches) to 2 sachines (14 feet) apart, separated by bands of clay and hard sandstone. Their dip is 45 degs. to 53 degs., and their direction 8 degs. to N. They are worked from below upwards. The price of the coal at the pit top is 5 to 6 copecs per pud (6s. 2½d. to 7s. 5½d. per ton).

In 1886-87, 3,447,758 puds (55,606 tons) were raised from this colliery. The coal is very hard, and has a slaty fracture of a greyish colour, and has the following composition:—C = 75·96 per cent., H = 5·31 per cent., O and N = 10·42 per cent., ash = 8·31 per cent. It gives 56 per cent. of a grey, brittle coke.

The upper Goobahinsky Colliery is situated near the station Goobaha, on the Looneffsky branch line. Three seams are worked, the Nicholas, the Varvara, and the Alexandra. The Nicholas Seam is worked by two adit levels, and is about 1 sachine (7 feet) thick; its general direction is 7 degs. to 8 degs. N.W., and occurs in a sandstone strata. The yearly output of this seam is 2,200,000 puds (35,484 tons). The coal is black, with a greasy lustre, it gives 63·55 per cent. of a compact coke, and has the composition C = 70·84 per cent., H = 5·00 per cent., ash = 14·56 per cent. It is a bituminous coal, and has a calorific capacity of 7·686·

The Varvara Seam is no longer worked, and the Alexandra Seam is only lately opened out. The roof is clay-slate, and the floor sandstone. Its direction is 10 degs. N.W., and its dip 30 degs. to E.

The Coal-fields on the East side of the Urals.—The Kamensky collieries are near the Kamensky Iron Works, and have only been lately opened out. The best coal is won from No. 2 and No. 6 seams. No. 2 is, in parts, 1½ sachine (10½ feet) thick, but on the average 4½ feet. No. 6 is 2½ feet thick. The coal from No. 6 easily cokes, and contains 23·6 per cent. of ash. No. 2 coal is very black, with a greasy lustre; in coking it swells to two or three times its volume, giving 78·3 per cent. of coke. It has the following composition:—S = 7 per cent., C = 82·5 per cent., H = 4·8 per cent., ash = 7·01 per cent., calorific power 8·044 units.

The Ferdinando-Leopoldoffsky Colliery. The coal seams have an exceedingly high inclination, 70 degs. to 80 degs. to W., with a direction N.W. 345 degs. The coal is very soft, and gives 85·24 per cent. of a pulverulent coke; it has the following composition:—C = 81·52 per cent., H = 3·72 per cent., ash = 9·86 per cent., moisture = 0·45 per cent.

The Egorshinsky collieries are not in work. The coal is anthracitic, and contains no sulphur, but much gas; it has the following composition:—C = 88·29 per cent., H = 3·44 per cent., W = 1·11 per cent., O = 4·02 per cent., ash = 3·14 per cent.

The Fadinsky coal is not worked; it contains a very high proportion of carbon, having the following composition:—C = 78·16 per cent., H = 2·77 per cent., O and N = 6·77 per cent., ash = 12·30 per cent.　　　　　　　　　　　J. H. M.